AF358830

Characterization of Conventional and Unconventional Hydrocarbon Reservoirs

Editors

Yuming Liu
Bo Zhang

Basel • Beijing • Wuhan • Barcelona • Belgrade • Novi Sad • Cluj • Manchester

Editors
Yuming Liu
China University of
Petroleum
Beijing
China

Bo Zhang
University of Alabama
Huntsville, AL
USA

Editorial Office
MDPI AG
Grosspeteranlage 5
4052 Basel, Switzerland

This is a reprint of articles from the Special Issue published online in the open access journal *Energies* (ISSN 1996-1073) (available at: https://www.mdpi.com/journal/energies/special_issues/ hydrocarbon_reservoirs).

For citation purposes, cite each article independently as indicated on the article page online and as indicated below:

Lastname, A.A.; Lastname, B.B. Article Title. *Journal Name* **Year**, *Volume Number*, Page Range.

ISBN 978-3-7258-1683-5 (Hbk)
ISBN 978-3-7258-1684-2 (PDF)
doi.org/10.3390/books978-3-7258-1684-2

Contents

Article

Linearized Frequency-Dependent Reflection Coefficient and Attenuated Anisotropic Characteristics of Q-VTI Model

Yahua Yang [1], Xingyao Yin [1,*], Bo Zhang [2], Danping Cao [1] and Gang Gao [3]

[1] School of Geosciences, China University of Petroleum (East China), Qingdao 266580, China; yyh@s.upc.edu.cn (Y.Y.); caodp@upc.edu.cn (D.C.)

[2] Department of Geological Science, The University of Alabama, Tuscaloosa, AL 35487, USA; Bzhang33@ua.edu

[3] School of Geophysics and Petroleum Resources, Yangtze University, Wuhan 430100, China; dragon_china316@163.com

* Correspondence: Xyyin@upc.edu.cn

Abstract: Seismic wave exhibits the characteristics of anisotropy and attenuation while propagating through the fluid-bearing fractured or layered reservoirs, such as fractured carbonate and shale bearing oil or gas. We derive a linearized reflection coefficient that simultaneously considers the effects of anisotropy and attenuation caused by fractures and fluids. Focusing on the low attenuated transversely isotropic medium with a vertical symmetry axis (Q-VTI) medium, we first express the complex stiffness tensors based on the perturbation theory and the linear constant Q model at an arbitrary reference frequency, and then we derive the linearized approximate reflection coefficient of P to P wave. It decouples the P- and S-wave inverse quality factors, and Thomsen-style attenuation-anisotropic parameters from complex P- and S-wave velocity and complex Thomsen anisotropic parameters. By evaluating the reflection coefficients around the solution point of the interface of two models, we analyze the characteristics of reflection coefficient vary with the incident angle and frequency and the effects of different Thomsen anisotropic parameters and attenuation factors. Moreover, we realize the simultaneous inversion of all parameters in the equation using an actual well log as a model. We conclude that the derived reflection coefficient may provide a theoretical tool for the seismic wave forward modeling, and again it can be implemented to predict the reservoir properties of fractures and fluids based on diverse inversion methods of seismic data.

Keywords: fluids-bearing fractured reservoirs; Q-VTI effective medium model; seismic attenuated anisotropic characteristics; AVOF reflection coefficient

Citation: Yang, Y.; Yin, X.; Zhang, B.; Cao, D.; Gao, G. Linearized Frequency-Dependent Reflection Coefficient and Attenuated Anisotropic Characteristics of Q-VTI Model. *Energies* **2021**, *14*, 8506. https://doi.org/10.3390/en14248506

Academic Editor: Pål Østebø Andersen

Received: 9 November 2021
Accepted: 13 December 2021
Published: 16 December 2021

Publisher's Note: MDPI stays neutral with regard to jurisdictional claims in published maps and institutional affiliations.

1. Introduction

Development of seismic acquisition and processing technology makes it possible to sufficiently employ useful information embedded in seismic data, e.g., amplitude variation with offset, azimuth and frequency (AVO, AVAz, AVF), to estimate fluids and fractures. Recently, many studies revealed that seismic wave exhibit velocity dispersion and anisotropy while propagating in attenuated fractured media and attenuated finely layered media [1–9]. The seismic wave velocity dispersion refers to the phenomenon that the velocity varies with the frequency, and it accompanies with the seismic wave amplitude attenuation, which means amplitude decreases with the increase of distance. Therefore, the modeling of frequency-dependent attenuation and anisotropy of seismic waves, and the inversion for attenuation factors and anisotropic parameters using frequency-dependent seismic amplitude data, may help improve the reliability of the detection of fractured reservoirs and infilling fluids [10,11].

Under the assumption of static equivalent effective medium model, the rock physics models are employed to model how fractures induce the frequency-independent anisotropy, e.g., the linear slip model proposed by Schoenberg [12], the isolated fracture model of Hudson [13], the uniform pore model of Thomsen [1], and the model combining the linear slip model and anisotropic Gassmann equation proposed by Gurevich [14]. Meanwhile, Thomsen [1] demonstrates that the exchange of fluids between pores and fractures during the seismic wave propagation can affect the anisotropic elastic properties. On the other hand, dynamic equivalent medium models are proposed to describe how seismic wave propagates in fractured rocks in the case of considering the effect of frequency variation [2–9]. Typically, Chapman [15] proposed a model which considers coupled fluid motion on both the grain scale and fracture scale, which concludes that frequency-dependent anisotropy and strong anisotropic attenuation can occur in the seismic frequency band when large fractures are present, and it reveals that fracture and fluid properties can be estimated from frequency-dependent seismic data.

To model how the seismic amplitude varies with incident angle and frequency, we consider the effects of the parameters of anisotropy and attenuation on the reflection coefficient. Under the assumption of slight changes in properties across the reflection boundary, Aki and Richards [16] proposed linearized reflection coefficients which are the analytical solutions of the Zoeppritz equations. However, it is complicated to solve the Zoepprtitz equations that are extended to viscoelastic anisotropic media. Based on the Born approximation, Shaw and Sen [17] presented an approach to derive linearized reflection coefficients for arbitrary anisotropic media using the perturbation in stiffness matrix of anisotropic media. Following them, Zong [18] derived the seismic wave scattering coefficient in terms of P-wave and S-wave quality factors in a viscoelastic medium, Moradi and Innanen [19,20] derived the expressions for scattering potentials of PP wave and proposed a frequency-independent linearized reflection coefficient in the attenuated VTI medium. Chen [21,22] presented a linearized azimuthal and frequency-dependent PP-wave reflection coefficient in terms of dry rock elastic properties, dry fracture weaknesses and a new indicator of oil-bearing fractured reservoirs. Pan [23] used Born formalism and first-order perturbation assumption to derive a matrix-fluid-fracture decoupled-based linearized PP-wave reflection coefficient for a fluid-saturated fractured porous medium.

In the present study, we focus on the case of Q-VTI medium with low-loss attenuation and weak anisotropy, which means we neglect the term proportional to higher orders of the attenuation factors and Thomsen anisotropic parameters, and we let P-wave, SV-wave and SH-wave propagate in the linear constant Q attenuation reference media. We express the PP wave scattering potentials and derive the linearized frequency-dependent reflection coefficient for the Q-VTI medium. Utilizing the reflection coefficients, we analyze the variation of reflection coefficients with the incident angle and angular frequency in two reservoir models, and we also model how the attenuation factors and Thomsen anisotropic parameters affect the reflection coefficients. We conclude that, combining the rock physics effective model, the derived reflection coefficient may provide a theoretical tool to model how pore-, fracture-, and fluid-related parameters (e.g., porosity, fracture density, fluid modulus) affect the seismic wave amplitude, and can also be employed to estimate these parameters from incident angel- and frequency-dependent seismic data.

2. Theories and Methods

2.1. Approximation of Frequency-Dependent Complex Stiffness Tensors for Q-VTI Model

Seismic wave velocity in viscoelastic media is expressed as a function of v_0, a phase velocity at an arbitrary reference frequency ω_0, and Q, a quality factor describing absorption and attenuation. Kjartansson [24] derives the complex and frequency-dependent phase velocity $\tilde{v}$ based on the linear constant Q model as,

$$\tilde{v}(\omega) = v_0 \left(i\frac{\omega}{\omega_0} \right)^{\frac{1}{\pi}Q^{-1}}, \tag{1}$$

where, the accent mark '~' indicates the complex velocity in viscoelastic medium. Using Equation (1), the quality factor is computed as $Q = \tilde{v}^{\mathrm{Re}}/\tilde{v}^{\mathrm{Im}}$, where $\tilde{v}^{\mathrm{Re}}$ and $\tilde{v}^{\mathrm{Im}}$ are the real and imaginary parts of the complex velocity $\tilde{v}$.

We approximate the complex velocity using the Maclaurin series expansion of the exponential function, and preserve the first two terms of the expansion. The complex velocity is given by,

$$\widetilde{v}(\omega) \approx v_0\left[1 + \frac{1}{\pi}Q^{-1}\ln\left(i\frac{\omega}{\omega_0}\right)\right].\tag{2}$$

Similar to the derived approximate complex velocity, we express the complex stiffness tensor $\widetilde{c}_{IJ}$ as,

$$\widetilde{c}_{IJ}(\omega) = c_{IJ}^0\left[1 + \frac{2}{\pi}Q_{IJ}^{-1}\ln\left(i\frac{\omega}{\omega_0}\right)\right],\tag{3}$$

where, c_{IJ}^0 is the elastic stiffness tensor at an arbitrary reference frequency ω_0, and Q_{IJ}^{-1} is the corresponding inverse quality factors.

A consistent description of P-wave property in VTI medium with weak anisotropy is given in terms of Thomsen anisotropic parameters [25,26]. For Q-VTI media, the complex Thomsen parameters are given by,

$$\widetilde{\varepsilon} = \frac{\widetilde{c}_{11}-\widetilde{c}_{33}}{2\widetilde{c}_{33}}$$

$$\widetilde{\gamma} = \frac{\widetilde{c}_{66}-\widetilde{c}_{44}}{2\widetilde{c}_{44}}\tag{4}$$

$$\widetilde{\delta} = \frac{(\widetilde{c}_{13}+\widetilde{c}_{44})^2-(\widetilde{c}_{33}-\widetilde{c}_{44})^2}{2\widetilde{c}_{33}(\widetilde{c}_{33}-\widetilde{c}_{44})}$$

where, the Thomsen-style attenuation-anisotropic parameters ε_Q, γ_Q and δ_Q are given by Zhu and Tsvankin [27,28] as,

$$\varepsilon_Q = \frac{Q_{11}^{-1}-Q_{33}^{-1}}{Q_{33}^{-1}}$$

$$\gamma_Q = \frac{Q_{66}^{-1}-Q_{44}^{-1}}{Q_{44}^{-1}}\tag{5}$$

$$\delta_Q = \frac{\left(Q_{13}^{-1}+Q_{44}^{-1}\right)^2-\left(Q_{33}^{-1}-Q_{44}^{-1}\right)^2}{2Q_{33}^{-1}\left(Q_{33}^{-1}-Q_{44}^{-1}\right)}$$

The parameters ε_Q and γ_Q represent the difference between the horizontal and vertical attenuation coefficients of P- and SH-waves, respectively, however, δ_Q is defined through the second derivative of the P-wave attenuation coefficient in the symmetry direction, which refers to the coupling between the attenuation and velocity anisotropy.

We stress that in this study we consider the Q-VTI medium with constant attenuation and weak anisotropy (i.e., $|\varepsilon|$, $|\delta|$, $|\gamma| \ll 1$), which means the second and higher orders of quality factors and Thomsen parameters are neglected in the approximation process of the complex tensors. Consequently, the components of frequency-dependent complex stiffness tensor $\widetilde{c}_{IJ}(\omega)$ are expressed in terms of two inverse quality factors, three Thomsen anisotropy parameters and corresponding Thomsen-style attenuation-anisotropic parameters,

$$\widetilde{c}_{11} = \rho v_p^2(1+2\varepsilon) + \rho v_p^2 Q_p^{-1}(1+2\varepsilon+\varepsilon_Q)I_\omega$$

$$\widetilde{c}_{13} = \rho v_p^2(1+\delta) - 2c_{55} + \rho v_p^2 Q_p^{-1}(1+\delta+\delta_Q)I_\omega - 2\rho v_s^2 Q_s^{-1}I_\omega$$

$$\widetilde{c}_{33} = \rho v_p^2 + \rho v_p^2 Q_p^{-1}I_\omega\tag{6}$$

$$\widetilde{c}_{55} = \rho v_s^2 + \rho v_s^2 Q_s^{-1}I_\omega$$

$$\widetilde{c}_{66} = \rho v_s^2(1+2\gamma) + \rho v_s^2 Q_s^{-1}(1+2\gamma+\gamma_Q)I_\omega$$

where, $I_\omega = \frac{2}{\pi} \ln \frac{\omega}{\omega_0} + i$. The approximate results above decouple the effective factors from the complex tensors and distinguish the real and imaginary parts, elasticity and attenuation parameters.

2.2. Approximation of Frequency-Dependent Reflection Coefficient for Q-VTI Model

A scattering model of seismic wave interaction in an attenuated anisotropic medium is shown in Figure 1, which consists of a homogeneous reference medium and perturbations described by eleven properties. Taking density as an example, the term $\Delta \rho = \rho - \rho_0$ in Figure 1, represents the perturbation in density, which represents the difference between the rock density ρ and the reference density ρ_0.

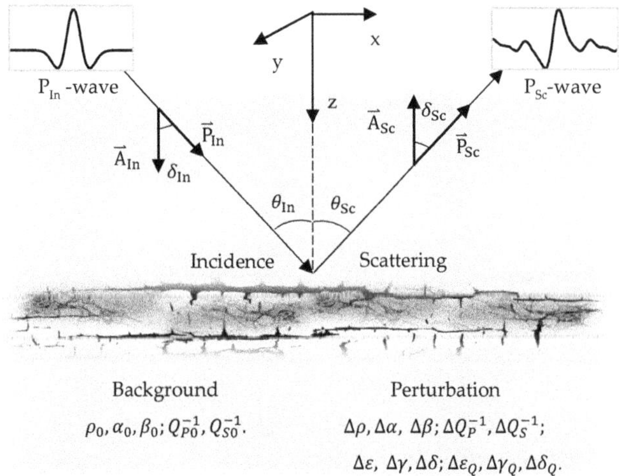

Figure 1. Schematic of seismic wave propagate in an attenuation anisotropy medium based on the perturbation theory. It is characterized by three elastic parameters P-wave velocity α, S-wave velocity β and density ρ; two viscoelastic parameters P-wave quality factor Q_P and S-wave quality factor Q_S; three anisotropic Thomsen parameters ε, γ, δ and corresponding attenuation Thomsen parameters ε_Q, γ_Q, δ_Q. Note that the subscript '0' stands for the properties of background (reference medium) and the mark 'Δ' stands for the properties of small perturbation.

Hence, the complex stiffness matrix of the Q-VTI medium can be re-expressed as the sum of the anisotropic perturbation and the stiffness matrix of a homogeneous isotropic background based on the perturbation theory. It has been shown in the Appendix A Equation (A1).

Since the quasi-Zoeppritz equation of Q-VTI medium is very complicated, we aim to derive the approximation reflection coefficient for P-to-P wave based on the Born approximation. A relationship between the reflection coefficient and the scattering functions is given by Shaw and Sen [29], and we extend it to the attenuated anisotropic medium in the present study,

$$\widetilde{R}(\theta, \omega) = \frac{1}{4\rho_0 \cos^2 \theta} \widetilde{S}(r_0), \tag{7}$$

where, ρ_0 is the density of the background medium, and $\widetilde{S}(r_0)$ is the scattering function related to the perturbations of stiffness tensors and density, which is given by,

$$\widetilde{S}(r_0) = \Delta\rho\xi + \Delta\widetilde{c}_{IJ}\eta_{IJ}, \tag{8}$$

where, $\xi = t_m t'_m \big|_{r=r_0}$, $\eta_{IJ} = t'_m p'_n t_k p_l \big|_{r=r_0}$. t and p are the polarization and the slowness vectors, respectively, which are given in the Equation (A2). $\Delta\rho$ and $\Delta\widetilde{c}_{IJ}$ represent the perturbation in density and complex elastic stiffness, respectively. The position vector r_0 is the point on a horizontal interface separating two weak anisotropic media, where Snell's law of reflection for a source-receiver pair is satisfied. The subscripts I and J refer to Voigt's concise notation.

The Einstein summation convention over repeated indices applies to Equation (8), and the scattering function for the frequency-dependent Q-VTI medium is written as,

$$\widetilde{S}(r_0) = \Delta\rho\cos 2\theta + \frac{1}{\alpha_0^2}\left[\Delta\widetilde{c}_{11}\sin^4\theta + 2(\Delta\widetilde{c}_{13} - 2\Delta\widetilde{c}_{55})\sin^2\theta\cos^2\theta + \Delta\widetilde{c}_{33}\cos^4\theta\right]. \quad (9)$$

Substituting the Equation (10) into Equation (7), we finally obtain the linearized approximate incident angle and frequency dependent PP-wave reflection coefficient (AVOF) for the Q-VTI medium,

$$\widetilde{R}_{PP}^{QVTI}(\theta,\omega) = \widetilde{A}(\omega) + \widetilde{B}(\omega)\sin^2\theta + \widetilde{C}(\omega)\sin^2\theta\tan^2\theta, \quad (10)$$

where,

$$\widetilde{A}(\omega) = \tfrac{1}{2}\left[\left(\tfrac{\Delta\rho}{\rho_0} + \tfrac{\Delta\alpha}{\alpha_0}\right) + \tfrac{1}{2}\Delta Q_P^{-1}I_\omega\right]$$

$$\widetilde{B}(\omega) = \tfrac{1}{2}\left[\tfrac{\Delta\alpha}{\alpha_0} - 4\tfrac{\beta_0^2}{\alpha_0^2}\left(\tfrac{\Delta\rho}{\rho_0} + 2\tfrac{\Delta\beta}{\beta_0}\right) + \Delta\delta + \tfrac{1}{2}\left(\Delta Q_P^{-1} - 8\tfrac{\beta_0^2}{\alpha_0^2}\Delta Q_S^{-1} + 2Q_{P0}^{-1}\Delta\delta_Q\right)I_\omega\right] \quad (11)$$

$$\widetilde{C}(\omega) = \tfrac{1}{2}\left[\tfrac{\Delta\alpha}{\alpha_0} + \Delta\varepsilon + \tfrac{1}{2}\left(\Delta Q_P^{-1} + Q_{P0}^{-1}\Delta\varepsilon_Q\right)I_\omega\right]$$

in which, the first term $\widetilde{A}(\omega)$ denotes the amplitude of P-wave at zero offset or normal incidence, the second term $\widetilde{B}(\omega)\sin^2\theta$ characterizes reflection coefficient at intermediate angles, and the third term $\widetilde{C}(\omega)\sin^2\theta\tan^2\theta$ describes the result approached to critical angle. Similar to the analysis of amplitude versus offset (AVO) in the isotropic elastic medium, the coefficient $\widetilde{A}(\omega)$ is called intercept, $\widetilde{B}(\omega)$ is called gradient and the third coefficient $\widetilde{C}(\omega)$ is called curvature. The derived reflection coefficients involve three elastic parameters P-wave velocity α, S-wave velocity β and density ρ; two attenuation parameters P-wave inverse quality factor Q_P^{-1} and S-wave inverse quality factor Q_S^{-1}; two Thomsen anisotropic parameters ε, δ; and two Thomsen-style attenuation-anisotropic parameters ε_Q, δ_Q. The subscript '0' stands for the properties of background (reference medium) and the mark 'Δ' stands for the properties of small perturbation. We normally take the average value and the difference value of two layers as the background and perturbation properties, respectively.

Note that, if we neglect frequency dispersion and attenuation, Equation (11) becomes the linearized reflection coefficient for elastic VTI medium derived by Rüger [30,31]. If we let the perturbation of anisotropy be zero, Equation (11) is exactly the same as the linearized PP-wave reflection coefficient for elastic isotropic media given by Shuey [32].

In addition, we obtain the form of reflectivity of each parameters using the Equation (11) to exhibits their contributions,

$$\widetilde{R}_{PP}^{QVTI}(\theta,\omega) = R_{PP}^{ISO}(\theta) + R_{PP}^{QISO}(\theta,\omega) + R_{PP}^{ANI}(\theta) + R_{PP}^{QANI}(\theta,\omega), \quad (12)$$

where,

$$R_{PP}^{ISO}(\theta) = \sec^2\theta R_P - 8g\sin^2\theta R_S + (1 - 4g\sin^2\theta)R_D$$

$$R_{PP}^{QISO}(\theta,\omega) = \tfrac{1}{4}\sec^2\theta I_\omega\Delta Q_P^{-1} - 2g\sin^2\theta I_\omega\Delta Q_S^{-1}$$

$$R_{PP}^{ANI}(\theta) = \tfrac{1}{2}\sin^2\theta\Delta\delta + \tfrac{1}{2}\sin^2\theta\tan^2\theta\Delta\varepsilon$$

$$R_{PP}^{QANI}(\theta,\omega) = \tfrac{1}{2}Q_{P0}^{-1}\sin^2\theta I_\omega\Delta\delta_Q + \tfrac{1}{4}Q_{P0}^{-1}\sin^2\theta\tan^2\theta I_\omega\Delta\varepsilon_Q$$

$$(13)$$

where, $R_P = \tfrac{1}{2}\tfrac{\Delta\alpha}{\alpha_0}$, $R_S = \tfrac{1}{2}\tfrac{\Delta\beta}{\beta_0}$, $R_D = \tfrac{1}{2}\tfrac{\Delta\rho}{\rho_0}$, $g = \tfrac{\beta_0^2}{\alpha_0^2}$, $I_\omega = \tfrac{2}{\pi}\ln\tfrac{\omega}{\omega_0} + i$.

Similar to the approximate formula of elastic isotropic reflection coefficient proposed by Aki-Richards [14], this is the mathematical bridge and basis for obtaining all elastic, attenuated and anisotropic parameters of Q-VTI model through simultaneous inversion.

3. Test and Analysis

3.1. Characteristics of Reflection Coefficients for Q-VTI Model

To analyze the characteristics of reflection coefficients, we compute the reflection coefficients around the solution point of the interface of two models using the derived reflection coefficient equation. Parameters of two models are shown in Tables 1 and 2 separately. For Model 1, we take the mud shale as the upper layer and oil shale as the lower layer, and for Model 2, we take the mud shale as the upper layer but calcareous sandstone as the lower layer. The properties of background and perturbation are the average value and the difference value of two layers, respectively. The elastic and anisotropic parameters of two models come from the compiled table of Thomsen [25] about the measured anisotropy in sedimentary rocks.

Figures 2 and 3 show the variation of reflection coefficients with the incident angle θ and frequency f for Model 1 and 2, respectively. We consider four cases of (1) elastic isotropy, (2) elastic anisotropy, (3) attenuated isotropy, and (4) attenuated anisotropy to compute the reflection coefficients using the derived reflection coefficient equation. The results show in sub-Figure (a,b), (c,d), (e,f) and (g,h), respectively and the value of colors represent in corresponding colorbars, where sub-Figure (a,c,e,f) and (b,d,f,h) exhibit the real part and the imaginary part of reflection coefficients separately. We stress that all parameters of the same properties on the vertical axis have the same scale so that the different degree of various influence can be observed directly.

In the case of elastic isotropic assumption, the derived reflection coefficient becomes the linearized P-P reflection coefficient given by Aki and Richards [16]. Therefore, the reflection coefficients are real numbers and controlled only by P-wave velocity α, S-wave velocity β and density ρ. In Figure 2a, we observe the real parts of reflection coefficients increase with the incident angle but frequency-independent. It exhibits the fourth AVO type in the case of the interface separating the mud shale and oil shale model. In Figure 2b, we observe that the imaginary parts of reflection coefficients are equal to zero.

Table 1. The parameters of attenuated anisotropic model 1.

Layer	α (km/s)	β (km/s)	ρ (g/cm^3)	ε	δ	Q_P^{-1}	Q_S^{-1}	ε_Q	δ_Q
Mud-shale	5.073	2.998	2.68	0.010	0.012	0.001	0.001	0.001	0.001
Oil-shale	4.231	2.539	2.37	0.200	0.100	0.205	0.118	0.046	0.025

Table 2. The parameters of attenuated anisotropic model 2.

Layer	α (km/s)	β (km/s)	ρ (g/cm^3)	ε	δ	Q_P^{-1}	Q_S^{-1}	ε_Q	δ_Q
Mud-shale	5.073	2.998	2.68	0.010	0.012	0.001	0.001	0.001	0.001
Calcareous Sandstone	5.460	3.219	2.69	0.000	−0.264	0.177	0.056	−0.025	0.050

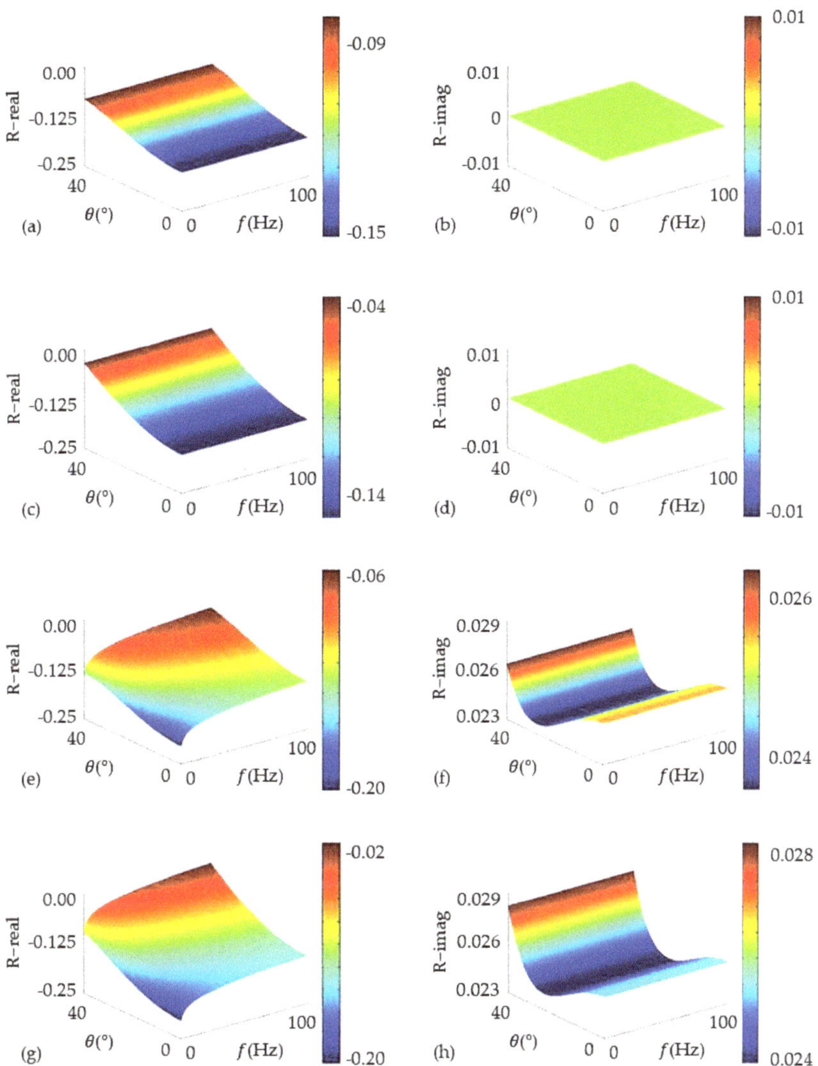

Figure 2. The variation of reflection coefficient with the incident angle θ and frequency f for model 1. (**a**,**b**), (**c**,**d**), (**e**,**f**) and (**g**,**h**) show the real part and imaginary part of reflection coefficients in four cases of assumption: (1) elastic isotropy, (2) elastic anisotropy, (3) attenuated isotropy, and (4) attenuated anisotropy separately. Different reflection coefficient values represent in corresponding colorbars.

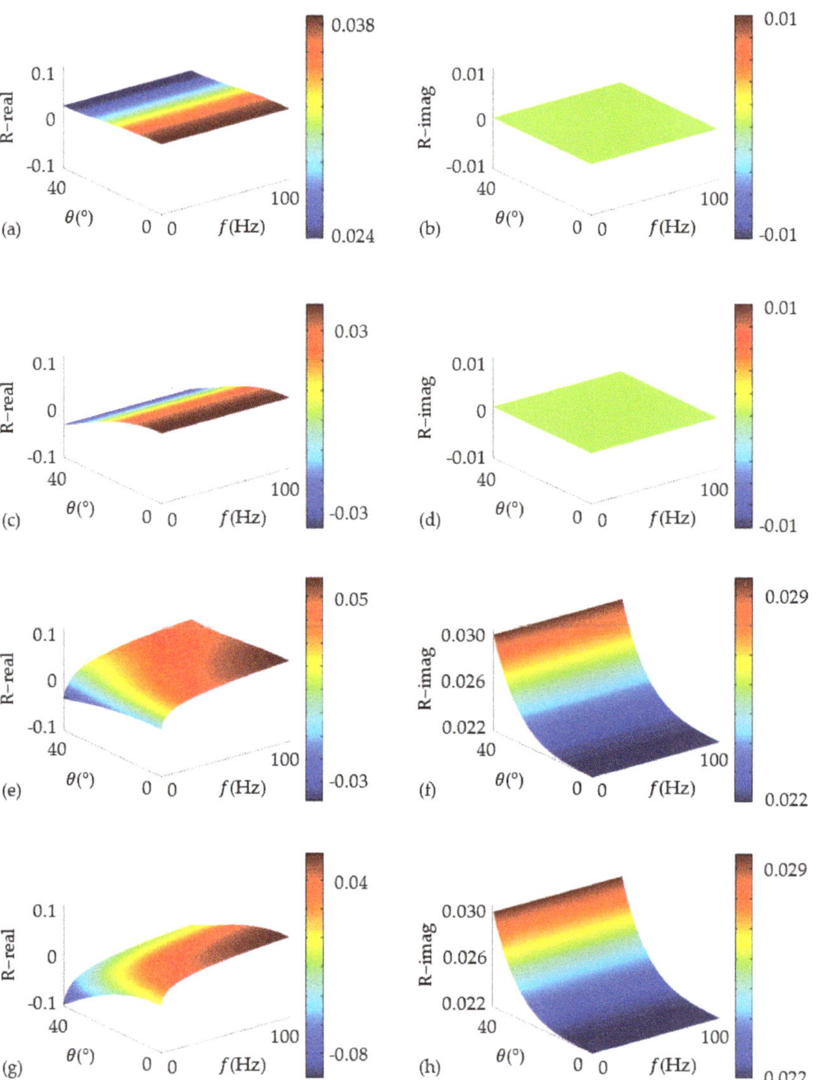

Figure 3. The variation of reflection coefficient with the incident angle θ and frequency f for model 2. (**a**,**b**), (**c**,**d**), (**e**,**f**) and (**g**,**h**) show the real part and imaginary part of reflection coefficients in four cases of assumption: (1) elastic isotropy, (2) elastic anisotropy, (3) attenuated isotropy, and (4) attenuated anisotropy separately. Different reflection coefficient values represent in corresponding colorbars.

The derived reflection coefficient is exactly the same as the reflection coefficient proposed by Rüger [30,31] when we only take the effect of anisotropy into consideration. The reflection coefficients are frequency-independent, as shown in Figure 2c, and the imaginary parts are also zero, as shown in Figure 2d. We conclude that the anisotropic parameters ε, δ just affect the value of reflection coefficients, however, the AVO type is the same as Figure 2a.

It shows slightly difference in the case of attenuated isotropic assumption. The reflection coefficients become complex numbers, and vary with both incident angle and frequency, as shown in Figure 2e. We emphasize that the AVO type has not been changed by the attenuation parameters Q_p^{-1}, Q_s^{-1}. In Figure 2f, we observe the imaginary part of reflection coefficients vary with incident angle but frequency-independent.

Figure 2g,h show the characteristics of the complex reflection coefficients in the case of attenuated anisotropic assumption, which are the similar to the results of reflection coefficients for the attenuated isotropic assumption. We observe the Thomsen-style attenuation-anisotropic parameters ε_Q, δ_Q have litter effect in larger incident angles and higher frequency on the reflection coefficients.

Then, we compute the reflection coefficients around the solution point of the interface separating the mud shale and calcareous sandstone. The same characteristics appear in this model, as shown in Figure 3. We observe the reflection coefficients decrease with the incident angle but frequency-independent in Figure 3a. It exhibits the second AVO type in the case of model 2 due to P-wave velocity α, S-wave velocity β and density ρ. The anisotropic parameters ε, δ also only affect the value of reflection coefficients and the AVO type doesn't change, as shown in Figure 3c. Figure 3e–h show that the reflection coefficients are complex numbers and the real parts vary with frequency caused by the attenuation. However, the Thomsen-style attenuation-anisotropic parameters ε_Q, δ_Q contribute much smaller to the reflection coefficient than Q_p^{-1} and Q_s^{-1} because they exist in the terms of high order.

In the following, we focus on the effect of the crucial parameters of anisotropy ε, δ and attenuation Q_p^{-1}, Q_s^{-1}. We proceed to the analysis of how perturbations in anisotropic parameters and attenuation factors affect reflection coefficients. The P-wave velocity α, S-wave velocity β and density ρ are set up as the same as the model 1 (Table 1), and nine groups of perturbations in anisotropic $\Delta\delta$, $\Delta\varepsilon$ and six groups of perturbations in attenuation ΔQ_p^{-1}, ΔQ_s^{-1}, as shown in Tables 3 and 4. Using the derived reflection coefficient equation, we obtain the reflection coefficients variation with incident angle and frequency in the case of different perturbations.

Table 3. The effect of elastic anisotropic perturbation on the reflection coefficient.

Anisotropic Perturbation	1	2	3
$\Delta\delta$	0.3	0	-0.3
$\Delta\varepsilon$	0.2, 0, -0.2	0.2, 0, -0.2	0.2, 0, -0.2

Table 4. The effect of attenuated isotropic perturbation on the reflection coefficient.

Attenuated Perturbation	1	2	3
ΔQ_P^{-1}	0	0.02	0.2
ΔQ_S^{-1}	0	0, 0.012	0, 0.012, 0.12

We first consider the effect of perturbations in anisotropy $\Delta\delta$ and $\Delta\varepsilon$ on the reflection coefficients, as shown in Figure 4. For this case, the derived linearized reflection coefficient is equal to the reflection coefficient proposed by Rüger [30,31] because the attenuation parameters are neglected. In Figure 4a we observe the reflection coefficients are frequency-independent, and in Figure 4b we observe the imaginary part of reflection coefficients are equal to zero. In Figure 4a, the intercept of reflection coefficients is a constant, and equals to the result computed for the isotropic model (the red solid line). By comparing three sets of the same type of lines (dashed lines, solid lines and doted dashed lines), respectively, we observe the gradients of curves vary with $\Delta\delta$. For example, the gradients of blue dashed line, red dashed line and black dashed line increase with $\Delta\delta$ when $\Delta\varepsilon$ is equal to a constant 0.2, but the curvature of them are the same. In the meanwhile, we observe the curvatures vary with $\Delta\varepsilon$ by comparing three sets of same color of lines (black lines, red lines, and blue lines), respectively. For example, the curvatures of black doted dashed line, black solid line and black dashed line increase with $\Delta\varepsilon$ when $\Delta\delta$ is equal to a constant 0.3, but the gradient of them is a constant. It appears the same characteristics in the rest groups.

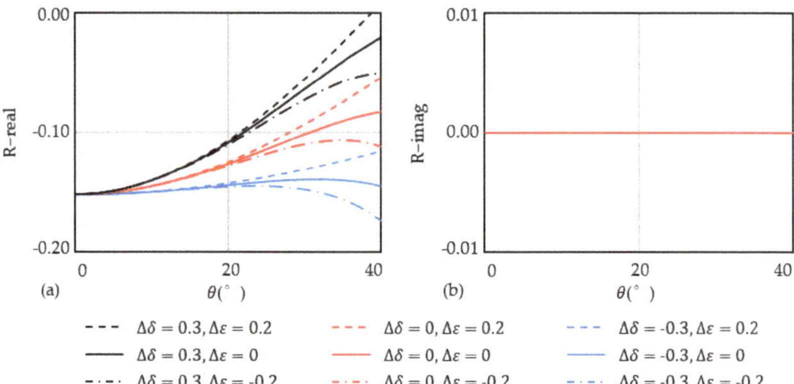

Figure 4. The effect of anisotropic parameters $\Delta\varepsilon$ and $\Delta\delta$ on the reflection coefficients. At this point, the reflection coefficients are real numbers and frequency-independent. (**a**) shows the real part of it in nine combinations of anisotropy parameters, and (**b**) shows the imaginary part equal to zero.

Next, we analyze the effect of perturbations in P- and S-wave attenuation factors ΔQ_p^{-1}, ΔQ_s^{-1} on the reflection coefficients, as shown in Figure 5. For this case, we neglect the parameters of anisotropy and attenuation anisotropy. Figure 5a–f show the real and imaginary parts of reflection coefficients computed using six combinations of attenuation parameters presented in legend in three case of frequency: (1) 5 Hz, (2) 25 Hz and (3) 65 Hz. We mention that the reference frequency is 25 Hz. In Figure 5, we observe that the reflection coefficients are complex numbers, and the real part is frequency-dependent; however, the imaginary part is frequency-independent.

Figure 5a shows the effect of ΔQ_p^{-1}, ΔQ_s^{-1} on the real part of reflection coefficients when the frequency is equal to 5 Hz. The red solid line represents the result computed for the elastic isotropic model because the attenuation parameters are equal to zero. By comparing the solid line of red, black and blue, the intercept, gradient and curvature of them all decrease with the inverse quality factor of P-wave ΔQ_p^{-1}. Three blue lines illustrate the inverse quality factor of S-wave ΔQ_s^{-1} only affects the gradient of the real parts and increases it. Figure 5c shows the real part of reflection coefficients are equal to the result of elastic isotropic assumption in 25 Hz since the natural logarithm of frequency term becomes zero when the frequency we took is equal to reference frequency. Figure 5e shows the effect of ΔQ_p^{-1}, ΔQ_s^{-1} on the real part of reflection coefficients when the frequency is equal to 65 Hz. By comparing the solid line of red, black and blue, the intercept, gradient and curvature of them all increase with the inverse quality factor of P-wave ΔQ_p^{-1}. Three blue lines illustrate the inverse quality factor of S-wave ΔQ_s^{-1} only affects the gradient of the real parts but decreases it.

We conclude the imaginary part of reflection coefficients are frequency- independent, as shown in Figure 5b,d,e. By comparing the solid line of red, black and blue, the intercept, gradient and curvature of the imaginary parts all increase with the inverse quality factor of P-wave ΔQ_p^{-1}. However, the gradient of the imaginary part decreases with the inverse quality factor of S-wave ΔQ_s^{-1} which is illustrated by three blue lines.

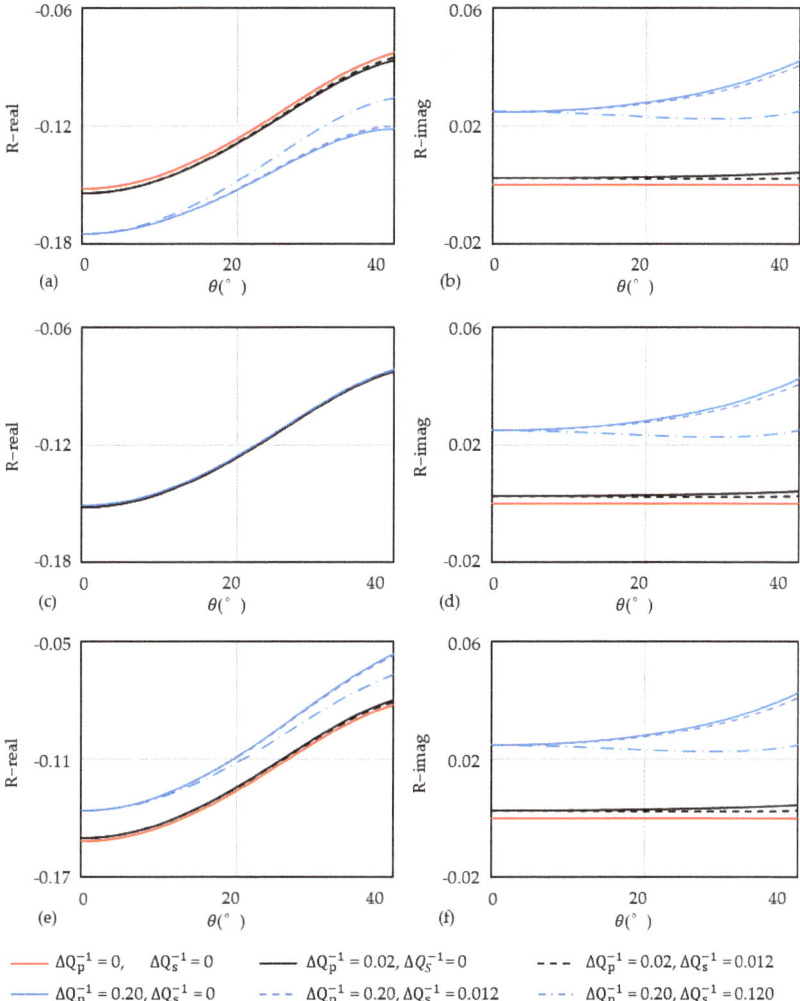

Figure 5. The effect of attenuation parameters ΔQ_p^{-1} and ΔQ_s^{-1} on the reflection coefficients. At this point, the reflection coefficients are complex numbers and frequency-dependent. (**a**–**f**) show the real part and imaginary part of reflection coefficients in three cases of 5 Hz, 25 Hz and 65 Hz separately. Different lines exhibit the results of six combinations of attenuation parameters. The reference frequency is set to 25 Hz.

3.2. Inversion Test for Q-VTI Model

We use the synthetic seismic data to verify the feasibility of the proposed equation for inversion. At first, we choose a well logging data to build a fractured model which is shown in Figure 6. We find that the places with high calcite content developing pores and fractures bearing fluids, and there is no good correspondence of the basis elastic parameters P- and S-wave velocity with it.

Then, we calculate the complex stiffness matrix using Chapman model [4–6] and further obtain the P- and S-wave velocity, density, inverse quality factors, anisotropic parameters and Thomsen-style attenuated anisotropic parameters. As is shown in Figure 7, we acquire these new parameters which is vary with different frequency and incident angle. The P- and S-wave velocities exhibit in this figure are their real parts, and the P- and S-wave velocity we estimated match well with the measured value except some differences in the place of pore, fracture and fluid anomaly. Significantly, the inverse quality factors of P- and S-wave show obvious differences in both frequency and incident angle especially in pore, fracture and fluid anomaly. These small differences are apparent because they are orders of magnitude smaller. They represent the ratio of the imaginary and real parts of P- and S-wave velocity, which is reflect the attenuation characteristics of P- and S-wave. What's more, the trends of Q_p^{-1}, Q_s^{-1} and ε are more consistent with that of porosity, fracture density and fluid saturation than α and β.

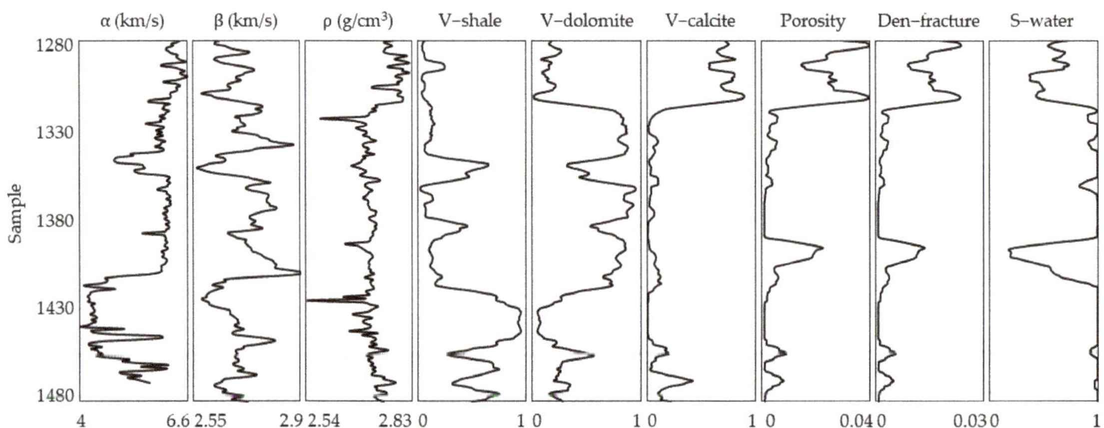

Figure 6. The well logging data of a fractured model which is ready to calculate the attenuated and anisotropic parameters.

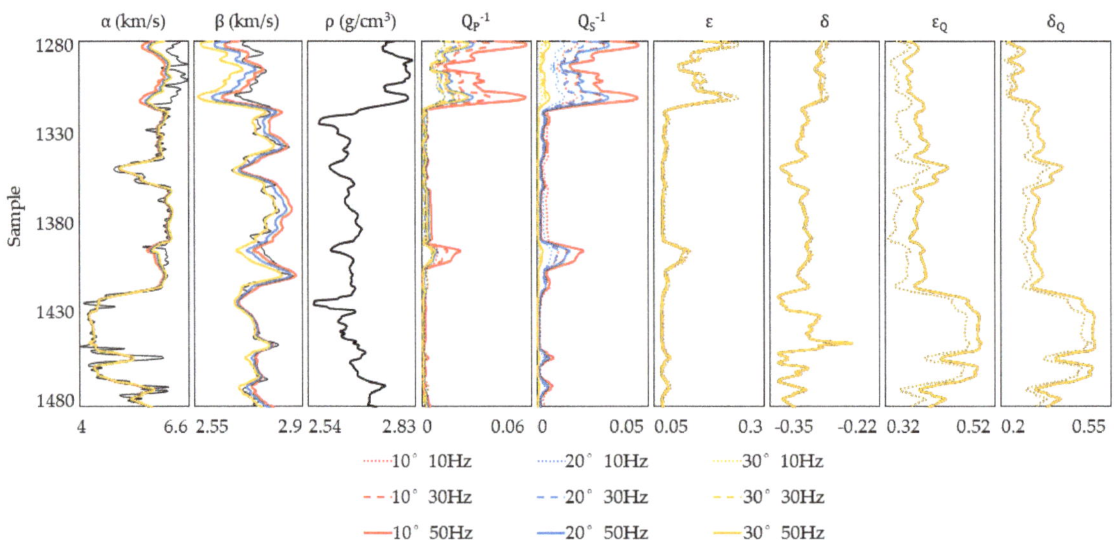

Figure 7. The estimated results of the attenuated and anisotropic parameters using Chapman model in different frequency and incident angle. The black lines are the measured values from well-logging.

Next, we generate synthetic seismic data in small incident angles utilizing Ricker wavelets with different frequencies, as is shown in Figure 8. We add Gaussian random noise into the synthetic seismic data to generate noisy seismic data of signal-to-noise ratio being 5. The result is shown in Figure 9 and used as the observed seismic data for inversion through the Equation (12).

Figure 10 plots comparisons between true values of model and inversion results of each parameter. We take the average value estimated by the Chapman model as the true value for comparison, and calculate the relative error of each inverted parameter, as shown in Figure 11. We observe a close match between inversion results and true values given data with a moderate noise.

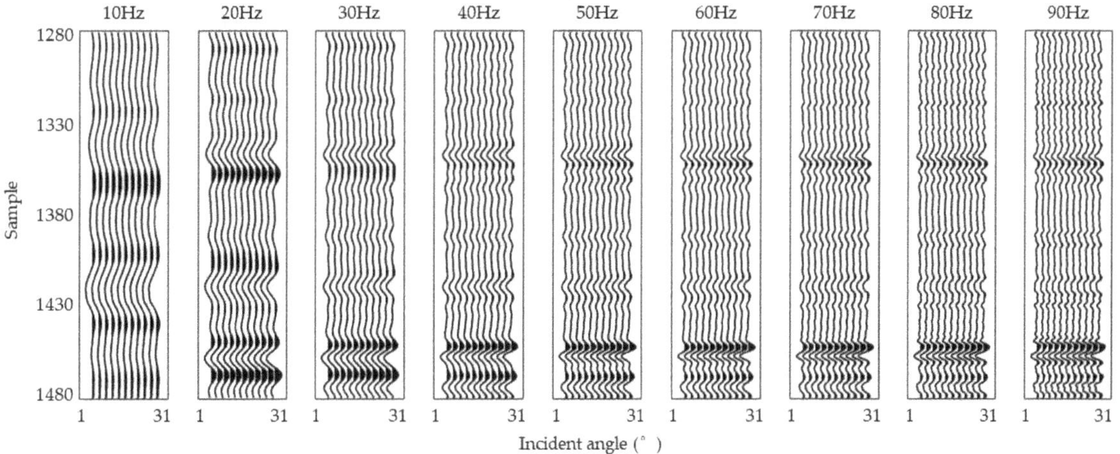

Figure 8. The synthetic seismic data for a set of frequency in small incident angles.

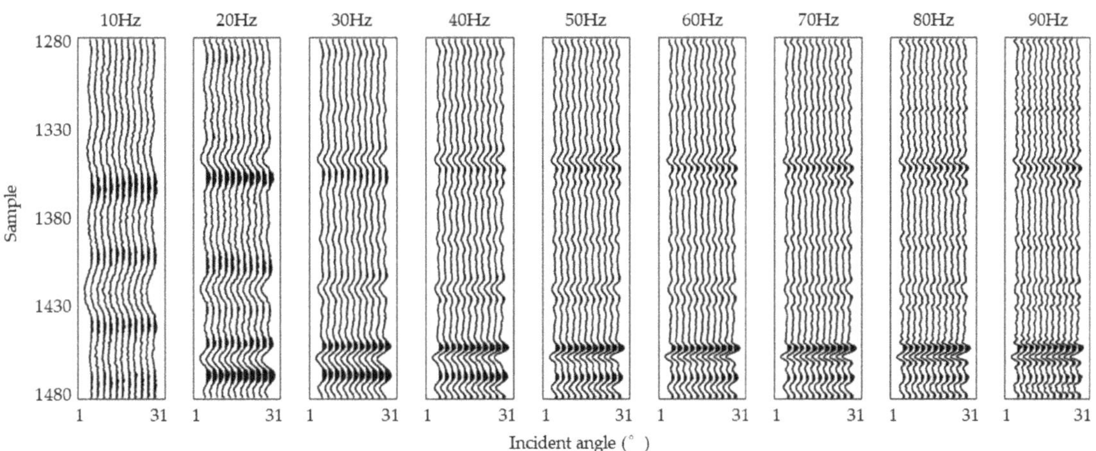

Figure 9. The synthetic seismic data which added Gaussian random noise with S/N of 5. These results will be used as the observed seismic data for inversion. The lower frequency data affected more by noises.

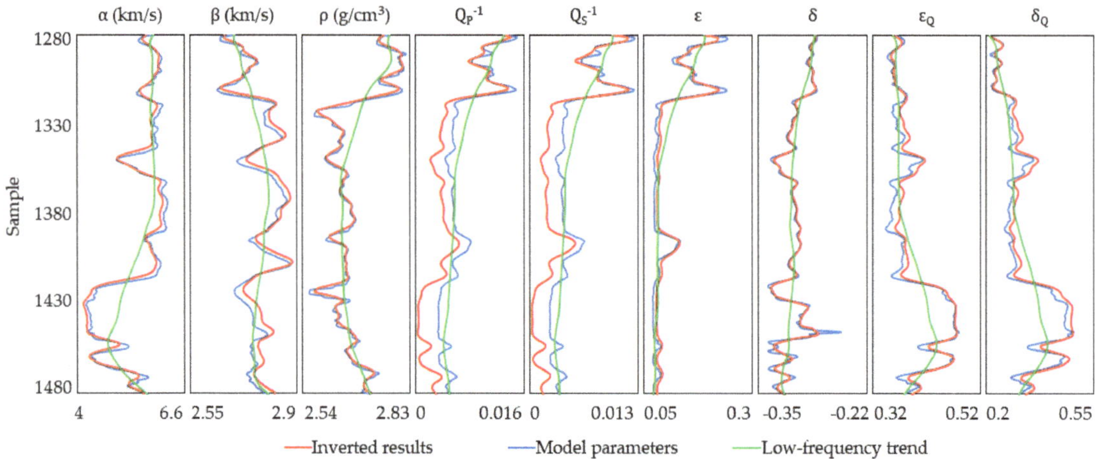

Figure 10. Comparison of the inverted results and true values of the P- and S-wave velocity, density, inverse quality factors, anisotropic parameters and Thomsen-style attenuated anisotropic parameters.

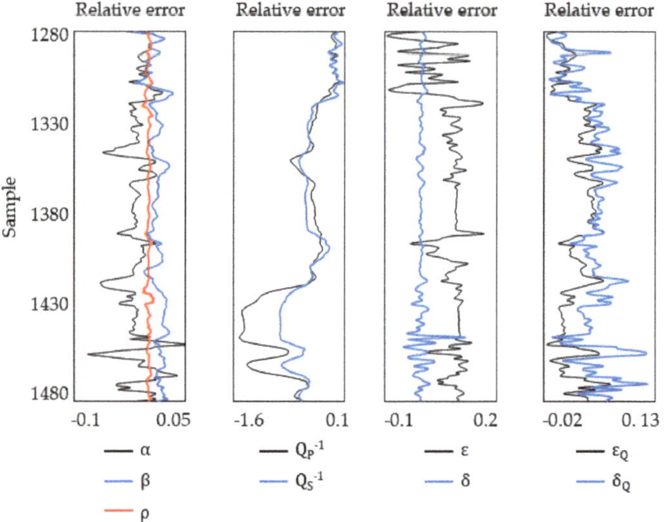

Figure 11. The relative errors between inverted result and model value of each parameter.

4. Result and Discussion

It is possible to obtain relatively simple approximations for the PP-wave reflection coefficient of the linear constant Q-VTI model under the assumption of weak anisotropy and low-loss attenuation of velocities. One of the advantages is the Q-VTI model considers frequency dispersion and anisotropy of velocities at the same time. Another advantage is the equation decouple the inverse quality factors of P- and S-wave and the Thomson-style attenuation anisotropic factors from the complex velocities and Thomson anisotropy parameters. Based on the derived reflection coefficient, we analyze some characteristics of reflection coefficients in the case of different reservoirs and groups of parameters.

The type of AVO is determined by the differences in P-wave velocity, S-wave velocity and density across the interface. Both the anisotropic parameters and the attenuation parameters only change the value of the reflection coefficient; however, they don't affect the type of AVO. The perturbations of anisotropy $\Delta\delta$ and $\Delta\varepsilon$ contribute to the gradient and the curvature. The perturbations in attenuated anisotropy $\Delta\delta_Q$ and $\Delta\varepsilon_Q$ also contribute to the gradient and the curvature, but the contribution is much smaller than that of $\Delta\delta$ and $\Delta\varepsilon$. The attenuation factor of P-wave ΔQ_p^{-1} affects the intercept, gradient and curvature at the same time, however, the attenuation factor of S-wave ΔQ_s^{-1} only affects the gradient.

The reflection coefficients become complex numbers when we consider the effect of the attenuation, in which the real part represents the amplitude and the imaginary part represents the phase. The real parts of reflection coefficients vary with the frequency caused by the attenuation terms, however, the imaginary parts are frequency-independent. The real part of reflection coefficient is inversely proportional to ΔQ_p^{-1} but proportional to ΔQ_s^{-1} because the natural logarithm of frequency term is negative when the frequency is less than the reference frequency. In the same way, the real part of reflection coefficient is proportional to ΔQ_p^{-1} but inversely proportional to ΔQ_s^{-1} because the natural logarithm of frequency term is positive when the frequency is greater than the reference frequency. In particular, the reflection coefficient to be equivalent to the elastic cases because the natural logarithm of frequency term becomes zero when the frequency we took is equal to the reference frequency.

The contribution of the attenuation anisotropic term that the high order to the reflection coefficient is very small compared with other properties. In the meanwhile, the value of third term which affects the reflection coefficient at a large incident angle is much smaller than the first two terms when the P-wave incident at a small angle.

We observe that the relative error of inverse quality factors seems to be large, which is still caused by their relatively small order of magnitude than other parameters, but the trend of their inversion results is completely consistent with the true values. Thus, we still regard the inversion test shows a well result to verify the feasibility of the proposed equation. In addition, the attenuated and anisotropic parameters are not only the better indicators of pores, fractures and fluids than the P- and S-wave velocity, but also enable us to avoid further inversion of physical parameters such as porosity, fracture density and fluid saturation.

5. Summary and Conclusions

Under the assumption of low-loss attenuation and weak anisotropy of velocities, we derive the linearized approximate frequency-dependent reflection coefficient based on the linear constant Q-VTI model. We observe that the reflection coefficient is related to the parameters of anisotropy and attenuation simultaneously, and varies with both the incident angle and frequency. It appears that the analysis of the AVOF characteristics may guide us to identify the characteristics of anisotropy and attenuation in the real working area, and we can take full advantage of the seismic data of different incident angles and frequencies to predict the fluid-filled pores and fractures in the reservoirs using our derived reflection coefficient equation.

Moreover, the attenuated anisotropic medium is more suitable for modeling how seismic wave propagates in underground layers than that proposed under the assumption of elastic isotropic or anisotropic medium. Focusing on the attenuated anisotropic medium, we consider the effects of Thomsen parameters that are related to anisotropy caused by fractures and the attenuation factors that are sensitive to attenuation caused by intrinsic attenuation and the fluid-filled in pores and fractures on the reflection coefficient, which may provide more useful information for detecting fractures and fluids using the observed seismic data. We conclude that we present a valuable expression of reflection coefficient, which can be employed for the analysis of seismic wave response modeling for different types of reservoirs, and the derived reflection coefficient can also guide the inversion for the properties that are related to fractures and fluids using frequency components of seismic amplitudes.

Author Contributions: Conceptualization, Y.Y. and X.Y.; methodology, Y.Y. and G.G.; software, Y.Y. and G.G.; validation, Y.Y., X.Y., G.G., D.C. and B.Z.; formal analysis, Y.Y.; investigation, Y.Y.; resources, D.C.; data curation, D.C.; writing—original draft preparation, Y.Y.; writing—review and editing, B.Z.; visualization, Y.Y.; supervision, X.Y.; project administration, X.Y.; funding acquisition, X.Y. All authors have read and agreed to the published version of the manuscript.

Funding: This research was funded by the National Natural Science Foundation of China (42030103, 42004092).

Acknowledgments: The authors would like to acknowledge the many current and past researchers who have contributed to the foundational work for this study. We also thank the anonymous reviewers and editors for their constructive suggestions.

Conflicts of Interest: The authors declare no conflict of interest.

Appendix A

The complex stiffness matrix of the Q-VTI medium can be re-expressed as the sum of the anisotropic perturbation and the stiffness matrix of a homogeneous isotropic background based on the perturbation theory.

$$
\begin{aligned}
\tilde{c}_{QVTI} &=
\begin{bmatrix}
\tilde{c}_{11} & \tilde{c}_{11} - 2\tilde{c}_{66} & \tilde{c}_{13} & 0 & 0 & 0 \\
\tilde{c}_{11} - 2\tilde{c}_{66} & \tilde{c}_{11} & \tilde{c}_{13} & 0 & 0 & 0 \\
\tilde{c}_{13} & \tilde{c}_{23} & \tilde{c}_{33} & 0 & 0 & 0 \\
0 & 0 & 0 & \tilde{c}_{55} & 0 & 0 \\
0 & 0 & 0 & 0 & \tilde{c}_{55} & 0 \\
0 & 0 & 0 & 0 & 0 & \tilde{c}_{66}
\end{bmatrix} \\[4pt]
&=
\begin{bmatrix}
\tilde{c}_{33}^{0} & \tilde{c}_{33}^{0} - 2\tilde{c}_{55}^{0} & \tilde{c}_{33}^{0} - 2\tilde{c}_{55}^{0} & 0 & 0 & 0 \\
\tilde{c}_{33}^{0} - 2\tilde{c}_{55}^{0} & \tilde{c}_{33}^{0} & \tilde{c}_{33}^{0} - 2\tilde{c}_{55}^{0} & 0 & 0 & 0 \\
\tilde{c}_{33}^{0} - 2\tilde{c}_{55}^{0} & \tilde{c}_{33}^{0} - 2\tilde{c}_{55}^{0} & \tilde{c}_{33}^{0} & 0 & 0 & 0 \\
0 & 0 & 0 & \tilde{c}_{55}^{0} & 0 & 0 \\
0 & 0 & 0 & 0 & \tilde{c}_{55}^{0} & 0 \\
0 & 0 & 0 & 0 & 0 & \tilde{c}_{55}^{0}
\end{bmatrix} \\[4pt]
&+
\begin{bmatrix}
\Delta\tilde{c}_{33} & \Delta\tilde{c}_{33} - 2\Delta\tilde{c}_{55} & \Delta\tilde{c}_{33} - 2\Delta\tilde{c}_{55} & 0 & 0 & 0 \\
\Delta\tilde{c}_{33} - 2\Delta\tilde{c}_{55} & \Delta\tilde{c}_{33} & \Delta\tilde{c}_{33} - 2\Delta\tilde{c}_{55} & 0 & 0 & 0 \\
\Delta\tilde{c}_{33} - 2\Delta\tilde{c}_{55} & \Delta\tilde{c}_{33} - 2\Delta\tilde{c}_{55} & \Delta\tilde{c}_{33} & 0 & 0 & 0 \\
0 & 0 & 0 & \Delta\tilde{c}_{55} & 0 & 0 \\
0 & 0 & 0 & 0 & \Delta\tilde{c}_{55} & 0 \\
0 & 0 & 0 & 0 & 0 & \Delta\tilde{c}_{55}
\end{bmatrix} \\[4pt]
&+
\begin{bmatrix}
\tilde{c}_{11} - \tilde{c}_{33} & \tilde{c}_{12} - \tilde{c}_{33} + 2\tilde{c}_{55} & \tilde{c}_{13} - \tilde{c}_{33} + 2\tilde{c}_{55} & 0 & 0 & 0 \\
\tilde{c}_{12} - \tilde{c}_{33} + 2\tilde{c}_{55} & \tilde{c}_{11} - \tilde{c}_{33} & \tilde{c}_{13} - \tilde{c}_{33} + 2\tilde{c}_{55} & 0 & 0 & 0 \\
\tilde{c}_{13} - \tilde{c}_{33} + 2\tilde{c}_{55} & \tilde{c}_{13} - \tilde{c}_{33} + 2\tilde{c}_{55} & 0 & 0 & 0 & 0 \\
0 & 0 & 0 & 0 & 0 & 0 \\
0 & 0 & 0 & 0 & 0 & 0 \\
0 & 0 & 0 & 0 & 0 & \tilde{c}_{66} - \tilde{c}_{55}
\end{bmatrix}
\end{aligned}
\tag{A1}
$$

where, the stiffness coefficients in the square brackets on the right-hand side represent isotropic attenuated background, isotropic attenuated perturbations and anisotropic attenuated perturbations, respectively.

The polarization vectors and slowness vectors corresponding the incident and reflected waves are given by,

$$t = [\sin\theta\cos\varphi, \sin\theta\sin\varphi, \cos\theta]$$

$$t' = [-\sin\theta\cos\varphi, -\sin\theta\sin\varphi, \cos\theta]$$

$$p = (1/\tilde{v})[\sin\theta\cos\varphi, \sin\theta\sin\varphi, \cos\theta] \tag{A2}$$

$$p' = (1/\tilde{v})[-\sin\theta\cos\varphi, -\sin\theta\sin\varphi, \cos\theta]$$

References

1. Thomsen, L. Elastic anisotropy due to aligned cracks in porous rock. *Geophys. Prospect.* **1995**, *43*, 805–829. [CrossRef]
2. Carcione, J.M. Constitutive model and wave equations for linear, viscoelastic, anisotropic media. *Geophysics* **1995**, *60*, 537–548. [CrossRef]
3. Carcione, J.M. *Wave Fields in Real Media: Wave Propagation in Anisotropic, Anelastic, Porous and Electromagnetic Media*, 2nd ed.; Elsevier: Amsterdam, The Netherlands, 2007.
4. Chapman, M. Frequency-dependent anisotropy due to meso-scale fractures in the presence of equate porosity. *Geophys. Prospect.* **2003**, *51*, 369–379. [CrossRef]
5. Chapman, M.; Liu, E.; Li, X. The influence of fluid-sensitive dispersion and attenuation on AVO analysis. *Geophys. J. Int.* **2006**, *167*, 89–105. [CrossRef]
6. Chapman, M. Modeling the effect of multiple sets of mesoscale fractures in porous rock on frequency-dependent anisotropy. *Geophysics* **2009**, *74*, D97–D103. [CrossRef]
7. Ba, J.; Cao, H.; Yao, F.; Nie, J.; Yang, H. Double-porosity rock model and squirt flow in the laboratory frequency band. *Appl. Geophys.* **2008**, *5*, 261–276. [CrossRef]
8. Guo, J.; Shuai, D.; Wei, J.; Ding, P.; Gurevich, B. P-wave dispersion and attenuation due to scattering by aligned fluid saturated fractures with finite thickness: Theory and experiment. *Geophys. J. Int.* **2018**, *215*, 2114–2133. [CrossRef]
9. Guo, J.; Gurevich, B.; Shuai, D. Frequency-dependent P-wave anisotropy due to scattering in rocks with aligned fractures. *Geophysics* **2020**, *85*, MR97–MR105. [CrossRef]
10. Thanh, H.V.; Sugai, Y.; Nguele, R.; Sasaki, K. A New Petrophysical Modeling Workflow for Fractured Granite Basement Reservoir in Cuu Long Basin, Offshore Vietnam. In Proceedings of the 81st EAGE conference and exhibition, London, UK, 3–6 June 2019; pp. 1–5. [CrossRef]
11. Thanh, H.V.; Sugai, Y.; Nguele, R.; Sasaki, K. Integrated workflow in 3D geological model construction for evaluation of CO_2 storage capacity of a fractured basement reservoir in Cuu Long Basin, Vietnam- Sciencedirect. *Int. J. Greenh. Gas Control* **2019**, *90*, 102826. [CrossRef]
12. Schoenberg, M. Elastic wave behavior across linear slip interfaces. *J. Acoust. Soc. Am.* **1980**, *68*, 1516–1521. [CrossRef]
13. Hudson, J.A. Wave speeds and attenuation of elastic waves in material containing cracks. *Geophys. J. Int.* **1981**, *64*, 133–150. [CrossRef]
14. Gurevich, B. Elastic properties of saturated porous rocks with aligned fractures. *J. Appl. Geophys.* **2003**, *54*, 203–218. [CrossRef]
15. Chapman, M.; Maultzsch, M.; Liu, E.; Li, X. The effect of fluid saturation in an anisotropic multi-scale equant porosity model. *J. Appl. Geophys.* **2003**, *54*, 191–202. [CrossRef]
16. Aki, K.; Richards, P.G. *Quantitative Seismology: Theory and Methods*; W.H. Freeman: San Francisco, CA, USA, 1980; p. 932.
17. Shaw, R.K.; Sen, M.K. Use of AVOA data to estimate fluid indicator in a vertically fractured medium. *Geophysics* **2006**, *71*, C15–C24. [CrossRef]
18. Zong, Z.; Yin, X.; Wu, G. Complex seismic amplitude inversion for P-wave and S-wave quality factors. *Geophys. J. Int.* **2015**, *202*, 564–577. [CrossRef]
19. Moradi, S.; Innanen, K.A. Born scattering and inversion sensitivities in viscoelastic transversely isotropic media. *Geophys. J. Int.* **2017**, *211*, 1177–1188. [CrossRef]
20. Moradi, S. Scattering of Seismic Waves from Arbitrary Viscoelastic-Isotropic and Anisotropic Structures with Applications to Data Modelling, FWI Sensitivities and Linearized AVO-AVAz Analysis. Ph.D. Thesis, University of Calgary, Calgary, AB, Canada, 2017.
21. Chen, H.; Innanen, K.A.; Chen, T. Estimating P- and S-wave inverse quality factors from observed seismic data using an attenuative elastic impedance. *Geophysics* **2018**, *83*, R173–R187. [CrossRef]
22. Chen, H.; Li, J.; Innanen, K.A. Inversion of differences in frequency components of azimuthal seismic data for indicators of oil-bearing fractured reservoirs based on an attenuative cracked model. *Geophysics* **2020**, *85*, 1MJ-Z13. [CrossRef]
23. Pan, X.; Zhang, G.; Cui, Y. Matrix-fluid-fracture decoupled-based elastic impedance variation with angle and azimuth inversion for fluid modulus and fracture weaknesses. *J. Pet. Sci. Eng.* **2020**, *189*, 106974. [CrossRef]
24. Kjartansson, E. Constant Q-wave propagation and attenuation. *J. Geophys. Res. Atmos.* **1979**, *84*, 4737–4748. [CrossRef]

25. Thomsen, L. Weak elastic anisotropy. *Geophysics* **1986**, *51*, 1954–1966. [CrossRef]
26. Thomsen, L. Reflection seismology over azimuthally anisotropic media. *Geophysics* **1988**, *53*, 304–313. [CrossRef]
27. Zhu, Y.; Tsvankin, I. Plane-wave propagation in attenuative transversely isotropic media. *Geophysics* **2006**, *71*, T17–T30. [CrossRef]
28. Zhu, Y.; Tsvankin, I.; Dewangan, P.; Wijk, K.V. Physical modeling and analysis of P-wave attenuation anisotropy in transversely isotropic media. *Geophysics* **2007**, *72*, D1–D7. [CrossRef]
29. Shaw, R.K.; Sen, M.K. Born integral, stationary phase and linearized reflection coefficients in weak anisotropic media. *Geophys. J. Int.* **2004**, *158*, 225–238. [CrossRef]
30. Rüger, A. Reflection Coefficients and Azimuthal AVO Analysis in Anisotropic Media. Ph.D. Thesis, Colorado School of Mines, Denver, CO, USA, 1996.
31. Rüger, A. Variation of P-wave reflectivity with offset and azimuth in anisotropic media. *Geophysics* **1998**, *63*, 935–947. [CrossRef]
32. Shuey, R.T. A simplification of the Zoeppritz equations. *Geophysics* **1985**, *50*, 609–614. [CrossRef]

Article

A Multi-Point Geostatistical Seismic Inversion Method Based on Local Probability Updating of Lithofacies

Zhihong Wang [1,*], Tiansheng Chen [2], Xun Hu [3], Lixin Wang [4,5] and Yanshu Yin [4,5,*]

[1] Research Institute of Petroleum Exploration & Development, PetroChina, P.O. Box 910,20#, Xueyuan Road, Beijing 100083, China
[2] SINOPEC Petroleum Exploration and Production Research Institute, 31 Xueyuan Road, Haidian District, Beijing 100083, China; chents.syky@sinopec.com
[3] College of Geosciences, China University of Petroleum (Beijing), 18 Fuxue Road, Changping, Beijing 102249, China; 2020310040@student.cup.edu.cn
[4] Key Laboratory of Oil and Gas Resources and Exploration Technology, Ministry of Education, Yangtze University, Wuhan 430100, China; 201571323@yangtzeu.edu.cn
[5] School of Geosciences, Yangtze University, 111 University Road, Caidian District, Wuhan 430100, China
[*] Correspondence: wzh2331@sina.com (Z.W.); yys@yangtzeu.edu.cn (Y.Y.)

Abstract: In order to solve the problem that elastic parameter constraints are not taken into account in local lithofacies updating in multi-point geostatistical inversion, a new multi-point geostatistical inversion method with local facies updating under seismic elastic constraints is proposed. The main improvement of the method is that the probability of multi-point facies modeling is combined with the facies probability reflected by the optimal elastic parameters retained from the previous inversion to predict and update the current lithofacies model. Constrained by the current lithofacies model, the elastic parameters were obtained via direct sampling based on the statistical relationship between the lithofacies and the elastic parameters. Forward simulation records were generated via convolution and were compared with the actual seismic records to obtain the optimal lithofacies and elastic parameters. The inversion method adopts the internal and external double cycle iteration mechanism, and the internal cycle updates and inverts the local lithofacies. The outer cycle determines whether the correlation between the entire seismic record and the actual seismic record meets the given conditions, and the cycle iterates until the given conditions are met in order to achieve seismic inversion prediction. The theoretical model of the Stanford Center for Reservoir Forecasting and the practical model of the Xinchang gas field in western China were used to test the new method. The results show that the correlation between the synthetic seismic records and the actual seismic records is the best, and the lithofacies matching degree of the inversion is the highest. The results of the conventional multi-point geostatistical inversion are the next best, and the results of the two-point geostatistical inversion are the worst. The results show that the reservoir parameters obtained using the local probability updating of lithofacies method are closer to the actual reservoir parameters. This method is worth popularizing in practical exploration and development.

Keywords: local updating; permanent updating ratio of probability; multi-point geostatistical inversion; cyclic iteration; correlation coefficient; Xinchang gas field

Citation: Wang, Z.; Chen, T.; Hu, X.; Wang, L.; Yin, Y. A Multi-Point Geostatistical Seismic Inversion Method Based on Local Probability Updating of Lithofacies. *Energies* **2022**, *15*, 299. https://doi.org/10.3390/en15010299

Academic Editor: Alexei V. Milkov

Received: 29 October 2021
Accepted: 25 December 2021
Published: 2 January 2022

Publisher's Note: MDPI stays neutral with regard to jurisdictional claims in published maps and institutional affiliations.

1. Introduction

Seismic inversion is an important approach to lithology identification and oil–gas interpretation. It converts conventional seismic reflection records into acoustic impedance properties and reservoir parameters in order to give them a more definite geological meaning. It is a common concern of oil and gas geophysicists and geologists to directly apply seismic inversion methods to fine reservoir characterization and modeling. However, due to the noise of seismic data, the finite frequency of seismic waves, and the incomplete mapping of geological attributes to the seismic physical parameters, the inversion and interpretation of seismic records into reservoir attributes are not unique and pose great challenges. Some scholars have conducted a lot of research on eliminating the impact of noise, Ghaderpour [1] proposed a method of seismic data regularization and random noise attenuation via least-squares spectral analysis in frequency wavenumber domain, due to the accuracy of the estimated wavenumbers, the total number of iterations of the method is significantly reduced and the efficiency is significantly improved. However, there are still many problems in the process of connecting seismic property with geology. The design and development of advanced seismic inversion methods that integrate geological, rock geophysics, and even the production of dynamic data, have been important topics for exploration geophysicists, and two types of inversion methods, namely, deterministic inversion and (geological) statistical inversion [2], have gradually formed. Deterministic inversion obtains the maximum posteriori probability model through an optimization algorithm and minimizes the error. Although strong reflector information can be recovered well and the inversion results have a good lateral continuity, the resolution of the inversion results can only reach the resolution of the seismic data due to the limited bandwidth of the seismic data [3]. In order to improve the resolution, the consensus is that it is necessary to integrate various geological (logging) information into the reservoir inversion using spatial reservoir correlation [2]. In geological modeling, this spatial correlation is mainly represented by the variogram function. Journel and Huijbregt [4] first developed the reservoir geological modeling method integrating seismic data, which laid a solid theoretical foundation for seismic stochastic inversion. In 1994, Hass and Dubrule [5] proposed stochastic inversion based on sequential simulation in the First break, which is the prototype of the geostatistical inversion method. Since the spatial correlation is characterized by the vertical variogram function of the borehole data, the planar continuity is obtained from the seismic data. Therefore, the inversion effectively makes use of the vertical resolution of the well data, makes up for the limitation of the seismic bandwidth, and improves the inversion resolution [2,4–8]. In addition, the inversion probabilities are inferred using the Kriging method, and the Markov chain Monte Carlo (MCMC) method is used for sampling posterior probabilities [9–11], which satisfy the needs of statistical inversion uncertainty analysis and evaluation. Azevedo and Demyanov [12] have also conducted research on multi-scale uncertainty evaluation in geostatistical seismic inversion, this method combines geostatistical seismic inversion with a stochastic adaptive sampling and Bayesian inference of the metaparameters to provide more accurate and realistic uncertainty prediction without being limited by a large number of assumptions of large-scale geological parameters. Pereira [13] proposed iterative geostatistical seismic inversion combined with local anisotropy, this method adopts a random sequence simulation and joint simulation method, which can deal with the information of spatial variation, and uses local and independent variogram models to reduce the spatial uncertainty related to underground characteristics. Therefore, the geostatistical inversion method has been widely used and has achieved good results in practical applications.

With the development of geological modeling research, more and more modelers have pointed out that the variogram-based method is difficult to integrate more information in order to describe a complex curved reservoir morphology, and it cannot fully reveal the spatial variability [6,14–18]. It is necessary to combine the spatial distribution of multiple points to determine the reservoir's characteristics. Based on this idea, Guardiano and Srivastava [19] proposed the concept of a spatial multi-point joint distribution to represent complex reservoir structures, and they obtained the multi-point probability through repeated scanning of a training image (a quantitative grid-based reservoir lithofacies model) and data samples (i.e., the spatial multi-point combination model) and applied it to the prediction of the points to be estimated. Strebelle [15] improved this method by designing a search tree to store and access the multi-point probability, which improved the simulation efficiency. Multi-point geostatistics were formally introduced into actual reservoir modeling [15] and gradually replaced the traditional two-point geostatistics method based on the variogram function. This has also aroused the attention of geostatistical inversion scholars.

Gonzalez [8], who first attempted to apply multi-point geostatistics to reservoir inversion, used the improved Simpat method to obtain the lithofacies distribution, sampled the seismic attributes through the relationship between the lithofacies and seismic attributes, and finally used the likelihood function to determine the optimal matching elastic parameters. Their method emphasizes the control of the relative sedimentary facies quality; that is, the spatial continuity of the elastic parameter field and its sampling are controlled by a specific geological lithofacies model. They named the method mSIMPAT. However, the calculation efficiency of the mSIMPAT is low in the process of updating facies, which creates difficulties in actual seismic inversion. Jeong [20] replaced mSIMPAT with the direct sampling method, which they combined with the adaptive spatial resampling (ASR) method to improve the operation efficiency. However, the ASR method retains the optimal matching facies data and adds conditional data to guide the multi-point geostatistical facies modeling. The inverted elastic parameters were obtained through integral iteration without local lithofacies updating. Especially in lithofacies modeling, the elastic parameters obtained during previous iterations cannot be used as constraints. Liu [21] replaced mSIMPAT with the SNESIM method and combined it with the probability perturbation method (PPM) to accelerate the inversion iteration efficiency. The updating of the lithofacies model is conducted by disturbing the entire geological model using the probabilistic perturbation method without updating the local lithofacies. Although this disturbance satisfies the actual seismic observation data through annealing optimization, it is likely to be at the expense of disturbing the local specific deposition patterns. Because the specific lithofacies model plays an important role in the inversion, it not only determines the inversion's efficiency, but also the accuracy of the inversion [22–25]. Therefore, it is necessary to reconsider the local probability updating in facies modeling.

In this study, the iterative inversion method of Gonzalez [8] is revised. In the iterative process, the theory of the permanent probability updating ratio is used to integrate the early elastic parameters for the local lithofacies prediction. In addition, the inversion results of the current iteration are not only evaluated but are also compared with the previously partially retained lithofacies and the elastic parameters to determine whether to update. The theoretical model tests reveal that the improved method can reflect the distribution of the reservoir lithofacies and the elastic parameters better, and its calculation efficiency is high. Practical inversion of the Xinchang gas field data in China also demonstrates that the improved method has a higher inversion accuracy. The results of this research provide technical support for oil and gas exploration and development.

2. Principle and Methods

2.1. Inversion Principle and Multi-Point Geostatistical Inversion Method

All inversion processes can be regarded as the process of obtaining synthetic seismic records of the elastic parameters in a certain way and matching the real seismic records within an allowable error range, the principle of which can be expressed by the Bayesian formula [26].

$$\sigma_M(m) = c\gamma_M(m)\gamma_D(g(m)), \tag{1}$$

where c is the correction parameter and is a constant, $\gamma_M(m)$ is the prior probability, and $\gamma_D(g(m))$ is the likelihood function. M is the simulation region, m is the initial model or pattern group, $g(m)$ is the forward operator, and $\sigma_M(m)$ is the posterior probability.

Inversion is an inference process in which the prior probability is updated and made faithful to the actual seismic data, and the maximum posterior probability is the core objective. $\gamma_D(g(m))$ is used to measure the matching degree between the forward simulation record and the actual observed seismic track. Its elastic parameters are generally obtained from the prior probability sampling, and the wavelet comes from the actual seismic working area. Therefore, the core of the inversion lies in the method of obtaining the prior probability $\gamma_M(m)$ [17].

Haas and Dubrule [5] used a sequential Gaussian simulation to obtain the impedance data, in which the prior probability of the impedance was predicted using the variogram function, which also constituted the most initial geostatistical inversion. Subsequently, different scholars discussed the influence of the prior information on the Bayesian inversio. Accurate prediction of the prior probability is the key to improving the accuracy of the seismic inversion. Considering that multi-point statistics can obtain higher-order prior statistics from training images and can integrate more information than the second-order statistics of the variogram function, using multi-point geostatistics to predict the prior impedance information is a potential development direction. However, multi-point geostatistics is mainly applicable to discrete variables, and it is difficult to predict continuous variables. In seismic inversion, it is often necessary to establish statistical rock physics models; that is, the statistical relationship between the elastic parameters m_{elas} (such as the impedance and velocity) and the reservoir properties m_{res} (such as the lithofacies). According to the chain rule of conditional probability, the prior probability can be written as

$$\gamma_M(m_{res}, m_{elas}) = P_{prior}(m_{res}, m_{elas}) = P_{petro}(m_{elas}|m_{res})P_{prior}(m_{res}). \tag{2}$$

Thus, the prior probability of the lithofacies can be predicted using multi-point statistics, and the current joint prior probability distribution of the elastic parameters–lithofacies can be obtained from the lithofacies and elastic parameter probability [8].

The likelihood function $\gamma_D(g(m))$ is used to measure the error between the forward simulated record and the actual observed seismic trace. Selecting a specific likelihood function is essential to determining what is a good enough fit. It can be based on the distribution of the measurement errors, or it can be assessed subjectively, for example, using the seismic root mean square error or correlation coefficient. The likelihood function is generally expressed as the sum of the residuals between the forward simulated record and the actual seismic data (assuming that the seismic noise has a Gaussian random distribution with mean value of 0 and a variance of σ_e):

$$\gamma_D(g(m)) = \frac{1}{(2\pi\sigma_e^2)^2} \exp\left[-\sum \frac{(d - g(m))^2}{2\sigma_e^2}\right] \tag{3}$$

where D is the observed seismic trace, and $g(m)$ is the synthetic seismic trace. By combining Equations (1)–(3), the posterior probability can be expressed as

$$\sigma_M(m) = \gamma_M(m_{res}, m_{elas}|d) = c\left[\frac{1}{(2\pi\sigma_e^2)^2} \exp\left[-\sum \frac{(d - g(m))^2}{2\sigma_e^2}\right]\right]\left[P_{petro}(m_{elas}|m_{res})P_{prior}(m_{res})\right]. \tag{4}$$

Once the a posteriori probability distribution is calculated, it can be used several times for sampling and characterization of the a posteriori probability of the reservoir model. Each model in the model set is consistent with the geological knowledge and the prior information in the training image. Lithofacies and the actual seismic data have a better matching relationship. This sampling is usually achieved using MCMC sampling. However, it takes a long time for the Markov chain to visit all the state spaces, and it converges slowly to a stationary distribution. Gonzalez [8] cleverly designed the internal and external double iteration method to achieve an inversion effect using a limited number of iterations. Its two core processes of this method are preprocessing and inversion. Preprocessing is the preparation of the information required for the inversion, including training images, statistical relationship between lithofacies and elastic parameters, and well data. Inversion is an iterative process. First, a random path is defined, the prior probability of the lithofacies is obtained through multi-point scanning of the training images, and the geological model library is established. The selection of different geological models can be regarded as the external iteration. Then, according to relationship between lithofacies and elastic parameters, the attribute values, such as the acoustic velocity and density are extracted, which is the internal iteration. According to the attribute values obtained from the simulation, the reflection coefficient sequence is obtained, and the forward simulation record is obtained through seismic wavelet convolution and is compared to the actual seismic record. If the error between them meets the set condition, the attribute value of the point to be estimated is retained; otherwise, it is extracted and simulated again. If the internal iteration is completed, and the best matching geological model is not found, the cycle is broken out and a new geological model is searched from the model library. The above steps are repeated until the given conditions are met in order to achieve seismic inversion and reservoir prediction.

2.2. Method Improvement

Gonzalez [8] introduced multi-point statistical inversion (mSIMPAT), which has been widely applied and studied. Because the mSIMPAT method is used to search for the best matching deposition pattern, the entire training image must be scanned repeatedly each time. When the size of the training image and the data sample is slightly larger, the overall scanning will seriously increase the computational load. In the process of internal and external double iteration, the optimal elastic parameters and lithofacies data obtained from the previous external iteration inversion do not provide information and constraints for the next inversion iteration cycle, resulting in each iteration cycle being independent. Thus, it is difficult to iteratively update the local lithofacies model. To solve the above problems, the iterative inversion algorithm was improved.

In view of the low computational efficiency of the mSIMPAT method, scholars replaced it with the direct sampling (DS) method and the SNESIM method. The DS method is a direct matching method [27]. Since it does not need to store the multi-point conditional probability, it avoids the storage problem when the probability of the training image scanned is larger. Because of the non-integral scanning, its computational efficiency is significantly better. Local areas can be selected during scanning, which can ensure the reproduction of the local characteristics of the sedimentary model and reflect the non-stationary reservoir structure to a certain extent, and it is more suitable for reservoir prediction involving complex changes. Therefore, the DS method is a natural choice to replace the mSIMPAT method as the prior probability method [20]. However, the DS method is still difficult to implement in terms of local updating under synthetic elastic parameter constraints. In contrast, the SNESIM method has a high computational efficiency because it stores all of the multi-point probabilities through one scan. Single point prediction more easily integrates multiple sources of information, especially the elastic parameters obtained in the previous iteration. Therefore, in this study, the SNESIM method was chosen to replace the mSIMPAT method.

In view of the local updating of the lithofacies in the inversion process, the statistical relationship between the elastic parameters and the lithofacies is attained using the permanent ratio of the updating theory in the inner cycle [28]:

$$P(A|B,C) = \frac{1}{1+x} = \frac{a}{a+bc} \in [0,1], \tag{5}$$

$$a = \frac{1-P(A)}{P(A)} = \frac{P(\tilde{A})}{P(A)} \in [0,+\infty), \tag{6}$$

$$b = \frac{1-P(A|B)}{P(A|B)} = \frac{P(\tilde{A}|B)}{P(A|B)}, \tag{7}$$

$$c = \frac{1-P(A|C)}{P(A|C)} = \frac{P(\tilde{A}|C)}{P(A|C)}, \tag{8}$$

$$x = \frac{1-P(A|B,C)}{P(A|B,C)} = \frac{P(\tilde{A}|B,C)}{P(A|B,C)} \geq 0. \tag{9}$$

$P(A|B,C)$ is the current joint statistical probability of the training images and the elastic parameters. $P(A|B)$ is the multi-point probability under the condition of only lithofacies data. $P(A \mid C)$ is probability under the condition of the optimal elastic parameters of the previous inversion, which is known from the elastic parameters–lithofacies statistical probability. $P(A)$ is the lithofacies statistical probability obtained from the geological analysis; hence, a in Equations (6) and (10) can be interpreted as a prior distance to the event A occurring. Likewise, the values b and c in Equations (7), (8) and (10) state the uncertainty about occurrence of A, given information B and C, respectively. x is the uncertainly when knowing both B and C.

To describe the relationship between B and C, the τ factor is introduced:

$$\frac{x}{b} = \left(\frac{c}{a}\right)^{\tau(B,C)}, \tau(B,C) \geq 0. \tag{10}$$

$\tau(B,C)$ is an evaluation of the correlation degree between the seismic elastic parameters and the lithofacies, and it indicates whether the seismic elastic parameters reflect the type and distribution of the lithofacies, and it is generally obtained through trial and error.

According to Equations (5) and (10), the elastic parameters obtained from the previous iteration inversion can be used to constrain the local lithofacies prediction and update the local lithofacies model. In order to determine the optimal elastic parameters in the local inversion, the current forward simulation records are compared with the previous forward simulation records, including the optimal records retained in the earlier stage of the outer cycle.

In the outer cycle, the current overall inversion results are compared with the actual error to decide whether to retain the inversion results of the elastic parameters and repeat the cycle iteration. This continues until the local elastic parameter inversion and the global inversion satisfy the given conditions. Then, the cycle terminates and the inversion results are output.

2.3. Inversion Steps

Based on the above improvements, a multi-point geostatistical inversion method based on the local probability updating method for the inversion of lithofacies (LPUMI) was developed in this study. The main steps are as follows (Figure 1).

Step 1: Preprocessing

a. Check the data. Check whether the seismic data and well data are complete, including lithology, density, p-wave velocity, and s-wave velocity information.
b. Statistical analysis of the data. When the shear wave information cannot be obtained from the logging data, it can be estimated using empirical formulas. The probability density functions of the different elastic parameters of the lithofacies are established to provide a basis for the subsequent elastic parameter sampling. The plot of the lithofacies versus the elastic parameters is established to provide a basis for the fluid prediction.
c. The attribute values of the initial reservoir elastic parameters are given. According to the statistical well data, the initial elastic parameter attribute values, including the density, p-wave velocity, and s-wave velocity, are assigned to the simulation grid.
d. Build training images. Commonly, unconditional modeling methods such as object-based stochastic modeling, sedimentary process modeling, multi-point simulation results, outcrop and modern deposition models, digital geological sketches, and physical simulation interpretation are used to confirm the working area's reservoir characteristics for the training images.
e. Scan the training images to establish a search tree. Only the data events that actually appear in the training image are saved in the search tree. In order to limit the geometric configuration of the data events and prevent it from being too large, the maximum number of searched data needs to be defined. Build a search tree based on the sample of the largest search data.

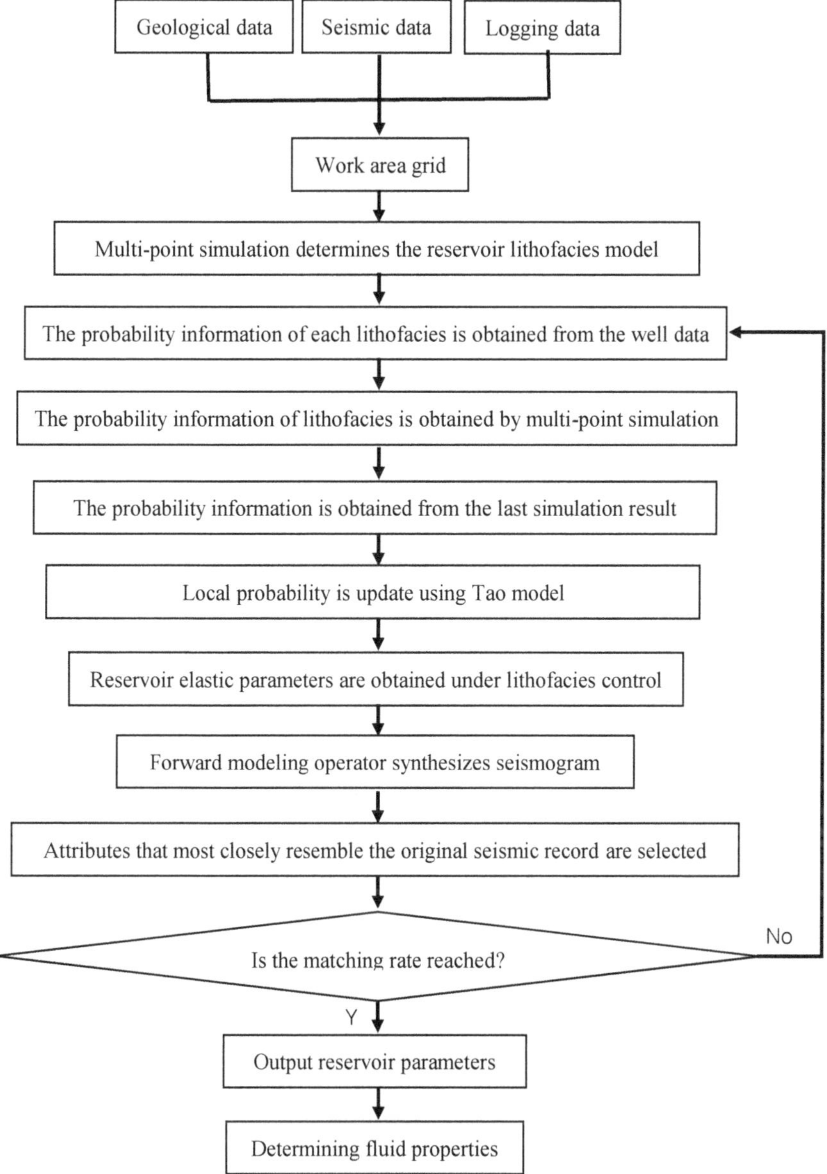

Figure 1. New multi-point geostatistical inversion flow chart considering local updating.

Step 2: SNESIM simulation using LPUMI

i.   Griding and assignment of the well data and elastic parameters. Each conditional data point is assigned to the nearest grid node in the simulation grid. If multiple conditional data points are assigned to the same grid node, the nearest one is assigned to the center of the grid node.

ii.   Define the path through the remaining nodes of the simulated grid. A path is a vector that contains all of the indexes of the grid nodes to be simulated in sequence. Random, one-way (i.e., the nodes are accessed in a regular order starting from one side of the grid), or any other path can be used. The simulation path is from a dense well area to a sparse well area and finally to a no well area.

iii. Search for domains that simulate node X. They consist at most of n nodes $\{x_1, x_2, \ldots, x_N\}$ that have recently been assigned to or simulated in the simulation grid. If the field of X is not found in the first iteration (such as the first unconditionally simulated node), a node Y is randomly selected in the TI, and its value (Z(y) to Z(x)) is assigned in the simulation grid. Then, proceed to the next node of the path.

iv. Determine the search tree's conditional probability $P(A \mid B)$.

v. Determine whether there is a point at which in the previous simulation, the elastic parameters were reserved. If there is, using the permanent ratio of the updating theory, probability $P(A \mid B)$ will update to $P(A \mid B,C)$. Otherwise, the update is still the conditional probability $P(A \mid B)$.

Step 3: Prestack inversion

According to the reservoir's elastic parameters obtained from the logging data, the density and p-wave velocity are uniformly sampled in the suggested data mode to obtain the p-wave impedance Z_P of the sample.

According to the relationships between the p-wave impedance Z_P and the s-wave impedance Z_S and the p-wave impedance Z_P and the density ρ given by Hampson and Russell (2005), in general, Z_S and ρ can be expressed as follows:

$$\ln(Z_S) = k \ln(Z_P) + k_c + \Delta L_S, \tag{11}$$

$$\ln(\rho) = m \ln(Z_P) + m_c + \Delta L_D. \tag{12}$$

They are looking for deviations away from a linear fit in logarithmic space. k and m are the corresponding slop. k_c and m_c are the corrsponding intercept. The deviations away from this straight line, ΔL_S and ΔL_D, are desired fluild anomalies. The seismic forward modeling record calculation of the proposed elastic parameters in the proposed data model is conducted as follows:

$$g(\theta) = \tilde{c_1} W(\theta) DL_P + \tilde{c_2} W(\theta) D\Delta L_S + W(\theta) D\Delta L_D, \tag{13}$$

where $\tilde{c_1} = \frac{1}{2}c_1 + \frac{1}{2}kc_2 + mc_3$, $\tilde{c_2} = \frac{1}{2}c_2$, $c_1 = 1 + tan^2\theta$, $c_2 = -8\gamma^2 tan^2\theta$, $c_3 = -0.5tan^2\theta + 2\gamma^2 sin^2\theta$, $\gamma = V_S/V_P$. $W(\theta)$ is the incident angle of the wavelet, D is the differential operator, $L_P = \ln(Z_P)$, $L_S = \ln(Z_S)$, $L_D = \ln(\rho)$, and $g(\theta)$ is the seismic forward modeling record.

The likelihood function Equation (3) and the posteriori probability Equation (4) are determined from the forward simulation record and the actual seismic record. Gonzalez's (2008) method is adopted to select the elastic inversion parameters that retain the maximum likelihood function as the results; or according to the Metropolis–Hasting optimization criterion, a large number of implementations of the lithofacies and elastic parameters are generated from the posterior function, and these implementations represent the probability distribution of the posterior function. The acceptance criteria of the model are proposed to determine the optimal inversion elastic parameters.

$$P_{accept} = min\left(1, \frac{P(f^*)}{P(f)} \cdot \exp\left[\sum \frac{\left(m - \mu_f\right)^2 - \left(m^* - \mu_f\right)^2}{2\sigma_f^2} + \sum \frac{(d - g(m))^2 - (d - g(m^*))^2}{2\sigma_e^2}\right]\right). \tag{14}$$

In consideration of the computational efficiency and algorithm continuity, Gonzalez's [8] method was adopted in this study to select the optimal matching inversion results through iterative comparison of the multiple sampling (generally 25–30).

Step 4: Iteration

All of the simulation grids are visited to realize a single inversion.

According to the matching degree of the synthetic seismic records and the actual records, it is judged whether the iteration should be terminated. If the conditions are not met, start again from Step 2 for the next external iteration. Usually, after six iterations, the average correlation coefficient of the seismic data is greater than 85% and the inversion results are output.

3. Model Testing

3.1. Theoretical Model Testing

The meandering river model with a low curvature in the first layer of the Stanford VI-E reservoir was taken as the test object, which is a $150 \times 200 \times 80$ model. The lithofacies were subdivided into point bar, channel, and floodplain mudstone deposits (Figure 2). The different microfacies have different elastic parameter distributions (Figure 3). By designing 68 virtual wells, the seismic inversion method was tested based on the given elastic parameters and the lithofacies interpreted from the well data. In order to verify the accuracy of the method, only 63 wells were selected as the condition wells, and the remaining five wells were used as the test wells to analyze the inversion results.

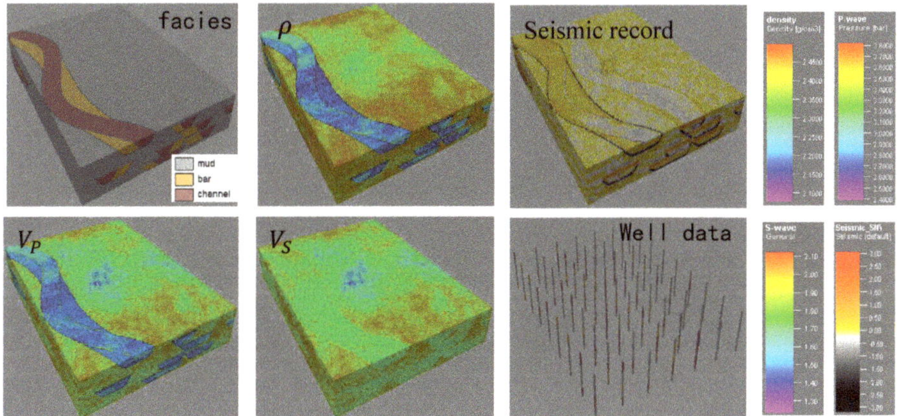

Figure 2. Stanford VI-E theoretical model.

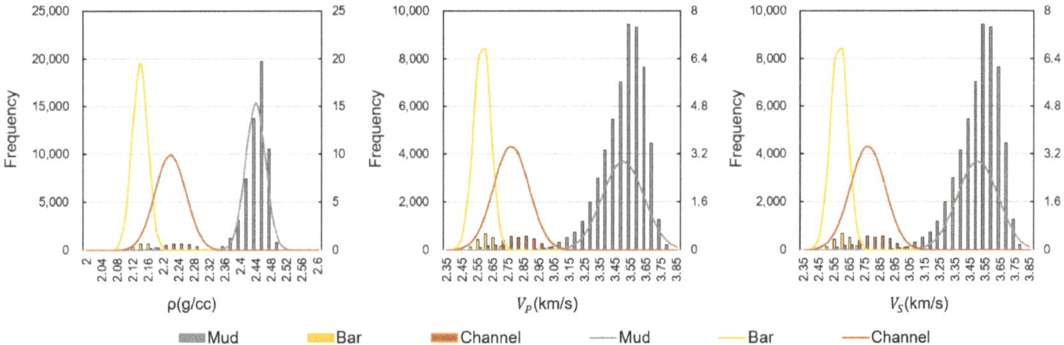

Figure 3. Statistical distribution of the elastic parameters of the different microfacies.

The theoretical lithofacies model was selected as the training image, and the tests were carried out using the traditional two-point statistical inversion method (TPI), the conventional multi-point statistical inversion method (MPI), and the multi-point statistical inversion with local probability updating method (LPUMI). The results show that compared with two-point statistical inversion, multi-point statistical inversion can reproduce the reservoir lithofacies better, and the inversion results are more consistent with the theoretical model. The synthetic seismogram is more similar to the actual seismogram (Figures 4–6). The average matching rate of the multi-point statistical inversion is 83.5%, while that of the two-point statistical inversion is 81.5%, indicating that the multi-point statistical inversion produced a more accurate prediction of the inter-well reservoir properties (Figure 7). According to the correlation coefficient of the seismic record calculated via the inversion, the correlation coefficient increases gradually as the number of iterations increases. After six iterations, the correlation between the inverted synthetic seismic track and the actual seismic track is close to 80%. The LPUMI has the largest correlation coefficient, reaching 0.78; the correlation coefficient of the MPI is in the middle (0.76); and the TPI has the lowest correlation coefficient (0.75) (Figure 8). The results show that the reservoir parameters obtained using the LPUMI are closer to the actual reservoir parameters. This shows that the proposed method is more reasonable and can be applied to actual reservoir inversion prediction.

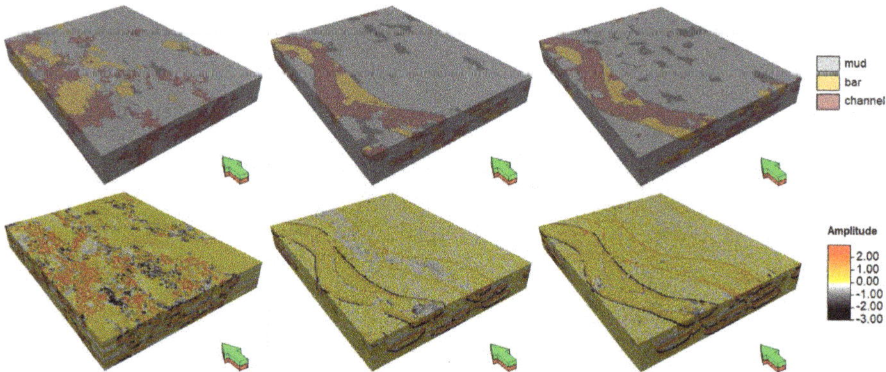

Figure 4. Lithofacies and forward modeling records of the inversion (**left**) TPI, (**middle**) MPI, and (**right**) LPUMI.

3.2. Real Reservoir Testing

The Xinchang gas field is located in the western part of the Sichuan Basin, China. The main gas-bearing horizon is the second member of the Xujiahe Formation, and the main sandbodies are braided delta front distributary channels and mouth bars. The thicknesses of the sand bodies are large and their horizontal distributions are wide. The horizontal distribution of the sweet spot reservoir is not uniform, which causes difficulties in the exploration and development of the gas reservoir.

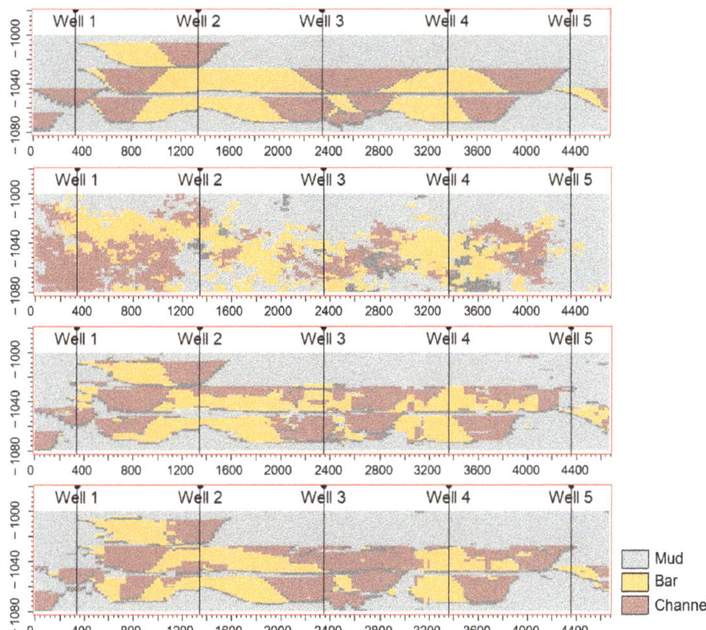

Figure 5. Lithofacies profiles of the five validation wells (from **top** to **bottom**, real reservoir, TPI, MPI, and LPUMI).

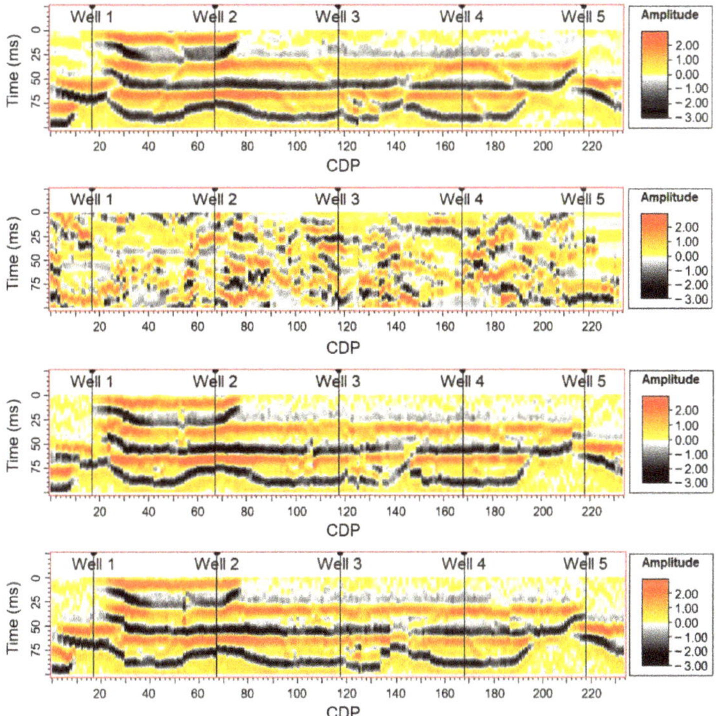

Figure 6. Seismic records obtained through well inversion (from **top** to **bottom**, real reservoir, TPI, MPI, and LPUMI).

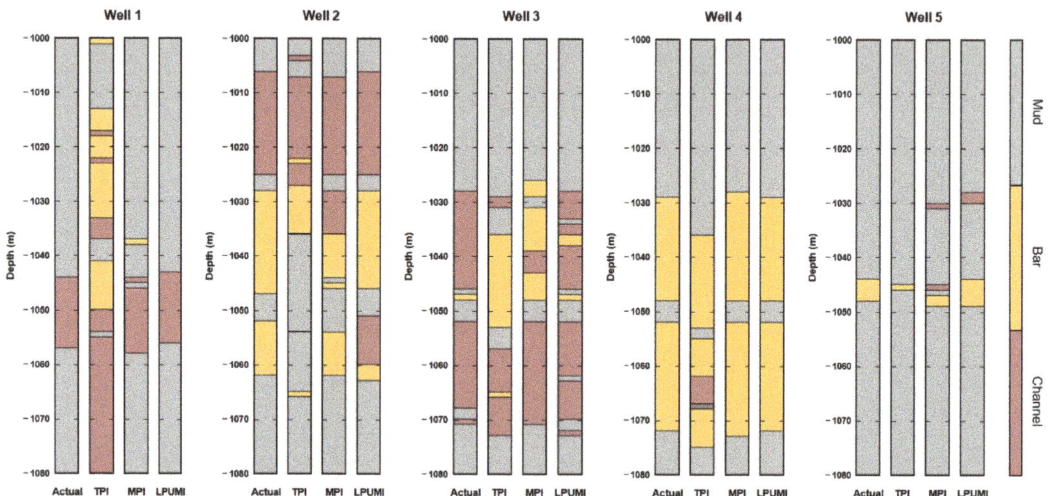

Figure 7. Comparison of the lithofacies distribution in five wells obtained using the different methods.

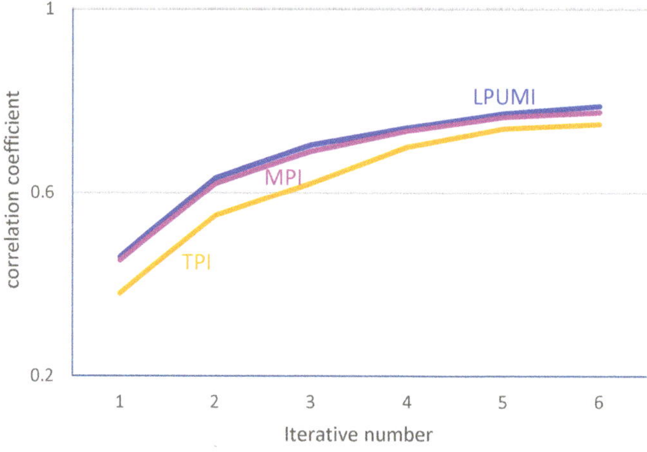

Figure 8. Correlation coefficients between the inversion trace and actual trace.

In this study, three methods, including the TPI, MPI, and LPUMI, were applied to the prediction of the $TX_2^3 - TX_2^7$ sand formation in the study area. Because this was mainly undertaken to test the proposed inversion method and compare it to the two existing methods, the inversion prediction was not performed for the entire region. Instead, a relatively regular local area with a relatively simple stratigraphic structure and no-fault development was selected to carry out the study. The total thickness of the vertical direction of the intercepted area is about 235 m, and the length in the I direction and J direction on the plane is about 4000 m (Figure 9). The grids were 100×100 m in the plane and 2 m in the vertical direction, and the total number of simulated grids was $40 \times 40 \times 118 = 188,800$. Figure 10 shows the pre-stack track sets at different angles (5°, 15°, and 25°) in the test block. Figure 11 shows the spatial distribution and attribute interpretation for 11 wells. The analysis shows that the main body of the channel is composed of sand and silt, with little mud. The p-S wave velocity has an obvious linear relationship, with a small p-S wave velocity ratio and a low gamma ray (GR) value. The interchannel region is mainly composed of clay deposits with a small amount of silt and fine sand. The p-S wave velocity has an obvious linear relationship, with a high p-S wave velocity ratio and a high GR value. The mouth bar is composed of fine and silty sand, with fine sorting and a pure quality, and it has a small S-wave velocity ratio and low GR value, which is similar to the main body of the channel (Figure 12). The statistical analysis of the elastic parameter–lithofacies was conducted based on the data for these 11 wells, and its probability distribution was established for the elastic parameter sampling under lithofacies control during the subsequent inversion (Figure 13). The seismic wavelets from different angles were extracted based on the seismic records of the sidewalks (Figure 14). After comprehensive analysis, 25 Hz theoretical Rick wavelets were selected for the inversion seismic record synthesis. Based on geological analysis, a three-dimension training image of the study area was established (Figure 15), which was used to calculate the two-point variogram function and to extract the multi-point prior probability.

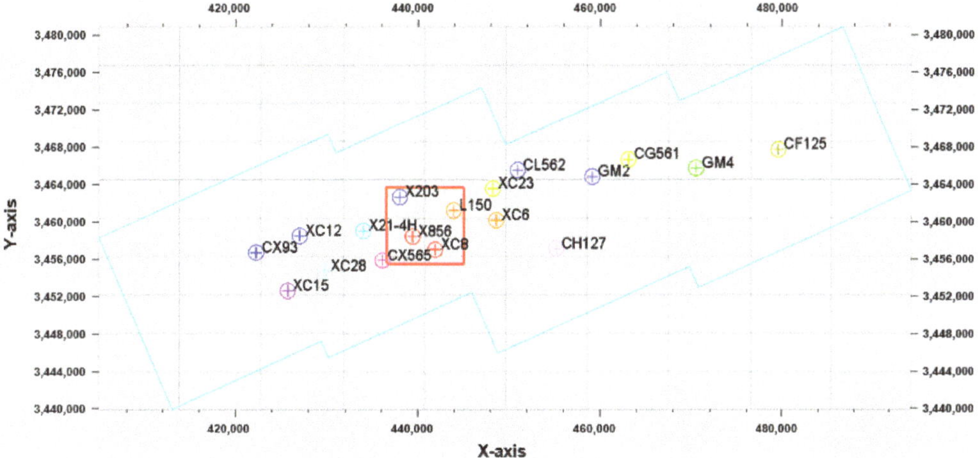

Figure 9. Drilling distribution map of the actual working area and the location of the inversion block.

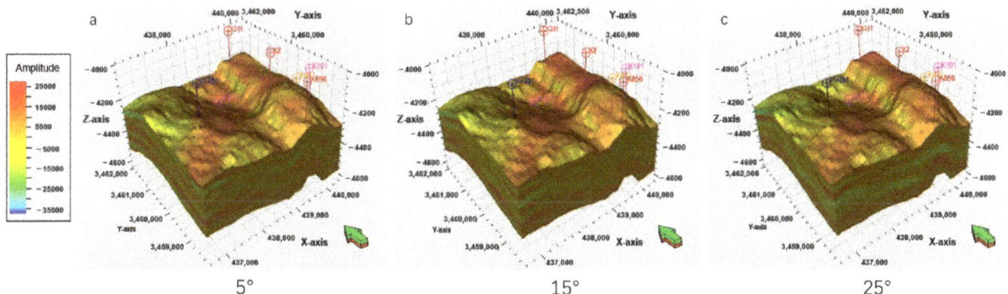

Figure 10. Pre-stack seismic track set for the inversion block: (**a**) 5°, (**b**) 15° and (**c**) 25°.

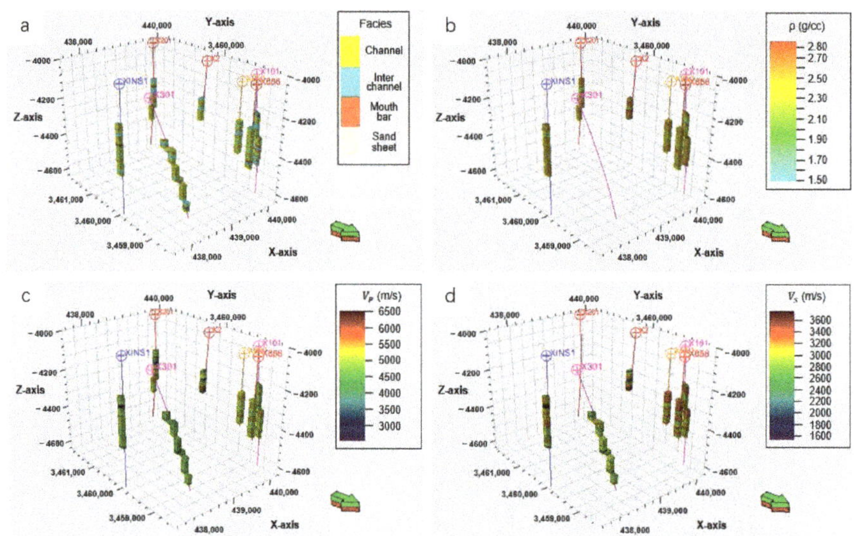

Figure 11. Data interpretation results for the 11 wells in the inversion block: (**a**) lithofacies, (**b**) density, (**c**) p-wave velocity, and (**d**) s-wave velocity.

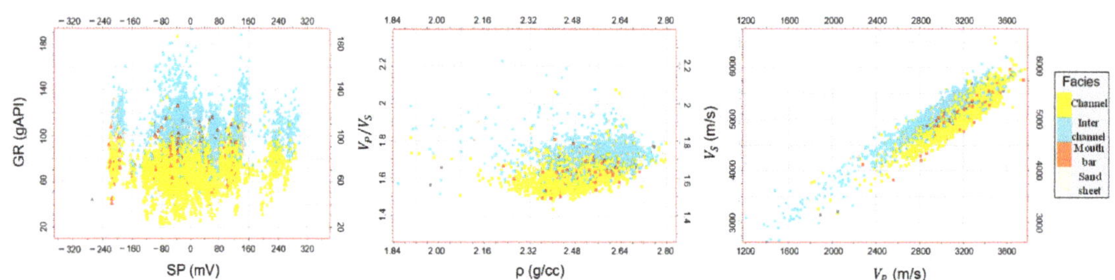

Figure 12. Statistical diagram showing the relationship between the elastic parameters and the lithofacies.

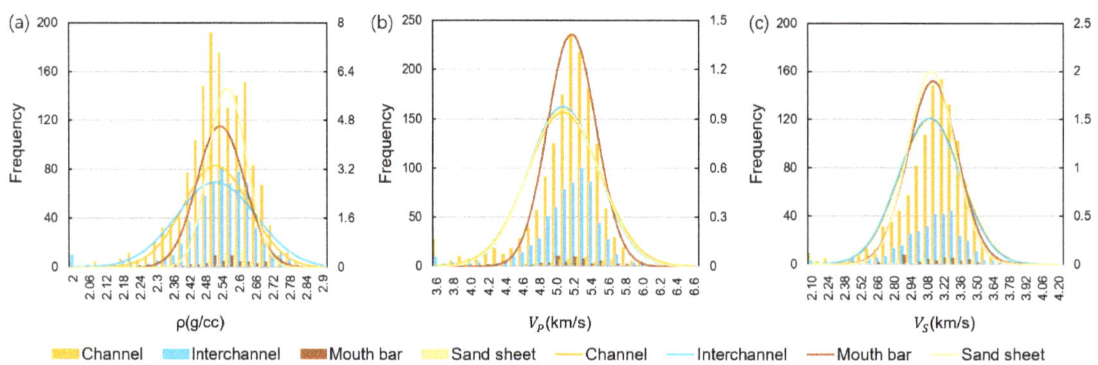

Figure 13. Analysis of the elastic parameters of the different lithofacies in the wells: (**a**) density, (**b**) p-wave velocity, and (**c**) shear wave velocity.

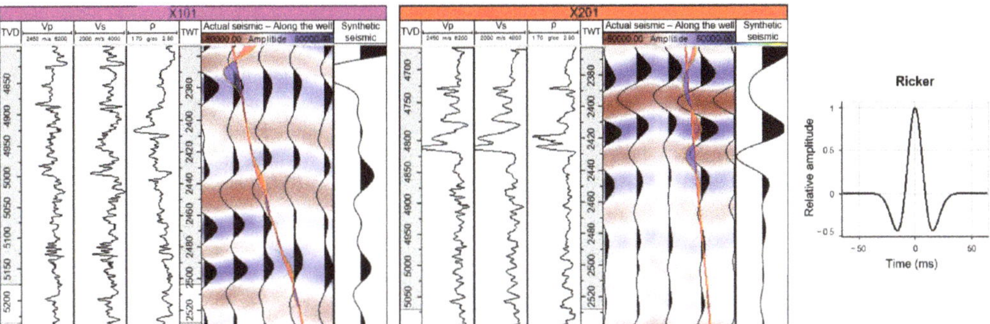

Figure 14. The synthetic and observed seismic traces.

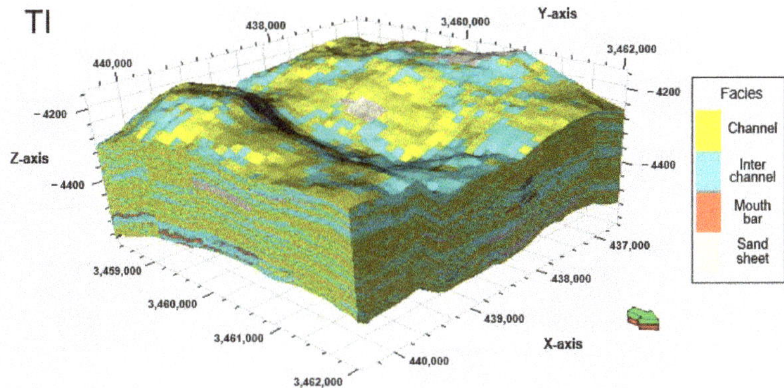

Figure 15. Training image of the inversion block.

The inversion profile results obtained using the different method were captured for wells Xins1–X201 for comparison (Figure 16). As can be seen from the profiles, overall, all of the inversion lithofacies profiles are mainly composed of channel sand bodies. The mouth bar deposits are locally developed, and the mudstone deposits are relatively scarce and are mainly developed in the upper part. The lithofacies inversion is relatively continuous and the distribution of the channel sand body is reflected well. However, in terms of the structure, the sand body continuity of the TPI is too good to reflect the complex heterogeneity. The distributions of the MPI and LPUMI are highly variable and are connected locally, and the overall structure is close to that of the actual reservoir. In terms of the seismic track records, the inversion seismic records of the MPI and LPUMI are close to the actual seismic records, while the TPI exhibits chaotic reflection characteristics, which are quite different from the actual continuous lithofacies distribution. The results show that the MPI and LPUMI are able to reflect the sand body and elastic parameters better.

Gonzalez [8] pointed out that the accuracies of the inversion iterations can be compared using the absolute error recorded by the forward simulation or the seismic trace correlation. Due to the possible errors in the time–depth conversion, direct comparison may cause large errors. However, the underground reservoir prediction is more likely to reveal the sand body and the spatial structure of the interlayer. If the reflected structure is similar, the overall similarity of the seismic records will increase. Therefore, the correlation between the forward simulation records and the actual records was used to compare the results of the different methods. The correlations between the forward modeling records and the actual seismic track were calculated. The results show that the correlation coefficient of the TPI is 0.72. The correlation coefficient of the MPI is 0.74, and that of the LPUMI is 0.77.

This demonstrates that the LPUMI results are closer to the actual seismic record and have a higher accuracy.

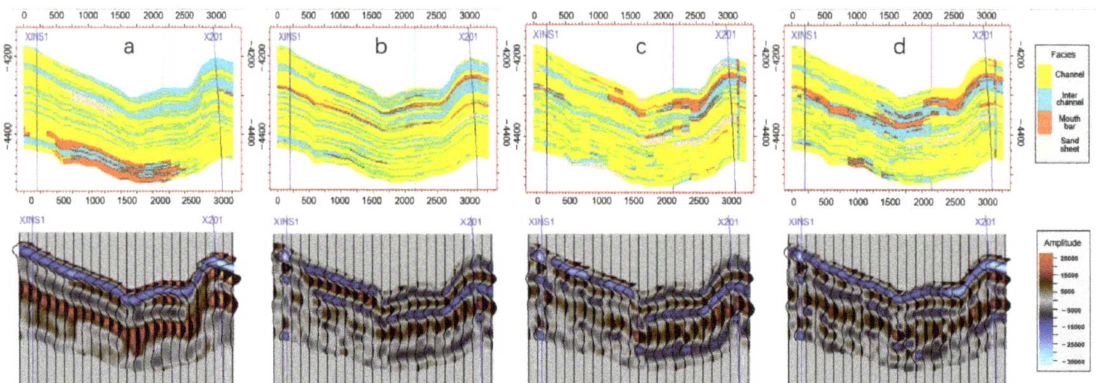

Figure 16. Inversion of the lithofacies model and synthetic seismic record through wells xins1–X201 for the training image and the different methods: (**a**) Training image, (**b**) TPI model, (**c**) MPI model, and (**d**) LPUMI model.

Furthermore, the cross-validation method was used to test and compare the methods. The cross section of well X853 was used to compare the prediction accuracies of the different methods. Both the MPI and LPUMI can reflect the characteristics of the upward transition of the mouth bar sand body, which is consistent with the migration trend of the lithofacies in the training image. However, the TPI can hardly reflect this characteristic. Compared with the actual seismic record, the synthetic seismic records of the MPI and LPUMI are closer to the actual seismic profile (Figure 17).

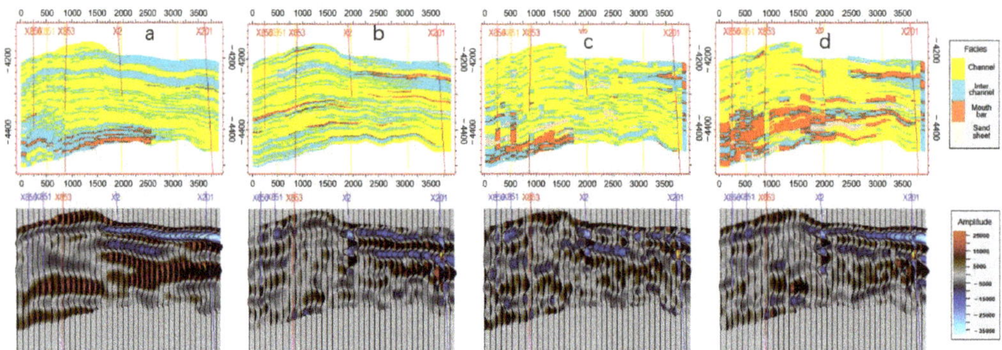

Figure 17. Training image and lithofacies model inversions obtained using the different methods and forward simulation records of well X853 through the drainage: (**a**) Training image, (**b**) TPI lithofacies model, (**c**) MPI lithofacies model, and (**d**) LPUMI lithofacies model.

According to the comparison of the elastic parameters across well X853 (Figure 18), the different inversion methods can reflect the variations in the elastic parameters in the area around the well successfully, with a good degree of matching. In terms of the elastic parameter errors, the TPI performed the best, followed by the MPI and the LPUMI. However, the differences are not significant. In terms of the correlation of the elastic parameters, the overall difference is not significant. The correlation of the LPUMI is 0.74, that of the MPI is 0.73, and that of the TPI is 0.72. However, the degrees of lithofacies matching are different. The results show that the best method is the LPUMI (0.862), followed by the MPI (0.856), and the TPI is the worst (only 0.78). This indicates that the LPUMI has more advantages.

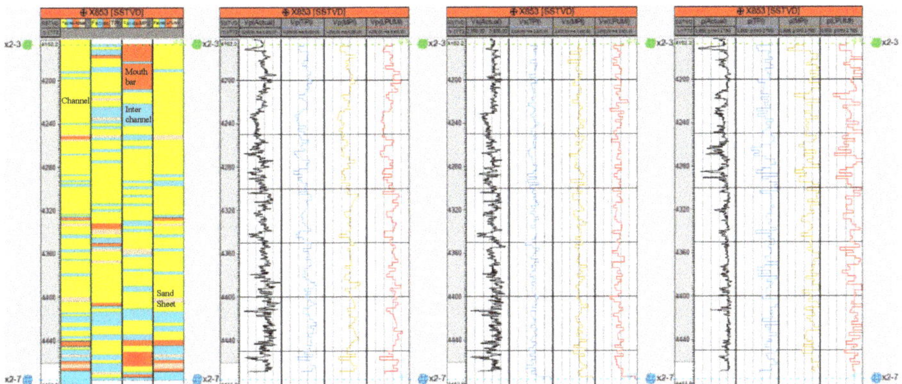

Figure 18. Comparison between the predicted results and the actual lithofacies and elastic parameters of the well.

4. Discussion

Seismic records are a comprehensive representation of subsurface lithofacies, physical properties, elastic parameters, and fluids. The essence of statistical inversion is to seek the optimal solution that reflects the underground reservoir parameters through convolution of the elastic parameters. The distribution of the elastic parameters is mainly revealed through rock physics modeling. In the field of geological modeling, due to the intrinsic relationship between the lithofacies and physical properties, developing facies-controlled reservoir parameter modeling method has become an important means of improving the accuracy of physical property modeling. Therefore, introducing the idea of facies control into seismic inversion can improve the inversion accuracy. In fact, Azevedo and Soares [29] compared the inversion results of the given lithofacies model with conventional inversion results and showed that the inversion elastic parameter distribution of the given lithofacies model is more reasonable and the iteration convergence is faster. Based on the theoretical model and practical model developed and tested in this study, the multi-point inversion method considering facies control is significantly better than the traditional two-point statistical inversion method without facies control. Therefore, making full use of lithofacies control in seismic inversion should be an important direction in the future. In a sense, the accuracy of the constrained lithofacies model determines the effectiveness of the final inversion effect.

In this improvement, the seismic elastic parameters are mainly used for local lithofacies updating. The SNESIM method is a single-grid point lithofacies forecasting method. When using elastic parameters to update the local lithofacies, only the information provided by the elastic parameters of the points to be estimated is used, which has no significant improvement effect compared with the conventional multi-point geostatistical inversion. This may be due to the fact that grid by grid updating of lithofacies does not significantly change the sedimentary facies model. In addition, probabilistic statistical sampling errors inevitably exist and are transmitted to the subsequent updates, resulting in limited improvement of the inversion accuracy. Another disadvantage of the SNESIM lithofacies prediction is that all reservoir predictions based on statistical methods require a stable lithofacies distribution model, but in reality, the lithofacies distribution is very complex and has non-stationary characteristics. This is one of the reasons why multi-point statistical methods such as the SIMPAT, Filtersim, and DS methods have been developed. This is also the reason why Gonzalez [8] used SIMPAT as the lithofacies inversion method.

Arpat (2005) pointed out that seismic information can be integrated in SIMPAT geological modeling; that is, a seismic reference image with a good correspondence to the lithofacies can be constructed, and its contribution to the distance can be taken into account in the lithofacies matching to constrain and guide the selection of the optimal lithofacies mode:

$$s\langle\cdot,\cdot\rangle = s_h\left\langle dev_T(u), pat_T^k\right\rangle + s_s\left\langle sdev_T(u), spat_T^k\right\rangle, \tag{15}$$

where $s_h\left\langle dev_T(u), pat_T^k\right\rangle$ is the similarity between the model at the point to be estimated $dev_T(u)$ and the data model pat_T^k in the lithofacies training image $s_s\left\langle sdev_T(u), spat_T^k\right\rangle$ and the similarity between the seismic attribute model $sdev_T(u)$ at the corresponding point to be estimated and the seismic attribute model $spat_T^k$ in the training image. It should be noted that the contribution of the seismic attributes (i.e., soft data) is different from that of well data (i.e., hard data), so it is necessary to effectively measure the contribution of the seismic attributes. The similarity value of the seismic training images is multiplied by a weight to represent the contribution of the seismic attributes in the similarity calculation. In addition, because the scale of the seismic attributes is not consistent with that of the facies attributes, the seismic attributes must be normalized before the similarity calculation in the application of the seismic data in order to avoid the absolute superiority of the seismic similarity in the entire similarity due to the different scales.

Lithofacies training images can be obtained from the geological anatomy and through sedimentary simulation. However, seismic attribute training image is often difficult to obtain. Based on rock physics modeling, the forward modelling can be conducted many times and the optimal matching seismic attributes can be calculated, which can be applied to Equation (15) to update the overall local lithofacies and improve the accuracy of the inversion.

The efficiency of the mSIMPAT algorithm is relatively low, and adding seismic attribute constraints will further increase the computational burden. Therefore, using parallel computing and deep learning theory to accelerate the inversion iteration is an important direction in future research.

5. Conclusions

This paper proposed a new multi-point geostatistical inversion through local iterative updating rock facies using the constrains of elastic parameters. An internal and external double cycle iteration mechanism was adopted to execute the iteration and updating. During the internal cycle iteration, the optimal elastic parameters obtained in the previous external cycle were combined with the statistical probability of the lithofacies and elastic parameters, and the elastic parameters were combined with the permanent ratio of the updating theory to achieve local lithofacies updating. Based on this, inversion prediction of the lithofacies and elastic parameters was carried out. In the outer loop, the current global inversion results were compared with the actual error to determine whether the inversion results of the elastic parameters meet the conditions, and the cycle iteration was carried out again until local elastic parameter inversion and global inversion satisfy the given conditions. Then, the cycle was terminated and the inversion results were output.

Both the theoretical and practical model tests conducted confirm that the correlation between the actual seismic track and the synthetic seismic track obtained using the LPMUI is the best, and the degree of lithofacies matching is the highest. The results of the MPI are the next best, and the results of the TPI are the worst, indicating that the reservoir parameters obtained using the LPMUI are closer to the actual reservoir parameters. This method is worth popularizing in practical exploration and development.

The calculation efficiency of double iteration in the LPMUI is much better than traditional MPI; however, it is still lower than the TPI. In future, the deep learning method or parallel computing method can be introduced to improve the calculation efficiency. Another improvement may exist in the use of the elastic parameters for rock lithofacies updating. Here, only the elastic property in the un-simulated grid was used to update the rock lithofacies in the same grid, a multi-point geostatistical simulation for sedimentary pattern reproduction and updating was not conducted, which may have caused the failure of the reproduction of a continuous geobody. How to use elastic parameters in a multi-point data template to update rock lithofacies patterns, is still a challenge for future work.

Author Contributions: Conceptualization, Y.Y. and T.C.; methodology, X.H. and L.W.; software, X.H.; validation, Z.W.; formal analysis, Z.W.; investigation, Z.W.; resources, T.C.; data curation, Z.W.; writing—original draft preparation, Z.W.; writing—review and editing, Z.W., X.H. and Y.Y.; visualization, X.H.; supervision, Z.W. and Y.Y.; project administration, Y.Y.; funding acquisition, Y.Y. All authors have read and agreed to the published version of the manuscript.

Funding: This work is supported by the National Natural Science Foundation of China (Grant Numbers 42130813 and 41872138).

Conflicts of Interest: The authors declare no conflict of interest.

Abbreviations

The following abbreviations are used in this manuscript:

LPUMI	a multi-point geostatistical inversion method based on the local probability updating method for the inversion of lithofacies.
MCMC	Markov chain Monte Carlo.
ASR	adaptive spatial resampling.

References

1. Ghaderpour, E. Multichannel antileakage least-squares spectral analysis for seismic data regularization beyond aliasing. *Acta Geophys.* **2019**, *67*, 1349–1363. [CrossRef]
2. Bosch, M.; Mukerji, T.; Gonzalez, E.F. Seismic inversion for reservoir properties combining statistical rock physics and geostatistics: A review. *Geophysics* **2010**, *75*, 165–176. [CrossRef]
3. Zhang, F.; Xiao, Z.; Yin, X. Bayesian stochastic inversion with seismic data constraints. *Oil Geophys. Prospect.* **2014**, *49*, 176–182.
4. Journel, A.G.; Huijbregts, C.J. *Mining Geostatistics*; Academic Press: New York, NY, USA, 1978.
5. Haas, A.; Dubrule, O. Geostatistical inversion-a sequential method of stochastic reservoir modelling constrained by seismic data. *First Break.* **1994**, *12*, 561–569. [CrossRef]
6. Deutsch, C.V. *GSLIB Geostatistical Software Library and User'S Guide*; Oxford University Press: Oxford, UK, 1992.
7. Bortoli, L.J.; Alabert, F.A.; Hass, A.; Journel, A.G. *Constraining Stochastic Images to Seismic Data*; Geostatistics Tróia'92; Spring: Dordrecht, The Netherlands, 1993; pp. 325–337. Available online: https://www.semanticscholar.org/paper/Constraining-Stochastic-Images-to-Seismic-Data-Bortoli-Alabert/20d8acea057bcc1d7f875a204689c2ee1243ad73 (accessed on 21 December 2021).
8. González, E.F.; Mukerji, T.; Mavko, G. Seismic inversion combining rock physics and multiple-point geostatistics. *Geophysics* **2008**, *73*, 11–21. [CrossRef]
9. Yang, P.J.; Yin, X.Y. Non-linear quadratic programming Bayesian prestack inversion. *Chin. J. Geophys.* **2008**, *51*, 1876–1882. (In Chinese)
10. Zhang, G.; Wang, D.; Yin, X. Seismic parameters estimation using MCMC method. *Oil Geophys. Prospect.* **2011**, *46*, 605–609.
11. Wang, P.; Li, Y.; Zhao, R. Lithology inversion using post-stack MCMC method. *Prog. Geophys.* **2015**, *30*, 1918–1925.
12. Azevedo, L.; Demyanov, V. Multi-scale uncertainty assessment in geostatistical seismic inversion. *Geophysics* **2019**, *84*, R355–R369. [CrossRef]
13. Pereira, P.; Calçôa, I.; Azevedo, L.; Nunes, R.; Soares, A. Iterative geostatistical seismic inversion incorporating local anisotropies. *Comput. Geosci.* **2020**, *24*, 1589–1604. [CrossRef]
14. Caers, J.; Srinivasan, S.; Journel, A.G. Geostatistical quantification of geological information for a Fluvial-Type North Sea Reservoir. In Proceedings of the SPE Annual Technical Conference and Exhibition, SPE-56655-MS. Houston, TX, USA, 3–6 October 1999.
15. Strebelle, S.; Journel, A.G. Reservoir modeling using multiple-point statistics. In Proceedings of the SPE Annual Technical Conference and Exhibition, SPE-71324-MS. New Orleans, LA, USA, 30 September–3 October 2001.
16. Yin, Y.; Zhang, C.; Li, J.; Shi, S. Research progress and Prospect of multi-point geostatistics. *J. Palaeogeogr.* **2011**, *13*, 245–253.
17. Yang, P. Geostatistical inversion: From two points to multiple points. *Prog. Geophys.* **2014**, *29*, 2293–2300.
18. Yang, P. Multi-point geostatistical random simulation based on pattern clustering and matching. *Prog. Geophys.* **2018**, *33*, 279–284.
19. Guardiano, F.B.; Srivastava, R.M. Multivariate Geostatistics: Beyond Bivariate Moments. In *Geostatistics Tróia '92, Quantitative Geology and Geostatistic*; Kluwer Academic Publications: Dordrecht, The Netherlands, 1993; pp. 133–144.
20. Jeong, C.; Mukerji, T.; Mariethoz, G. A fast approximation for seismic inverse modeling: Adaptive spatial resampling. *Math. Geosci.* **2017**, *49*, 845–869. [CrossRef]

21. Liu, X.; Li, J.; Chen, X.; Li, C.; Guo, K.; Zhou, L. A stochastic inversion method integrating multi-point geostatistics and sequential Gaussian simulation. *Chin. J. Geophys.* **2018**, *61*, 2998–3007.
22. Liming, S.; Wuyang, Y.; Fengchang, Y.; Xingyao, Y.; Xueshan, Y. Review and prospect of seismic inversion technology. *Oil Geophys. Prospect.* **2015**, *50*, 184–202.
23. Ling, Y.; Xiao, Y.; Sun, D.; Lin, J.; Gao, J. Influence factors analysis and seismic attribute interpretation of post-stack thin reservoir inversion. *Geophys. Prospect. Pet.* **2008**, *47*, 531–558.
24. Azevedo, L.; Nunes, R.; Soares, A.; Mundin, E.C.; Neto, G.S. Integration of well data into geostatistical seismic amplitude variation with angle inversion for facies estimation. *Geophysics* **2015**, *80*, 113–128. [CrossRef]
25. Azevedo, L.; Soares, A. *Geostatistical Methods for Reservoir Geophysics*; Springer International Publishing: Cham, Switzerland, 2017.
26. Tarantola, A. *Inverse Problem Theory and Methods for Model Parameter Estimation*; Society for Industrial and Applied Mathematics: Philadelphia, PA, USA, 2005; pp. 269–285.
27. Mariethoz, G.; Renard, P.; Straubhaar, J. The direct sampling method to perform multiple-point geostatistical simulations. *Water Resour. Res.* **2010**, *46*, 1–14. [CrossRef]
28. Journel, A.G. Combining knowledge from diverse sources: An alternative to traditional data independence hypotheses. *Math. Geol.* **2002**, *34*, 573–596. [CrossRef]
29. Arpat, B.G. Sequential Simulation with Patterns. Doctoral Dissertation, Stanford University, Stanford, CA, USA, 2005.

Article

A Technique of Hydrocarbon Potential Evaluation in Low Resistivity Gas-Saturated Mudstone Horizons in Miocene Deposits, South Poland

Anita Lis-Śledziona * and Weronika Kaczmarczyk-Kuszpit

Oil and Gas Institute—NRI, Lubicz Street 25A, 31-503 Krakow, Poland; kaczmarczyk-kuszpit@inig.pl
* Correspondence: lis-sledziona@inig.pl

Abstract: The petrophysical properties of Miocene mudstones and gas bearing-heteroliths were the main scope of the work performed in one of the multihorizon gas fields in the Polish Carpathian Foredeep. Ten boreholes were the subject of petrophysical interpretation. The analyzed interval covered seven gas-bearing Miocene horizons belonging to Sarmatian and Badenian deposits. The water saturation in shaly sand and mudstone intervals was calculated using the Montaron connectivity theory approach and was compared with Simandoux water saturation. Additionally, the Kohonen neural network was used for qualitative interpretation of four PSUs (petrophysically similar units), which represent the deposits of comparable petrophysical parameters. This approach allowed us to identify the sediment group with the highest probability of hydrocarbon saturation. Then, the spatial distribution of PSUs and reservoir parameters was carried out in Petrel. The resolution of the model was selected to reflect the variability of log-derived parameters. The reconstruction of the spatial distribution of shale volume, porosity, and permeability was performed with standard parametric modeling procedures using the Gaussian random function simulation stochastic algorithm, while PSU distribution and hydrocarbon saturation (SH) required a separate approach. The distribution into PSU groups was carried out by facies classification. Predefined ranges of clay volume, effective porosity, and permeability were used as discriminators to achieve spatial distribution of the PSU groups. The spatial distribution of hydrocarbon saturation was performed by creating the meta-attribute of this parameter and then reducing the derived pseudo-saturation model to physical values. Results included the creation of maps of hydrocarbon saturation that show the preferable areas with the highest hydrocarbon saturation for each gas horizon.

Keywords: shaly-sand; low resistivity gas reservoir; Montaron equation; clastic reservoir evaluation; Miocene sediments; heteroliths

Citation: Lis-Śledziona, A.; Kaczmarczyk-Kuszpit, W. A Technique of Hydrocarbon Potential Evaluation in Low Resistivity Gas-Saturated Mudstone Horizons in Miocene Deposits, South Poland. *Energies* **2022**, *15*, 1890. https://doi.org/10.3390/en15051890

Academic Editors: Yuming Liu and Bo Zhang

Received: 11 February 2022
Accepted: 2 March 2022
Published: 4 March 2022

Publisher's Note: MDPI stays neutral with regard to jurisdictional claims in published maps and institutional affiliations.

1. Introduction

Research on hydrocarbons in Miocene sediments of the Carpathian Foredeep conducted in recent years has led to the discovery of many gas horizons located in thin-layer heteroliths, mudstones, and sandstones. Many geophysical and geological research has been performed to recognize the possibilities of hydrocarbon accumulation in Miocene sediments [1–7]. The main influence on the formation of gas accumulations was the structural and facies factors, which define the horizons lithological boundaries [1]. There are three main sedimentary complexes in the study area: deltaic deposits, submarine fan sediments and fine-rhytmic turbidyte sediments of basin plain. The lowest part of the upper Baden-Sarmatian is dominated by heterolithic deposits formed in the environment of the basin plain. They consist of claystones and siltstones covered with thin layers of fine and medium-grained sandstones, the thickness of these sediments can be up to about 200 m. Higher in the profile the turbidite sediments of submarine fans can be observed. They are spread under the Carpathian thrust and along its present border. The thickness of these sediments can reach several meters and are the result of sedimentation within the distribution channels of the upper fan. Finer turbidite sediments are associated with sedimentation along the riverbed shafts. These sediments are about 300 m thick. Towards the top of the profile, the series of submarine fans turbidite deposits are gradually replaced by deltaic deposits of rather constant thickness. Deltaic deposits are characterized by a clay-sandy lithology and generally high collateral continuity. The sandstones of deltaic sediments, especially occurring in thin-bed heterolithic and mudstone lithofacies, are characterized by low textural and mineral maturity. The sandstones grain skeleton is composed of a very fine to medium-grained fraction [1–3].

The interesting is that despite there is sandstones complex with thickness of even 250 m, formed as submarine fans sediments with the excellent porosity and permeability, the existing hydrocarbon horizons mainly occurs in sandstones up to 50 m of thickness. It is probably due to dispersion of hydrocarbons in layers of large thickness [1]. There are not enough sealing units to prevent migration of hydrocarbons. The morphology of Precambrian basement had the main impact for developing of the deepest located gas accumulations mainly within fine-rhytmic turbidyte heteroliths being the subject of the interpretation in the presented work. The detail recognition and reconstruction of sedimentary environment and building a proper structural model is a main task to predict the stratigraphic traps which constitute favorable areas for hydrocarbon accumulations [1]. Seismostratigraphic works performed in the area of Wielkie Oczy—Graben and Markowce-Lubliniec elevation allowed to subdivide the Machow formation into seven genetic sequences composed mostly of deltaic deposits. The work allowed to detect the upplaping pinchouts that constitute structural-stratigraphic traps for gas generated in front-prodelta heteroliths [6]. In the area seismic-scale deltaic clinoforms can be observed on the south, these forms developed a shelf-to-basin floor relief of over 300 m [3].

However, the detection of hydrocarbon-saturated intervals still remains the greatest challenge. Resistivity logs measured in shaly-sand intervals show low values, as the occurrence of clay minerals constitute additional conductive components apart from water present in the rock. Shaly-sand formations are usually related to the high content of capillary water occupying the matrix micropores. Due to the high volume of clay minerals, there are many intervals with ambiguous saturation characteristics. Thus, a novel approach is required to detect low-resistivity gas-saturated zones. Mudstones are "non-Archie rock", where the saturation exponent (n) is not a constant, but its values change throughout the reservoir. It is expected that the saturation exponent in mudstones is low, as these sediments are likely to be wet. Thus, the Montaron equation [8] was adopted to calculate water saturation, where rock wettability is expressed through the water connectivity index (WCI). Additionally, the artificial neural network approach was used to identify the unit of the best gas accumulation properties and particularly to separate gas-saturated and water-saturated intervals. The process of water saturation estimation was supported by the modified qualitative Passey's [9] and Bowman's [10] methods primary dedicated the calculation of TOC content. A successful result of the analyses carried out in the profiles of the boreholes consequently raises the issue of the spatial distribution of these reservoir parameters. Thus, the authors attempt to reliably recreate the spatial continuity of the reservoir's parameters obtained from well logs through the use of geostatistical analyzes, multiple linear regression and neural network techniques in the face of the availability of seismic data. As seismic data are available, the results were integrated into the spatial seismic response through the use of supervised neural network methods. Finally, an attempt was made to identify the areas preferable for planning future wells with high gas accumulation potential.

2. Materials and Methods

2.1. Study Area

The location of the study is an area of multihorizon gas deposits typical of Miocene formations (Figure 1A). The analyzed area is in the central part of the Carpathian Foredeep, a short distance from the edge of the Carpathian overthrust, and is characterized by an uncomplicated geological structure, both in terms of stratigraphy and tectonics, which consists of three main complexes: (1) Precambrian basement with a strongly eroded morphological surface; (2) autochthonous Miocene sediments subdivided into Sarmatian and Badenian; and (3) Quaternary cover. Eroded Precambrian basement, consisting of quartzite sandstone and shales, is unconformably overlain by Badenian and Sarmatian sediments where many gas-bearing horizons were discovered. The Badenian formation consists of clays, marly clays, marls with thin laminas of mudstones, sandstones, and anhydrites. The Upper Badenian deposits are represented in the entire area by clay and clay–calcareous shales, sometimes interbedded with sandstone. The Sarmatian sediments consist of sandstone, mudstone, and claystone layers of similar thickness. The Sarmatian claystones are represented by gray, dark gray, and marly clayey shales with inserts and interlayers of light gray and gray fine-grained calcareous sandstones. The upper part of the Lower Sarmatian is associated with the presence of silty sands and claystones, while more sandstone layers are observed at greater depths. The Quaternary cover, up to 30 m, consists of clay, silt, sand, and gravel, from which gas exhalations from the underlying unsealed deposits or uninsulated boreholes were observed. This is related to the nature of hydrocarbon migration processes resulting from pressure differences, capillary forces, diffusion phenomena, etc. [11–13]. Gas-bearing horizons associated with the Lower Sarmatian are formed as sandstone and mudstone layers with good reservoir properties, which have an anticlinal structure. The series of sandstone and mudstone were also found to dip at small angles (1–2 degrees). Their average thickness is approximately 11 m (from 1 to 25 m effective thickness) with an average gas saturation of 63% (gas saturation from 52 to 83%). The average porosity and permeability of the gas horizons of this area are, respectively, 7.4% and 26 mD, with the maximum recorded values of these parameters equal to 30% and 2700 mD. Reservoir properties decrease due to compaction and the increase in calcite in the lithologic composition. The interbeds of claystones constitute an impervious barrier to the migration of gas, which is characterized by high methane content (not less than 95.77% CH_4) [14]. The individual gas accumulation is determined by the water–gas contact or lithological barrier related to changes in the sedimentary depositional environment. There are also gas accumulations related to fractures and cracks in the upper part of the Precambrian deposits, isolated by impervious anhydrite layers, but these gas occurrences are not a subject of this study.

2.2. Input Data

The petrophysical interpretation was carried out using data from 10 boreholes within several hundred meters of Miocene sediments covering 7 gas-bearing horizons. The gas deposit formed as anticline with the location of the wellbores is presented in Figure 2.

K-6, and K-9 wells, drilled in the 1970s and 1980s, represent a much lower technical level than the wells drilled in recent years (2011–2015). Thus, the analyzed wells differ significantly in quality, technical conditions, and available well log data. The standard dataset from the 1970s and 1980s included the following well logs: gamma ray (GR), neutron porosity (NPHI), resistivity logs (EN64, EN16, EL18), diameter (CALI), and spontaneous potential (PS).

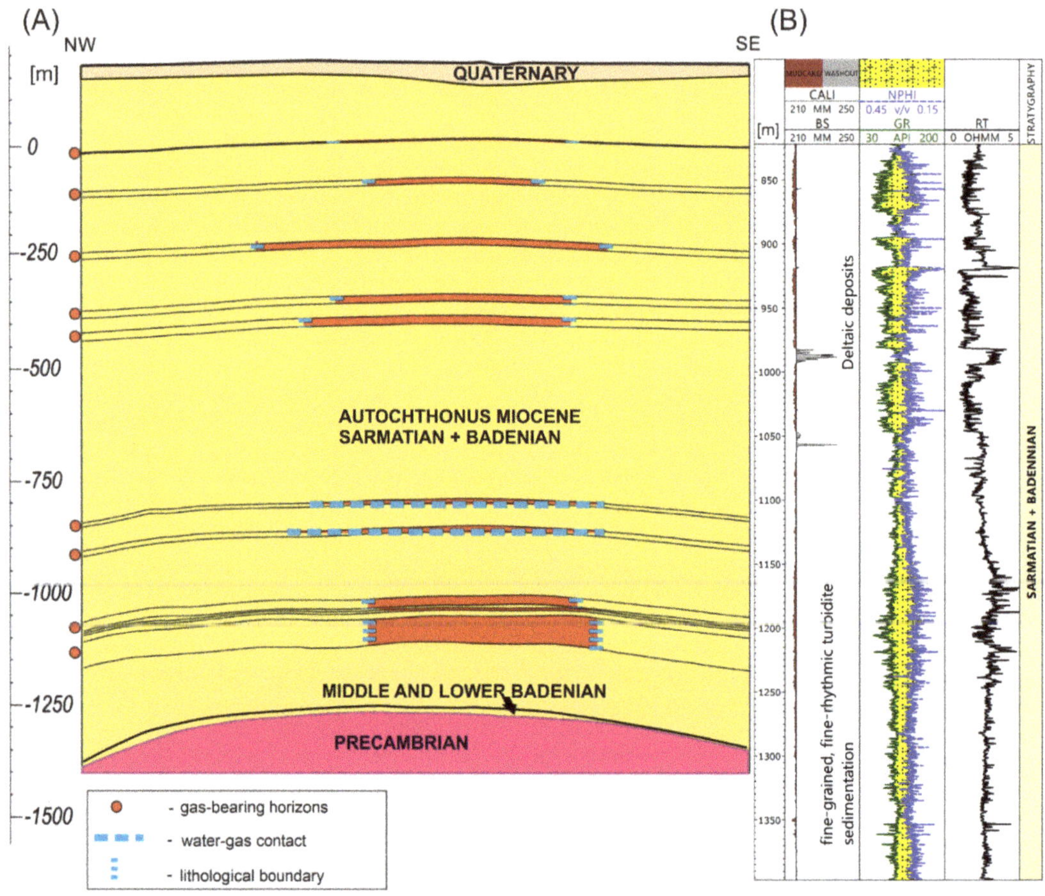

Figure 1. (**A**) Scheme of a typical deposit complex in the Miocene formations of the Carpathian Foredeep. (**B**) Well log identification of different Miocene sedimentary environments.

The first stage of the work was to collect the available laboratory and stratigraphic data, core descriptions, reservoir test information, and well logs available from individual boreholes. For wells K-17, K-18, K-19, K-20, and K-21K, the following laboratory measurements were performed: porosity measurement by MIP (mercury injection porosimetry) method, permeability measurements, NMR, XRD, and measurement of the Archie parameter (cementation exponent (m)). These boreholes were used to establish the relationships between well logs and laboratory data. They were the basis for the interpretation of older wells characterized by inferior technical conditions and limited sets of well logs. Laboratory analyses carried out on the cores were the basis for the calibration of effective porosity (correlation between laboratory-measured bulk density and effective porosity), the definition of the relationship between porosity (PHIE) and permeability (K), and calculation of the irreducible water content (Swi_Kapilar) based on NMR data. Moreover, the XRD data enabled us to calibrate the calculated clay volume. Datasets of new boreholes additionally include the following logs: PE (photoelectric factor), DT (compressional slowness), bulk density (RHOB), and medium (ILM) and deep (ILD) induction resistivity logs. Table 1 contains the input data used in the interpretation, and Figure 3 is a graphical presentation of the input data from the K-17 well. In boreholesK-25K and K-27, there were also available measurements of potassium (POTA), thorium (THOR), and uranium (URAN) concentration.

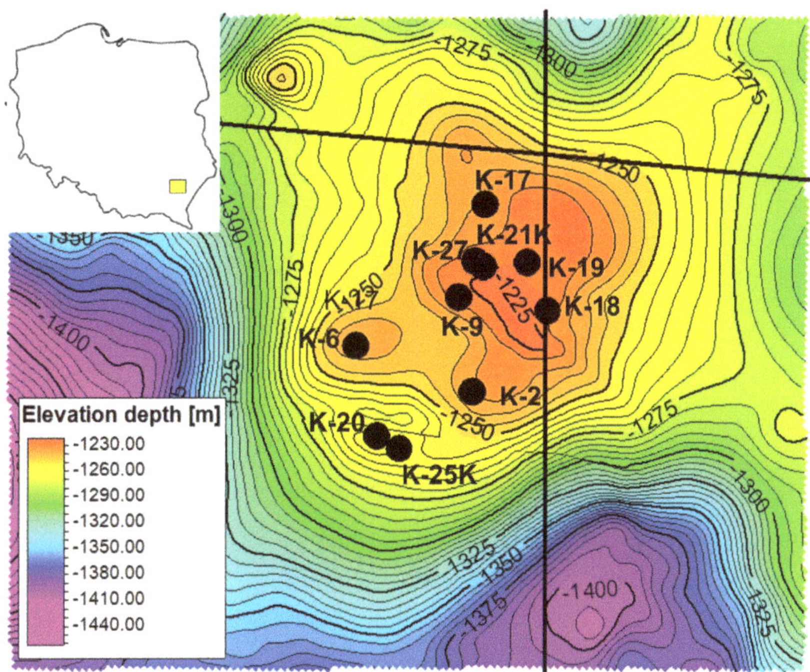

Figure 2. Interpreted gas deposit formed as anticline with the location of wellbores.

Table 1. Well log data available for the interpreted wells.

Well Name	Available Well Logs	Laboratory Core Measurements
K-2	CALI, GR, SP, NPHI, EL18, EN16, EN64	PHIE, Perm., RHOB
K-6	CALI, GR, SP, NPHI, EL18, EN16, EN64	PHIE, Perm., RHOB
K-9	CALI, GR, SP, NPHI, EL18, EN16, EN64	
K-17	CALI, GR, PE, NPHI, DT, RHOB, PE, ILM, ILD	PHIE, NMR, XRD
K-18	CALI, GR, PE, NPHI, DT, RHOB, PE, ILM, ILD	PHIE, NMR, XRD
K-19	CALI, GR, PE, NPHI, DT, RHOB, PE, ILM, ILD	PHIE, NMR, XRD
K-20	CALI, GR, PE, NPHI, DT, RHOB, PE, ILM, ILD	PHIE, NMR, XRD
K-21K	CALI, GR, PE, NPHI, DT, RHOB, PE, ILM, ILD, URAN, POTAS, THOR	
K-25K	CALI, GR, PE, NPHI, DT, RHOB, PE, ILM, ILD, URAN, URAN, THOR	
K-27	CALI, GR, PE, NPHI, DT, RHOB, PE, ILM, ILD	

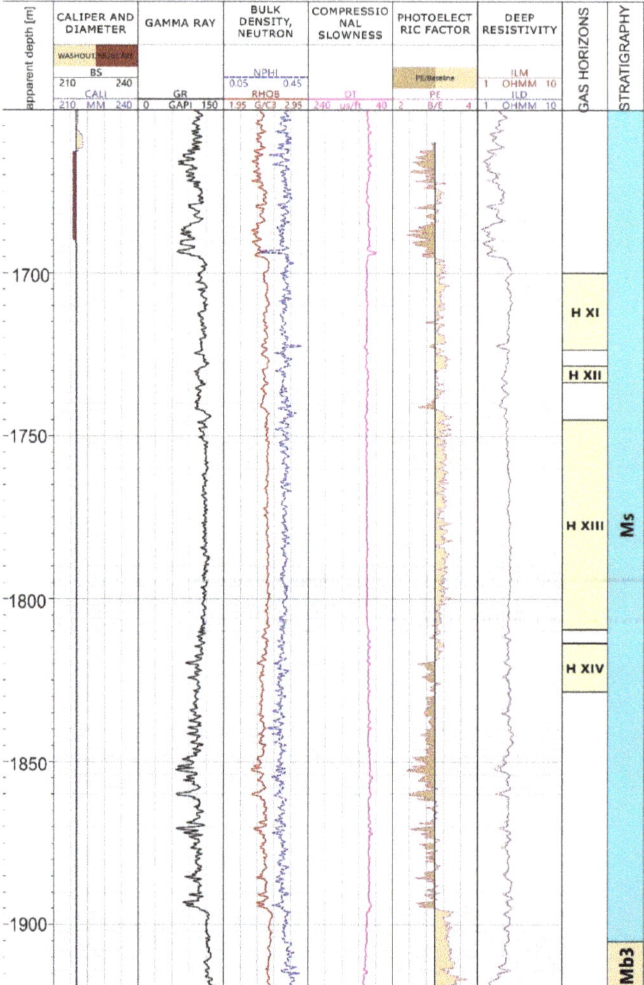

Figure 3. Well log data used in the interpretation of the K-17 well.

2.3. Evaluation of Reservoir Parameters from Well Logs (Vcl, Phie, Swi)

The shaly-sand reservoir, due to high clay volume content and rather low porosity, belongs to unconventional gas accumulation, and the applied methodology of water saturation differs from the classical methods used in the analysis of conventional deposits. The approach to calculate other reservoir parameters such as clay volume (Vcl), effective porosity (Phie), and irreducible water saturation (capillary-bound water) (Swi_kapilar) was based on well logging and laboratory measurements from cored intervals together with the results of well tests. In order to evaluate the key reservoir parameters, a simple petrophysical model was built. The results of X-ray diffraction laboratory measurement and crossplots of Potassium and Thorium concentration indicate show that the dominant clay minerals are: illite and montmorillonite, while rock matrix consists of quartz with admixtures of K-Feldspar, Plagioclase, Calcite and Dolomite.

$$\text{clay volume} + \text{matrix} + \text{porosity}$$

The best calibration of the volumetric content of clay minerals with the results of laboratory measurements of total clay mineral content using the X-ray diffraction method was obtained using the Stieber model for Miocene and Pliocene deposits in the form of Equation (1).

$$\text{Vcl} = \frac{\text{GRindex}}{3 - 2 \times \text{GRindex}}$$ (1)

where

$$\text{GRindex} = \frac{\text{GR} - \text{GRmatrix}}{\text{GRcl} - \text{GRmatrix}},$$

and GRcl is the gamma ray for clay, and GRmatrix is the gamma ray value for the sandstones.

The calculations assumed values of GR, namely, 30–40 API for sandstones and 120–160 API for claystones. These values were determined during optimization procedures. Moreover, the correlation between GR and laboratory XRD measurements of total clay mineral content was also performed. The following relationship was obtained Equation (2):

$$\text{Vcl} = 0.455 \times \text{GR} - 29.95$$ (2)

This dependence was used to estimate the clay volume in individual wells. The effective porosity values (Phie) were calculated using two approaches from neutron density log crossplots and using laboratory-measured effective porosity (PHIE) and bulk density (RHOB) (Figure 4A). The neutron porosity of clay, NPHI_Vcl = 0.4, and bulk density of clay, RHOB_Vcl = 2.45, were empirically derived. The total porosity was estimated on the assumption that the clay porosity was Phie_{cl} = 0.2. The assumption of clay porosity was based on the relationship between total clay volume obtained from the X-ray diffraction analysis technique (XRD) and clay-bound water saturation from NMR data.

$$\text{PhiT} = \text{Phie} + \text{Vcl} \times \text{Phi}_{cl}$$ (3)

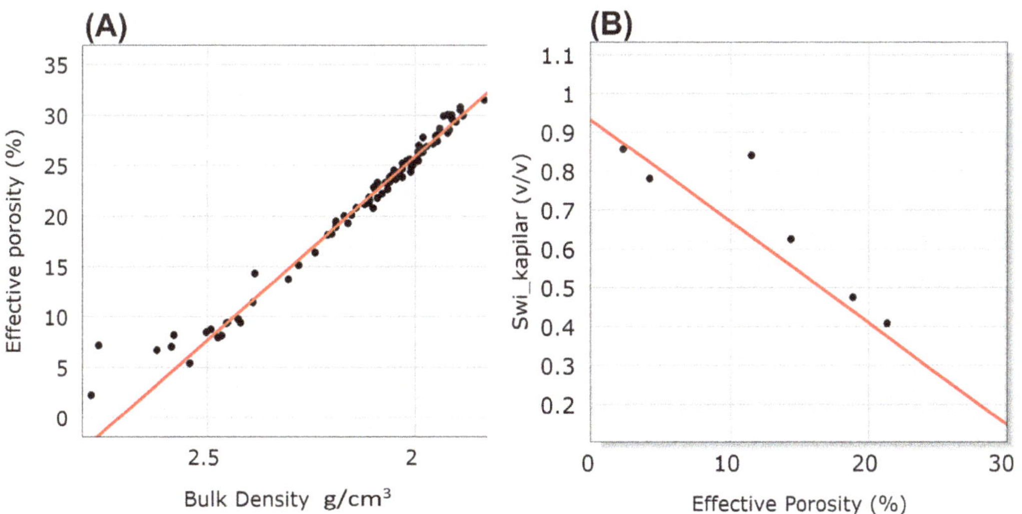

Figure 4. (**A**) Correlation of the measured effective porosity and the bulk density; (**B**) correlation of laboratory-measured porosity and capillary water content measured by NMR.

Calculation of capillary water content (Swi_Kapilar), water bound in small pores in the matrix, was also performed with the use of NMR laboratory measurements. A trend line was determined between capillary water saturation and effective porosity (NMR) (Figure 4B), and based on this relationship, Swi_Kapilar was calculated. Each equation based on the use of well logs required the assumptions of certain coefficients; some of these parameters are often unknown and challenging to calculate, especially in shaly-sand or mudstone formations. Errors during the stage of calculating the effective porosity undoubtedly have a large effect on errors in gas resource estimation. More particularly, if the reservoir has moderate to low effective porosity, then much attention should be given to the precise estimation of this parameter.

2.4. Qualitative and Quantitative Methods of Identifying Perspective Gas-Saturated Zones

The most important step during petrophysical interpretation is detection of perspective gas-saturated intervals (qualitative methods), which is followed by calculation of water saturation. Interpretation of gas accumulation present in mudstones is difficult due to high clay volume, which, due to high conductivity, lowers the values of recorded resistivities. Clays, similar to water, are additional conductive media in the formation. This usually leads to underestimation of gas resources.

According to Archie's assumption [15], the presence of hydrocarbons in rock is related to the difference between the resistivity of the unflushed zone (Rt) and the water-saturated zone (R0). Then, knowing the values of the saturation coefficient (n), one can determine the values of Sw. However, when the interpretation of water saturation is carried out through a formation with a thickness of several hundred meters, consisting of thin layers of sandstone, mudstones, and claystones, proper determination of the saturation exponent (n) is very difficult. The values of this parameter are closely related to the irreducible water content, size of the pores, and wettability of the rock. The resistivity of the water-saturated zone (R0) can be calculated based on formation water resistivity (R_w) and Archie parameters: cementation exponent (m), and saturation exponent (n). Calculation of R_w values requires information about the temperature in the borehole and the salinity. While the temperature in the borehole can be determined based on the results of temperature logs performed in some boreholes in evaluated gas fields or calculated based on temperature gradients for a given area, the salinity/mineralization of water is often unknown. Even the laboratory-measured mineralization of a water sample collected from the evaluated borehole does not guarantee that this sample is not a mixture of mud filtrate and formation water. A more reliable method to determine R0 could be based on well logs measured in situ that reflect the reservoir conditions. The analyzed interval consists of gas-saturated sandstones/mudstones and water-saturated sandstones/mudstones. On the crossplot of bulk density (RHOB) and formation resistivity (Rt), the separation between the gas-saturated zone and the water-saturated zone is clearly visible, especially in the intervals with low clay volume. Thus, the resistivity of the water-saturated zone R0 can be calculated with the use of exponential Equation (4):

$$R0 = a \times RHOB^b \tag{4}$$

where a and b are empirically derived constants; for this study, a = 0.0008 and b = 8.8554.

If the bulk density log is not available, the neutron porosity log can also be used.

The presence of hydrocarbons can be defined by the RI (resistivity index) Equation (5):

$$RI = \frac{Rt}{R0} = \frac{Rt}{a \times RHOB^b} \tag{5}$$

Then, the RI obtained can be calibrated with laboratory-measured Sw values, determined, for example, by the Dean–Stark method. During Dean–Stark extraction, a fresh core sample is weighed and subjected to fluid extraction by boiling solvent. Then water is condensed and collected. These laboratory measurements are highly recommended for this type of formation in order to calibrate the water saturation. Unfortunately, there were no laboratory water saturation data available in the evaluated gas field. In this case, the values of Sw were calculated with the use of the generalized connectivity Equation (6) [16,17]:

$$RI = \left(\frac{1 - Sc}{Sw - Sc} \right)^{\mu} \tag{6}$$

where Sc is critical water saturation.

Assuming as a conductive medium: irreducible water bound in clay minerals (CBW), irreducible capillary water held by capillary forces in matrix micropores (KBW) and free water (FFW), and with the assumption that CBW has different resistivity than KBW and free water (FFW), the following equations can be derived [18]:

$$\sigma = [(CBW + KBW + FFW) \times PhiT - WCIsh]^{\frac{1}{\mu}} \tag{7}$$

$$\begin{aligned} WCI_{sh} &= (CBW + KBW + FFW) \times PhiT \\ &- \left[\frac{CBW \times PhiT \times \sigma_{cw}^{\frac{1}{\mu}} + KBW \times PhiT \times \sigma_{w}^{\frac{1}{\mu}} + FFW \times PhiT \times \sigma_{w}^{\frac{1}{\mu}}}{\sigma_{w}^{\frac{1}{\mu}}} \right] \end{aligned} \tag{8}$$

$$\begin{aligned} WCI_{sh} &= (CBW + KBW + FFW) \times PhiT \\ &- \left[\frac{CBW \times PhiT \times \sigma_{cw}^{\frac{1}{\mu}}}{\sigma_{w}^{1/\mu}} + \frac{KBW \times PhiT \times \sigma_{w}^{\frac{1}{\mu}}}{\sigma_{w}^{1/\mu}} + \frac{FFW \times PhiT \times \sigma_{w}^{\frac{1}{\mu}}}{\sigma_{w}^{1/\mu}} \right] \end{aligned}$$

Then,

$$WCI_{sh} = CBW \times PhiT \left[1 - \left(\frac{\sigma_{cw}}{\sigma_{w}} \right)^{1/\mu} \right]$$

$$WCI_{sh} = CBW \times PhiT \left[1 - \left(\frac{R_{w}}{R_{cw}} \right)^{1/\mu} \right]$$

$$Sc = \frac{WCI}{PhiT}$$

$$Sc = -CBW \times \left[\left(\frac{R_{w}}{R_{cw}} \right)^{\frac{1}{\mu}} - 1 \right] \tag{9}$$

WCI—water connectivity index
CBW—clay-bound water, water bound in clay minerals
KBW—capillary-bound water, water bound in the micropores of the rock matrix
FFW—free fluid water, moveable water
σ_{w}—conductivity of free water and KBW
σ_{cw}—conductivity of CBW
R_{w}—free fluid water (FFW) and capillary-bound water (KBW) resistivity
R_{cw}—clay-bound water resistivity
μ—conductivity exponent
Swir = CBW $\times$ PhiT and Swi_Kapilar = KBW $\times$ Phie

Thus, the final water saturation (Sw) can be calculated as

$$Sw = \frac{(1 - Sc) + \sqrt[\mu]{RI} \times Sc}{\sqrt[\mu]{RI}} \tag{10}$$

The values of the conductivity coefficient μ are generally constant, and their range of variation is small (1.6–2).

2.5. Determination of Critical Water Saturation (Sc)

Although RI index can be calculated based on in situ measurement, the estimation of Sc values may require the values of R_w and R_{cw}. Thus, formation/free water resistivity (R_w) was calculated on the basis of the salinity of the water measured in the laboratory and the temperature measured in the borehole. The (R_w) values decreased with depth. In the interval below 1000 m, the R_w values ranged from 0.02 to 0.058 ohm. Resistivity of clay-bound water was calculated as follows Equation (11):

$$R_{cw} = \frac{CBW \times PhiT}{m} \times \frac{Rvcl}{a} \tag{11}$$

The cementation exponent (m) was calculated based on the relationship between laboratory-measured values of m and measured porosity, performed in well K-17 (12).

$$m = 0.089 \times \log(Phie) + 1.952 \tag{12}$$

However, if there are difficulties in cementation exponent calculation, no measurements of m are performed in the evaluated gas field. The simple equation proposed by Peeters &Holmes [19] can be used to calculate conductivity (C_{cw}) and resistivity (R_{cw}) of clay-bound water (CBW) Equation (13):

$$C_{cw} = \frac{C_{cl}}{Phi_{cl}^2}, \; R_{cw} = \frac{1000}{C_{cw}} \tag{13}$$

where C_{cl} is conductivity of clay, and Phi_{cl} is clay porosity.

The calculated water saturation values are SwT, which represents water saturation in total porosity. The following equation can be used to calculate effective water saturation (Sw) Equation (14):

$$Sw = 1 - \left[\frac{PhiT}{Phie} \times (1 - SwT) \right] \tag{14}$$

2.6. The Use of Artificial Neural Networks

Artificial neural networks are often used when conventional methods of analysis fail. In qualitative analyses of the presence of hydrocarbons in rock, appropriate sets of measured data are often used that allow one to observe so-called crossover, for example on a neutron–acoustic or density–neutron crossplot. This allows qualitative analysis to be performed and perspective zones to be distinguished [20–24]. There are two types of neural network: supervised and unsupervised. This research used unsupervised self-organizing maps (SOMs) [25] which are especially suitable for data surveys. This method creates a set of vectors to represent the input data and carries out topology preserving projection of the prototypes from the input space onto a low-dimensional grid [26]. The neurons and connections that transfer information are the basic elements. In this study, the input was the well logging measurements, and the output was the lithological classification in the form of PSU units. During the learning phase, an input–output correlation is established. The SOM is a fully connected neural network, where the output is generally organized into a 2D arrangement of neurons. SOMs are based on soft competition between neurons in the output layer. Training of the network is about finding similarities among input data and can be performed without a priori information, which is called unsupervised learning. Kohonen [25] described three complimentary processes—competition, cooperation, and synaptic adaptation—that are involved in the SOM algorithm. Neurons in the SOM are connected to adjacent neurons by neighborhood relations. In the training phase, vector x from the input is chosen, and the activation function is used to activate each unit. Usually, the activation function is expressed by the Euclidian distance between the weight vector (w_i) and input vector (x) [25]. If assumed that M is the size of SOM array, the unit number i ranges from 1 to M and adjacent units on the grid are celled neighbours; the neuron with the weight vector closest to the input vector x is called the best-matching unit (BMU). It has the smallest Euclidian distance and is described by Equation (15).

$$c_k = \text{argmin}||x_k - w_i|| \qquad (15)$$

where:

c_k is a winner index on the SOM for a data snapshot k, and c ranges from 1 to M. The "arg" states for "index". The weight vector of the winner is moved toward the presented input data as indicated by a time-decreasing learning rate α. While function h modifies the weight vectors of the neighboring units [27]. The rule of the learning is described by Equation (16) [28].

$$w_i(t+1) = w_i(t) + \alpha(t)h_{ci}(t)[x(t) - w_i(t)] \qquad (16)$$

where:

t—is learning iteration
x—states for an input pattern
$w_i(t)$ is the weight vector indicating the output unit's location in the data space at time t;
$x(t)$ is an input vector drawn from the input dataset at time t;
$\alpha(t)$ is the learning rate at time t;
h—spatial-temporal neighborhood function
$h_{ci}(t)$ is the neighborhood kernel around the 'winner' unit c.

In this study, the calculations were performed using Techlog Schlumberger software with the ISPOM module, which enables electrofacies detection based on Kohonen SOMs.

The presence of hydrocarbons, especially gas, affects not only resistivity logs but also porosity logs, such as RHOB, DT, and NPHI. Thus, the presence of hydrocarbons causes an increase in the interval time (DT) and a decrease in the values of bulk density and neutron porosity. While the presence of hydrocarbons in sandstone intervals significantly affects the recorded values of porosity logs, changes might be slight and not always unambiguous in mudstones. The decrease in neutron porosity and the increase in resistivity may also be associated with a higher content of calcite or dolomite in a given interval; however, in this case, we also observe an increase in bulk density values and a decrease in compressional slowness values.

Neural networks were used to classify data into PSUs (petrophysically similar units). They did not constitute a facies classification, but indicated intervals with similar petrophysical properties. Based on the knowledge of the geology in a given area and information concerning the influence of individual minerals on the different well logs, it indicates the hydrocarbon potential of each group and constitutes an alternative tool for qualitative identification of water-saturated and gas-saturated intervals. In the research area selected, the variability of the input data was small, related to the rather monotonous structure of Miocene sediments constituting a series of interspersed layers, namely sandstones, mudstones, and claystones of various thicknesses and variable carbonate content. These sediments were characterized by the lack of evident differentiation on gamma rays, causing significant difficulties in the interpretation of these types of deposits, which are the result of fine-grained, fine-rhythmic turbidite sedimentation [1] (Figure 1B). All curves used as input were normalized. The classification into petrophysically similar units (PSUs) was made in 10 wells covering the interval of seven gas-bearing horizons. The selected interval lies at a depth of about 1000–1500 m and includes mostly thin-layer sediments. In seven of ten interpreted wells, six well logs were used as input: gamma ray (GR), compressional slowness (DT), bulk density (RHOB), neutron porosity (NPHI), photoelectric factor (PE), and deep induction resistivity log ILD (Rt). Artificial neural networks based on the Kohonen [29] algorithm available in Techlog Schlumberger software were used for data classification. The fuzzy logic classification method was chosen. The network was trained on data from the K-19 well and the results validated for K-17. The network dimension of 10×10 was assumed. The trained network was applied to the wellbores K-17, K-18, K-20, K-21K, K-25K, and K-27.

In three archival wells where only three well logs (GR, NPHI, and Rt) were available, the supervised neural network method was used, with the reference to PSUs from well K-19. Despite that fact, the identified PSU groups corresponded to these from new wells, where six input data were used. Table 2 presents the mean values of the input data within the four PSUs and descriptions of their petrophysical properties.

Table 2. Average values of petrophysical parameters for individual groups.

	Units	PSU 1. Mudstone/Claystone with Higher Carbonate Content		PSU 2. Porous Sandstone, Mostly Water Saturated		PSU 3. Mudstones Mostly Water-Saturated		PSU 4. Gas-Saturated Shaly-Sands/ Mudstones	
Variables		Mean	Variance	Mean	Variance	Mean	Variance	Mean	Variance
Compressional Slowness	μs/m	326.329	28.985	322.360	36.055	339.215	24.054	323.570	51.090
Bulk Density	g/cm^3	2.495	0.000	2.362	0.001	2.455	0.001	2.444	0.001
Gamma Ray	API	106.158	8.848	83.506	43.401	104.500	20.555	96.040	17.988
Formation Resistivity	ohm.m	3.661	0.012	2.870	0.583	3.248	0.069	3.973	0.084
Neutron Porosity		0.263	0.000	0.243	0.001	0.283	0.000	0.243	0.000
Photoelectric Factor	b/elec	3.182	0.009	2.711	0.009	3.127	0.008	2.971	0.006

PSU 1 represents claystone and mudstones with carbonates, sediments with high clay-bound water content, and an average bulk density of 2.5 g/cm^3. This group was comparable to PSU 3 (mudstone); however, the values of neutron porosity for the first group were lower, and the bulk density and resistivity were higher. This was due to the admixture of carbonates present in this electrofacies. XRD measurements indicated between 20 and 30% of carbonates (calcite and dolomite) for this group, which is a group of sediments that constitute sealing layers; it has low effective porosity and is almost impermeable. However, the analysis of high resolution microresistivity logs from well K-19 showed several-meters-thick interbeds of gas-saturated sandstones present within this group of sediments. Unfortunately, these layers were too thin to be properly resolved by conventional well logs. A petrophysical approach to interpretation of this thin-bed formation required the use of high-resolution logs [30–35]. At this stage of interpretation, PSU 1 was treated as a sealing unit, as applied methods of interpretation did not allow the hydrocarbon potential of this unit to be properly estimated. The photoelectric factor values were the highest for this group, on average 3.18 b/elec. PSU 2 corresponded to porous sandstone or sandstone/mudstone heterolites. The average gamma ray value for this group was 83.5 API. The bulk density values in this group were the lowest, on average 2.36 g/cm^3. The average value of the photoelectric factor was equal to 2.71 b/elec. Average neutron porosity was 0.24. However, the resistivity of this group was very low at 2.87 ohm.m, which suggests that sandstones were mostly water saturated or were in the transition zone with the presence of both gas and water. This unit does not constitute the dominant group associated with gas accumulation. Effective porosity of sandstone layers was high and reached up to 23%. PSU 3 consisted of siltstones/mudstone with the highest neutron porosity of 0.28, compressional slowness value of 339 us/m, and an average resistivity equal to 3.2 ohm.m. These features may suggest that mudstones are organic-rich. However, the uranium content measurements available in the K-25K and K-21K wells were very low and generally did not exceed 4 ppm. The increase in uranium content did not show a relationship with PSU 3. In that case, this group probably is related to mudstones with high irreducible water content; it also might consist of higher volumes of swelling clays such as smectite [36] and low carbonate content. PSU 4 corresponded to layers with the highest hydrocarbon potential. It was characterized by the highest resistivity values of 3.97 ohm.m; the average value of neutron porosity was 0.24 and was similar to that of sandstones of PSU 2. The average gamma ray was 96 API, higher than for sandstones (83.5 API) and lower than for mudstones PSU 3 (104.5 API). The intervals corresponding to PSU 4 coincided with the intervals where calculated water saturation was low. PSU 3 and 4 showed similar parameters, both representing mudstones, but low values of neutron porosity and high values of formation resistivity definitely indicate gas saturation in PSU 4. These petrophysical interpretation results were used in Petrel to model the spatial distribution of clay volume, effective porosity, and hydrocarbon saturation (SH_FIN) within defined PSUs.

2.7. Spatial Distribution of Petrophysical Parameters

The spatial distribution of facies or petrophysical (clay volume, porosity, permeability, hydrocarbon saturation), geomechanical, and geochemical parameters can be presented through 3D models. Recognition of the variability of reservoir properties within the research area (e.g., a gas field) is an important part of hydrocarbon exploration. On the basis of spatial distributions or maps of individual parameters, one can make a decision that helps to minimize the exploitation costs and maximize production. The maps of the reservoir properties in the gas field, in addition to the structural model, are the basic elements of planning the exploration works, carrying out the calculations of resources, and preparing development projects for the discovered hydrocarbon accumulation. In addition to planning the location of boreholes, the spatial form of mapping the parametric variability of the analyzed reservoir also permits the selection of the well-type (vertical, horizontal). In order to spatially distribute the results of the PSU classification obtained in the profiles of the analyzed boreholes, a number of realizations were generated to reduce the geological uncertainty of the model assessment for the petrophysical PSUs, porosity, and hydrocarbon saturation model. The first stage of work was related to the construction of the structural model. The authors used the structural model created by K. Sowiżdzał. In the next step, definition of the vertical cell size based on the vertical heterogeneity of the log resolution was established. High resolution increased the probability of capturing important changes in reservoir parameters. Applying modern methods of data grouping also allowed us to capture details related to the reservoir and variability of petrophysical properties revealed in the wellbore profiles.

The final structural model consisted of seven gas-bearing horizons (marked in yellow) separated by layers that had not been subjected to the spatial modeling process (gray layers) (Figure 5). Due to the thin-bed nature of the Miocene deposits, the individual gas-bearing horizons, being the subject of the analysis, had an average vertical resolution of the grid of about 1 m (Figure 6). The total number of model cells in the intervals covered by the analysis was 6,925,200 (Table 3).

Table 3. Quantitative characteristics of the individual intervals that were the subject of the research.

Horizon	Average Thickness (m)	Number of Grid Cells
XI	22.2	875,600
XII	9.6	398,000
XIII	64.5	2,587,000
XIV	31.8	1,273,600
XIVa	14	597,000
XV	15.4	636,800
XVI	13.5	557,200

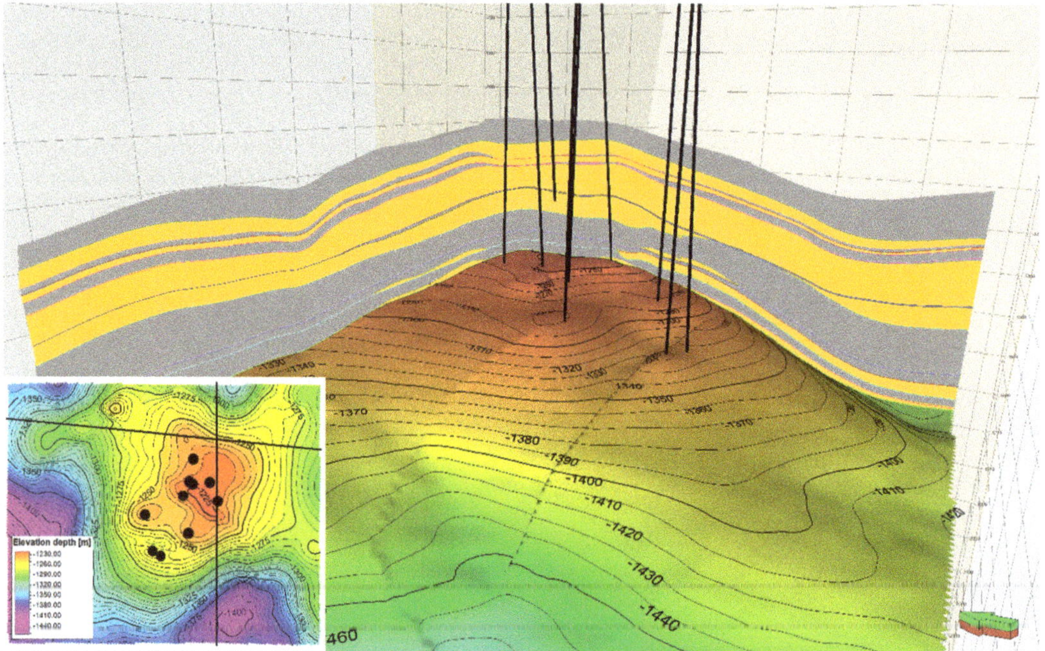

Figure 5. Cross-section through the structural model of the analyzed deposit complex (view from the west side) with a location map of the wells and cross-sectional lines.

The final model showed the spatial distribution of individual petrophysical groups through the variability of reservoir parameters in each PSU group. The range of variability for reservoir parameters in each PSU group was defined based on the interpretation performed for the boreholes (Table 4).

Table 4. The range of variability of the most important reservoir properties for each PSU group.

PSU	Vcl *(v/v)*	Phie *(v/v)*	K (mD)	Swi_Kapilar *(v/v)*	SH-Hydrocarbon Saturation (1-Sw)
1	0.55–0.7	0.03–0.05	0.01–0.1	0.8–0.9	1—sealing unit
2	0.1–0.35	0.15–0.23	10–20	0.3–0.4	SH
3	0.4–0.5	0.05–0.12	0.05–1	0.4–0.6	SH
4	0.35–0.45	0.08–0.15	0.1–10	0.4–0.5	SH

For hydrocarbon saturation, no predefined values for each group were applied (except PSU1, which was treated as a sealing unit), but the continuous curve of SH was used. First, the data analysis was carried out (defining the physical values of the modeled parameter) and the parametric modeling process itself.

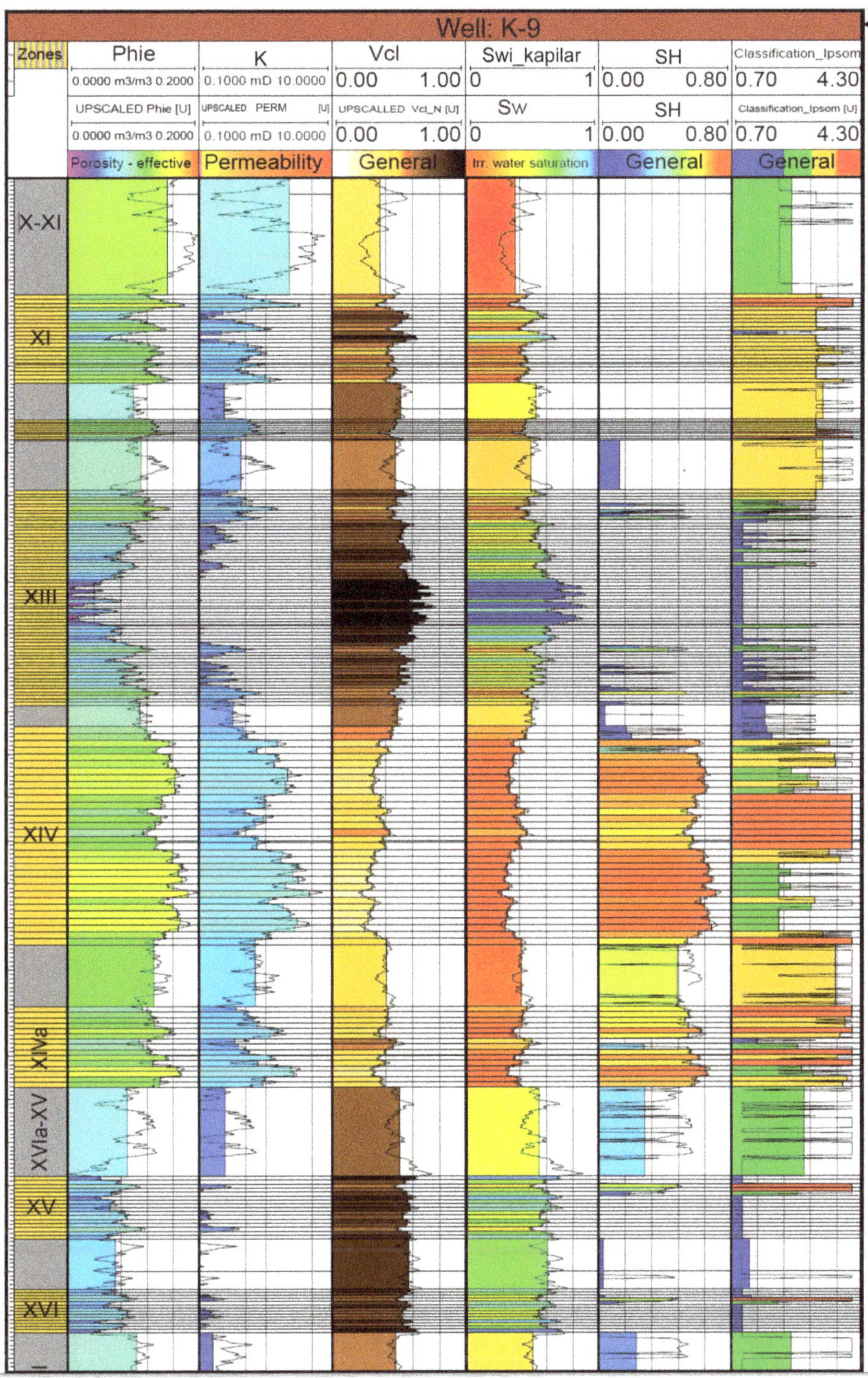

Figure 6. Well log data (solid black line) averaged at the vertical resolution of the grid.

2.8. Defining the Direction and Extent of Variograms

The analysis of the semivariogram function was carried out on the basis of a graph in which the x-axis showed the distance between the tested locations of the analyzed variable, the y-axis corresponded to 12 variances (i.e., semivariances), and the points were the values of the experimental variogram. Variogram analysis consists of adjusting the type of theoretical model to the nature of the variability observed in the data population, represented by the points of the experimental variogram by determining a number of parameters. The results are applied at the stage of parametric modeling of individual parameters

2.9. Parametric Modeling

At the parametric modeling stage, for each gas-bearing horizon, the results of data analyses carried out in the previous step were applied, and a computational algorithm was selected. For clay volume, effective porosity, and permeability, a stochastic Gaussian random function simulation algorithm was selected.

For each of the modeled parameters, 25 equally probable realizations of the spatial distribution were generated, which were finally subjected to the averaging process.

Parametric modeling results were verified by comparison of the histograms of the input data (well logs) with the data averaged in the resolution of the grid (upscaled logs) and models.

Based on the parametric models constructed in the previous step (Phie, Vcl), the reconstruction of the spatial distribution of reservoir parameters within the analyzed gas-bearing horizons was carried out. The procedure of spatial electrofacies classification consisted of the application of parametric models and the classifier formula in the form of an interpretation of electrofacies (the results of the PSU classification carried out with the use of the IPSOM module on well log data). The classification algorithm defined the membership of each grid cell to the PSU showing the highest degree of probability, also showing the value of this probability.

Finally, reconstruction of the spatial distribution of hydrocarbon saturation (SH) was performed. Neural networks were used to recreate the meta-attributes (pseudo-attributes) of this parameter (parameter with nonphysical values). Even so, pseudo-attributes can still be valuable information of a qualitative nature. The quantitative nature of the data was ensured by the procedure related to data analysis. At this stage, ranges of values (minimum, average, maximum) and a distribution curve for the value of a given parameter were defined based on information from the borehole.

3. Results

The results of well log data interpretation of the selected wells and spatial distribution of reservoir properties and PSUs in the form of property maps are presented below. Table 5 is a detailed description of each column's contents from Figures 7 and 8.

Table 5. Description of mnemonics and interpretation results of individual paths from Figures 6 and 7.

Track Number	Mnemonic	Description
1	MD	Measured depth
2	perforations	Perforated interval
3	GR	Normalized gamma ray
4	PE, NPHI, DT, RHOB	Normalized: photoelectric factor, neutron porosity, compressional slowness, bulk density
5	Rt	Normalized formation resistivity
6	Vcl	Clay volume
7	Phie, PhiT, PHIE	Effective porosity, total porosity, and laboratory-measured porosity (PHIE)

Table 5. *Cont.*

Track Number	Mnemonic	Description
8	XI, XII, XII, XIV	Horizon names
9	PSU groups	1—claystones/mudstones, 2—sandstones, 3—mudstones, 4—mudstones with high hydrocarbon potential
10	probability	Probability of occurrence of each facie at a specific depth
11	NPHI_base, NPHI	Qualitative method of detection of gas-saturated zones based on modified Passey's method: yellow—gas-saturated zones; brown—water saturated zones
12	Rt, R0_RHOB	Qualitative method of detection of gas-saturated zones based on difference between RT and R0
13	Swe_SIM, Swi_kapilar	Effective water saturation calculated from the Simandoux [37] formula (Swe_SIM), irreducible water saturation—Swi_kapilar
14	Swe_mont, Swi_kapilar	Effective water saturation calculated from the modified Montaron equation (Swe_mont), irreducible water saturation—Swi_kapilar
15	WCI	Water connectivity index

Calculated water saturation from the modified Montaron method indicates higher gas saturation in sandstones intervals comparing to Simandoux results. Simandoux model is basically dedicated shaly formations but if there is high variability of shale volume content as it is in heteroliths reservoirs there will be a necessity to subdivide reservoir into high number of thin beds and use different saturation model in each kind of facie, which may be confused and will lead to discontinuity, rapid changes in water saturation profile. Additionally, Simandoux method is only resistivity sensitive, it does not account for i.e., neutron porosity or bulk density changes, which may lead to underestimating water saturation in intervals where increasing resistivity is not related to the presence of hydrocarbon, for example, higher calcite content. Moreover, if there is little contrast or no contrast between gas saturated and water saturated zone Simandoux model will consequently show gas saturation through the whole interpreted interval or water saturation depending on assumed claystone resistivity. Involving RHOB or NPHI logs at stage of Sw calculation will give extra indicator that may confirm or not the presence of gas. The more input data taken into account, the lower the uncertainty of the calculated Sw. The advantage of Montaron model is also the fact of using measured "in situ" well logs as an input. The method does not rely on calculated parameters as Vcl or Phie that may have a high level of uncertainty. Via spatial propagation of petrophysical properties from boreholes, one can determine the favorable areas potentially gas saturated. The distribution of reservoir parameters was performed within the defined PSU's. Moreover, even the sole spatial propagation of PSU's might be a good qualitative hydrocarbon indicator, as PSU 4 is highly related to existing gas accumulations. This may give a first look at the perspective of the area before petrophysical interpretation is carried out.

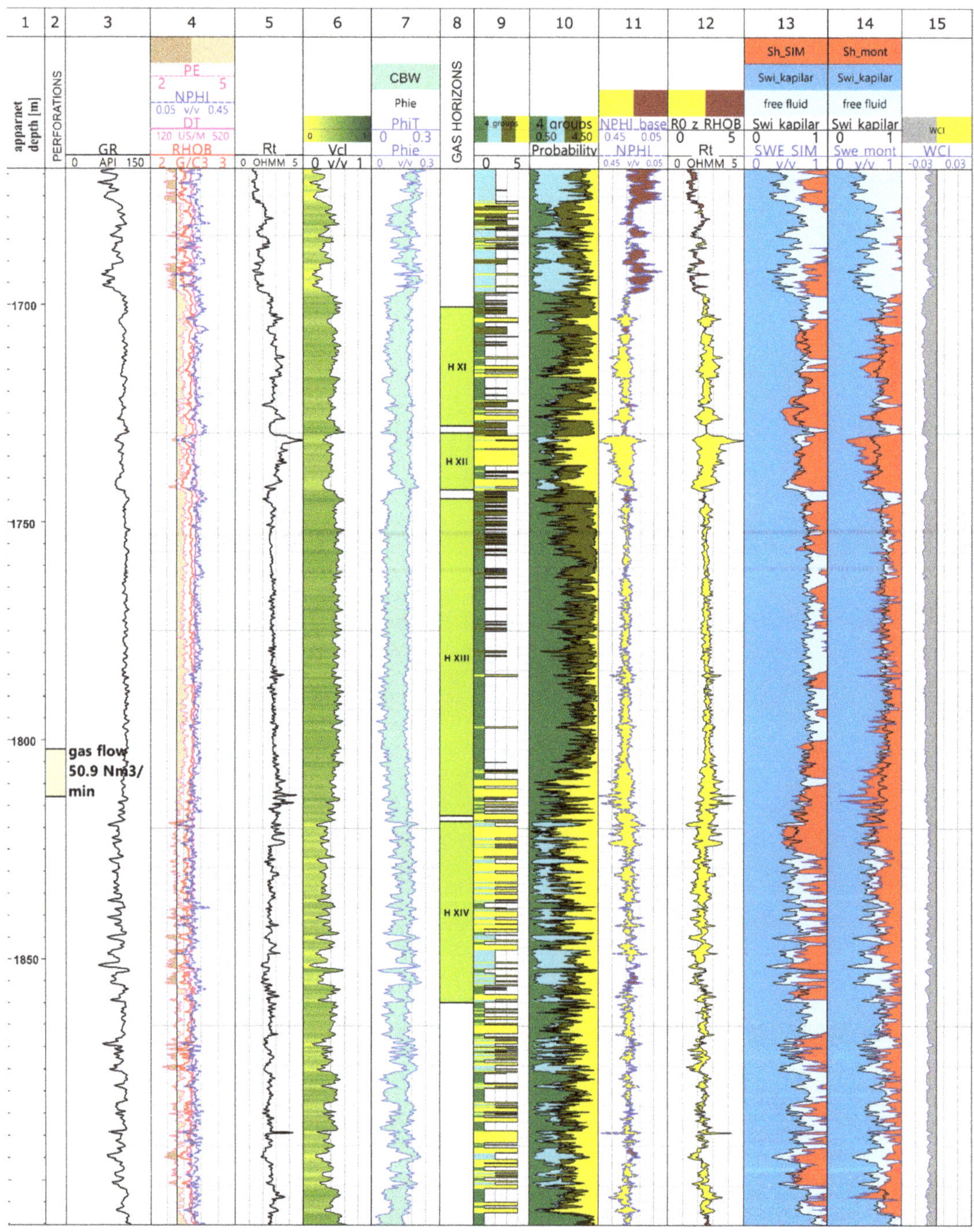

Figure 7. The results of petrophysical well log interpretation for well K-21K.

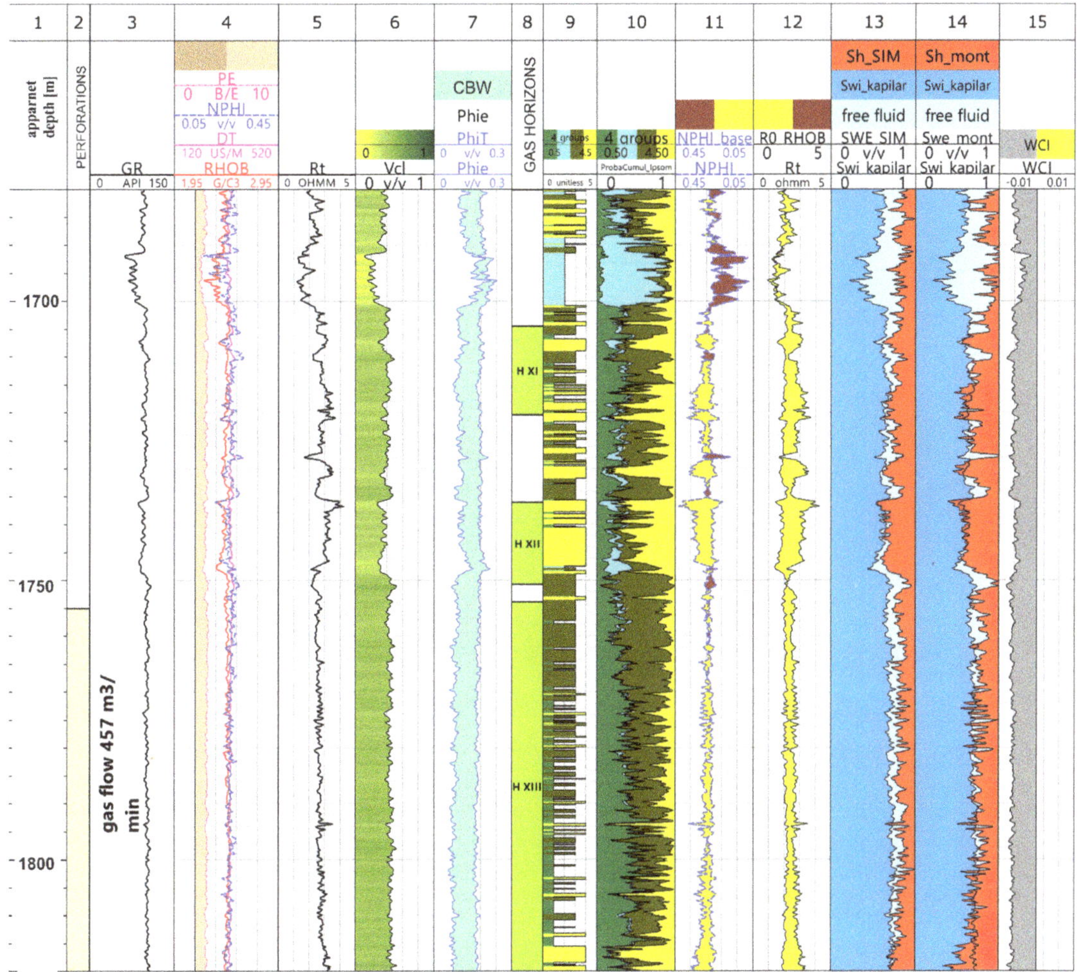

Figure 8. The results of petrophysical well log interpretation for well K-27.

High resolution of the model allowed for more reliable spatial variability of the individual layers that constitute the Miocene formation. Figure 9 present the effective porosity within the modeled horizons while Figure 10 show the spatial distribution of PSUs in horizon XV.

Figures 11 and 12 present the distribution of the hydrocarbon saturation (SH) in individual gas-bearing horizons. Finally, the works presented in this paper allowed for reconstruction of hydrocarbon saturation in the analyzed area and the generation of maps of average values of this parameter for each of the analyzed gas horizons.

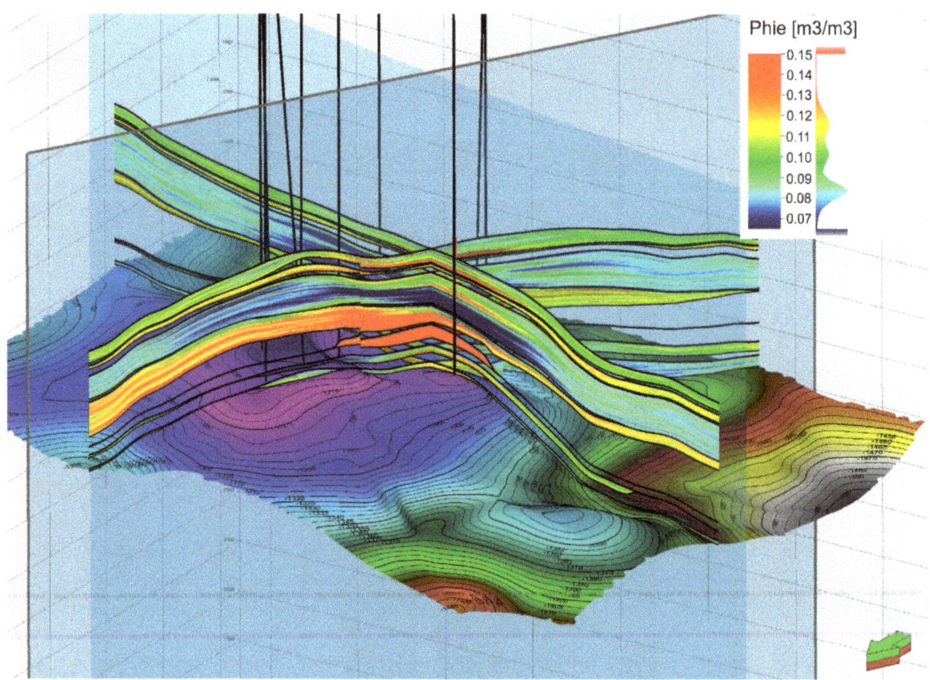

Figure 9. Final spatial distribution of effective porosity.

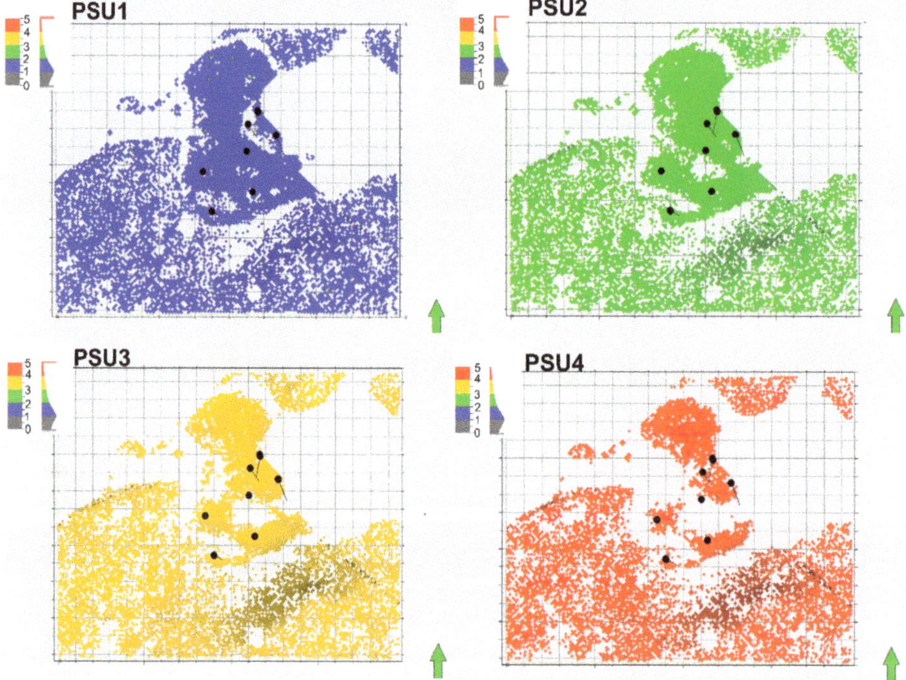

Figure 10. Spatial distribution of PSUs in gas horizon XV.

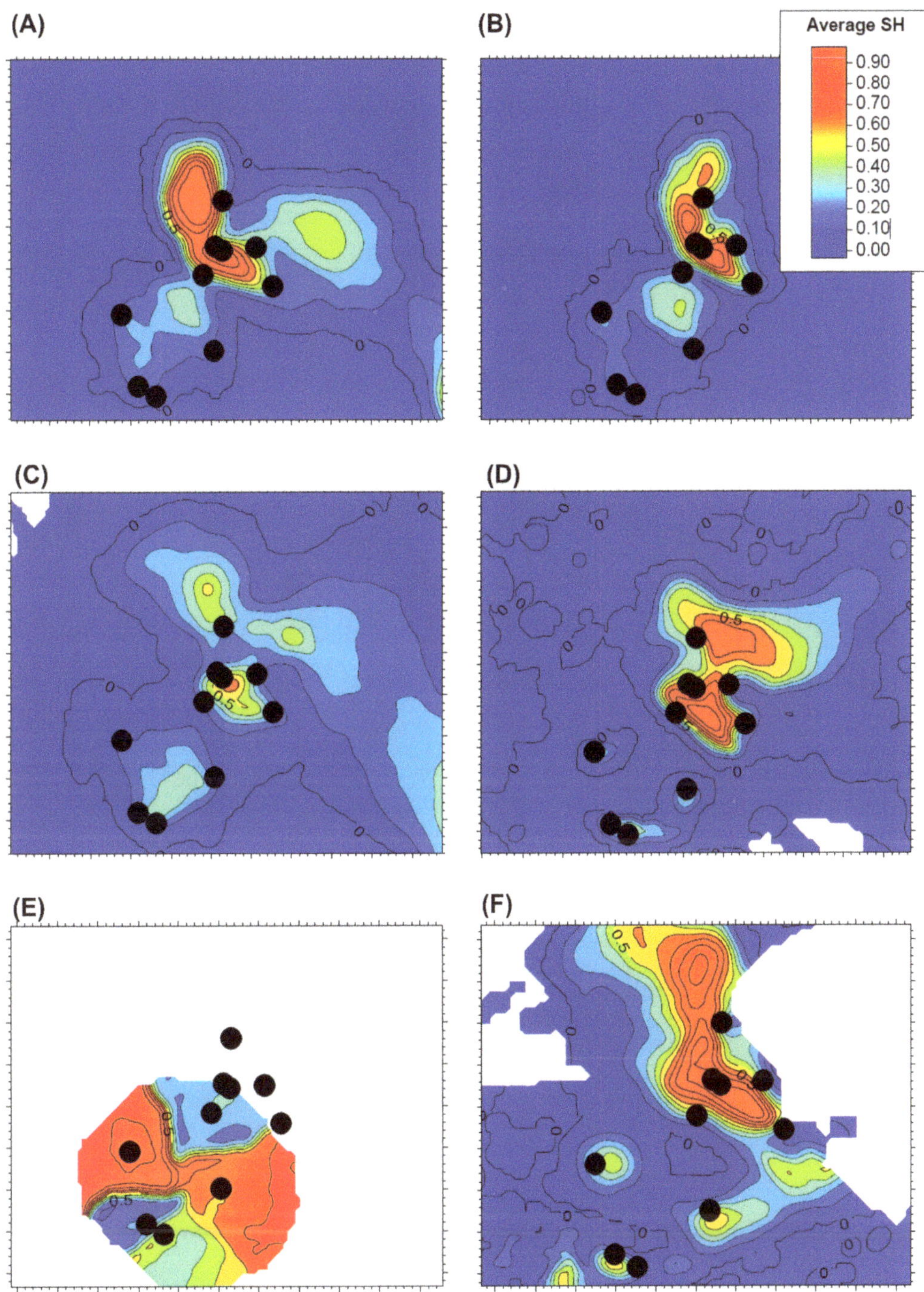

Figure 11. Maps of average hydrocarbon saturation in gas-bearing horizons: (**A**) XI, (**B**) XII, (**C**) XIII, (**D**) XIV, (**E**) XIVa, and (**F**) XV. White represents areas beyond the lithological barrier of the horizon.

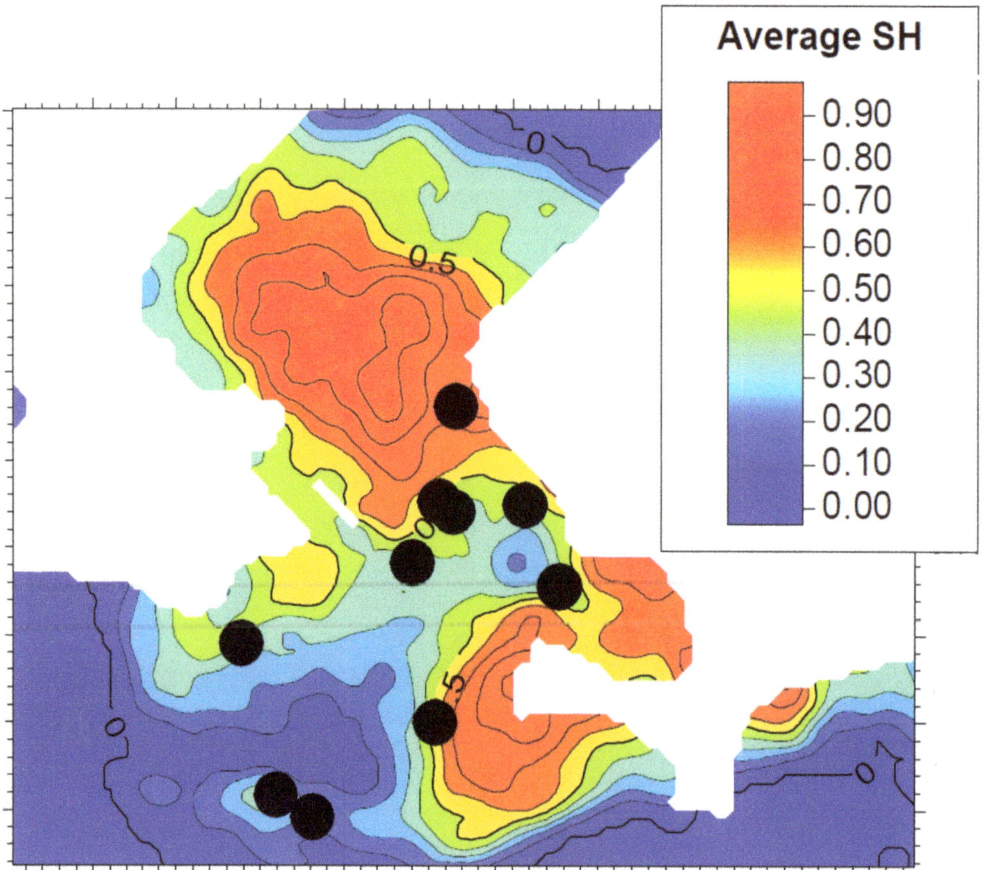

Figure 12. Maps of average hydrocarbon saturation in gas-bearing horizon XVI. White represents areas, beyond the lithological barrier of the horizon (Precambrian basement).

The analyses of the results presented in the form of maps of average hydrocarbon saturation indicate best perspectivity for horizons XV and XVI showed high hydrocarbon saturation in the north-west part of the anticline; in addition, the thickness of several dozen meters make these horizons attractive for the future exploitation. Additionally, the morphology of Precambrian basement constitutes the traps which preserve hydrocarbons from migration. In turn horizon XIVa showed high hydrocarbon saturation in the central and south part of the evaluated gas field, but the coverage of this horizon was significantly limited by the lithological changes, the increase in clay volume, and decrease in effective porosity. In addition, horizons XI and XIV indicate gas saturation in the northern and north-east part of the anticline that pointing out that area favorable for the exploration and drilling works. Horizon XIII was dominated by PSU 1 with low porosity and permeability. Detail interpretation of perspectives of this horizon requires the availability of high resolution well logs and much higher vertical resolution of the model.

4. Discussion

The interpretation of the Miocene multihorizons gas field allowed us to characterize the reservoir parameters of sandy-clay and silt sediments, in which most gas accumulation can be observed. Simultaneously, qualitative methods of detecting gas-saturated intervals were proposed based on well log data. An alternative modified connectivity equation for shaly-sands was applied to calculate water saturation. The model enabled us to calculate the value of Sw in the deposits formed as mudstones or sandstones with clay content higher than 35%, without the need to determine Archie's saturation exponent (n). This method is mainly based on the results of well log data that reflect the reservoir properties. The results of the Sw estimation are consistent; water saturation based on the Montaron method shows correlation with the Sw obtained from the Simandoux equation and clearly indicates intervals with high hydrocarbon potential. However, the Montaron method, which also uses the porosity logs (RHOB or NPHI) in the intervals with low resistivity contrast (LRC) zones, shows lower values of water saturation compared to Sw from the Simandoux equation. This indicates that porosity logs (RHOB, NPHI, DT) could constitute valuable additional input at the stage of water saturation calculation in formations of high clay content. On the basis of the interpretation, it is possible to define the properties of the sediments in which most of the gas accumulation is observed. The effective porosity of shaly gas horizons ranges from 5 to 15%, and the water saturation ranges from 26 to 50%. The effective porosity of these sediments is reduced by the irreducible water, the capillary water (Swi_kapilar), the content of which ranges from 40 to 70%. The qualitative analysis of mudstone and claystone heteroliths classified as PSU 1 did not clearly show the possibility of gas saturation in these formations, as the water saturation coefficient is very high, at 80–90%. However, the detailed analysis of high-resolution microresistivity measurements in the K-19 well confirmed the presence of thin layers of gas-saturated sandstones within these sediments (horizon XI and XIII). Proper calculation of the reservoir parameters within these rare sandstone inserts due to their thin-layer nature requires a separate interpretation approach based on high-resolution microresistivity logs. Moreover, spatial modeling of reservoir parameters within this group of sediments requires a significant increase in the resolution of the model.

Propagation of the petrophysical properties of individual PSU groups was carried out based on the well log interpretation and predefined ranges of variability of reservoir parameter for the individual group. The reconstruction of the spatial distribution of clay volume and porosity was performed with standard parametric modeling procedures using the Gaussian random function simulation stochastic algorithm, while for the two key parameters for this study, PSU grouping and SH hydrocarbon saturation, different approaches were used. The distribution of the classification into PSU groups was carried out by the facies classification procedure, which enabled us to define the membership of each grid cell to the group (1, 2, 3, 4) showing the highest degree of probability of its occurrence. The spatial distribution of hydrocarbon saturation was based on the use of the SH parameter calculated on the basis of the Montaron method and the petrophysical classification of PSU. The results of petrophysical interpretation indicate the presence of hydrocarbons, including those beyond predefined gas horizons. Thus arises the necessity of verification and redefinition of horizon boundaries.

5. Conclusions

Presented work confirmed the nature of gas deposits developed in the Miocene formation. As was noted in the previous works, the favorable conditions for gas accumulations provide the thin layers of sandstones and mudstones sealing by the laminas of impermeable claystones with admixtures of calcite and dolomite. The thick massive sandstones with effective porosity of 25–29% usually cannot keep hydrocarbons as there are not enough thick impermeable layers that will prevent the hydrocarbons from migration. Although the existing gas accumulation in the heteroliths of basin plain deposits have moderate reservoir properties porosity of several percent and higher level of clay volume, their thickness can reach even 100 m. That makes these multihorizons, consist of thin laminas of gas saturated mudstones and sandstones layered by impermeable claystone, economically important. The perforations documented the gas flow from these horizons of several dozen cubic meters per minute. Unfortunately, we still suffer from the small amount of laboratory measurements performed on core samples. There is need for an accurate recognition of lithological composition of these horizons that may be provide both by the X-ray diffraction tomography measurements of core sample and new well logging profiling. The most important from exploration point of view become the accurate calculation and core calibration of irreducible and free fluid volume. Miocene deposits are often thin-bedded thus new well logging technology should be used to collect data with higher resolution. Very useful in pay zone detection could be measurements of vertical (Rv) and horizonal (Rh) resistivities. Xaminer Multicomponent Induction (MCI) is dedicated to evaluation anisotropic thin-bedded formations and is a tool that measures formation resistivity both vertically and horizontally at different depths of investigation. When run with a directional tool, it also provides structural dip and azimuth. However, horizontal microresistivity can also be measured with the use of imagers of borehole walls (XRMI). The paper shows that even using basic well logs we can roughly assess the potential of Miocene heteroliths. New looks at conventional well log data in combination with laboratory measurements and propagation within the reservoir allowed indication of potentially gas saturated areas. We can deal with low resistivity contrast between gas and water saturated intervals by involving other available logs, usually dedicated porosity calculation (NPHI, DT, RHOB), at the stage of water saturation calculation.

Author Contributions: Conceptualization, A.L.-Ś. and W.K.-K.; methodology, A.L.-Ś. and W.K.-K.; software, A.L.-Ś. and W.K.-K.; validation, A.L.-Ś. and W.K.-K.; formal analysis, A.L.-Ś. and W.K.-K.; investigation, A.L.-Ś. and W.K.-K.; resources, A.L.-Ś. and W.K.-K.; data curation, A.L.-Ś. and W.K.-K.; writing—original draft preparation, A.L.-Ś. and W.K.-K.; writing—review and editing, A.L.-Ś.; visualization, A.L.-Ś. and W.K.-K.; supervision, A.L.-Ś.; project administration, A.L.-Ś. All authors have read and agreed to the published version of the manuscript.

Funding: This research was funded by the Polish Ministry of Science and Higher Education, Grant No. DK-4100-0068/2021. The authors would like to express their gratitude to the Polish Ministry of Science and Higher Education for funding this research.

Institutional Review Board Statement: Not applicable.

Informed Consent Statement: Not applicable.

Data Availability Statement: The data supporting the findings of this research are available on reasonable request from the corresponding author.

Conflicts of Interest: The authors declare no conflict of interest.

References

1. Dziadzio, P.; Liszka, B.; Maksym, A.; Stryszak, G. Środowisko sedymentacji utworów miocenu autochtonicznego w brzeżnej strefie Karpat, a interpretacja geologiczno-złożowa w obszarze Husów–Albigowa–Krasne. *Nafta-Gaz* **1997**, *53*, 407–414. (In Polish)
2. Dziadzio, P. Depositional sequences in Badenian and Sarmatian deposits in the SE parts of the Carpathian Foredeep (SE Poland). *Przegląd Geol.* **2000**, *48*, 1124–1138. (In Polish)
3. Porębski, S.J.; Pietach, K.; Hodiak, R.; Steel, R.J. Origin and sequential development of Upper Badenian-Sarmatian clinoforms in the Carpathian Foredeep Ba sin, SE Poland. *Geol. Carpathica* **2002**, *54*, 119–136.
4. Kurovets, I.; Prytulka, G.; Shpot, Y.; Peryt, T.M. Middle Miocene Dashava Formation sandstones, Carpathian Foredeep. *Ukr. J. Pet. Geol.* **2004**, *27*, 373–388. [CrossRef]
5. Mastalerz, K.; Wysocka, A.; Krzywiec, P.; Kasiński, J.; Aleksandrowski, P.; Papiernik, B.; Ryzer-Siupik, J. Miocene succession at the Ryszkowa Wola High (Sieniawa-Rudka area), Carpathian Foredeep Basin: Facies and stratigraphic interpretation of wellbore and 3D seismic data. *Przegląd Geol.* **2006**, *54*, 333–342. (In Polish)
6. Pietsch, K.; Porębski, S.J.; Marzec, P. The use of seismostratigraphy for exploration of Miocene gas-geaing reservoirs in the NE part of the Carpathian Foreland Basin (Poland). *Geologia* **2010**, *36*, 173–186. (In Polish)
7. Oszczypko, N.; Krzywiec, P.; Popadyuk, I.; Peryt, T. Carpathian Foredeep Basin (Poland and Ukraine): Its sedimentary, structural, and geodynamic evolution. In *The Carpathians and Their Fore Land: Geology and Hydrocarbon Resources*; Golonka, J., Picha, F.J., Eds.; American Association of Petroleum Geologists: Tusla, OK, USA, 2006.
8. Montaron, B. Connectivity Theory—A new approach to modeling Non-Archie rocks. *Petrophysics* **2009**, *50*, 102–115.
9. Passey, Q.; Creaney, S.; Kulla, J.; Moretti, F.; Stroud, J. A pratical model for organic richness from porosity and resistivity logs. *Am. Assoc. Pet. Geol. Bull.* **1990**, *74*, 1777–1794.
10. Bowman, T. *Direct Method for Determining Organic Shale Potential from Porosity and Resistivity Logs to Identify Possible Resource Plays* Mantle Oil & Gas; LLC AAPG Annual Convention & Exhibition: New Orleans, LA, USA, 2010.

11. Dudek, J.; Dudek, L.; Klimek, P. Badania ekshalacji gazu w rejonie złoża Przeworsk. In Proceedings of the Geopetrol 2004: Konferencja Naukowo-Techniczna nt. Efektywne Technologie Poszukiwania i Eksploatacji Złóż Węglowodorów, Zakopane, Polish, 20–23 September 2004.
12. Dudek, J.; Zaleska-Bartosz, J. Ocena Możliwości Realizacji Inwestycji Budowlanych na Terenie Złoża Przeworsk, w Kontekście Występowania Zagrożeń Ekshalacji Gazu Ziemnego. *Nafta-Gaz* **2011**, *7*, 463–466.
13. Kołodziejak, G. Ocena Możliwości Minimalizacji Zagrożeń Powodowanych Ekshalacjami Gazu Ziemnego na Terenie Miasta Przeworsk. *Nafta-Gaz* **2010**, *66*, 591–596.
14. Myśliwiec, M. Exploration for gas accumulations in the Miocene deposits of the Carpathian Foredeep using direct hydrocarbon indicators—Verification of anomalies (southern Poland). *Prz. Geol.* **2004**, *52*, 307–314.
15. Archie, G.E. The electrical resistivity log as an aid in determining some reservoir characteristics. *J. Pet. Technol.* **1942**, *1*, 55–62. [CrossRef]
16. Montaron, B. A Quantitative Model for the Effect of Wettability on the Conductivity of Porous Rocks SPE. In Proceedings of the SPE Middle East Oil and Gas Show and Conference, Manama, Bahrain, 11–14 March 2007. [CrossRef]
17. Jarzyna, J.; Krakowska, P. Dobór Parametrów Petrofizycznych Węglanowych Skał Zbiornikowych w Celu Podwyższenia Dokładności Wyznaczenia Współczynnika Nasycenia Wodą. *Nafta-Gaz* **2010**, *7*, 547–556.
18. Lee Myung, W. *Connectivity Equation and Shaly-Sand Correction for Electrical Resistivity*; Scientific Investigations Report 2011-5005; Unites States Geological Survey: Liston, VA, USA, 2011.
19. Peeters, M.; Holmes, A. Review of Existing Shaly-Sand Models and Introduction of a New Method Based on Dry-Clay Parameters. *Petrophysics* **2014**, *55*, 543–553.
20. Puskarczyk, E.; Jarzyna, J.; Porębski, S.J. Application of multivariate statistical methods for characterizing heterolithic reservoirs based on wireline logs—Example from the Carpathian Foredeep Basin (Middle Miocene, SE Poland). *Geol. Q.* **2015**, *59*, 157–168.
21. Puskarczyk, E. Artificial neural networks as a tool for pattern recognition and electrofacies analysis in PolishPalaeozoic shale gas formations. *Acta Geophys.* **2019**, *67*, 1991–2003. [CrossRef]
22. Moss, B.; Seheult, A. Does principal component analysis have a role in the interpretation of petrophysical data? In Proceedings of the 28th Annual Logging Symposium Transactions, Society Professional Well Log Analysts, London, UK, 29 June–2 July 1987.
23. Szabó, N.P.; Dobroka, M.; Kavanda, R. Cluster analysis assisted float-encoded genetic algorithm for a moreautomated characterization of hydro-carbon reservoirs. *Intell. Control Autom.* **2013**, *4*, 362–370. [CrossRef]
24. Wawrzyniak-Guz, K.; Puskarczyk, E.; Krakowska, P.I.; Jarzyna, J.A. Classification of Polish shale gas formations from Baltic Basin, Poland based on well logging data by statistical methods. In Proceedings of the International Multidisciplinary Scientific GeoConference SGEM, Albena, Bulgaria, 30 June–6 July 2016; pp. 761–768, ISSN 1314-2704. ISBN 978-619-7105-57-5.
25. Kohonen, T. *Self-Organizing Maps*; Springer: Berlin/Heidelberg, Germany, 2001. [CrossRef]
26. Odesanya, I.; Ogbamikhumi, A.; Azi, O.S. Well Log Analysis for Lithology Identification Using Self-Organizing Map (SOM). *Int. J. Res. Appl. Phys.* **2016**, *2*, 25–32.
27. Liu, Y.; Weisberg, H.R.; Mooers, N.K.C. Performance evaluation of the self-organizing map for feature extraction. *J. Geophys. Res. Ocean.* **2006**, *111*, 1–14. [CrossRef]
28. Thomson, E.R.; Emery, W.J. *Data Analysis Methods in Physical Oceanography*, 3th ed.; Elsevier: Amsterdam, The Netherlands, 2014; ISBN 978-0-12-387782-6.
29. Kohonen, T. Self-Organized Formation of Topologically Correct Feature Maps. *Biol. Cybern.* **1982**, *43*, 59–69. [CrossRef]
30. Nooh, A.Z.; Moustafa, E.A.A. Comparison of Quantitative Analysis of Image Logs for Shale Volume and Net to Gross Calculation of a Thinly Laminated Reservoir between VNG-NERGE and LAGIA-EGYPT. *Egypt. J. Pet.* **2017**, *26*, 619–625. [CrossRef]
31. Eshimokhai, S.; Akhirevbulu, O.E.; Osueni, L. Evaluation of Thin Bed Using Resistivity Borehole and NMR Imaging Techniques. *Ethiop. J. Environ. Stud. Manag.* **2012**, *4*, 96–102. [CrossRef]
32. Ruhovets, N.; Rau, R.; Samuel, M.; Smith, H.; Smith, M., Jr. *Evaluating Thinly Laminated Reservoirs Using Logs with Different Vertical Resolution Halliburton Logging Services*; Pennsylvania State University: State College, PA, USA, 2016; Available online: http://sp.lyellcollection.org/ (accessed on 1 March 2021).
33. Saxena, K.; Tyagi, A.; Klimentos, T.; Morriss, C.; Mathew, A. *Evaluating Deepwater Thin-Bedded Reservoirs with Rt Scanner*; Petromin: Kuala Lumpur, Malaysia, 2006.
34. Kusuma, D.P.; Audinno, R.T.; Pratama, I.P.; Halim, A. Integrated Analysis of The-Low Resistivity Hydrocarbon Reservoir in the "S" Field. In Proceedings of the Conference: Indonesian Petroleum Association, Fortieth Annual Convention & Exhibition, Jakarta, Indonesia, 25–27 May 2016. [CrossRef]
35. Lis-Śledziona, A. Multiscale evaluation of a thin-bed reservoir. *Geol. Geophys. Environ.* **2021**, *47*, 5–20. [CrossRef]
36. Przelaskowska, A.; Łykowska, G.; Klaja, J.; Kowalska, S.; Gąsior, I. Application of the cation exchange capacity parameter (CEC) to the characterization of the swelling capacity of lower Paleozoic, Carpathian Flysch and Miocene Carpathian Foredeep clay rocks. *Nafta-Gaz* **2015**, *6*, 384–389.
37. Simandoux, P. Dielectric measurements in porous media and applications in Shaly formations. In *Revue de L'institute Francais de Petrole-SPWLA English Translation Volume Shaly Sand*; SPWLA: Huston, TX, USA, 1982.

Article

Characteristics and Formation Mechanism of the Lower Paleozoic Dolomite Reservoirs in the Dongying Depression, Bohai Bay Basin

Xuemei Zhang [1,2], Qing Li [1,2,*], Xuelian You [3], Lichi Ma [4], Anyu Jing [4], Wen Tian [4] and Lang Wen [1,2]

1   State Key Laboratory of Petroleum Resources and Prospecting, China University of Petroleum (Beijing), Beijing 102249, China; 2019210099@student.cup.edu.cn (X.Z.); 2019210100@student.cup.edu.cn (L.W.)
2   College of Geosciences, China University of Petroleum (Beijing), Beijing 102249, China
3   School of Ocean Science, China University of Geosciences (Beijing), Beijing 100083, China; youxuelian@cugb.edu.cn
4   Exploration and Production Research Institute of Shengli Oilfield Company, Sinopec, Dongying 257001, China; malichi.slyt@sinopec.com (L.M.); jinganyu.slyt@sinopec.com (A.J.); tianwen755.slyt@sinopec.com (W.T.)
*   Correspondence: liqing@cup.edu.cn

Abstract: The Lower Paleozoic carbonate strata experience multi-stage tectonic activity and post-depositional volcanic activity in the Dongying Depression, Bohai Bay basin. These tectonic and magmatic activities have caused the reservoir to undergo severe diagenesis, resulting in strong reservoir heterogeneity. This study aims to identify the characteristics of dolomite, various reservoir spaces' characteristics, the origin of different types of dolomite, and the porosity evolution. According to crystal size and morphology, dolomites can be divided into three kinds of matrix dolomites and four kinds of dolomite cements. The petrology and geochemistry of the dolomite suggests that matrix dolomite is formed from seawater. The medium-to-coarse-crystalline dolomite cement (D3) has a higher $^{87}Sr/^{86}Sr$ ratio (0.7119 to 0.7129) and a higher homogenization temperature (>125 °C), suggesting that the fluid for the precipitation of D3 is a mixed fluid formed by hydrothermal fluid eroding the ^{87}Sr-rich feldspar sandstone. The strikingly negative $\delta^{18}O$ values (-23.7 to -25.7‰ VPDB) of saddle dolomite (D4) indicate that D4 precipitated from hydrothermal fluids and the Mg^{2+} source may be due to dissolution of the host dolomite that formed in the evaporation environment. The reservoir spaces of the target strata in the study area mainly include fractures, dissolution vugs, intercrystalline pores, and moldic pores. Dissolution is the basis for forming high-quality dolomite reservoirs. The faults and fractures provided favorable conditions for dissolution. Hydrothermal fluid and organic acid were the main dissolution fluids for the dolomite reservoir, which were beneficial to the development of secondary pores. In the study area, organic acid dissolution was shown to contribute more than hydrothermal dissolution in the study area.

Keywords: buried hill reservoir; the lower Paleozoic; dolomitization; diagenesis; pore evolution

Citation: Zhang, X.; Li, Q.; You, X.; Ma, L.; Jing, A.; Tian, W.; Wen, L. Characteristics and Formation Mechanism of the Lower Paleozoic Dolomite Reservoirs in the Dongying Depression, Bohai Bay Basin. *Energies* **2022**, *15*, 2155. https://doi.org/10.3390/en15062155

Academic Editors: Yuming Liu and Bo Zhang

Received: 11 January 2022
Accepted: 11 March 2022
Published: 16 March 2022

Publisher's Note: MDPI stays neutral with regard to jurisdictional claims in published maps and institutional affiliations.

1. Introduction

Dolomite reservoirs are important hydrocarbon reservoirs. According to the statistics of 226 large–medium-scale carbonate hydrocarbon fields, dolomite reservoirs account for 90% of the world's carbonate hydrocarbon fields [1,2]. The development of the dolomite reservoirs is also related to the changes in climate that took place during the geological period. Older carbonate strata commonly have higher ratios of dolomite reservoirs [3]. Domestically, dolomite reservoirs are located in the Majiagou Formation in the northern Ordos Basin [4], in the Qixia and Dengying Formation and in the Sichuan Basin [5–9], in the Changxing–Feixian Formation in the Puguang Oilfield, and in the Lower Paleozoic in the Tarim Basin [10].

As for the genesis of dolomite reservoirs, most dolomite reservoirs are considered to have been initially formed by reflux dolomitization in an arid climate environment and capillary concentration [11,12]. Furthermore, assuming that there was enough Mg^{2+} and a dynamic mechanism, this microbial-induced dolomitization can also occur in carbonate formations [13,14]. Since the 1990s, hydrothermal dolomitization and buried dolomitization under the deep-buried environments have been proposed by scholars [15–19]. Additionally, dolomitization under high-temperature and deep-buried environments might improve the permeability and porosity of dolomite reservoirs [20].

Previous studies have suggested that carbonate rocks are more susceptible to the alteration of multi-type and multi-stage diagenesis during diagenetic evolution. The development of reservoir spaces and the quality of the late physical properties are significantly controlled by diagenesis [21,22]. Carbonate reservoirs that have been influenced by the diagenesis have more complicated pore evolution [23–25]. In recent years, the qualities of carbonate rock reservoirs have been the focus of many studies [26]. There are several types of diagenesis that affect the reservoir properties of carbonate rocks, including dolomitization, fracturing, dissolution, compaction, and cementation. Dolomitization, dissolution, and fracturing are the main constructive types of reservoir diagenesis [22,27,28], while cementation and compaction represent the destructive forms of reservoir diagenesis [29,30]. In recent years, research on carbonate diagenesis has mainly focused on the structurally controlled hydrothermal alteration of carbonate reservoirs, and studies have pointed out that the effects of tectonic-controlled hydrothermal alteration on reservoirs are multifaceted [15,31–33]. Therefore, it is difficult to clearly delineate the time boundaries of different types of diagenesis. Meanwhile, studies still present some doubts regarding the dissolution mechanism of carbonate minerals by deep hydrothermal fluids [34].

In the last few decades, many studies have been carried out on the carbonate reservoir in the Paleozoic strata of the Dongying Depression [35], many of which have focused on the reservoir characteristics of the buried hill zone in the western portion of the depression and the Caoqiao-Guangrao buried hill zone in particular [36], and these studies have proposed that carbonate reservoirs can be divided into the two following categories: weathering crust-type buried hills and inner dissolution-type buried hills [37]. The development of these carbonate reservoirs was mainly controlled by karstification and fracture development, and it has been suggested that the dolomite formations are more likely to form high-quality reservoirs [38]. Dolomite reservoirs have been explored in the Ordovician Badou Formation in the Zhuanghai area, which were previously predicted to be favorable reservoir areas [39]. In recent years, the dolomite reservoir of the Yeli-Liangjiashan Formation has also become a key exploration target in the Shengli Oilfield, especially in the Zhuanghai area [40]. Most studies have concluded that dissolution and dolomitization are the key factors for forming favorable reservoirs. However, there are insufficient studies on the lower Paleozoic carbonate reservoirs in the Gaoqing-pingnan area. There is little research on dolomite reservoirs and dolomite cement types in this area.

Although scholars have studied the southwest area of the Dongying Depression, most of these studies have focused on the influence of atmospheric fresh water on buried hill reservoirs. However, these ignored the fact that the matrix dolomite and dolostone cement are the most important factors influencing the formation of high-quality reservoirs. Therefore, this study focuses on the matrix dolomite and all types of dolostone cement of the Lower Paleozoic formation in the southwest of the Dongying Depression. The major goals of this study are to (1) analyze the genesis of the dolostone cements, diagenesis, and the pore evolution characteristics under the different diagenesis, utilizing petrographic, geochemical data; (2) carry out the study on the relationship between diagenesis and reservoir pore evolution in the study area. This study is conducive to understanding the reservoir characteristics and their distribution laws, providing some geological basis for later exploration.

2. Geological Setting

The Dongying Depression is located on the southeast of the Jiyang Sag, with an area of 5700 km^2 [41] and mostly faces northeast. The Dongying Depression belongs to a sub-tectonic unit in the Jiyang Sag, Bohai Bay Basin [42]. The study area mainly comprises the Gaoqing-Pingnan area in the southwest of the Dongying Depression (Figure 1a). The southern upper plate of the Gaoqing fault belongs to the Gaoqing uplift, and the eastern falling plate belongs to the Boxing depression [43].

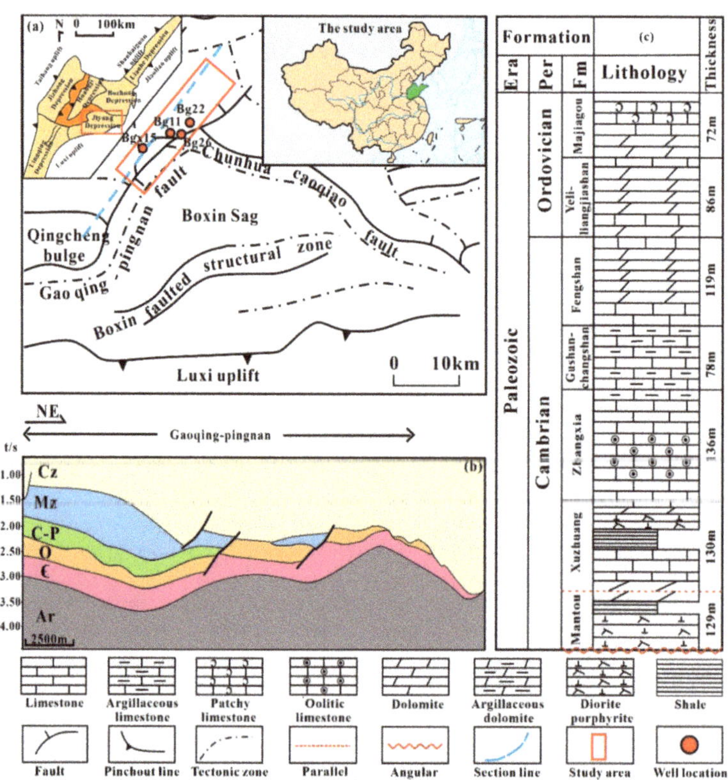

Figure 1. (**a**) Tectonic setting of the Dongying Depression, Bohai Bay Basin; (**b**) characteristics of formation development in NE-trending strata profile; section line is shown in (**a**); (**c**) stratigraphy of the Lower Paleozoic Formation in Dongying Depression, Bohai Bay Basin (modified after [44]). The map of China is quoted from [45].

In the study area, the Lower Paleozoic carbonate strata have experienced multiple periods of tectonic activity and post-depositional volcanic activity. Tectonic activity in the study area is divided into two main periods. The first large tectonic activity occurred during the Indosinian (Late Permian–Triassic period). The tectonic activity in the Indosinian period caused strong denudation of the Paleozoic strata in the relatively uplifted area of the study area. The second large-scale tectonic activity occurred in the Yanshan period (Jurassic–Early Cretaceous period). This tectonic activity was an extremely frequent and violent tectonic activity following the Indosinian period, and this tectonic activity was also accompanied by strong volcanic activity.

The objective strata consist of the Lower Paleozoic Ordovician and Cambrian carbonate reservoir [46]. The Cambrian strata consist of the Mantou Formation, Maozhuang Formation, Zhangxia Formation, Gushan Formation, Changshan Formation, and Fengshan Formation. The whole Cambrian is about 10–760 m thick. The Ordovician strata include the Yeli-Liangjiashan Formation, Majiagou Formation, and Badou Formation. The Ordovician strata is about 20–240 m thick [47]. The Badou Formation is missing from the study area due to uplift and denudation (Figure 1b,c). Due to the strata of uplift that took place during the tectonic activity period, Carboniferous and Permian strata are absent in the higher parts of the structure. Therefore, the Lower Paleozoic strata is in contact with the overlying Mesozoic unconformity [48].

The sedimentary facies in the Paleozoic portions of the Dongying Depression comprise clastic sedimentary facies series and carbonate sedimentary facies series according to sedimentary rock types. The clastic sedimentary facies series were formed in supratidal and intertidal flat environments. The clastic sedimentary facies series can be further divided into mud flat and sand beach sedimentary subfacies according to their environment and lithology. Due to the decrease in the terrigenous material supply, the study area is mainly composed of argillaceous dolomite or limestone deposits. In one section of the study area, shale and mudstone are interbedded with dolomite and limestone. Carbonate sedimentary facies series can be further divided into evaporative platform facies, restricted platform facies, and open platform facies. The sedimentary facies comprise supratidal and intertidal zones from the Mantou and Xuzhuang periods, the main lithology is composed of oolitic limestone interbedded with shale; high-energy subtidal shoals and open seas were widely developed during the Zhangxia and Gushan periods. Intertidal flat and restricted sea facies were widely developed during the Changshan and the Early Ordovician periods. The main rock types comprise argillaceous limestone, silty limestone, and dolomite. The deposited environment gradually changed to shallow-water supratidal and intertidal zones and deeper-water subtidal zones during the Middle Ordovician period [44,48]

The Dongying Depression is a fault depression basin that was developed on the background of bedrock paleotopography. This basin went through stable lifting during the Paleozoic period, fold uplifting during the Triassic to middle Jurassic periods, preliminary fault depression between the late Jurassic and Cretaceous periods, and rifting and spreading in the Cenozoic periods, resulting in whole depression [44,49]. These complex tectonic activities laid the foundation for the formation of multiple diageneses. Additionally, the magmatic hydrothermal activity that accompanied these tectonic activities provided a source of materials for the development of hydrothermal cement.

3. Analytical Methods

All samples were collected from the southwest section of the Dongying Depression, specifically in the Gaoqing-pingnan area. Wells Bg 26, Bg 22, Bgx 15, and Bg 11 were included for sample collection (Figure 1a). A total of 70 thin sections were half-stained with a mixture of potassium ferricyanide and alizarin red to qualitatively discriminate between dolomite types, which was carried out according to the different petrographic characteristics of the different dolomites after they had been stained (Table 1).

Cathodoluminescence (CL) analysis was performed at the Analytical Laboratory of the Beijing Research Institute of Uranium Geology (BRIUG) to determine the cement generations.

The oxygen and carbon stable isotope analyses were carried out at the Analytical Laboratory of the Beijing Research Institute of Uranium Geology (BRIUG). Carbonate powders were reacted with 100% phosphoric acid for 4 h at 25 °C for calcite and at 50 °C for dolomite, and the resultant CO_2 was measured to determine its oxygen and carbon isotopic ratios on a Delta plus mass spectrometer. The isotope values were determined relative to the Vienna Peedee Belemnite standard (VPDB). Reported in the standard notation relative to standard VPDB for carbonate ratios and VSMOW for oxygen ratios. $\delta^{18}O$ (VSMOW) values were converted to $\delta^{18}O$ (VPDB) values. The reproducibility values of the isotopic measurements for both C and O were better than $\pm0.06\permil$ and $\pm0.08\permil$, respectively.

Table 1. Information on the samples used in this study.

Well	Sample Depth (m)	Formation	Lithology Description	Sample Analysis
Bg26	2515.6	Yeli-Liangjiashan	Limestone	Microscopic observation
Bg26	2790.5	Zhangxia	Oolitic limestone	Microscopic observation
Bg26	2791.2	Zhangxia	Bioclastic limestone	C, O isotope; $^{87}Sr/^{86}Sr$
Bg26	2791.5	Zhangxia	Bioclastic limestone	Microscopic observation
Bg26	2793.2	Zhangxia	Bioclastic limestone	C, O isotope; $^{87}Sr/^{86}Sr$
Bg26	2795.4	Zhangxia	Bioclastic limestone	Microscopic observation
Bg26	2795.5	Zhangxia	Grain limestone	Cathodoluminescence
Bg26	2798.2	Zhangxia	Dolomitic bioclastic limestone	C, O isotope; $^{87}Sr/^{86}Sr$
Bg26	2801.9	Zhangxia	Calcareous dolomite	C, O isotope; $^{87}Sr/^{86}Sr$; Cathodoluminescence
Bg26	2803.5	Zhangxia	Algal limestone	Microscopic observation
Bg22	2211.17	Ordovician	Microcrystalline dolomite	Microscopic observation
Bg22	2219.27	Ordovician	Microcrystalline dolomite	Microscopic observation
Bg22	2233.04	Ordovician	Microcrystalline dolomite	C, O isotope; $^{87}Sr/^{86}Sr$
Bg22	2236.4	Ordovician	Microcrystalline dolomite	C, O isotope; $^{87}Sr/^{86}Sr$
Bg22	2236.76	Ordovician	Argillaceous dolomite	Microscopic observation
Bg22	2246.89	Ordovician	Microcrystalline dolomite	Microscopic observation
Bg22	2348.95	Ordovician	Microcrystalline dolomite	C, O isotope
Bg22	2398	Ordovician	Fine-crystalline dolomite	Microscopic observation
Bg22	2399.1	Ordovician	Fine-crystalline dolomite	C, O isotope
Bgx15	2238.9	Ordovician	Microcrystalline dolomite	C, O isotope
Bg11	2233.04	Ordovician	Microcrystalline dolomite	$^{87}Sr/^{86}Sr$; Cathodoluminescence
Bg11	2434.5	Ordovician	Argillaceous limestone	C, O isotope
Bg11	2435.24	Ordovician	Argillaceous limestone	C, O isotope; $^{87}Sr/^{86}Sr$
Bg11	2449.5	Ordovician	Argillaceous limestone	Cathodoluminescence

The $^{87}Sr/^{86}Sr$ isotope ratios were determined for selected matrix dolomite and dolomite cements using a Thermal Ionization Mass Spectrometer (TIMS, Phoenix) at the analytical Laboratory of the Beijing Research Institute of the Uranium Geology (BRIUG). The analytical precision of the individual runs was determined to be 0.00005 (2σ). The mean standard error of the mass spectrometer performance was ± 0.00003 for standard GB/T 17672-1999. The measured $^{87}Sr/^{86}Sr$ isotope values were normalized using the following formula

$$(^{87}Sr/^{86}Sr)_{St} = (^{87}Sr/^{86}Sr)_{Sa} \times (1 + 2f) \tag{1}$$

$$f = [(^{87}Sr/^{86}Sr)_{St}/(^{87}Sr/^{86}Sr)_{Sa}] \div 2 \tag{2}$$

$$(^{87}Sr/^{86}Sr)_{Nor} = (^{87}Sr/^{86}Sr)_{Sa} \times (1 + f) \tag{3}$$

where $(^{87}Sr/^{86}Sr)_{St}$ is the standard $^{87}Sr/^{86}Sr$ values, $(^{87}Sr/^{86}Sr)_{St} = 8.37521$; $(^{87}Sr/^{86}Sr)_{Sa}$ is the measured value of samples; $(^{87}Sr/^{86}Sr)_{Nor}$ is the normalized value of samples.

In this work, the primary fluid inclusions of the dolomite and calcite in the fractures and vugs were selected for systematic microscopic temperature measurement. Fluid inclusion microthermometry measurements were carried out using a microscopic heating and cooling stage (Linkam THMSG600) at the Institute of Geology Chinese Academy of Geological Science, and the measurements were taken within a temperature range of -196 to $+600\ ^{\circ}C$ and at a test accuracy between $\pm 0.5\ ^{\circ}C$ and $\pm 2\ ^{\circ}C$. The temperatures of the fluid inclusions were obtained by freezing and warming. First, liquid nitrogen was used to cool the fluid inclusions. The changes that took place in the fluid inclusions during the temperature drop were observed. The inclusions were slowly warmed back up after freezing. When performing the homogenization temperature measurements, the heating rate was $5\ ^{\circ}C/min$, and the changes in the two gas–liquid phases were observed during the heating process. The heating rate was controlled to $1\ ^{\circ}C/min$ in order to accurately record the homogenization temperature when the first phase was close to disappearing.

4. Results

4.1. Petrography

Based on the core, thin section observation, the dolomites were able to be divided into two types: matrix dolomite and dolomite cements. The matrix dolomite mainly included argillaceous dolomite (M1), microcrystalline dolomite (M2), and fine-crystalline dolomite (M3). The dolomite cement filling the fractures and vugs included powder-to-fine-crystalline dolomite cement (D1), fine-crystalline ferroan dolomite cement (D2), medium-to-coarse-crystalline dolomite cement (D3), and saddle dolomite (D4).

4.1.1. Argillaceous Dolomite (M1)

In this type of dolomite, the crystals were smaller in size, less than 10 μm, with a euhedral–subhedral texture. There was no development of intercrystalline pores or intercrystalline dissolution pores between the dolomite crystals. The fractures were filled with seepage silt or clay minerals during uplift periods (Figure 2a,b). Argillaceous dolomite is commonly found in Middle Ordovician strata.

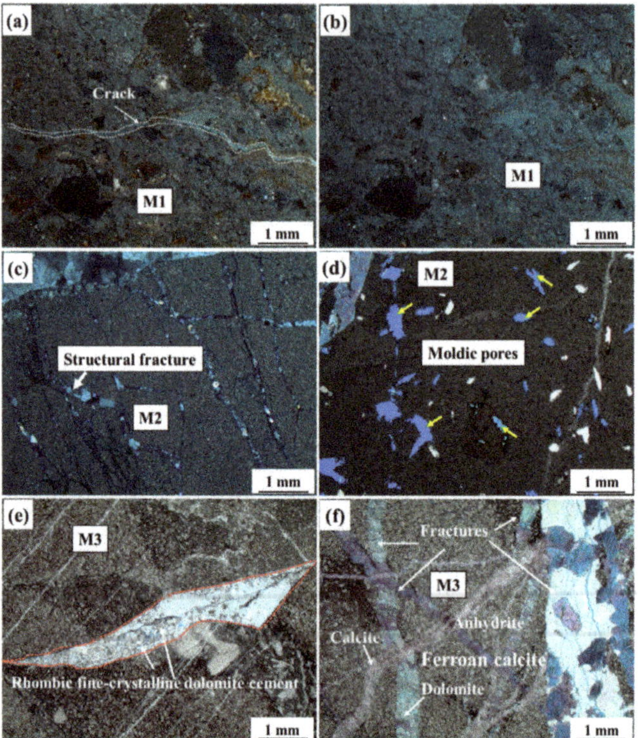

Figure 2. Petrological characteristics of various matrix dolomites in the Lower Paleozoic strata in the Dongying Depression, Bohai Bay Basin. (**a**) Argillaceous dolomite (M1), Well Bg22 2236.67 m, Ordovician formation, plane-polarized light; (**b**) same field of view as A, orthogonal light; (**c**) microcrystalline dolomite (M2), Well Bg22 2211.17 m, Ordovician formation, the structural fractures developed and filled by dolomite, orthogonal light; (**d**) microcrystalline dolomite (M2), Well Bg22 2233.04 m, Ordovician formation, dissolution vugs are developed, plane-polarized light; (**e**) fine-crystalline dolomite (M3), Well Bg22 2398 m, Ordovician formation, plane-polarized light; (**f**) multi-stage fractures are developed in fine-crystalline dolomite, Well Bg22 2399.1 m, Ordovician formation, plane-polarized light.

4.1.2. Microcrystalline Dolomite (M2)

This type of dolomite is defined as having a dolomite mineral of more than 90% in the dolomite rocks. This type of dolomite also shows the development of fractures, which are filled by medium-coarse-crystalline dolomite cements (Figure 2c). As the burial depth increases, selective dissolution vugs can be observed in the microcrystalline dolomite (Figure 2d). Microcrystalline dolomite mainly developed in the middle to lower parts of the Ordovician strata.

4.1.3. Fine-Crystalline Dolomite (M3)

This type of dolomite is defined as having a dolomite mineral content of more than 90% in the dolomite rocks, and the dolomite mainly comprises euhedral–subhedral crystals. Fine-crystalline dolomite is larger than 100 μm. Anhydrite cements can be seen, and part of the rhombic fine-crystalline dolomite is dispersed in the anhydrite cements (Figure 2e), indicating that the dolomite may have been formed by evaporation in an arid environment. Structural fractures developed in the fine-crystalline dolomite, which were successively filled with dolomite, calcite, and anhydrite (Figure 2f).

4.1.4. Powder-to-Fine-Crystalline Dolomite Cements (D1)

This type of dolomite cement is different from the fine-crystalline matrix dolomite. According to the different occurrence characteristics of the powder-to-fine-crystalline dolomite cements (D1), it can be divided into three types: (1) Most of the powder-to-fine-crystalline dolomites are distributed along stylolite (Figure 3a–c). Powder-to-fine-crystalline dolomite is about 46 μm in size. Powder-to-fine-crystalline dolomite cements (D1) are commonly rhomboid crystals and tend to have a euhedral structure, allowing them to form more intercrystalline pores. This type of dolomite is commonly nonluminous under the CL (Figure 4b). (2) The other type of powder-to-fine-crystalline dolomite mainly comprises the dolomite patches found in limestone and has euhedral crystals and poorer sorting (Figure 3e). This dolomite ranges from 39 μm to 88 μm in size. (3) Powder-to-fine-crystalline dolomite cements fill in the oolites. The early fluid selectively dissolved the interior of the oolite, and the late Mg-rich fluid recrystallized into the interior of the oolite to form rhombic dolomite crystals (Figure 3f). This dolomite ranges from 69 to 218 μm in size. Its crystals are usually characterized by "mist core bright edges" and are commonly nonluminous under the CL, only faint dull red cathodoluminescence can be seen at the edges of these dolomite crystals (Figure 4d). This type of dolomite is mainly distributed in the Zhangxia Formation of the Cambrian strata.

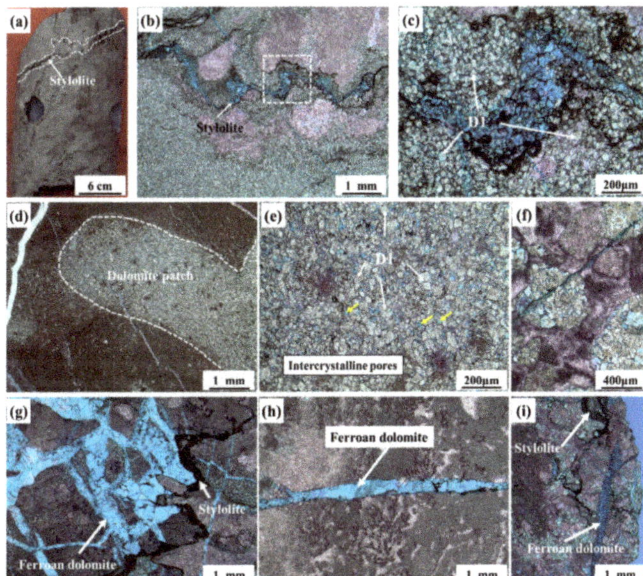

Figure 3. Petrological characteristics of the powder-fine dolomite and dolomite cements (D1). (**a**) Lower angle stylolite in the core, Well Bg26 2801.9 m, Zhangxia formation; (**b**) the characteristics of the powder-to-fine-crystalline dolomite cements (D1) around the stylolite. Stylolite is filled with black organic matter, Well Bg26 2801.9 m, Zhangxia formation; (**c**) local amplification of b; (**d**) the characteristics of the dolomite patch, the outside of the dolomite patch is argillaceous limestone, Well Bg26 2515.6 m, Yeli-Liangjiashan formation; (**e**) local amplification of d, intercrystalline pores (yellow arrow); (**f**) the characteristics of the powder-to-fine-crystalline dolomite cements (D1) interior of the oolite, Well Bg26 2790.5 m, Zhangxia formation; (**g**) the petrological characteristics of the fine-crystalline ferroan dolomite cement (D2) and the relationship between the fracture and stylolite, Well Bg26 2801.9 m, Zhangxia formation; (**h**) algal limestone. The petrological characteristics of the fine-crystalline ferroan dolomite cement (D2), Well Bg26 2803.5 m, Zhangxia formation; (**i**) dolomitic bioclastic limestone. The petrological characteristics of the fine-crystalline ferroan dolomite cement (D2), Well Bg26 2803.7 m, Zhangxia formation.

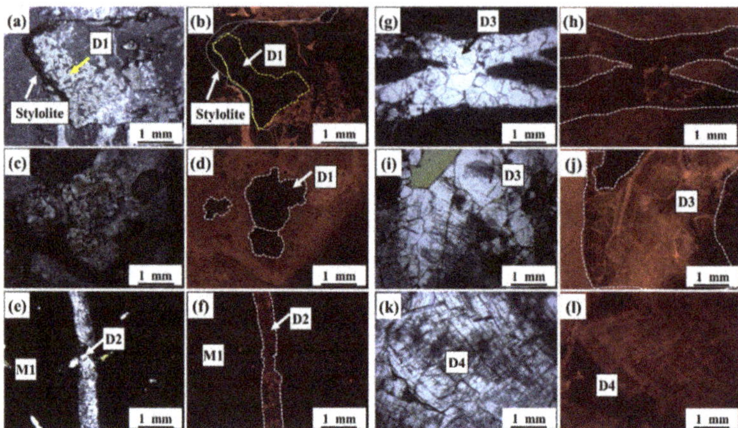

Figure 4. Thin section and the cathodoluminescence photomicrographs showing different kinds of dolomite cements in the Lower Paleozoic strata in the Dongying Depression, Bohai Bay Basin. (**a**) The powder-to-fine-crystalline dolomite cements (D1) along with the stylolite, Well Bg11 2449.5 m, Ordovician formation, plane-polarized light; (**b**) characteristics of cathodoluminescence in the same field of view as a. The powder-to-fine-crystalline dolomite cements within the yellow dotted line, CL; (**c**) the powder-to-fine-crystalline dolomite cements (D1) filled in the oolite and have the characteristics of "mist core bright edges", Well Bg26 2795.5 m, Zhangxia formation, plane polarized light; (**d**) the powder-fine-crystalline dolomite cements (D1) filled in the oolite within the white dotted line. The powder-fine-crystalline dolomite cements (D1) filled in the oolite in the same field of view as c, within the white dotted line, CL; (**e**) fractures developed in the microcrystalline dolomite (M2), which were filled in with fine-crystalline ferroan dolomite cements (D2), Well Bg22 2233.04, Ordovician formation, plane-polarized light; (**f**) characteristics of cathodoluminescence in the same field of view as e, and there are fine-crystalline ferroan dolomite cements (D2) in the white dotted line showing the weak dark red cathodoluminescence characteristics, CL; (**g**) the medium-to-coarse-crystalline dolomite cements (D3) filled in the fractures, Well Bg26 2801.9 m, Zhangxia formation, plane-polarized light; (**h**) characteristic of cathodoluminescence of the medium-to-coarse-crystalline dolomite cement (D3) filled in the fractures in the same field of view as g, and showing the weak cathodoluminescence, CL; (**i**) the characteristics of the medium-to-coarse-crystalline dolomites (D3) filled in the fractures, Well Bg26 2795.5 m, Zhangxia formation, plane-polarized light; (**j**) characteristics of cathodoluminescence of the medium-to-coarse-crystalline dolomite cements (D3) in the same field of view as i. The medium-to-coarse-crystalline dolomite cements with the obvious luminous bands, CL; (**k**) the characteristics of the saddle dolomite (D4), Well Bg22 2233.04 m, Ordovician formation, plane-polarized light; (**l**) characteristics of cathode luminescence of the saddle dolomites (D4) in the same field of view as k, CL. CL is a cathode luminescence characteristic.

4.1.5. Fine-Crystalline Ferroan Dolomite Cements (D2)

Fine-crystalline ferroan dolomite cements mainly filled in the fractures. It can be inferred that the fine-crystalline ferroan dolomite cement was formed after the stylolite (Figure 3g,h). This kind of dolomite cement is dyed blue by alizarin red mixed with potassium ferric hydroxide (Figure 3g–i). The crystal size of the ferroan dolomite depends on the fracture scale. In general, fractures with a width less than 0.25 cm are highly likely to be filled with fine-crystalline ferroan dolomite cements (D2). This type of dolomite cement commonly has a weak dull red color under CL (Figure 4f).

4.1.6. Medium-to-Coarse-Crystalline Dolomite Cements (D3)

Medium-to-coarse-crystalline dolomite cements consist of 250 μm to 500 μm crystals, with some being larger than 500 μm. The dolomite crystals mostly display a planar texture (Figure 5b,c,e,f). This type of dolomite cement fills in dissolution or structure fractures

(Figure 5a,d). The dolomite crystals were partially dissolved by late organic acid fluid, forming dissolution pores (Figure 5b,c,e). This type of dolomite is mainly present in the Cambrian strata and in smaller amounts in the Ordovician strata. The medium-to-coarse-crystalline dolomite cements have red cathodoluminescence characteristics, with obvious luminous bands in the dolomite crystals (Figure 5h–j)

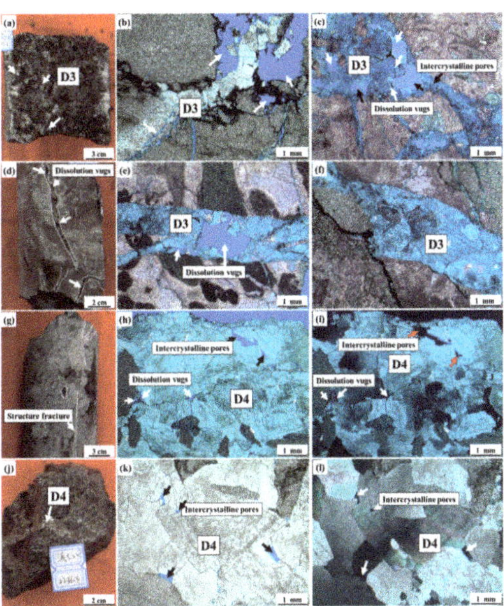

Figure 5. Petrological characteristics of the medium-to-coarse-crystalline dolomite cements (D3) and saddle dolomite (D4). (**a**) The medium-to-coarse-crystalline dolomite cements (D3) filled in the vertical fractures, and there is an obvious hydrocarbon charging phenomenon, Well Bg26 2793.2 m, Zhangxia formation, core; (**b**) microcrystalline dolomite; the developed dissolution fractures are filled in with medium-to-coarse-crystalline dolomite cements. The hydrocarbon dissolved the dolomite crystals and formed the dissolution vugs, shown by the white arrow, Well Bg22 2246.89 m, Ordovician formation, plane-polarized light; (**c**) bioclastic limestone; the medium-to-coarse-crystalline dolomite cements are filled in the structure fracture and formed the intercrystalline pores, shown by the black arrow. Some of the dolomite crystals were dissolved by the hydrocarbon and formed the dissolution vugs, shown by white arrow, Well Bg26 2793.2 m, Zhangxia formation, plane-polarized light; (**d**) the vertical fracture dissolved by the fluid and formed the dissolution vugs (shown by the white arrow), which are filled by coarse dolomite crystals, Well Bg26 2795.1 m, Zhangxia formation, core; (**e**) bioclastic limestone; structure fracture is developed and is filled in with medium-to-coarse-crystalline dolomite cements. Additionally, dolomite crystals are dissolved by fluid, forming the dissolution vugs (shown by white arrow), Well Bg26 2795.4 m, Zhangxia formation, plane-polarized light; (**f**) bioclastic limestone; medium-to-coarse-crystalline dolomite cements are tightly filled in the dissolution fractures. The dolomite crystals at the edge of the fracture are mainly medium-crystalline dolomite cements, while the size of the dolomite crystals at the center of the fracture increases gradually, Well Bg26 2798.2 m, Zhangxia formation, plane-polarized light; (**g**) microcrystalline dolomite; structural fractures developed, Well Bg22 2211.17 m, Ordovician formation, core; (**h**) microcrystalline dolomite; saddle dolomites are filled in the fracture, Well Bg22 2211.17 m, Ordovician formation, plane-polarized light; (**i**) saddle dolomites in the same view as h, perpendicular polarized light; (**j**) microcrystalline dolomite; high-angle structural fractures are filled by the saddle dolomites, Well Bg22 2236.4 m, Ordovician formation, core; (**k**) microcrystalline dolomite; characteristic of saddle dolomites are filled in fractures, and intercrystalline pores are formed by saddle dolomite crystals (the black arrow), Well Bg22 2246.89 m, Ordovician formation, plane-polarized light; (**l**) saddle dolomite in the same view as k, perpendicular polarized light.

4.1.7. Saddle Dolomite (D4)

Saddle dolomites are coarse crystalline, are much larger than 500 µm, and exhibit curved crystal faces. Under perpendicular polarized light, the saddle dolomites show wavy extinction (Figure 5i,l). Small-scale intercrystalline pores are developed between the saddle dolomites (Figure 5h,k). Parts of the edge of the saddle dolomites are dissolved and form the dissolution vugs (Figure 5h). Saddle dolomite has been observed in fractures (Figure 5g,j) and is mainly present in pores or cavities in the Ordovician strata. Previous studies have shown that the saddle dolomite is related to fractures and is possibly due to hydrothermal events [31,50].

4.2. Dolomite Reservoir Characteristics

There are various reservoir spaces in the dolomite reservoir of the Lower Paleozoic strata in the Dongying Depression. Reservoir spaces can be divided into pores and fractures according to the observations of the cores and thin sections.

4.2.1. Pore

The pores types include moldic pores, intercrystalline pores, and dissolution vugs.

Moldic Pore

Moldic pores are formed by the selective dissolution of early calcite or gypsum cements by meteoric water [51], and the morphology of the dissolved particles is retained. The pore diameter (long axis) of the moldic pores in the study area is between 211 µm and 655 µm (Figure 2d).

Intercrystalline Pore

Intercrystalline pores occupy a large proportion of all types of reservoir space. These intercrystalline pores are between mineral crystals [52]. Dolomite crystals with different structures form intercrystalline pores of different sizes. The powder-to-fine-crystalline dolomite cements (D1) mainly form micro intercrystalline pores (Figure 3e). Saddle dolomite (D4) overgrowth also forms small-scale intercrystalline pores (average size 288 µm) (Figure 4h,i,k,l and Figure 6a). Medium-to-coarse-crystalline dolomite cement (D3) tends to form relatively large intercrystalline pores (Figures 4c and 6b).

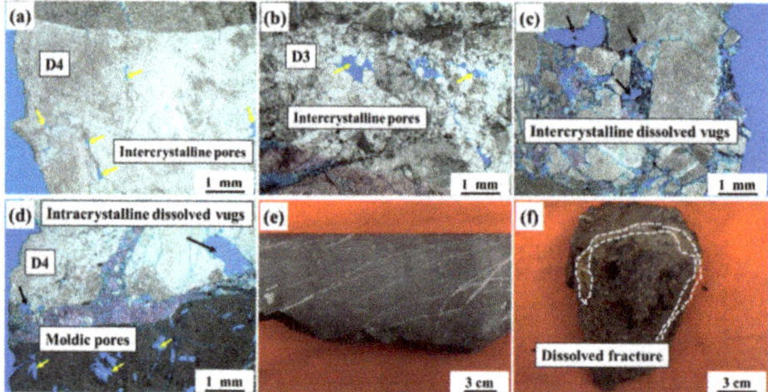

Figure 6. The reservoir space characteristics of the dolomite reservoir in the Dongying Depression, Bohai Bay Basin. (**a**) The intercrystalline pores (yellow arrow) between the saddle dolomites (D4), Well Bg22 2246.89 m, Ordovician formation, plane-polarized light; (**b**) the intercrystalline pores (yellow arrow) between the medium-to-coarse-crystalline dolomite cements, Well Bg26 2793.2 m, Zhangxia formation, plane-polarized light; (**c**) intercrystalline-dissolved vugs (black arrow) between the saddle dolomites (D4), Well Bg22 2236.4 m, Ordovician formation, plane-polarized light; (**d**) intracrystalline dissolved vugs (black arrow) within the saddle dolomites (D4), Well Bg22 2233.04 m, Ordovician formation, plane-polarized light; (**e**) microcrystalline dolomite (M2) high-angle structural fractures, Well Bg22 2348.4 m, Ordovician formation, core; (**f**) microcrystalline dolomite (M2), dissolved fractures (white dotted line), the saddle dolomites fill in the fracture, Well Bg22 2235.84 m, Ordovician formation, core.

Dissolution Vugs

Dissolution vugs include intercrystalline-dissolved vugs and intracrystalline-dissolved vugs. Dolomite were dissolved and enlarged by supergene and late diagenetic dissolution, forming enlarged dissolution pores [53]. Dissolution vugs commonly have harbor shapes. The dissolution fluids are mainly composed of hydrothermal fluid and/or organic acid fluid. Black bitumen can be seen in or around the edge of the dissolution vugs.

4.2.2. Fracture

Fractures can be divided into structural fractures and dissolution fractures according to their genesis [54,55]. Tectonic stress promotes the formation of structural fractures, including low-angle fractures, oblique fractures, and high-angle fractures, which are relatively straight (Figure 6e). Since then, diagenetic fluids have mainly dissolved (expand) the structural fractures or unstable rocks, causing the fracture edge to be curved and smooth. Hydrothermal minerals and black residual bitumen filled within the dissolution fractures, indicating that the dissolution fluid may be hydrothermal fluid or organic acid fluid (Figures 4b and 6f).

4.3. Isotope Data

4.3.1. Stable Isotopes

The results of the O isotope VPDB and C-isotope VPDB values are presented in Figure 7. The argillaceous dolomites (M1) have $\delta^{18}O$ VPDB values of about $-6‰$, and the $\delta^{18}O$ values of the microcrystalline dolomite (M2) range from $-9.3‰$ to $-6.2‰$. The fine-crystalline dolomite (M3) has $\delta^{18}O$ values of about $-9.3‰$. The dolomite cements filling in the fractures or dissolution vugs have more negative $\delta^{18}O$ values than the matrix dolomites do. The $\delta^{18}O$ values of the powder-to-fine-crystalline dolomite cements (D1) range from $-10‰$ to $-5.2‰$, the $\delta^{18}O$ value of the fine-crystalline ferroan dolomite cements (D2) is $-14.1‰$. The $\delta^{18}O$ values of the medium-to-coarse-crystalline dolomite cements (D3) range from $-15.7‰$ to $-12.7‰$. The saddle dolomites have more negative $\delta^{18}O$ values than other types of the dolomite in the study areas, ranging from $-25.7‰$ to $-23.7‰$. The $\delta^{13}C$ values of the dolomites range from $-1.9‰$ to $0.5‰$.

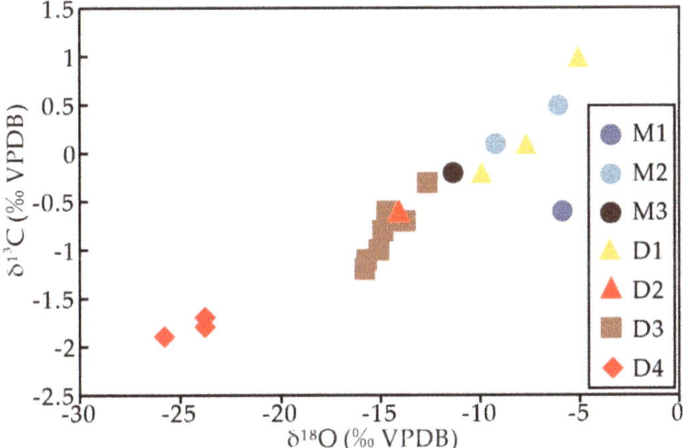

Figure 7. Stable isotope values for the dolomites of the Lower Paleozoic strata in the Dongying Depression, Bohai Bay Basin. M1: argillaceous dolomite; M2: microcrystalline dolomite; M3: fine-crystalline dolomite; D1: powder-to-fine-crystalline dolomite cement; D2: fine-crystalline ferroan dolomite cement; D3: medium-to-coarse-crystalline dolomite cement; D4: saddle dolomite.

4.3.2. Strontium Isotope

The Sr isotope ratios of the dolomites are displayed in Figure 8. The microcrystalline dolomite (M2) $^{87}Sr/^{86}Sr$ ratio is about 0.7091, within the range of the Latest Cambrian to Middle Ordovician seawater [56,57]. The $^{87}Sr/^{86}Sr$ ratio of the powder-to-fine-crystalline dolomite cements (D1), fine-crystalline ferroan dolomite cements (D2), and the medium-to-coarse-crystalline dolomite cements (D3) range from 0.7119 to 0.7129, which are higher than those of the microcrystalline dolomite. The $^{87}Sr/^{86}Sr$ ratio of the saddle dolomite (D4) has a wide range, from 0.7093 to 0.7112.

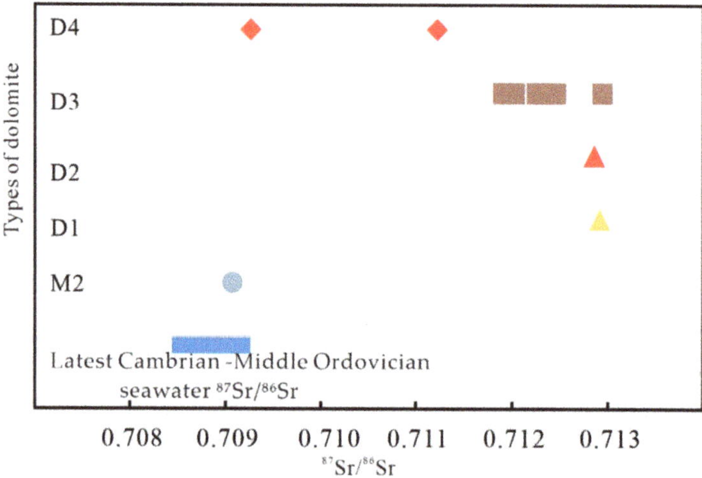

Figure 8. Comparison of radiogenic strontium isotope ratios between Lower Paleozoic dolomite and Latest Cambrian to Middle Ordovician seawater. The $^{87}Sr/^{86}Sr$ ratios of Latest Cambrian to Middle Ordovician seawater from [56,57]. M2: microcrystalline dolomite; D2: fine-crystalline ferroan dolomite cements; D3: medium-to-coarse-crystalline dolomite cements; D4: saddle dolomite.

5. Discussion

5.1. Diagenesis of the Dolomite Reservoir

Carbonate rocks may undergo a variety of diagenesis mechanisms after deposition, changing the physical properties of carbonate reservoirs. According to the thin sections and core study, the dolomite reservoirs in the study areas mainly experienced six types of diagenesis: dolomitization, dissolution, fracturing, cementation, compaction, and pressure solution.

5.1.1. Dolomitization

Dolomitization is widely developed in the limestone at the bottom of the Ordovician formation and the oolitic limestone in the Cambrian formation. The $^{87}Sr/^{86}Sr$ ratios of the fine-crystalline dolomite cements have similar $^{87}Sr/^{86}Sr$ ratio values to those of argillaceous dolomite (M1) and microcrystalline dolomite (M2), and the Latest Cambrian to Middle Ordovician seawater [56,57]. D1 are associated with stylolite and anhydrite. These characteristics indicate that powder-to-fine-crystalline dolomite cements (D1) may have been formed by the evaporation and condensation of Cambrian–Ordovician seawater in the early stage.

5.1.2. Dissolution

In the area around where dissolution occurs, the residual black bitumen can be seen around the edges of the dolomite crystals and in some dissolution fractures (Figure 5b) and dissolution vugs (Figure 6c), indicating the charging of the organic matter fluid. The margin of the dissolution fracture is not straight and is an embay in shape, and the local enlargement of the fractures appeared (Figure 5c). Pyrite, saddle dolomite, quartz, and other hydrothermal minerals have been observed in the dolomite reservoirs, indicating hydrothermal fluid activity. Additionally, these hydrothermal minerals are accompanied by dissolution pores (Figure 9a). This evidence indicates that organic acid fluids and hydrothermal fluids are the main dissolution fluids for the dolomite, leading to the dissolution of dolomite and the formation of dissolution fractures or dissolution vugs.

Figure 9. Diagenesis types of dolomite reservoirs in the Dongying Depression, Bohai Bay Basin. (**a**) Microcrystalline dolomite; hydrothermal dissolution vugs. The pyrite filled in the vugs (yellow arrow), and medium-to-coarse-crystalline dolomite cement was dissolved and formed intragranular dissolution pores, Well Bg22 2246.89 m, Ordovician formation, backscattered image; (**b**) microcrystalline dolomite; low-angle fracture, filled with calcite cements; Well Bg26 2515.2 m, Yeli-Liangjiashan formation, core; (**c**) oblique fractures (white arrow) formed before high-angle fractures (black arrow), Well Bg26 2803.7 m, Zhangxia formation, core; (**d**) microcrystalline dolomite; vertical fracture (yellow arrow), filled with saddle dolomite; Well Bg22 2236.4 m, Ordovician formation, core; (**e**) the stylolite (white arrow) was formed by compaction and pressure solution, and filled argillaceous material; Well Bg22 2219.27 m, Ordovician formation, thin section; (**f**) the stylolite (red arrow) formed by compaction and pressure solution, and filled with organic matter; Well Bg22 2348.95 m, Ordovician formation, thin section. D3: medium-to-coarse-crystalline dolomite cements; D4: saddle dolomite.

5.1.3. Fracturing and Cementation

Tectonic stresses cause rocks to fracture and can form a series of fractures. The main types of fractures include low-angle fractures, high-angle fractures, and vertical fractures. The sequence of fractures can be distinguished according to the intersecting relationships between fractures. In the early stages, low-angle fractures and oblique fractures were mainly formed and were filled by the calcite cements (Figure 9b,c). In the late stages, high-angle fractures and vertical fractures were formed and were filled by the medium-to-coarse-crystalline dolomite cements (D3) and saddle dolomites (D4) (Figure 9d).

5.1.4. Compaction and Pressure Solution

A large number of stylolites were found in the study area, indicating that the reservoir underwent strong compaction and pressure solution (Figure 9e,f).

5.2. Origins of the Different Types of Dolomite

5.2.1. Matrix Dolomites

Matrix dolomites mainly include argillaceous dolomite (M1), microcrystalline dolomites (M2), and fine-crystalline dolomites (M3) from the restricted and tidal flat face and tend to be mainly microcrystalline to fine crystals with a euhedral–subhedral crystal texture. M1 and M2 have the smallest crystal size. This indicates that M1 and M2 grow during the early stages of diagenesis [58]. The $\delta^{13}C$ values of the matrix dolomites (0.5~−1‰, VPDB) are similar to those of contemporaneous limestone (−1.5~−1.1‰). The $\delta^{18}O$ values of M1 and M2 range from −9.3 to −6‰ VPDB. Using the dolomite–water oxygen isotope fractionation equation $1000Ln\alpha_{(dolomite-water)} = 3.14 \times 10^6 T^{-2} - 2.0$ proposed by Land [59], and assuming a temperature of 25 °C, the $\delta^{18}O$ values of the water present during the growth of M1 and M2 were calculated. The results of the calculations show that the $\delta^{18}O$ values of the water present during the growth of M1 and M2 ranged from −12.1 to −8‰ VSMOW. The $\delta^{18}O$ values of the brachiopods from the Cambrian–Ordovician strata commonly range from −10 to −3‰ VPDB [60]. Using the calcite–water oxygen isotope fractionation equation $1000Ln\alpha_{(calcite-water)} = 2.78 \times 10^6 T^{-2} - 2.89$ proposed by Friedman and O'Neil [61], the $\delta^{18}O$ values of seawater during the growth of brachiopods were calculated. The $\delta^{18}O$ values of the seawater during brachiopod growth are between −9.2 and −2.1‰ VSMOW (Figure 10). It can be seen that the $\delta^{18}O$ values of the argillaceous dolomite (M1) and microcrystalline dolomite (M2) were more negative than the $\delta^{18}O$ value of the Cambrian–Ordovician seawater (−9.2 to −2.1‰, VSMOW). However, the $^{87}Sr/^{86}Sr$ ratio of microcrystalline dolomite (M2) is about 0.7091, falling within the $^{87}Sr/^{86}Sr$ range of Latest Cambrian to Middle Ordovician seawater. These indicate that seawater may be the main diagenetic fluid for M1 and M2 formation. M1 and M2 have more negative $\delta^{18}O$ values than that of seawater and are likely to be affected by the hydrothermal fluid or meteoric water. Fine-crystalline dolomite (M3) have a larger crystal size than M1 and M2, indicating that M3 were formed relatively later than M1 and M2. Moreover, early diagenetic gypsum (now transformed to anhydrite) is present in the dolomite strata. This supports sabkha capillary zone dolomitization [62,63] and reflux dolomitization models [64]. Anhydrite cements have been observed in the formation where M3 are developed in the study area. Additionally, the main sedimentary environments in the study area are the gypsum dolomite flat with strong evaporation and supratidal–intertidal flat during the deposition period of Majiagou formation [65]. Therefore, M3 could have been formed by the seepage reflux of seawater in an evaporative environment.

5.2.2. Powder-to-Fine-Crystalline Dolomite Cements (D1)

Power-to-fine-crystalline cements (D1) are commonly planar rhombic textures, indicating that D1 were formed at relatively low temperatures during the early shallow burial stage [66]. D1 are distributed along the stylolite in limestone samples, indicating that the formation of D1 is related to stylolite. D2 is generally distributed on both sides of the stylolite. Some dolomites are dissolved and have uneven crystal edges (Figure 3b,c). Some studies have reported that the dissolution of the limestone along the stylolite may accelerate dolomitization through the flow of Mg-rich fluids [67]. Therefore, the powder-to-fine-crystalline dolomite cements associated with stylolite may be formed at the same time as or after stylolite. Stylolite was formed by the chemical compaction and is of various scales and sizes. Some scholars have considered that pressure dissolution can occur at the depth of 610~914 m, forming stylolite [68], and can also occur at the deep burial environment [69]. In recent years, pioneers have proposed that stylolite forms at a depth of between 500 m and 1 km [70]. According to the burial history in the study areas. D1 may have been formed by metasomatic calcite from lower Ordovician dolomites. The $\delta^{18}O$ values of the powder-to-fine-crystalline dolomite cements (D1) ranged from −5.2‰ to −10‰ (VPDB). Using the dolomite–water oxygen isotope fractionation equation proposed by Land $1000Ln\alpha_{(dolomite-water)} = 3.14 \times 10^6 T^{-2} - 2.0$, the $\delta^{18}O$ values of fluid forming dolomite precipitates were calculated. The calculation results show that the $\delta^{18}O$ values of the fluid-forming dolomite precipitates are between −8 and −3.8‰ VSMOW. The $\delta^{18}O$ values of fluid forming dolomite precipitates are similar to the oxygen isotope values of the Paleozoic seawater (−9.2 to −2.1‰, VSMOW) [71]. This suggests that seawater is the main dolomitic fluid. However, Figure 8 shows that the $^{87}Sr/^{86}Sr$ ratio of the D1 dolomite is greater than that of the Latest Cambrian to Middle Ordovician seawater, further indicating that D1 may be influenced by the continental formation of water rich in radioactive Sr in the shallow burial environment [72]. In some samples, unstable calcite carbonate filled in ooids replaced by D1 (Figure 3f). Calcite that filled in the ooids is dissolved to form moldic pores, the dolomitizing fluid then enters the moldic pores. When the fluid is highly saturated, recrystallization development occurs to form the powder-to-fine-crystalline dolomite cements (D1) [58,66]. Sparry calcite cements developed in the intercrystalline pores (Figure 3f), suggesting that the powder-to-fine-crystalline dolomite cements were formed relatively early in diagenetic history.

5.2.3. Fine-Crystalline Ferroan Dolomite Cements (D2) and Medium-to-Coarse-Crystalline Dolomite Cements (D3)

D2 and D3 dolomite cements have a larger crystal size than D1 dolomite cements, indicating that D2 and D3 dolomite cements are formed in higher-temperature conditions [58,73]. The most obvious difference between the D2 and (D3) dolomite cements and D1 dolomite cements is that the D2 and D3 dolomite cements are stained blue and exhibit the ferroan dolomite cement characteristics under the microscope. The iron-rich dolomite is the result of hydrothermal fluids action in a deep burial environment [74,75]. Dolomite crystals have obvious cathodoluminescence bands, suggesting that the D2 and D3 dolomite cements have multi-stage recrystallization [76]. The D2 and D3 dolomite cements filled in the fractures. Previous studies have analyzed the carbon and oxygen isotopes of calcite filling in high-angle fractures formed in the Yanshan period and found $\delta^{18}O$ values between $-19‰$ and $-15‰$ VPDB, $\delta^{13}C$ values between $-0.5‰$ and $-3.0‰$ VPDB [77], and the fluid inclusions homogenization temperatures are between 120 °C and 150 °C. Using the calcite–water oxygen isotope fractionation equation $1000Ln\alpha_{(calcite\text{-}water)} = 2.78 \times 10^6 T^{-2} - 2.89$ proposed by Friedman and O'Neil [61], the $\delta^{18}O$ values of fluid-forming calcite precipitates were calculated. The $\delta^{18}O$ values of fluid forming calcite precipitates ranged from -1.4 to $0.18‰$ VSMOW. The $\delta^{18}O$ values of D2 and D3 ranged from -15.8 to $-12.7‰$ VPDB. Using the ferroan dolomite–water oxygen isotope fractionation equation $1000Ln\alpha_{(dolomite\text{-}water)} = 2.78 \times 10^6 T^{-2} + 0.11$ proposed by Fisher and Land [78], the $\delta^{18}O$ values of fluid forming D2 and D3 precipitates were calculated (Figure 10). The results show that the $\delta^{18}O$ values of fluid forming D2 and D3 precipitates are between $-1.4‰$ and $0.2‰$ VSMOW. This result shows a high degree of overlap with the $\delta^{18}O$ values of the diagenetic fluids from the calcite in the Yanshan veins in the study area, indicating that the diagenetic fluid of D2 and D3 is likely to be similar to the vein calcite cements formed during the Yanshan period. The calcite cement and D2, and D3 in the veins, were formed from precipitation from hydrothermal fluids during the Yanshan period. The crust was folded and uplifted, and the platform was activated, forming a series of fault depressions and uplifting the fault blocks during the Late Triassic–Early Jurassic periods, forming the fracture systems [79]. The Yanshan Movement, which began at the end of the Early Jurassic period, was the initial stage of volcanic activity in the Bohai Bay Basin [80,81]. In particular, the second act of the Yanshan Movement was characterized by intense magmatic activity accompanied by intense faulting [82], providing a sufficient heat source and Fe^{2+} for iron-rich dolomite cements. Furthermore, the homogenization temperatures of fluid inclusions, which were shown to range in temperature from 120 to 181 °C, are higher than the maximum burial temperature (140 °C) in the study area [83], suggesting that D2 and D3 may be formed by hydrothermal fluid. D2 and D3 dolomite cements have higher $^{87}Sr/^{86}Sr$ ratios (0.7119 to 0.7129) than Cambrian–Ordovician seawater (Figure 8) and are close to the crust source of hydrothermal fluids [84]. A higher $^{87}Sr/^{86}Sr$ ratio indicates that ^{87}Sr-enriched fluids were involved in the diagenesis of the dolomite. ^{87}Sr-enriched fluid may derive from the fluid formed by the eroding of siliciclastic sediments containing argillaceous and/or feldspathic components. The Maozhuang and Xuzhuang formations are deposited with shale and sandstone with a total thickness of about 200 m. When the sandstone is dominated by feldspar, migrating hydrothermal fluids may be able to acquire radioactive Sr isotope. Therefore, D2 and D3 have the characteristics of a high temperature of hydrothermal fluid and high $^{87}Sr/^{86}Sr$ ratio of crust source.

5.2.4. Saddle Dolomite (D4)

Saddle dolomites in the Ordovician formation are filled with high-angle fractures, have largest crystals, curved crystal faces, and wavy extinction characteristics, suggesting that D4 may have been formed by the rapid precipitation of high temperatures of Mg-rich fluids during late diagenesis [17,63]. Previous studies and the high homogenization temperature of the inclusions (>200 °C) have indicated that D4 was formed by hydrothermal fluids.

The $\delta^{18}O$ values of dolomite can reflect the environmental conditions and temperature during the formation and diagenesis of carbonate rocks [85,86]. Saddle dolomites (D4) show that the $\delta^{18}O$ values (-23.7 to -25.7‰ VPDB) are more negative than other types of the dolomite cements (Figure 7). Utilizing the formula $10^3 Ln\alpha_{(dolomite-water)} = 3.14 \times 10^6 T^{-2} - 2.0$ proposed by Land [59], the $\delta^{18}O$ of the water present during the growth of D4 was between -6.7‰ and -5.0‰ VSMOW. The $\delta^{18}O$ values of the D4 are more negative compared to the $\delta^{18}O$ VSMOW values of the magmatic water (Figure 10). Additionally, this result is more negative than that of common hydrothermal saddle dolomites. The $\delta^{18}O$ value of D4 fluid is significantly lower than that of normal saddle dolomite, which may be caused by the following reasons: (1) The $\delta^{18}O$ values of the dolomite are controlled by temperature. As the temperature increases, ^{18}O are depleted, Therefore, dolomite has a lower $\delta^{18}O$. The homogenization temperature of the saddle dolomite inclusions in the lower Paleozoic in the southwest of the Dongying Depression is more than 200 °C (Figure 10), which is higher than that of common saddle dolomite inclusions (100~180 °C) [17,87]; (2) there may be ^{18}O depleted hydrothermal fluids injection during diagenesis. The host dolomites can be dissolved by the hydrothermal fluids. Saddle dolomites are mainly developed in the Ordovician microcrystalline dolomite strata. The main sedimentary environment of the Majiagou period is gypsum dolomite flat with strong evaporation and supratidal–intertidal flat in the study area [65]. The evaporation phase is enriched with light oxygen, so the host dolomites also have low $\delta^{18}O$ values. When the host dolomites with low $\delta^{18}O$ values were dissolved by hydrothermal fluids with low $\delta^{18}O$ values, saddle dolomite may have had extremely negative $\delta^{18}O$ values. The $^{87}Sr/^{86}Sr$ ratios of some host dolomites are similar to that of saddle dolomite. Which also supports this inference.

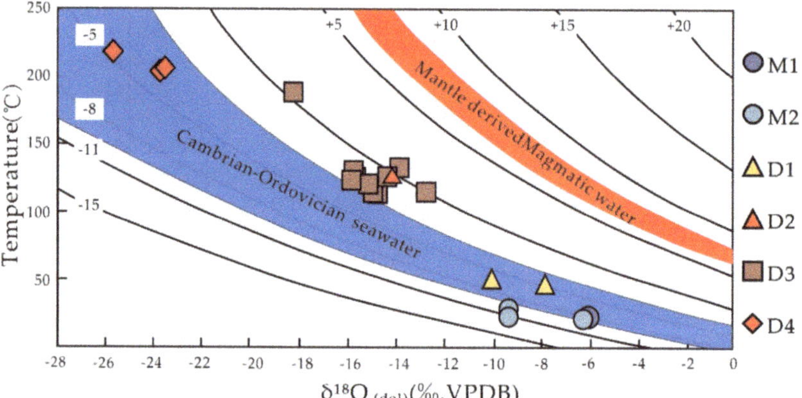

Figure 10. Crossplots of fluid inclusion homogenization temperature (Th) against oxygen signature for the matrix dolomite and various dolomite cements in the southwest of the Dongying Depression, Bohai Bay Basin. The data for mantle-derived magmatic water are quoted from [88]. M1: argillaceous dolomite; M2: microcrystalline dolomite; D1: powder-to-fine-crystalline dolomite cements; D2: fine-crystalline ferroan dolomite cements; D3: medium-to-coarse-crystalline dolomite cements; D4: saddle dolomite.

5.3. Implications for Porosity Development

5.3.1. Diagenetic Pore Evolution

On the basis of petrological observation, this study identified the diagenesis and diagenetic sequence of the reservoir and summarized the diagenetic pore evolution model of the reservoir (Figure 11).

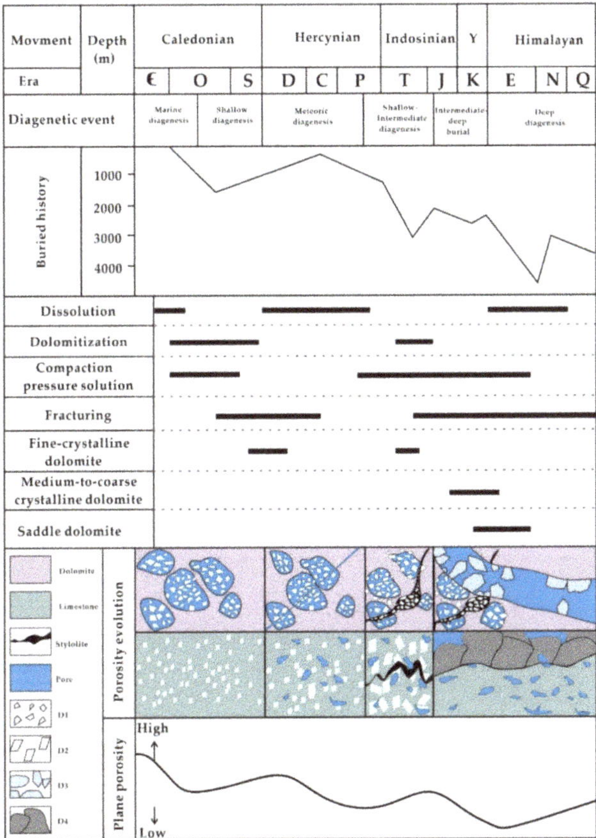

Figure 11. Diagenetic events and pore evolution of dolomite reservoir of the Lower Paleozoic strata in the Dongying Depression, Bohai Bay basin. Y represents Yanshan movement; D1: powder-to-fine-crystalline dolomite cements; D2: fine-crystalline dolomite cements; D3: medium-to-coarse-crystalline dolomite cements; D4: saddle dolomite.

In the early diagenetic stage, early calcite cements filled in the space between grains (oolites or rubble), reducing the primary porosity. The Cambrian–Ordovician strata were uplifted and affected by Caledonian movement [89], and the oolites were selectively dissolved by diagenetic fluids and formed moldic pores. At the moment, the powder-to-fine-crystalline dolomite cements (D1) formed by the seepage reflux mechanism are cemented in residual oolites. However, the plane porosity produced by the powder-to-fine-crystalline dolomite cements (D1) are commonly lower than 0.1%. Dolomitization at the early diagenesis is not the main diagenesis to increase the porosity. With the increase in the buried depth, the strata enters the subsidence stage, and pore reduction such as compaction occurs. Compaction reduces rock volume, dissolves grains, and decreases the proportion of intercrystalline pores, leading to a sharp decrease in porosity [90].

5.3.2. The Influence of Diagenesis on Reservoir Properties

Studies have shown that the fracture system, dissolution, and cementation play important roles in the formation of reservoir spaces [17,32,91,92]. The plane porosity of the rocks can reach 48% in the early stages of fracture formation. After the fault system is formed, hydrothermal fluids along with the fractures move upward, the matrix dolomite dissolves, and dissolution vugs and dissolution fractures begin to form. The saturability of the fluid changes as the fluids react with the host rocks. Subsequently, D3 and D4 dolomite cements begin to form during the intermediate-deep burial diagenesis. The D3 and D4 dolomite cements are fully filled in (or semi-filled in) the fractures. This cementation process causes serious damage to the previously formed fracture reservoir system [93]. After cementation, the plane porosity of the rocks is only 6% and can be even less than 1%. The residual pores after cementation are the main sites of late organic acid dissolution.

Petrological observations reveal high-temperature dolomite cements, pyrite, and secondary dissolution (irregular edges of pores and fractures), and the residual bitumen observed in the interior of the dissolution pores and fractures. In summary, the study area is extensively influenced by hydrothermal fluids and organic acids.

Fluid migration is controlled by faults [17,87,94]. The more developed the fault, the more obvious the fluid action. Dissolution is the main diagenesis to form effective reservoir spaces in the study area [40,95]. A series of needle-like dissolution vugs are formed as hydrothermal fluids migrate up the fractures and faults [91]. Dissolution is the key factor to form high-quality reservoirs. Due to the striking volcanic activity, hydrothermal dissolution is a main dissolution process during the intermediate-to-deep burial stages. When the reservoir is closer to the intrusions and faults, the dissolution phenomenon is more obvious, and more dense dissolution vugs are developed. The reservoir shows good productivity. Since the dissolution and precipitation of the hydrothermal fluid are usually accompanied and since fluid saturation increases, hydrothermal dissolution gradually turns into hydrothermal precipitation [16]. The cements formed by hydrothermal precipitation fill the fractures.

In general, the plane porosity of saddle dolomite is lower than that of medium-to-coarse-crystalline dolomite (Figure 12), which may be due to the joint influence of the crystal structure and mineral filling degree of the fractures. Saddle dolomite has larger crystals than medium-to-coarse-crystalline dolomite, and the fractures are usually fully filled in with saddle dolomite. Therefore, the rock is less affected by fluid dissolution, and the plane porosity of the rock is relatively low (Figure 12a,b). Medium-to-coarse-crystalline dolomite partially fills in fractures, and fractures with high residual pores. Additionally, the rocks that developed medium-to-coarse-crystalline dolomite have a higher degree of dissolution than the rocks that developed saddle dolomite (Figure 12). Magmatic activity mainly developed during the Yanshan movement, and some magmatic activity developed during the late Triassic period [79,96,97]. Residual bitumen is commonly filled between dolomite cements formed by high-temperature fluids. The oil and gas charging period represent the sedimentary period of the Shahejie Formation [98]. All of this evidence indicates that the dissolution of organic acids occurs after hydrothermal dissolution. Therefore, the dissolution vugs that are formed by organic acid dissolution can be better preserved [99]. Organic acid dissolution can form higher plane porosity than hydrothermal dissolution according to other plane porosity statistics (Figure 12). Additionally, the reservoirs are closer to the source layer or basin and the dissolution of organic acids is more intense.

Dolomite is more likely to form effective fractures than limestone due to its greater mechanical strength [100,101]. Thus, dolomite reservoirs are more likely to form high-density fractures as well as fractured reservoir systems. Additionally, the closer the formations and wells are to the fault plane, the higher the density of the structural fractures according to the statistic of the cores and thin sections (Figure 13). These fractures are interconnected to form a large fractured reservoir system, which plays a constructive role in reservoir reconstruction. The rocks with a high fracture density and low fracture filling degree are often subjected to more intense dissolution, and the rocks have higher plane porosity

(Figure 12c). On the contrary, the dissolution of rocks is not obvious and the plane porosity percentage is relatively low (Figure 12d).

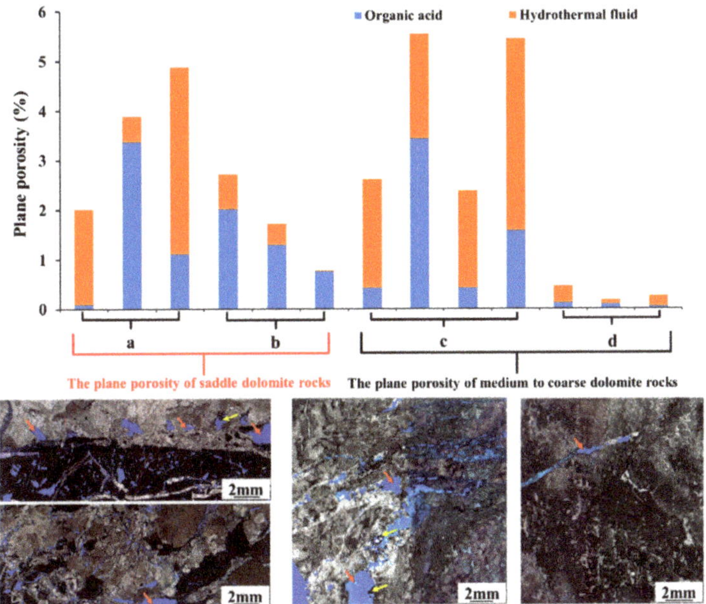

Figure 12. The total plane porosity of all kinds of dolomite cements with different textures and the rate of contribution of different fluids to the percent of area pores. (**a**) The contribution of dissolution to the plane porosity of the rocks that developed saddle dolomite and its microscopic characteristics when dissolution is strong; (**b**) when the dissolution is not obvious, the contribution of dissolution to the plane porosity of the rocks developed saddle dolomite and its microscopic characteristics; (**c**) the rock developed with a high-density fracture area, and the image shows the contribution of dissolution to the plane porosity of the rocks; (**d**) the rock developed a low-density fracture area, and the image shows the contribution of dissolution to the plane porosity of the rocks; red arrow: dissolution vugs; yellow arrow: residual bitumen.

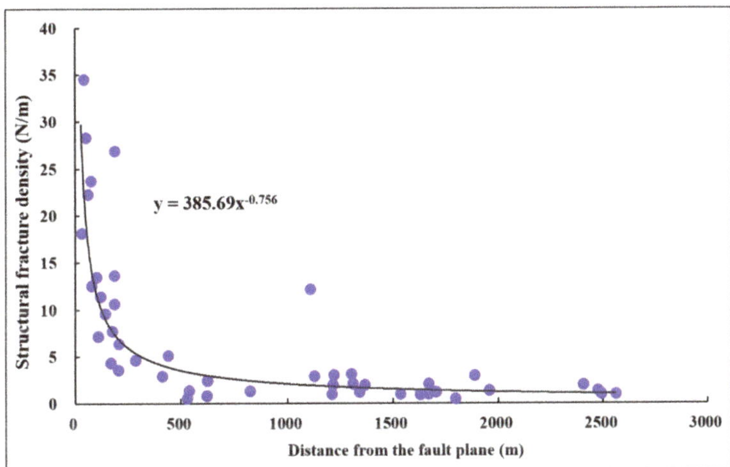

Figure 13. Relationship between fracture density and its distance to fault in the southwest of the Dongying Depression, Bohai Bay Basin. A part of data are taken from [46].

There are differences in the diagenetic process of different formations in a single well. The diagenesis observed in well Bg22 is relatively simple due to the relatively simple lithology of the host rock and the fact that the well did not drill through the Ordovician formation. Controlled by the physical properties of the original lithology, the fracturing widely developed, and the host rock is mostly breccia. Burial dissolution is also well developed in brecciated dolomite formation, which results in higher porosity. Well Bg22 coring is incomplete due to strong fracturing and dissolution. The main cements in well Bg22 are saddle dolomite and anhydrite. A small amount of fine-crystalline dolomite formed in deeper strata. Anhydrite cements formed in the evaporative phase can be observed in the fractures of the dolomite development strata (Figure 14).

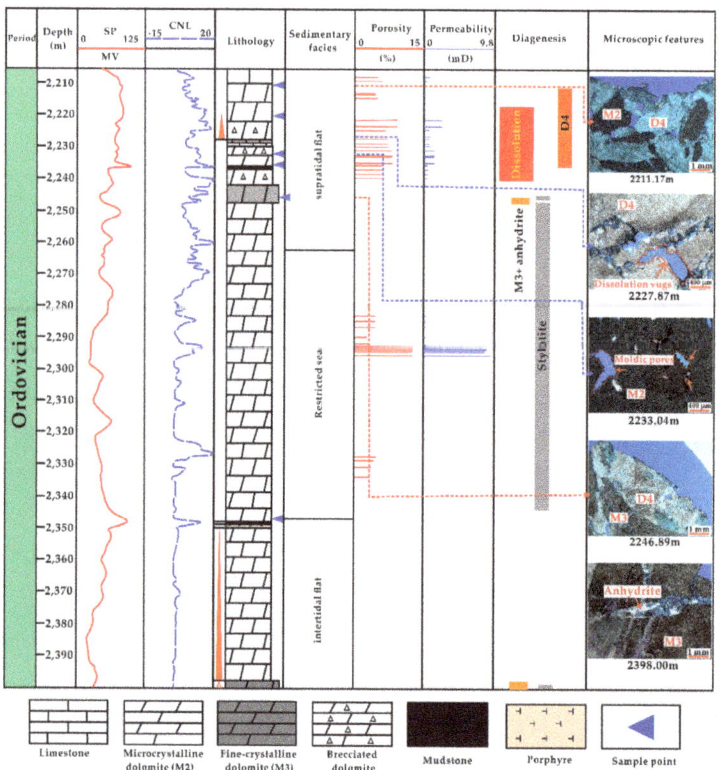

Figure 14. The stratigraphic log of well Bg22 and differences of diagenesis in different strata. M1: argillaceous dolomite; M2: microcrystalline dolomite; M3: fine-crystalline dolomite; D4: saddle dolomite.

In well Bg26, dissolution mainly occurs in the Ordovician strata close to unconformity and faults. In the Cambrian strata, oolitic limestone and bioclastic limestone with better original porosity also have obvious dissolution. Stylolites formed by pressure solution are developed in the Ordovician and Cambrian strata. However, the density of the stylolites in the Ordovician strata are higher than that of the Cambrian strata. Dolomitization is widely developed in well Bg26 and the main products are powder-to-fine dolomite cements (D1). With the increase in the burial depth, the properties of the powder-to-fine-crystalline dolomite cements (D1) change obviously. The crystal size gradually increases. D1 is usually associated with stylolites or formed dolomite patches in the Ordovician strata, while in the Cambrian strata, D1 is mainly filled in oolitic or bioclastics due to changes in the depositional environment. D2 and D3 are products of hydrothermals, so D2 and D3 are developed in Cambrian strata close to the intrusion, and are widely developed in the Gushan and Zhangxia formation (Figure 15).

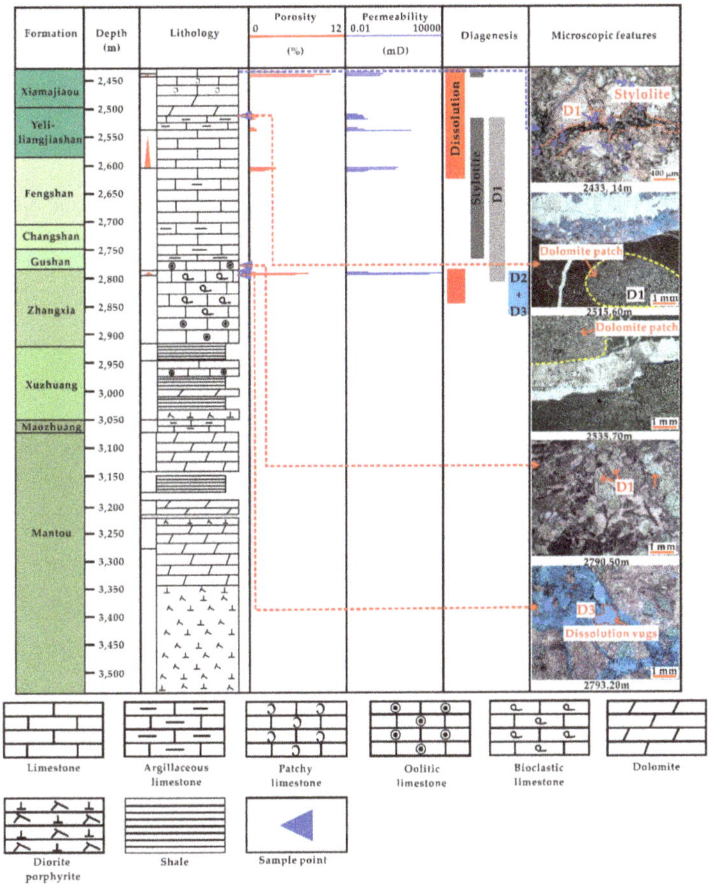

Figure 15. Differences of diagenesis in different strata in well Bg 26. D1: powder-to-fine-crystalline dolomite cements; D2: fine-crystalline ferroan dolomite cements; D3: medium-to-coarse-crystalline dolomite cements.

Four wells (Bg26, Bg22, Bgx15, Bg11) were selected for the study. Bg26 is the closest to the Gaoqing-pingnan fault and oil-generating zone, and suffered the most complicated diagenesis. Dolomitization products are power-to-fine-crystalline dolomite filling inside the oolitic or associated with the stylolites (Figures 15 and 16a). In the Yeli-Liangjiashan formation, dolomitization results in the formation of patchy dolomite in micritic limestone (Figure 16b). Well Bg26 was more easily affected by the hydrothermal fluids and organic acid. The hydrothermal fluid cement associated with dolomite are mainly fine-crystalline dolomite cement (D2) and medium-to-coarse-crystalline dolomite cement (D3) (Figure 16c). Organic acid dissolution is most obvious in well Bg26 due to it is closest to the oil-generating depression (Boxing depression). Well Bg22 did not drill through the Ordovician formation. Bg22 is located in a higher tectonic position than Bg26, and the sedimentary environment is dolomitic flat with strong evaporation for a period of time and forms dolomite with higher original porosity and brittleness. Additionally, the high tectonic position and development of the unconformity plane ensures the host dolomite is easily affected by meteoric water, and it is dissolved by meteoric water to form pores and fractures, which also makes the host dolomite have the geochemical characteristics of meteoric water. Then, hydrothermal fluids enter the dolomite reservoirs along these pores or fractures. Hydrothermal fluids react with the host dolomite and precipitate saddle dolomite cement (D4) with meteoric water geochemical characteristics in the fractures and pores (Figure 16d). Due to dolomite's high mechanical fracture strength, it is affected by both hydrothermal and tectonic processes. In well Bg22, the fracturing is obvious and the overall coring is incomplete. Well Bgx15 did not drill through the Ordovician formation. Selective dissolution pores are developed in the Ordovician microcrystalline dolomite, forming moldic pores (Figure 16e).

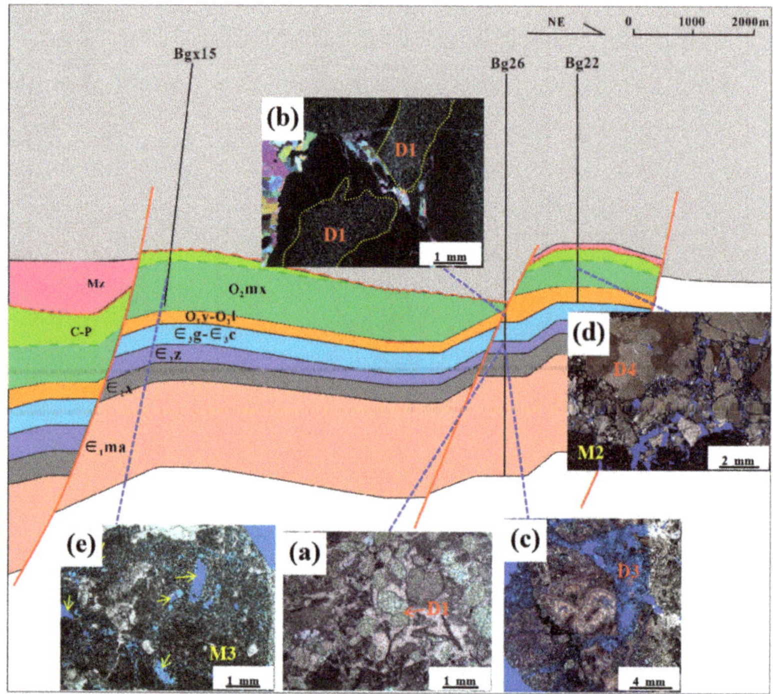

Figure 16. The difference of diagenesis between different wells. (**a**) Oolitic limestone, Well Bg26 2790.5 m, Zhangxia formation; (**b**) argillaceous limestone, well Bg26 2540 m, Yeli-Liangjiashan formation; (**c**) bioclastic limestone, well Bg26 2793.2 m, Zhangxia formation; (**d**) microcrystalline dolomite, well Bg22 2227.87 m Ordovician formation; (**e**) microcrystalline dolomite, well Bgx15 2238.9 m, Ordovician formation, moldic pores (yellow arrow); M2: microcrystalline dolomite; M3: fine-crystalline dolomite; D1: powder-to-fine-crystalline dolomite cement; D3: medium-to-coarse-crystalline dolomite cement; D4: saddle dolomite.

6. Conclusions

The following conclusions can be drawn based on petrographic and geochemical investigations of the dolomites at the Lower Paleozoic Formation in the southwest of the Dongying Depression, Bohai Bay Basin:

(1) Dolomites can be divided into the matrix dolomite and dolomite cements. The diagenetic fluid of matrix dolomite is mainly seawater. Dolomite cements can be divided into powder-to-fine-crystalline dolomite cement (D1), fine-crystalline ferroan dolomite cement (D2), medium-to-coarse-crystalline dolomite cement (D3), and saddle dolomite (D4). The powder-to-fine-crystalline dolomite cement (D1) could form by the seepage reflux of sea water or Mg-enriched pore fluids followed by recrystallization during the burial period. The fracture filling dolomite cements, such as D2 and D3 are most likely related to the mixed fluids formed by hydrothermal fluid erosion of the [87]Sr-enriched feldspar sandstone in the Maozhuang and Xuzhuang formation. Under the hydrothermal conditions, the host

dolomite formed by the evaporative phase was dissolved by hydrothermal fluids, then Mg-enriched fluid precipitation to form saddle dolomite, resulting in saddle dolomite with the negative $\delta^{18}O$ value.

(2) The results of the petrology show that there are six types of diagenesis: dolomitization, dissolution, fracturing and cementation, compaction, and pressure solution. Diagenesis in different diagenetic environments has different effects on reservoirs.

(3) Faults and fracture systems are key control factors for dolomite reservoirs and can be used as effective reservoir spaces. The stronger the tectonic activity is, the more faults develop, and the higher the fracture density, the more high-quality reservoirs can be formed. In addition, faults can be used as effective migration channels for fluids that are helpful for later dissolution.

(4) Hydrothermal fluid dissolution and organic acid dissolution result in dolomite reservoirs having better reservoir performance. Due to the late action time of organic acid dissolution, the dissolution vugs can be well preserved. In the study area, organic acid dissolution contributes more than hydrothermal dissolution.

Author Contributions: Conceptualization, X.Z. and Q.L.; methodology, X.Y.; software, L.W.; validation, A.J.; formal analysis, X.Y.; investigation, L.W.; resources, W.T.; data curation, A.J.; writing—original draft preparation, X.Z.; writing—review and editing, Q.L.; visualization, X.Z.; supervision, L.M.; project administration, Q.L.; funding acquisition, L.M. All authors have read and agreed to the published version of the manuscript.

Funding: This research was funded by the National Natural Science Foundation of China (grant number 41602137, 41972107), the National Natural Science Foundation of China (grant number U19B6003), the Strategic Cooperation Technology Projects of CNPC and CUPB (grant number ZLZX2020-02), and the Science Foundation of China University of Petroleum, Beijing (grant number 2462020YXZZ022).

Institutional Review Board Statement: Not applicable.

Informed Consent Statement: Not applicable.

Data Availability Statement: All data used in this research are easily accessible by downloading the various documents appropriately cited in the paper.

Acknowledgments: The authors greatly appreciate the Exploration and Production Research Institute of Shengli Oilfield Company for providing samples and data access and for permission to publish the results.

Conflicts of Interest: The authors declare no conflict of interest.

References

1. Bai, G.P. Distribution patterns of giant carbonate fields in the world. *J. Palaeogeogr. (Chin. Ed.)* **2006**, *8*, 241–250.
2. Halbouty, M.T. Giant Oil and Gas Fields of the Decade 1990–1999. AAPG Memoir. In Proceedings of the 4th AAPG Conference, Denver, CO, USA, 7–10 April 1991.
3. Zhao, W.Z.; Shen, A.J.; Zheng, J.F.; Qiao, Z.F.; Wang, X.F.; Lu, J.M. Discussion on pore genesis of dolomite reservoirs in Tarim, Sichuan and Ordos Basin and its guiding significance to reservoir prediction. *Sci. Sin.* **2014**, *44*, 1925–1939.
4. Zhang, J.T.; Jin, X.H.; Gu, N.; Bian, C.R.; Yang, J.Q.; He, Y.L. Differences and development patterns of karst reservoirs in Majiagou Formation, northern Ordos Basin. *Oil Gas Geol.* **2021**, *42*, 1159–1168.
5. Hao, Y.; Zhou, J.G.; Zhang, J.Y.; Ni, C.; Gu, M.F.; Xin, Y.G. Characteristics and controlling factors of dolomite reservoir of Middle Permian Qixia Formation in northwest Sichuan Basin. *Sediment. Geol. Tethyan Geol.* **2013**, *33*, 68–74.
6. Hu, A.P.; Pan, L.Y.; Hao, Y.; Shen, A.J.; Gu, M.F. Origin, Characteristics and Distribution of Dolostone Reservoir in Qixia Formation and Maokou Formation, Sichuan Basin, China. *Mar. Orig. Pet. Geol.* **2018**, *23*, 39–52.
7. Shu, X.H.; Zhang, J.T.; Li, G.R.; Long, S.X.; Wu, S.X.; Li, H.T. Characteristics and genesis of hydrothermal dolomites of Qixia and Maokou Formations in northern Sichuan Basin. *Oil Gas Geol.* **2012**, *33*, 442–448.
8. Zhu, Y.D.; Jin, Z.Y.; Sun, D.S.; Deng, Y.M.; Zhang, R.Q.; Yuan, Y.R. Hydrothermal dolomitization of Sinian Dengying Formation in south China and its influence on reservoir formation: A case study of Central Guizhou Uplift. *Chin. J. Geol. (Sci. Geol. Sin.)* **2014**, *49*, 161–175.
9. Ma, Y.S.; Guo, T.L.; Zhao, X.F.; Cai, X.Y. Formation mechanism of deep high-quality dolomite reservoir in Puguang Gasfield. *Sci. Sin.* **2007**, *37*, 43–52.

10. Zheng, H.R.; Wu, M.B.; Wu, X.W.; Zhang, T.; Liu, C.Y. Oil-gas exploration prospect of dolomite reservoir in the Lower Paleozoic of Tarim Basin. *Acta Pet. Sin.* **2007**, *28*, 1–8.

11. Sibley, D.F. *Climatic Control of Dolomitization, Seroe Domi Formation (Pliocene), Bonaire, NA*; The Society of Economic Paleontologists and Mineralogists: East Lansing, MI, USA, 1980; pp. 247–258.

12. Shinn, E.A.; Ginsburg, R.N.; Lloyd, R.M. Recent Supratidal Dolomite from Andros Island, Bahamas. In *Dolomitization and Limestone Diagenesis*; Pray, L.C., Murray, R.C., Eds.; SEPM Society for Sedimentary Geology: Tulsa, OK, USA, 1965; pp. 112–113.

13. Vasconcelos, C.; Bernasconi, S.; Grujic, D.; Tien, A.J.; Mckenzie, J.A. Microbial mediation as a possible mechanism for natural dolomite formation at low temperatures. *Nature* **1995**, *377*, 220–222. [CrossRef]

14. Vasconcelos, C.; Mckenzie, J.A. Microbial mediation of modern dolomite precipitation and diagenesis under anoxic conditions (Lagoa Vermelha, Rio de Janeiro, Brazil). *J. Sediment. Res.* **1997**, *67*, 378–390.

15. Jiang, L.; Pan, W.; Cai, C.; Jia, L.; Pan, L.; Wang, T.; Li, H.; Chen, S.; Chen, Y. Fluid mixing induced by hydrothermal activity in the ordovician carbonates in Tarim Basin, China. *Geofluids* **2015**, *15*, 483–498. [CrossRef]

16. Smith, L.B. Origin and reservoir characteristics of Upper Ordovician Trenton–Black River hydrothermal dolomite reservoirs in New York. *AAPG Bull.* **2006**, *90*, 1691–1718. [CrossRef]

17. Davies, G.R.; Smith, L.B. Structurally controlled hydrothermal dolomite reservoir facies: An overview. *AAPG Bull.* **2006**, *90*, 1641–1690. [CrossRef]

18. Spencer, C.C.; Mullis, J. Chemical study of tectonically controlled hydrothermal dolomitization: An example from the Lessini Mountains, Italy. *Geol. Rundsch.* **1992**, *81*, 347–370. [CrossRef]

19. White, T.; Al-Aasm, I.S. Hydrothermal dolomitization of the Mississippian Upper Debolt Formation, Sikanni gas field, northeastern British Columbia, Canada. *Bull. Can. Pet. Geol.* **1997**, *45*, 297–316.

20. Ronchi, P.; Masetti, D.; Tassan, S.; Camocino, D. Hydrothermal dolomitization in platform and basin carbonate successions during thrusting: A hydrocarbon reservoir analogue (Mesozoic of Venetian Southern Alps, Italy). *Mar. Pet. Geol.* **2012**, *29*, 68–89. [CrossRef]

21. Du, Y.L.; Li, S.Y.; Wang, B.; Zhao, D.Q.; Yang, D.D. Diagenesis of the Lower-Middle Permian Carbonate in the Wuwei-Chaohui area, Anhui Province. *Acta Geol. Sin.* **2011**, *85*, 543–556.

22. Giles, M.R.; De Boer, R.B. Origin and significance of redistributional secondary porosity. *Mar. Pet. Geol.* **1990**, *7*, 378–397. [CrossRef]

23. Ali, M.Y. Carbonate cement stratigraphy and timing of diagenesis in a Miocene mixed carbonate-clastic sequence, offshore Sabah, Malaysia: Constraints from cathodoluminescence, geochemistry, and isotope studies. *Sediment. Geol.* **1995**, *99*, 191–214.

24. Paradis, S.; Lavoie, D. Multiple-stage diagenetic alteration and fluid history of Ordovician carbonate-hosted barite mineralization, Southern Quebec Appalachians. *Sediment. Geol.* **1996**, *107*, 121–139. [CrossRef]

25. Heydari, E. Porosity loss, fluid flow, and mass transfer in limestone reservoirs: Application to the Upper Jurassic Smackover formation, Mississippi. *AAPG Bull.* **2000**, *84*, 100–118.

26. Esteban, M.; Taberner, C. Secondary porosity development during late burial in carbonate reservoirs as a result of mixing and/or cooling of brines. *J. Geochem. Explor.* **2003**, *78*, 355–359. [CrossRef]

27. Wang, L.; Shi, J.A.; Wang, Q.; Wang, J.P.; Zhao, X.; Sun, X.J.; Zhao, L.B. Analysis on main controlling factors of Ordovician carbonate reservoir in southwest margin of Ordos Basin. *Pet. Geol. Recovery Effic.* **2005**, *12*, 10–13.

28. Choquette, P.W.; Cox, A.; Meyers, W.J. Characteristics, distribution and origin of porosity in shelf dolostones; Burlington-Keokuk Formation (Mississippian), US Mid-Continent. *J. Sediment. Res.* **1992**, *62*, 167–189.

29. Xie, G.P. Diagenesis and Porosity Evolution of Crystal Garin Dolomite in the Upper Section of the 4th Member of Leikoupo Formation in the Western Sichuan Depression. *J. Yangtze Univ. (Nat. Sci. Ed.)* **2015**, *12*, 24–26.

30. Mo, J.; Wang, X.Z.; Xie, L.; Zhou, Z.; Lin, G.; Xiong, J.W. Diagenesis and Pore Evolution of Carbonate in Sinian Dengying Formation in Central Sichuan Province. *J. Oil Gas Technol.* **2013**, *35*, 32–38.

31. Lavoie, D.; Jackson, S.; Girard, I. Magnesium isotopes in high-temperature saddle dolomite cements in the lower Paleozoic of Canada. *Sediment. Geol.* **2014**, *305*, 58–68. [CrossRef]

32. Biehl, B.C.; Reuning, L.; Schoenherr, J.; Lüders, V.; Kulka, P.A. Impacts of hydrothermal dolomitization and thermochemical sulfate reduction on secondary porosity creation in deeply buried carbonates: A case study from the Lower Saxony Basin, northwest Germany. *AAPG Bull.* **2016**, *100*, 597–621. [CrossRef]

33. Ardiansyah, K.; Hilary, C.; Jack, S.; Peter, K.S.; Adrian, B.; Hamish, R.; Fiona, W.; Cathy, H. Evaluating new fault controlled hydrothermal dolomitization models: Insights from the Cambrian Dolomite, Western Canadian Sedimentary Basin. *Sedimentology* **2020**, *67*, 2945–2973.

34. Huang, S.J.; Wang, C.M.; Huang, B.B.; Zou, M.L.; Wang, Q.D.; Gao, X.Y. Scientific research frontiers and considerable questions of carbonate diagenesis. *J. Chengdu Univ. Technol. (Sci. Technol. Ed.)* **2008**, *35*, 1–10.

35. Liu, W.; Xiao, C.T.; Lv, Y.L. Analysis of Burid-hill Reservoir Forming Conditions in the East Section of South Slope of Dongying Depression. *J. Oil Gas Technol.* **2004**, *26*, 6–7.

36. Ge, X. Buried Hill Reservoir Forming Research of Ordovician System in Caoqiao Oilfield. Master's Thesis, Ocean University of China, Qingdao, China, 2015.

37. Qiu, Z.J. Study on the Geological Characteristics and Control Factors of Caoqiao Buried Hill Reservoir in Dongying Depression, Shandong. Master's Thesis, Kunming University of Science and Technology, Kunming, China, 2016.

38. Jiang, W. Characteristics of High-Quality Reservoir of the Palezoic Carbonate Buried Hills in the Donging Depression. Master's Thesis, China University of Petroleum (East China), Dongying, China, 2017.

39. Wei, X. Evaluation of Karst Reservoir in the Ordovician Badou Formation in the Zhuanghai Area, Jiyang Depression. Master's Thesis, Chengdu University of Technology, Chengdu, China, 2019.

40. Guo, Y.X. A study on constructive diagenesis of the dolomite reservoir within the Yeli-Liangjiashan Formation In the ZHuanghai area of Jiyang Depression. *Acta Mineral. Sin.* **2021**, *41*, 163–170.

41. Li, C.G. Control of fault systems on oil and gas distribution in Dongying Depression. *Oil Gas Geol.* **1994**, *15*, 87–93.

42. Xiong, Z.; Wang, L.S.; Li, C.; Shi, X.B.; Guo, S.P.; Wang, J. Distribution geotemperature in Dongying Depression, Shengli oil and gas field, North China basin. *Geol. J. China Univ.* **1999**, *5*, 312–321.

43. Jiang, Y.L.; Rong, Q.H. Formation pattern of oil-gas pools and distribution of hydrocarbon in Gaoqing area. *Pet. Geol. Exp.* **1998**, *20*, 14–19.

44. Chen, X.; Cao, Y.C.; Yuan, G.H.; Wang, Y.Z.; Zan, N.M. Origin and distribution model of the lower Paleozoic carbonata reservoirs in Pingfangwang-Pingnan buried hills, Dongying Sag. *J. China Univ. Pet. (Ed. Nat. Sci.)* **2020**, *44*, 1–14.

45. Li, G.Y. *Atlas of China's Petroliferous Basins*; Petroleum Industry Press: Beijing, China, 2002; pp. 58–276.

46. Fan, C.T.; Feng, Y.L.; Fu, J.P. Analysis on conditions and rules of reservoir forming of buried hill in Dongying sag. *Pet. Geol. Recovery Effic.* **2002**, *9*, 35–37.

47. Zan, N.M.; Wang, Y.Z.; Cao, Y.C.; Yuan, G.H.; Chen, X.; Jiang, W.; Zhai, G.H.; Song, M.S. Characteristics and development of reservoir space of the Lower Paleozoic buried hills in Dongying Sag, Bohai Bay Basin. *Oil Gas Geol.* **2018**, *39*, 355–365.

48. Lin, S.H.; Wang, H.; Zhang, G.X.; Wu, Y.X.; Chen, H.Y.; Wei, H.B. Pool features of buried hill in west part of Dongying Depression. *Oil Gas Geol.* **2000**, *21*, 360–363.

49. Tian, Y.M. Structural Evolution and Oil-Gas Accumulation Analysis of Buried Hill in the South of Dongying Sag, Bohaiwan Basin. Master's Thesis, Chengdu University of Technology, Chengdu, China, 2005.

50. Machel, H.G. Saddle dolomite as a by-product of chemical compaction and thermochemical sulfate reduction. *Geology* **1987**, *15*, 836–940. [CrossRef]

51. Dun, T.J. Reservoir research status and development trend. *Northwestern Geol.* **1995**, *16*, 1–15.

52. Xue, H.; Han, C.Y.; Xiao, B.Y.; Han, J.Y. Origin of Reservoirs in the Lower Cambrian Xiaoerbulak Formation, Tarim Basin. *Acta Sedimentol. Sin.* **2019**, *43*, 79–88.

53. Jin, C.G.; Liu, J.W. Characteristics of Ordovician Karst reservoir in southern Zhidan, Ordos Basin. *Ground Water* **2016**, *38*, 239–241.

54. Yang, N.; Lü, X.X.; Pan, W.Q. Feature of fracture development in Ordovician carbonate reservoir of Lunnan burial hill. *J. Xi'an Shiyou Univ. (Nat. Sci. Ed.)* **2004**, *19*, 40–42.

55. Zhao, J.L.; Gong, Z.W.; Li, G.; Feng, C.Y.; Bai, X.; Jian, J.; Fu, B.; Hong, Y. A review and perspective of identifying and evaluating the logging technology of fractured carbinate reservoir. *Prog. Geophys.* **2012**, *27*, 537–547.

56. Burke, W.H. Variation of seawater $^{87}Sr/^{86}Sr$ throughout Phanerozoic time. *Geology* **1982**, *10*, 516–519. [CrossRef]

57. Mcarthur, J.M.; Howarth, R.J.; Bailey, R.T. Strontium Isotope Stratigraphy: LOWESS Version 3: Best Fit to the Marine Sr-Isotope Curve for 0–509 Ma and Accompanying Look-up Table for Deriving Numerical Age. *J. Geol.* **2001**, *109*, 155–170. [CrossRef]

58. Gregg, J.M.; Shelton, K.L. Dolomitization and dolomite neomorphism in the back reef facies of the Bonneterre and Davis formations (Cambrian), southeastern Missouri. *J. Sediment. Res.* **1990**, *60*, 549–562.

59. Land, L.S. The application of stable isotopes to studies of the origin of dolomite and to problems of diagenesis of clastic sediments. In *Stable Isotope in Sedimentary Geology (SC10)*; Arthur, M., Anderson, T., Kaplan, I., Veiser, J., Land, L., Eds.; The Society of Economic Paleontologists and Mineralogists (SEPM): Tulsa, Oklahoma, USA, 1983; Chapter 4.

60. Veizer, J.; Bruckschen, P.; Pawellek, F.; Diener, A.; Ala, D. Oxygen isotope evolution of Phanerozoic seawater. *Palaeogeogr. Palaeocl.* **1997**, *132*, 159–172. [CrossRef]

61. Friedman, I.; O'Neil, J.R. *Compilation of Stable Isotope Fractionation Factors of Geochemical Interest*; United States Government Printing Office: Washington, DC, USA, 1977; Volume 440, Chapter KK.

62. Kinsman, D.J.J. Gypsum and anhydrite of recent age, Trucial Coast, Persian Gulf. *North. Ohio Geol. Soc. Clevel. Ohio* **1966**, *1*, 302–326.

63. Machel, H.G. Concepts and models of dolomitization: A critical reappraisal. *Geol. Soc. Lond. Spec. Publ.* **2004**, *235*, 7–63. [CrossRef]

64. Adams, J.E.; Rhodes, M.L. Dolomitization by seepage refluxion. *AAPG Bull.* **1960**, *44*, 1912–1920.

65. Wang, S.H.; Song, G.Q.; Xu, C.H.; Chen, L. Early Palaeozoic Sedimentary Facies in the Shengli Oil Province, North China Platform. *Sediment. Facies Palaeogeogr.* **1997**, *17*, 34–40.

66. Sibley, D.F.; Gregg, J.M. Classification of dolomite rock textures. *J. Sediment. Res.* **1987**, *57*, 967–975.

67. Merino, E.; Canals, À. Self-accelerating dolomite-for-calcite replacement: Self-organized dynamics of burial dolomitization and associated mineralization. *Am. J. Sci.* **2011**, *311*, 573–607. [CrossRef]

68. Dunnington, H.V. Stylolite development post-dates rock induration. *J. Sediment. Res.* **1954**, *24*, 27–49. [CrossRef]

69. Tang, J.C.; Chen, H.H.; Wang, J.H.; Chen, K.Q.; Qi, K.L. The Diagenesis of the Upper Paleozoic Carbonate Rocks in the Southeast Xiang Depression. *J. Southwest Pet. Univ. (Sci. Technol. Ed.)* **2007**, *29*, 43–46.

70. Beaudoin, N.; Koehn, D.; Lacombe, O.; Lecouty, A.; Billi, A.; Aharonov, E.; Parlangeau, C. Fingerprinting stress: Stylolite and calcite twinning paleopiezometry revealing the complexity of progressive stress patterns during folding—The case of the Monte Nero anticline in the Apennines, Italy. *Tectonics* **2016**, *35*, 1687–1721. [CrossRef]

signals, including how they are influenced by noise. The Wigner–Ville distribution, which is a time–frequency analysis method, is used along with multi-wavelet decompression–reconstruction technology to identify coal beds and to exclude the interference of coal beds. A nail-type wavelet and compressed sensing technology were used to remove the strong shield interference so that the seismic response of the target layer was highlighted, and the ability to predict sandstone reservoirs was improved [13]. Zhang et al. [14] successfully suppressed the subwave parametrization using subwave spectral shaping of the raw seismic data; by implementing this technique in combination with compressed sensing processing, they were also able to achieve rejection of strongly shielded signals. Gu et al. [15] adopted a new high-resolution inversion technology that makes full use of lateral seismic waveform space change information instead of a traditional variogram, and they achieved high-precision thin reservoir prediction even under a strong shielding effect. However, it is difficult to describe thin coal seams by seismic data alone due to the small thickness and random distribution of coal. Seismic forward modeling is a valid method of identifying the seismic response of coal seams [16–18]. Moreover, an approach based on AVO characteristics has advantages in detecting special lithology and fluids. Pre-stack seismic analysis highlights special lithological information better than post-stack seismic analysis [18–25]. Therefore, it is also necessary to investigate the reflection mechanism of coal seams so that the influence of coal seams on reservoir prediction can be reduced.

Based on petrophysical characteristics of the coal seams in the study area, we investigate the differences in elastic parameters between coal seams and sandstone or mudstone and identify the distinguishing parameters that can indicate coal seams. The seismic response of different sand–coal stratigraphic assemblages is simulated by seismic forward modeling. The modeling result indicates that the post-stack seismic response is dominated by coal bed, whereas the response of sandstone can hardly be recognized. In contrast, the difference between the pre-stack AVO response characteristics of coal seams and gas-bearing sandstones has been clarified based on the statistics pertaining to AVO characteristics of drilled wells. Therefore, we propose a method to reduce the interference of coal beds in sandstone reservoir prediction using far-gather seismic information. This method has significantly improved the accuracy of reservoir prediction and sand description in sand–coal coupled environments and has been applied successfully in the exploration of coal-rich strata in the Pingbei slope belt, Xihu Depression.

2. Geological Setting

The East China Sea Shelf Basin (ECSSB) is a Meso-Cenozoic superimposed basin lying in the east margin of the Eurasian continental plate. The basin is bounded by the Chinese mainland to the west and by the Okinawa Trough to the east. The Xihu Depression, which is located in the eastern part of the ECSSB and is predominantly NNE-striking, is the largest depression of the ECSSB [26,27]. The Xihu depression can be divided into three tectonic elements: The Western Slope Belt, the Central Inversional Structural Belt and the Eastern Half-graben Belt. The area studied in this paper, the Pingbei slope belt, is located in the north of the Western Slope Belt (Figure 1). The Xihu Depression has been developed since the end of the Mesozoic and is predominantly filled with Cenozoic clastic sediments. The basin's evolution is composed of two phases: the early syn-rift phase was due to extension from the end of the Cretaceous to the Paleocene, and the following post-rift phase was due to thermal subsidence from the Eocene to the Oligocene [28–30].

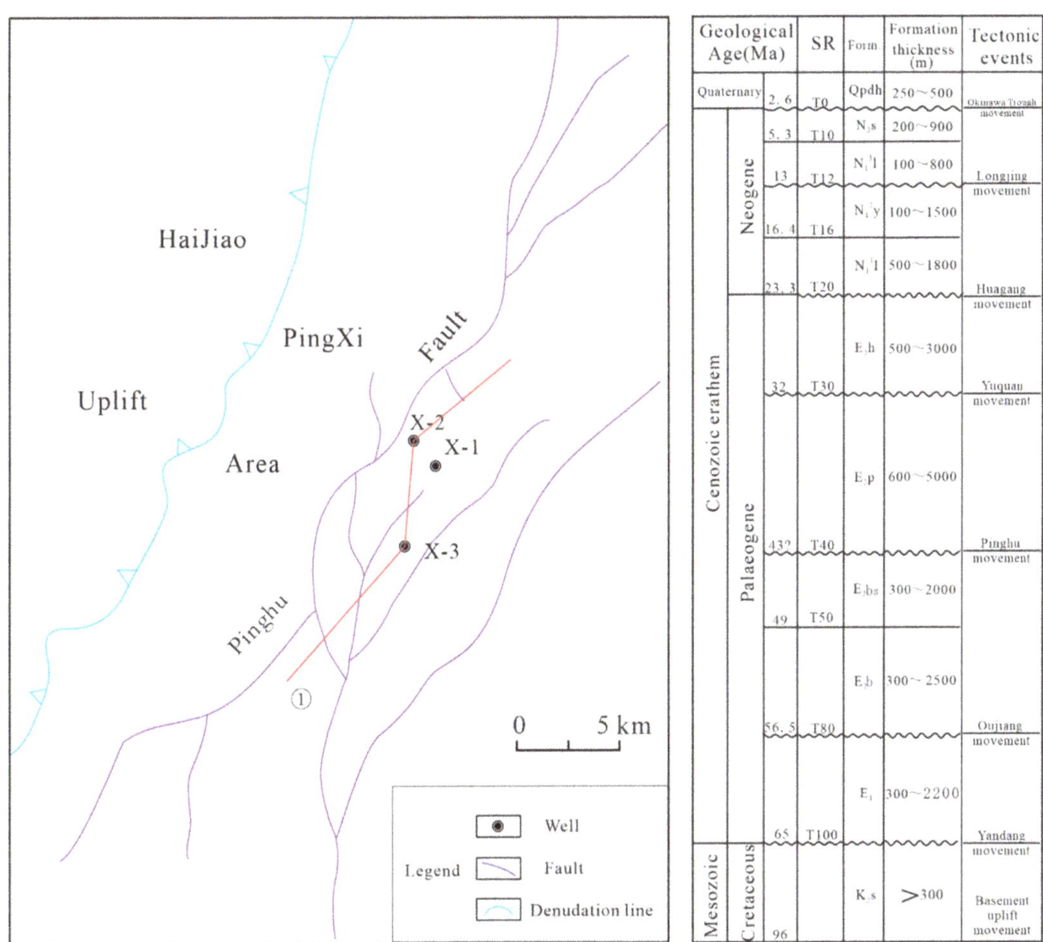

Geological Age (Ma)			SR	Form.	Formation thickness (m)	Tectonic events
Cenozoic erathem		Quaternary	2. 6 — T0	Qpdh	250~500	Okinawa Trough movement
Cenozoic erathem	Neogene		5. 3 — T10	N_2s	200~900	
Cenozoic erathem	Neogene		13 — T12	N_1^3l	100~800	Longjing movement
Cenozoic erathem	Neogene		16. 4 — T16	N_1^2y	100~1500	
Cenozoic erathem	Neogene		23. 3 — T20	N_1^1l	500~1800	Huagang movement
Cenozoic erathem	Palaeogene		32 — T30	E_3h	500~3000	Yuquan movement
Cenozoic erathem	Palaeogene		43? — T40	E_3p	600~5000	Pinghu movement
Cenozoic erathem	Palaeogene		49 — T50	E_2bs	300~2000	
Cenozoic erathem	Palaeogene		56. 5 — T80	E_2b	300~2500	Oujiang movement
Cenozoic erathem	Palaeogene		65 — T100	E_1	300~2200	Yandang movement
Mesozoic	Cretaceous		96	K_2s	>300	Basement uplift movement

Figure 1. Primary geological structures and the regional stratigraphy in the study area. Locations of wells and seismic profiles in Figure 9 are indicated.

The Pinghu Formation was deposited during the Eocene and comprises interbedded mudstone, siltstone and thin coal beds. Comprehensive analysis of data from the Pinghu Formation in the study area, including seismic phase, drilling core, and logging data, suggests that the coal seam developed mainly in the sedimentary environment of the tidally influenced deltaic plain, the tidally influenced deltaic foreshore divergent interfluve, the supratidal zone and the braided river deltaic foreshore divergent interfluve. The coal beds are interbedded with sandstone and mudstone in the Pinghu Formation. The interbedded layers have small thicknesses ranging from 0.5 to 2 m. The coal beds are difficult to recognize throughout the study area because their lateral change is radical (Figure 2).

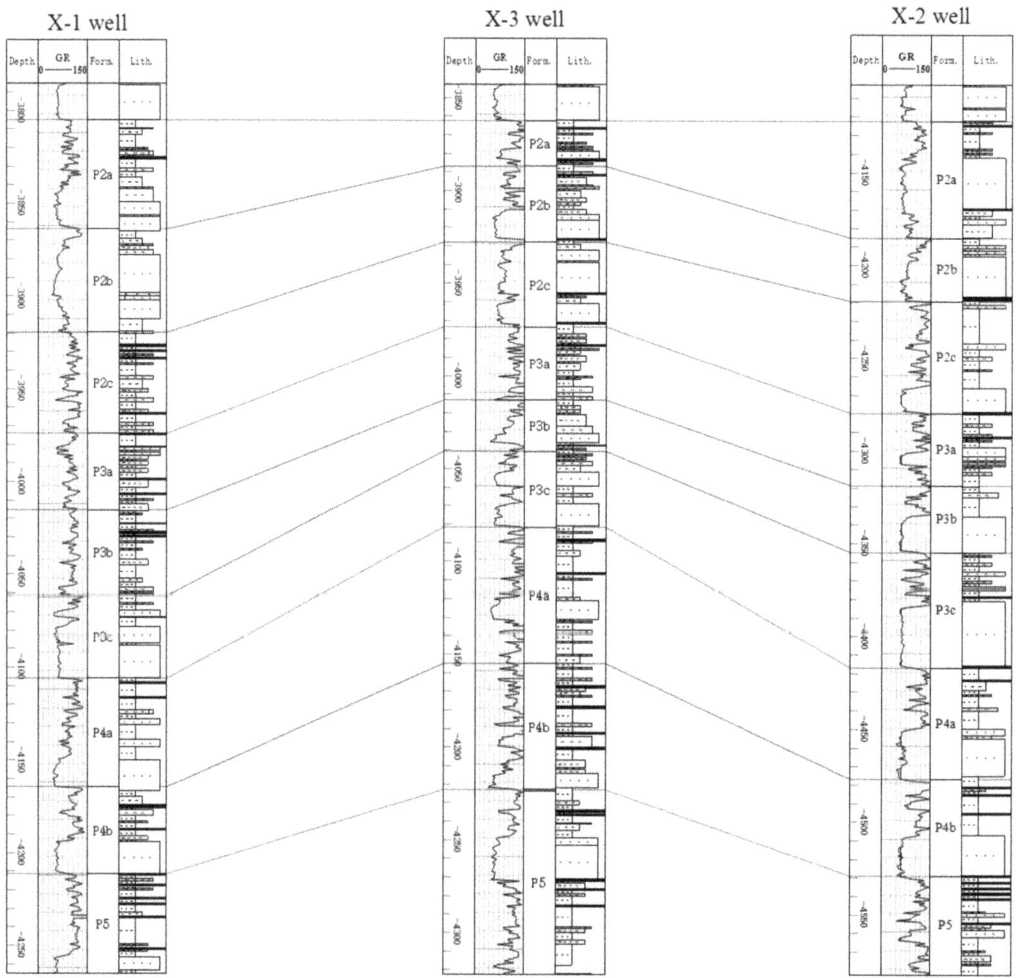

Figure 2. Litho-stratigraphic units of the Pinghu Formation in the Xihu Depression. For locations of the wells, see Figure 1.

3. Data and Methods

We investigate the distribution and reflective characteristics of coal-bearing strata, sandstone and mudstone by means of petrophysics, post-stack forward modeling and pre-stack wave equation forward modeling. First, the petrophysical properties of different lithology assemblages are clarified based on statistics. Second, a series of post-stack and pre-stack forward models are designed and simulated according to the petrophysical characteristics. The modeling result indicates that post-stack seismic analysis demonstrates ambiguity due to the interference of coal seams with the amplitude, frequency and phase of sandstone. Finally, we suppress the interference of coal seams using far-offset partial stacked seismic analysis, taking advantage of the fact that the energy of coal seams decreases as the offset angle increases.

3.1. Analysis of Petrophysical Characteristics

The target layers in the study area are generally buried deeper than 4000 m. The distribution characteristics of P-impedance differ based on lithology (Figure 3), with sandstone P-impedance being the largest, mudstone P-impedance the second largest, and coal seam

P-impedance the smallest. The sandstone P-impedance decreases when the sandstone contains gas, such that the sandstone and mudstone P-impedance are indistinguishable. Hence, only coal seam lithology can be effectively distinguished by P-impedance. In contrast, statistics concerning the ratio of compressional and shear wave velocity (hereafter referred to as Vp/Vs) of different lithologies show that Vp/Vs can effectively distinguish between the lithology of sand and mud (Figure 4). Vp/Vs of sandstone shows low-value characteristics, while those of mudstone and coal seam both show high-value characteristics. Thus, reservoir prediction and fine description of the sand body in this area can be conducted using Vp/Vs.

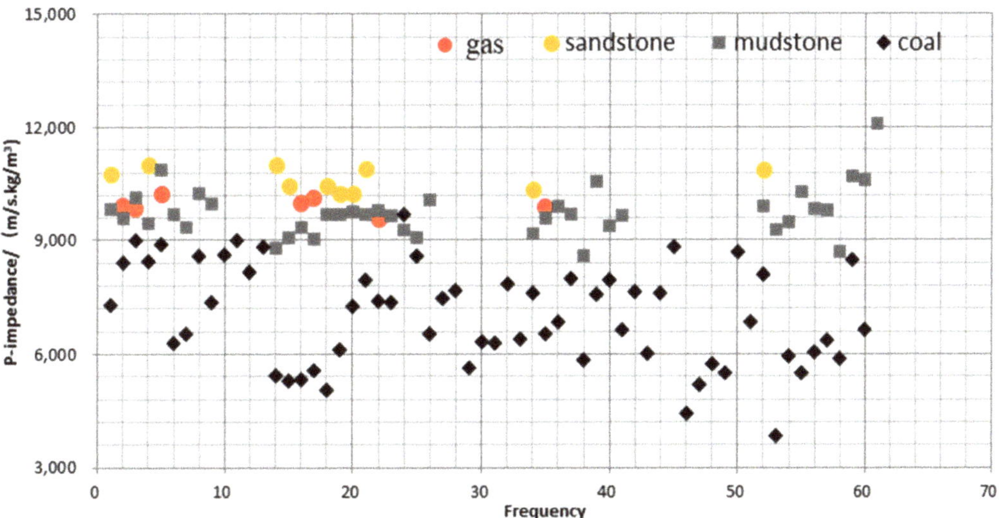

Figure 3. P-impedance distribution characteristics of different lithologies.

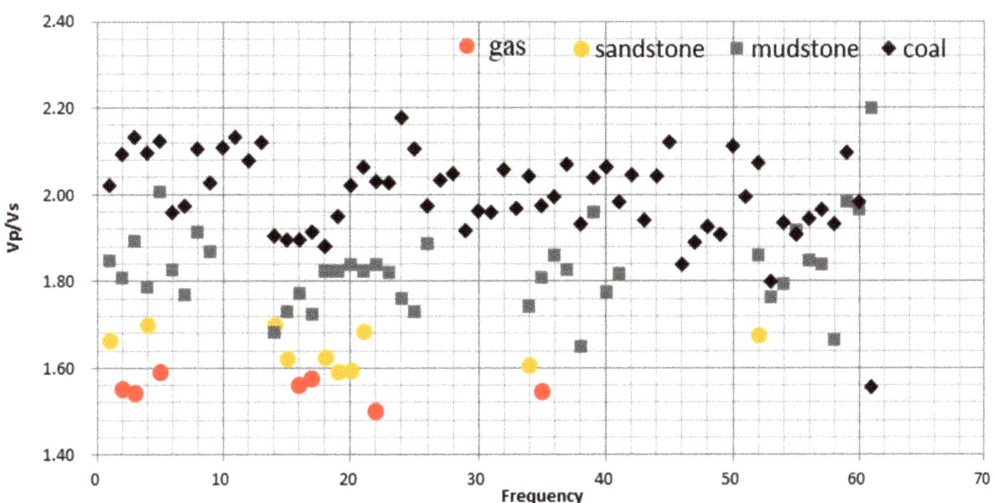

Figure 4. Distribution characteristics of the velocity ratios of compressional and shearing waves for different lithologies.

3.2. Analysis of Factors That Influence Seismic Amplitude

Based on the above petrophysical analysis, the coal seams are characterized by low P-wave impedance and a strong-amplitude trough response in post-stack seismic data. In order to analyze the influence of thin coal seams on the sandstone reservoir, forward models of sand–coal coupling with different numbers of coal seams and seam spacing are designed: the model parameters are shown in Table 1. The seismic forward models of coal seams and sand–coal coupling are compared (Figure 5). Both the coal seams and the gas-bearing sandstone mostly show the characteristics of strong trough bright spots on post-stack seismic data, while a small number shows the characteristics of weak amplitude response due to the influence of coal seam spacing. With the same coal seam spacing, the reflective amplitude of both coal seams and gas-bearing sandstone-top surfaces are stronger the larger the number of coal seams is. With the same number of coal seams, the reflective amplitude of the sandstone-top surfaces gradually increases when the sand–coal spacing decreases. The above analysis suggests that the seismic reflection from the sandstone top is influenced by both the number and spacing of coal seams. It is difficult to analyze the seismic response of sand and coal assemblages on post-stack seismic data because the distribution of coal seams, which is affected by tides, is random in number and spacing.

Table 1. Model parameters of thin gas-bearing sandstone with different AVO types.

Lithology	P Wave Velocity (m/s)	S Wave Velocity (m/s)	Density (g/cm^3)	Vp/Vs
Mudstone	4150	2220	2.63	1.87
Sandstone	4027	2430	2.43	1.65
Coal seam	2700	1350	1.90	2.0

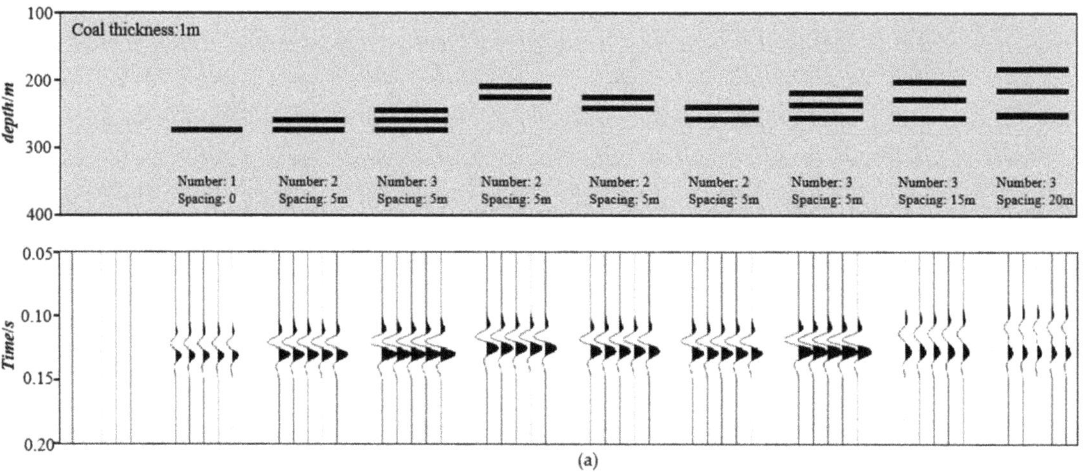

Figure 5. *Cont.*

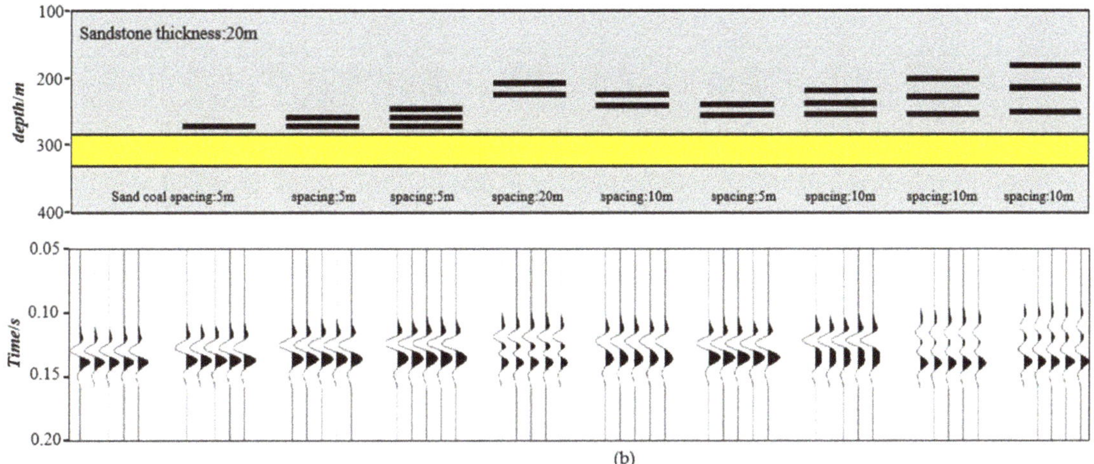

Figure 5. Forward models with coal seams and sand–coal assemblages. (**a**) Geological models with only coal seams; (**b**) geological models with both coal seams and sandstone (without coal seams, the sandstone-top surface shows trough reflection, while the sandstone–coal assemblage may show stronger or weaker amplitude and the response of sandstone cannot be recognized).

3.3. A Method to Suppress Coal Seam Interference Based on Partial Stacked Seismic Analysis

Given that the seismic response characteristics of sandstone–coal assemblages are complex, the response characteristics of coal seams and sandstone cannot be effectively distinguished by conventional seismic data alone, and other means of analysis are required. The statistics for all coal seams and gas-bearing sandstone in the study area (Tables 2 and 3) show that most of the coal seams exhibit low P-impedance, positive gradient and IV AVO type, while gas-bearing sandstone exhibits low P-impedance, negative gradient and II-III AVO type. In the near-gather seismic data, the difference between gas-bearing sandstone and coal seams is not obvious, which is the fundamental reason why their responses cannot be effectively distinguished using only post-stack seismic analysis. However, in the far-gather seismic data, coal seams gradually decrease in energy, while gas-bearing sandstone gradually increases in energy. Thus, the influence of coal seams can be eliminated through the far-gathers of partial post-stack seismic analysis.

Table 2. Statistics concerning elastic parameters and AVO characteristics of drilled coal seams in the Pingbei slope belt.

Lithology	P-Wave Velocity (m/s)	S-Wave Velocity (m/s)	Density (g/cm^3)	Intercept	Gradient	AVO Type
Mudstone	4150	2220	2.63	—	—	—
	2700	1350	2.00	−0.3477	0.4657	IV
	2639	1300	2.13	−0.3276	0.4525	IV
	2900	1378	2.10	−0.2894	0.4271	IV
	2698	1290	2.15	−0.3125	0.4504	IV
	2617	1300	2.14	−0.3293	0.4503	IV
	2747	1362	1.90	−0.3646	0.4873	IV
Coal	2956	1605	2.14	−0.2708	0.3237	IV
seam	2801	1600	2.00	−0.3301	0.3625	IV
	2900	1627	2.02	−0.3085	0.3461	IV
	2760	1529	1.80	−0.3885	0.4535	IV
	2849	1600	2.18	−0.2794	0.3124	IV
	2736	1654	2.07	−0.3245	0.3154	IV
	2619	1344	1.75	−0.4271	0.5417	IV
	2895	1337	1.80	−0.3655	0.5192	IV

Table 3. Statistics concerning elastic parameters and AVO characteristics of drilled sandstone in the Pingbei slope belt.

Lithology	P-Wave Velocity (m/s)	S-Wave Velocity (m/s)	Density (g/cm^3)	Intercept	Gradient	AVO Type
Mudstone	4150	2220	2.63	-	-	-
	4027	2518	2.43	−0.0546	−0.1309	II
	4114	2492	2.42	−0.0459	−0.1004	II
	4300	2402	2.39	−0.0115	−0.0414	II
	4273	2517	2.44	−0.0229	−0.0966	II
	4080	2485	2.47	−0.0399	−0.1148	III
	4361	2743	2.49	−0.0026	−0.2247	III
Sandstone	4136	2492	2.40	−0.0474	−0.0919	II
	4062	2575	2.45	−0.0461	−0.1643	III
	4000	2415	2.41	−0.0621	−0.0708	II
	4230	2473	2.46	−0.0239	−0.0838	II
	4157	2480	2.38	−0.0491	−0.0769	II
	4235	2550	2.42	−0.0314	−0.1151	III
	4070	2515	2.39	−0.0575	−0.1117	III

We simulated pre-stack forward modeling using the wave equation method in the frequency domain because thin coal seam assemblages do not meet the semi-infinite-space hypothesis condition of the Zoeppritz equation. The acoustic wave equation in the time domain can be expressed as

$$\frac{\partial^2 u(x,z,t)}{\partial x^2} + \frac{\partial^2 u(x,z,t)}{\partial z^2} - \frac{1}{v(x,z)^2}\frac{\partial^2 u(x,z,t)}{\partial t^2} = -f(x,z,t) \tag{1}$$

where $v(x,z)$ is the wave velocity in the media, which is a function of position when the media is anisotropic. f is the source function, which is usually set as Ricker wavelet. Take the Fourier transform of t to transform Equation (1) into the frequency domain. $u(x,z,t)$ is transformed into $u(x,z,\omega)$, and $f(x,z,t)$ is transformed into $f(x,z,\omega)$. Therefore, Equation (1) is transformed into

$$\frac{\partial^2 u(x,z,\omega)}{\partial x^2} + \frac{\partial^2 u(x,z,\omega)}{\partial z^2} - \frac{\omega^2}{v^2}u(x,z,\omega) = -f(x,z,\omega) \tag{2}$$

The study area is grided according to a geological model. A finite-difference operator is used to discretize the continuity equation at each grid element. The difference equation is solved, and an approximation of the solution at each grid element is obtained. Surface receivers are designed based on an actual observation system so that the real seismic reflection records can be obtained. Figure 6 shows a comparison of the forward modeling results as obtained by the Zoeppritz equation and the wave equation. The amplitude energy of coalbed-top surface decreases with an offset in the wave equation model, whereas it increases with an offset in the Zoeppritz equation model. These results indicate the wave equation model agrees better with the actual seismic response of coal beds than the Zoeppritz equation model.

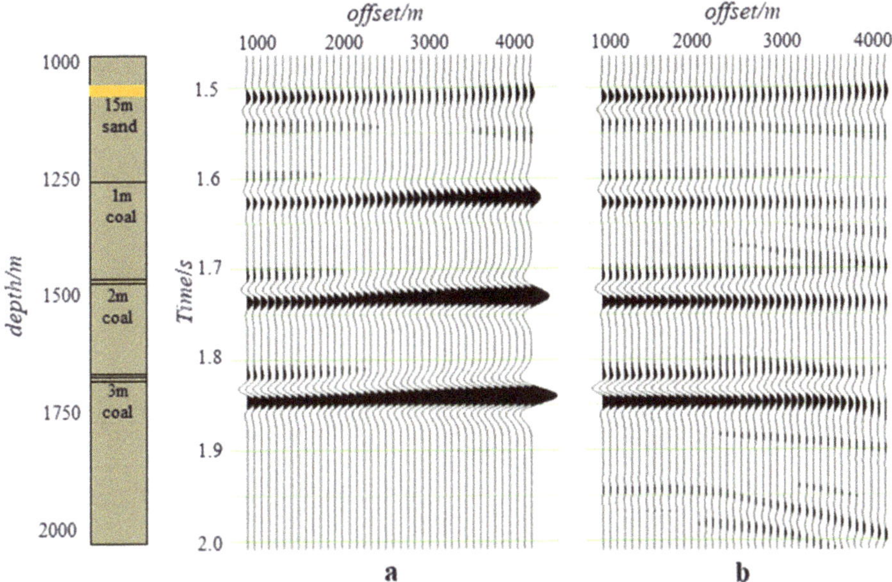

Figure 6. Comparison of the forward modeling results obtained using the Zoeppritz equation and the wave equation. (**a**) Zoeppritz equation modeling result; (**b**) wave equation modeling result.

In order to confirm that the far angle partial stacked seismic data can eliminate the interference of coal seams, forward models with different sand–coal spacing are designed. The parameters of the models are the same as in Table 1, where the thickness of each coal seam is 1 m, and the sand–coal spacing gradually changes from 1 m to 80 m. The finite-difference wave equation forward modeling method, using the aforementioned frequency domain and a Ricker wavelet of 25 Hz, is employed. The partial stacked seismic data with different angles (Figures 7 and 8) show that, in the near-gather stack seismic data, the trough energy of the sandstone top is enhanced by the influence of the coal seam, and the amplitude of the gas-containing sandstone increases by 1.7 times relative to the sandstone without a coal seam. With the increase in the partial stack angle, the influence of the coal seam on the amplitude of the sandstone-top surface becomes weaker and weaker. The amplitude of the sandstone with a coal seam is the same as that of the sandstone without a coal seam in the far-gather seismic data. Based on the above analysis, it is believed that a far-angle partial stacked seismic analysis can better eliminate the influence of the coal seam and improve the accuracy of sandstone prediction.

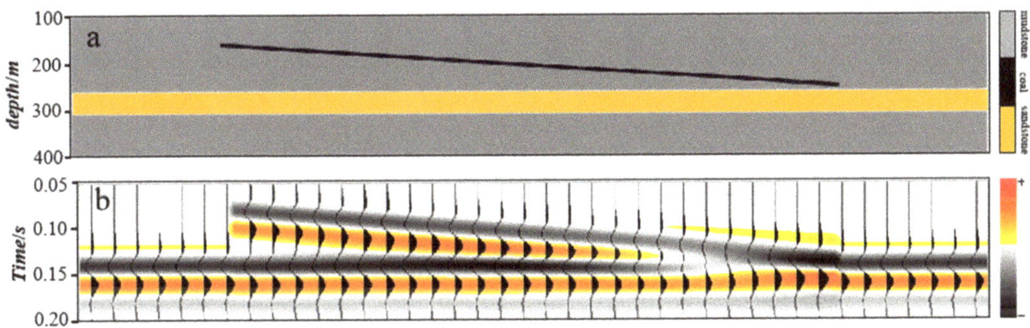

Figure 7. *Cont.*

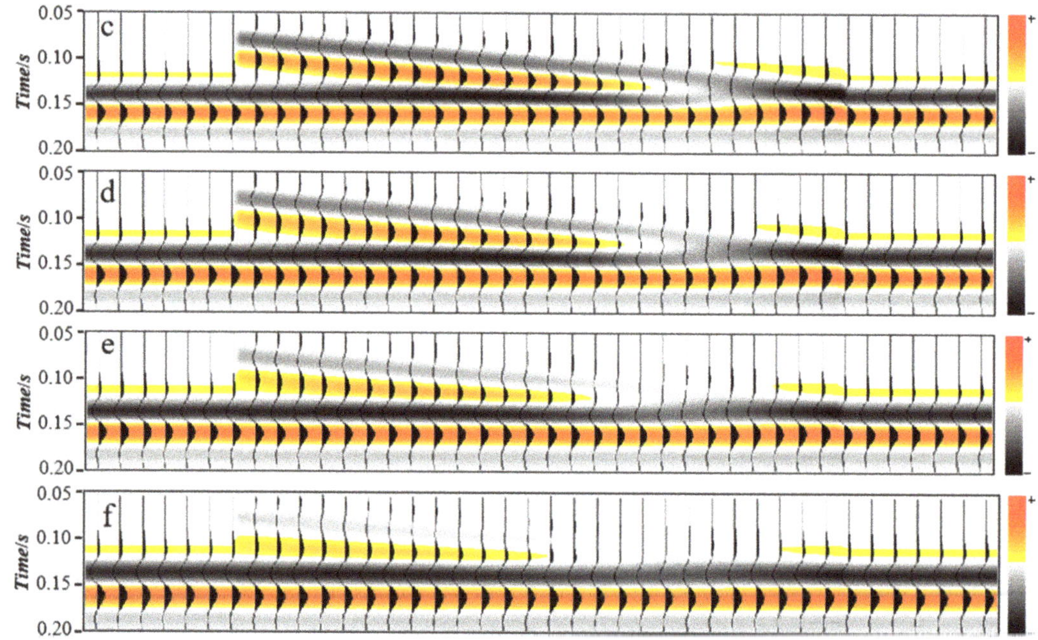

Figure 7. Seismic response of the partial stacked seismic analysis for sandstone–coal assemblage at different stack angles: (**a**) geologic model; (**b**) partial stacked seismic analysis 0–7°; (**c**) partial stacked seismic analysis 7–14°; (**d**) partial stacked seismic analysis 14–21°; (**e**) partial stacked seismic analysis 21–28°; (**f**) partial stacked seismic analysis 28–35°.

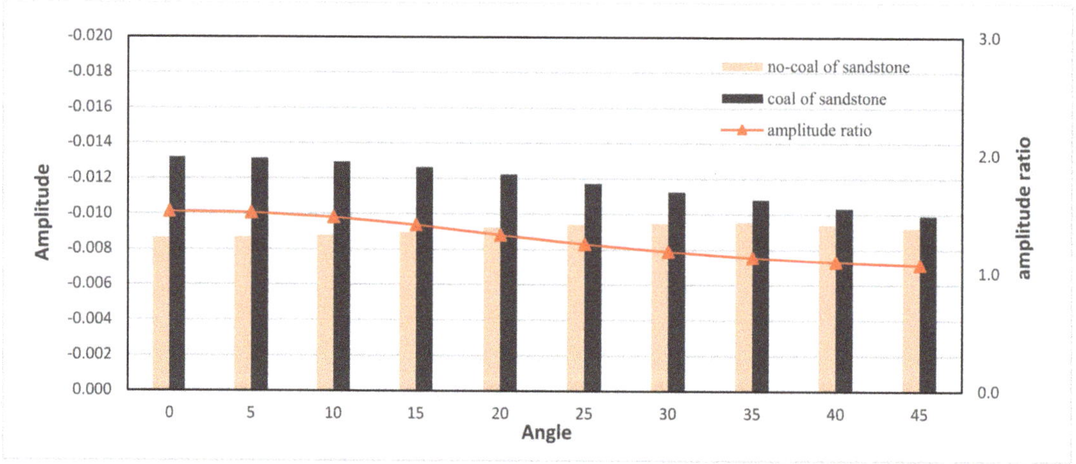

Figure 8. Partial stacked seismic amplitude of sandstone-top surface at different stack angles.

4. Results

Fault-block traps, which are dominated by faults, and structure-lithostratigraphic traps, which are dominated by faults and channel sands, are the main trap types in the study area, so it is important to recognize lithologic boundaries accurately. Sand–coal and mud–coal assemblages have been encountered during the drilling of many wells in the Pinghu Formation. Both the sand–coal and mud–coal assemblages show strong-amplitude bright spot reflections in seismic analyses due to the effect of coal seams. Consider the seismic profile across Wells X-2 and X-3 as an example (Figure 9a). The P4 layer of Well X-2 is mainly characterized by sand–coal coupling and shows strong-amplitude trough reflection in its seismic profile. The P8 layer of Well X-2 is mainly mud–coal coupling and shows strong-amplitude trough reflection. It is difficult to identify the distribution of sandstone channels in post-stack seismic analysis.

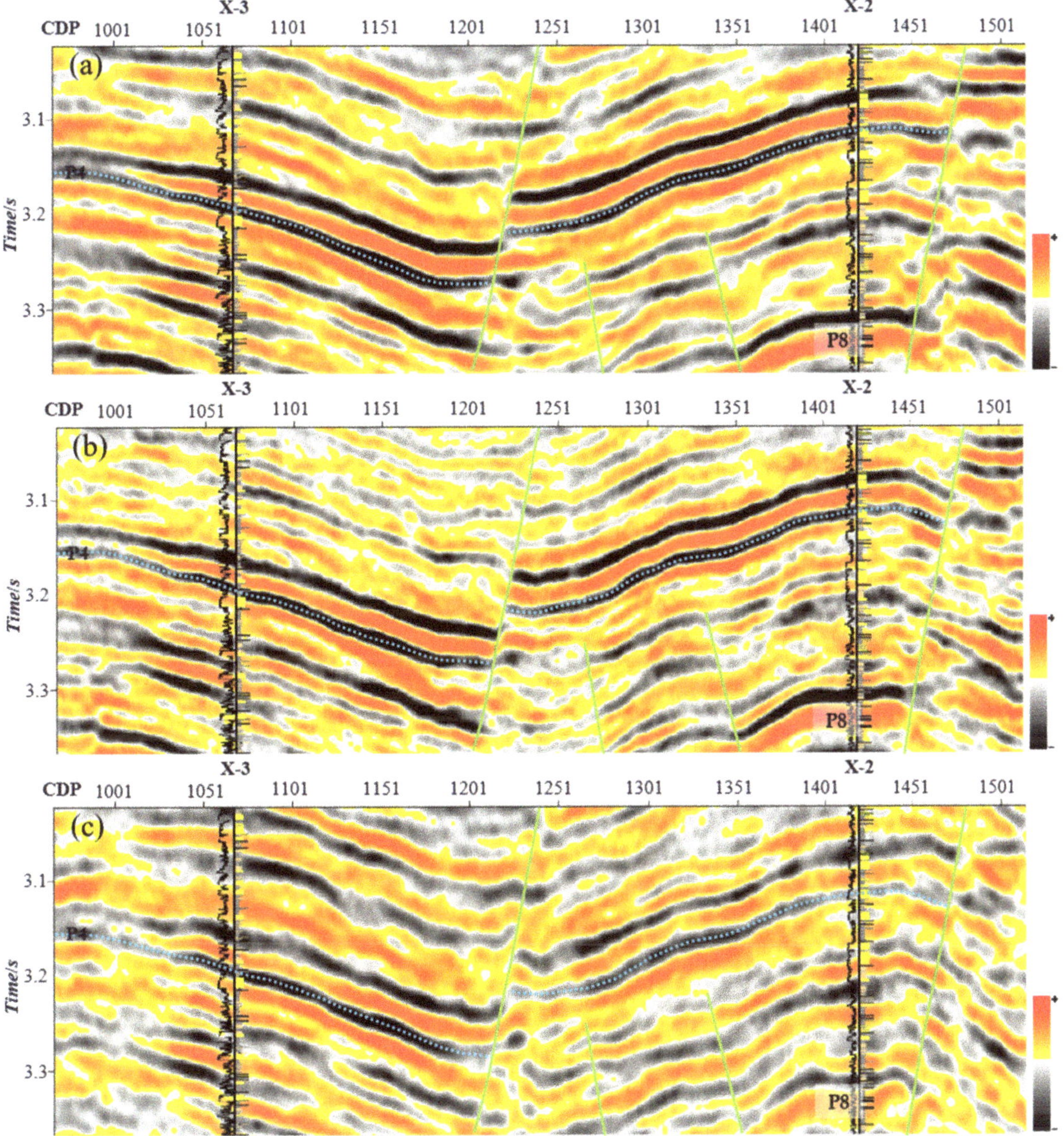

Figure 9. *Cont.*

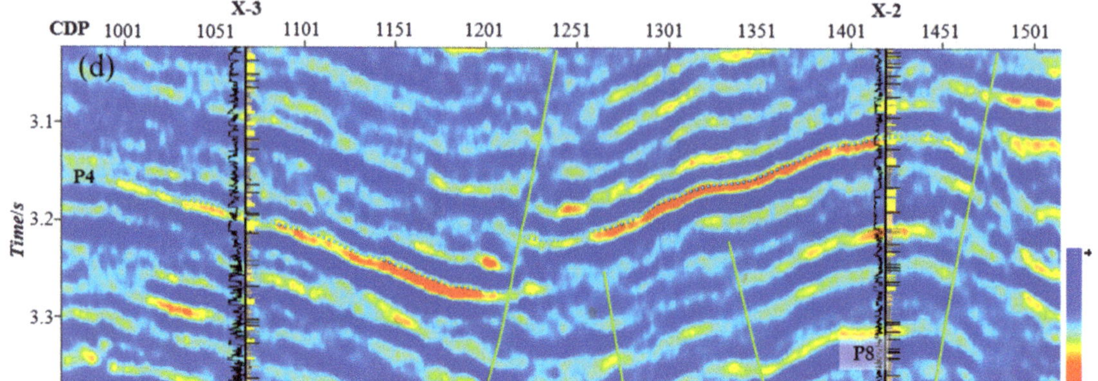

Figure 9. Post-stack and partial stacked seismic profiles across Wells X-2 and X-3. For locations of the profiles, see Figure 1. (**a**) The post-stack seismic profile; (**b**) the near offset partial stacked seismic profile; (**c**) the far-offset partial stacked seismic profile; (**d**) P + G hydrocarbon detection profile.

In order to eliminate the influence of the coal seam, the partial stack seismic profiles at different angles are analyzed (Figure 9b,c). It is found that the coal seam exhibits a very low P-impedance and a strong-amplitude reflection characteristic on the near-gather seismic data, which is a similar result to that of the post-stack seismic data, and is not conducive to a fine characterization of the sandstone. However, in the far-offset partial stack seismic profile, the coal seam shows a class-IV AVO signature with a strong amplitude in the near angle and a weak amplitude in the far angle, so the amplitude of the mudstone–coal stratigraphic assemblage (P8 in Well X-2) is significantly weakened in the far angle gather and in the P + G hydrocarbon detection profile (Figure 9d). In contrast, the sand–coal stratigraphic assemblage (P4 in Well X-2) shows a strong-amplitude reflection signature in the far-offset partial stack seismic profile and exhibits a stream channel in the downward direction of the strong trough. Moreover, gas-containing characteristics can be found in P4 in the P + G hydrocarbon detection profile (Figure 9d). Weak amplitude in the far-offset partial stack seismic analysis should be considered a predictor of a coal seam, while strong amplitude in the far-offset partial stack seismic analysis should be considered a predictor of gas-containing sandstone. The actual drilling result is reliably consistent with the predictions. The sand–coal assemblage can show either a strong or a weak amplitude in the minimum amplitude attribute of the P4 layer due to variation in sand–coal spacing and in the number and thickness of coal seams, and a large number of false bright spot reflections can also be seen (Figure 10a). In the minimum amplitude attribute of the far-offset partial stack seismic analysis, the false bright spot reflections have been effectively eliminated, and the pattern of braided channels is more clearly defined. Thus, it is evident that the far-offset partial stack seismic analysis is of great benefit in advanced reservoir prediction and description.

We evaluated a structure-lithostratigraphic trap and designed Well X-3 in the footwall based on the temporal and spatial relationships between the braided channels and faults. A 22 m thick gas-containing layer with 12.6% porosity in P4 was encountered by Well X-3, and the proven reserve of natural gas in the P4 layer increased by about 1.4 billion cubic meters. Three wells were successfully drilled, and the drilling success rate improved from 50% to 80% thanks to the method introduced in this paper. The far-gather seismic attribute was applied successfully and was proven helpful in the exploration of low-porosity, low-permeability reservoirs in the coal-bearing strata in the study area.

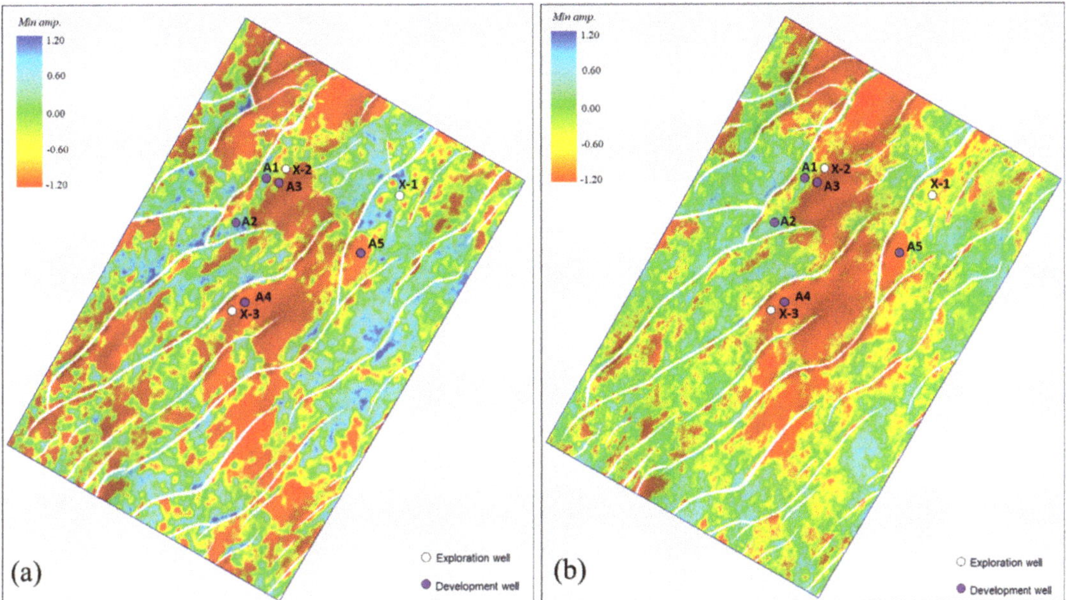

Figure 10. Minimum amplitude attribute of the P4 layer in the study area: (**a**) the minimum amplitude of post-stack seismic analysis; (**b**) the minimum amplitude of far-offset partial stacked seismic analysis.

5. Discussion

Although the interference of coal seams can be effectively reduced by far-offset partial stacked seismic analysis, multi-layer coal seams will cause residual energy illusions in far-gather analysis. In order to exclude the energy abnormality illusion and to demonstrate the applicability of the method described in this paper, we designed a pre-stack forward model of a multi-layer coal seam. The model uses a horizontally layered medium. Three types of strata are simulated: sandstone, a sandstone–coal assemblage (the thicknesses of the three coal seams are 1.0 m, 0.5 m and 1.0 m, and the distance between sandstone and coal is 5 m) and a coal seam (Figure 11).

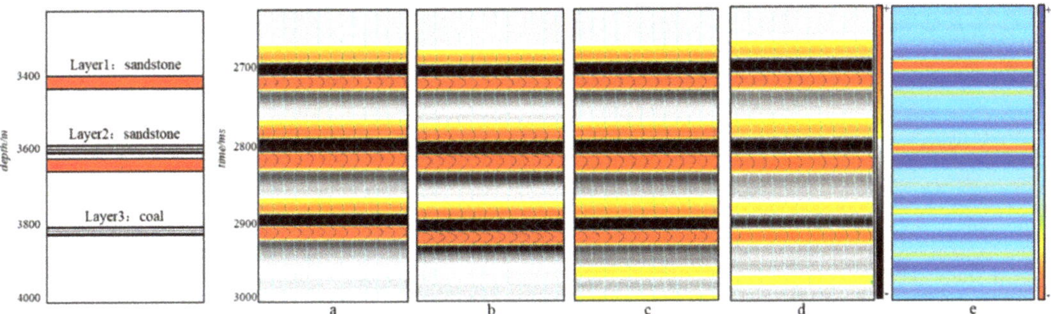

Figure 11. Seismic response of forward models with a sandstone–coal assemblage at different stack angles: (**a**) post-stack seismic analysis; (**b**) partial stacked seismic analysis 1–15°; (**c**) partial stacked seismic analysis 15–30°; (**d**) partial stacked seismic analysis 30–45°; (**e**) P + G hydrocarbon detection profile.

The simulation results are shown in Figure 10. The sandstone model exhibits a class-IIb AVO feature, and the trough energy increases as the offset angle increases. The sandstone–

coal assemblage model exhibits a class-IV AVO feature, and its trough energy and frequency are lower than the sandstone model. The coal seam model exhibits a typical class-IV AVO feature, and the trough energy decreases as the offset angle increases. There is a residual energy illusion in the far-gather of the coal seam model, which disturbs the prediction of sandstone, and the greater the thickness of coal seam, the stronger the interference. We analyzed the P + G attribute to exclude the energy abnormality of the multi-layer coal seam. The coal seam shows high Vp/Vs in the P + G attribute, whereas sandstone shows low Vp/Vs. Therefore, we can suppress the coal-bearing interference and improve the drilling success rate using the far-gather seismic attribute integrated with the P + G attribute.

6. Conclusions

We investigated the seismic response of sandstone–coal assemblages, both by analyzing the petrophysical and AVO characteristics of sandstone and coal seams and by forward modeling sandstone–coal assemblages. A method of suppressing coal seam interference to improve predictions of reservoirs using far-gather seismic attributes has been proposed. Based on these analyses and a successful case study, we draw the following conclusions:

(1) Coal seams are widespread within the Pinghu Formation and have small thicknesses and low impedance characteristics. P-wave impedance can only be used to detect coal seams, whereas Vp/Vs can be used to detect sandstone reservoirs.

(2) The finite-difference acoustic wave equation in the frequency domain can effectively eliminate the interference of the far-offset waveform distortion that is typical of coal seams, and the simulation results are consistent with the actual seismic response characteristics. The forward modeling of sandstone–coal assemblages indicates that the responses of coal seams and sandstone cannot be distinguished by post-stack seismic data, but the interference of coal seams can be eliminated by far-offset partial stacked seismic analysis.

(3) The far-gather seismic attribute has been applied successfully in reservoir prediction and sand body description in the study area, which contributed to the successful exploration of coal-rich strata in the Pingbei slope belt.

(4) The value of popularizing the method introduced in this paper has been proven in both the ECSSB and the Ordos Basin, where the accuracy of reservoir prediction in coal-bearing strata has been improved.

Author Contributions: Methodology, X.Z. and L.D.; software, C.Y.; investigation, Y.M.; data curation, R.Z. and Y.L.; writing—original draft preparation, Y.M.; writing—review and editing, X.W.; visualization, M.L. All authors have read and agreed to the published version of the manuscript.

Funding: This research received no external funding.

Institutional Review Board Statement: Not applicable.

Informed Consent Statement: Not applicable.

Data Availability Statement: Data available from the authors upon request.

References

1. Abbas, A.; Zhu, H.T.; Zeng, Z.W.; Zhou, X.H. Sedimentary facies analysis using sequence stratigraphy and seismic sedimentology in the Paleogene Pinghu Formation, Xihu Depression, East China Sea Shelf Basin. *Mar. Pet. Geol.* **2018**, *93*, 289–297. [CrossRef]
2. Lei, C.; Yin, S.Y.; Ye, J.R.; Wu, J.F. Geochemical Characteristics and Hydrocarbon Generation History of Paleocene Source Rocks in Jiaojiang Sag, East China Sea Basin. *Editor. Comm. Earth Sci. J. China Univ. Geosci.* **2021**, *46*, 3575–3587.
3. Liu, J.S.; Zhang, S.P. Natural gas migration and accumulation patterns in the central-north Xihu Sag, East China Sea Basin. *Nat. Gas Geosci.* **2021**, *32*, 1163–1176.
4. Liu, J.S.; Li, S.X.; Qin, L.Z.; Yi, Q.; Chen, X.D.; Kang, S.L.; Shen, W.C.; Shao, L.Y. Hydrocarbon generation kinetics of Paleogene coal in Xuhu sag, East China Sea Basin. *Acta Pet. Sin.* **2020**, *41*, 1174–1187.
5. Su, A.; Chen, H.H.; Wu, Y.; Lei, M.Z.; Li, Q.; Wang, C.W. Genesis, origin and migration-accumulation of low-permeable and nearly tight-tight sandstone gas in the central western part of Xihu sag, East China Sea Basin. *Acta Pet. Sin.* **2018**, *92*, 184–196.

6. Diessel, C.F. *Coal-Bearing Depositional Systems*; Springer: Berlin/Heidelberg, Germany, 1992.

7. Zou, F.; Xue, Y.J. Strong amplitude suppression of coal seam based on synchrosqueezed wavelet transform. *Prog. Geophys.* **2018**, *33*, 1198–1204.

8. Liu, L.; Zhang, Q.; Zhang, J.H.; Ban, L.; Li, J.L. Application of the Strong Shielding Peeling Technique Based on Matching Pursuit Algorithm in the Reservoir Prediction of Fan 159 Well Block. *Comput. Tomogr. Theory Appl.* **2016**, *25*, 331–337.

9. Liu, J.; Zhang, Z.T.; Liu, D.L. Sediment boundary identification and fluid detection for the seismic data with strong background reflections. *Geophys. Prospect. Pet.* **2016**, *55*, 142–149.

10. Han, W.G.; Zhang, J.G. Theoretical study on characteristic of weak signal and its identification. *Oil Geophys. Prospect.* **2011**, *46*, 232–236.

11. Ping, A. Application of Multi-Wavelet Seismic Trace Decomposition and Reconstruction to Seismic Data Interpretation and Reservoir Characterization. In Proceedings of the SEG/New Orleans Annual Meeting. 2006. Available online: https://onepetro.org/SEGAM/proceedings-abstract/SEG06/All-SEG06/SEG-2006-0973/93194?redirectedFrom=PDF (accessed on 10 January 2022).

12. Ping, A. Case Studies on Stratigraphic Interpretation and Sand Mapping Using Volume-Based Seismic Waveform Decomposition. In Proceedings of the SEG/New Orleans Annual Meeting. 2006. Available online: https://library.seg.org/doi/10.1190/1.2370307 (accessed on 10 January 2022).

13. Wang, X.X.; Yu, S.; Li, S.; Zhang, N.D. Two parameter optimization methods of multi-point geostatistics. *J. Petrol. Sci. Eng.* **2022**, *208*, 109724. [CrossRef]

14. Zhang, Y.Y.; Wei, X.W.; Tan, M.Y.; Gao, Q.J.; Zhu, D.R.; Li, S.X. Removal of seismic strong shield interface based on compressed sensing technology and its application. *Lithol. Reserv.* **2019**, *31*, 85–91.

15. Gu, W.; Zhang, X.; Xu, M.; Liang, H.; Zhang, D.J.; Luo, J.; Zheng, H. High precision prediction of thin reservoir under strong shielding effect and its application: A case study from Sanzhao Depression, Songliao Basin. *Geophys. Prospect. Pet.* **2017**, *56*, 439–448.

16. Wan, L.; Hurter, S.; Bianchi, V.; Li, P.; Wang, J.; Salles, T. The roles and seismic expressions of turbidites and mass transport deposits using stratigraphic forward modeling and seismic forward modeling. *J. Asian Earth Sci.* **2022**, *229*, 105110. [CrossRef]

17. Tomassi, A.; Trippetta, F.; Franco, R.; Ruggieri, R. From petrophysical properties to forward-seismic modeling of facies heterogeneity in the carbonate realm (Majella Massif, central Italy). *J. Pet. Sci. Eng.* **2022**, *211*, 110242. [CrossRef]

18. Wang, X.X.; Hou, J.G.; Li, S.H.; Dou, L.X.; Song, S.H.; Kang, Q.Q.; Wang, D.M. Insight into the nanoscale pore structure of organic-rich shales in the Bakken Formation, USA. *J. Pet. Sci. Eng.* **2019**, *176*, 312–320. [CrossRef]

19. Juan, C.M.; Ghisays, A.; Montes, L. AVO analysis with partial stacking to detect gas anomalies in the GÜEPAJÉ-3D project. *Geofís. Int.* **2013**, *52*, 249–260.

20. Ismail, A.; Ewida, H.F.; Al-Ibiary, M.G.; Zollo, A. Application of AVO attributes for gas channels identification, West offshore Nile Delta, Egypt. *Pet. Res.* **2020**, *5*, 112–123. [CrossRef]

21. Farfour, M.; Foster, D. New AVO expression and attribute based on scaled Poisson reflectivity. *J. Appl. Geophys.* **2021**, *185*, 104255. [CrossRef]

22. Yi, B.Y.; Lee, G.H.; Horozal, S. Qualitative assessment of gas hydrate and gas concentrations from the AVO characteristics of the BSR in the Ulleung Basin, East Sea (Japan Sea). *Mar. Pet. Geol.* **2011**, *28*, 1953–1966. [CrossRef]

23. Wang, X.C.; Pan, D.Y. Application of AVO attribute inversion technology to gas hydrate identification in the Shenhu Area, South China Sea. *Mar. Pet. Geol.* **2017**, *80*, 23–31. [CrossRef]

24. Liu, L.; Liu, L.H.; Wo, Y.J.; Sun, W. Pre-stack elastic parameter inversion of ray parameters. *J. Appl. Geophys.* **2019**, *162*, 13–21. [CrossRef]

25. Jiang, X.D.; Cao, J.X.; Hu, J.T. Pre-stack gather optimization technology based on an improved bidimensional empirical mode decomposition method. *J. Appl. Geophys.* **2020**, *177*, 104026.

26. Li, C.Y.; Wei, L.; Diao, H.; Cheng, X.; Hou, D.J. Hydrocarbon source and charging characteristics of the Pinghu Formation in the Kongqueting Structure, Xihu Depression. *Pet. Sci. Bull.* **2021**, *6*, 196–208.

27. Li, J.W.; Jiang, B.; Qu, Z.H.; Yin, S.; Xu, J.; Li, P. Tectonic evolution and control of coal in Donghai Xihu Sag. *Coal Geol. Explor.* **2016**, *44*, 22–27.

28. Ding, F.; Liu, J.S.; Jiang, Y.M.; Zhao, H.; Yu, Z.K. Source and migration direction of oil and gas in Kongqueting area, Xihu Sag, East China Sea Shelf Basin. *Mar. Geol. Quat. Geol.* **2021**, *41*, 156–165.

29. Li, J.J.; Jiang, Y.M.; Hou, G.W.; Xie, J.J.; Jiang, X. Constraints of slope break belt on oil and gas trapping-An example from the Pinghu Formation in the Kongqueting area of Pinghu Slope. *Mar. Geol. Quat. Geol.* **2021**, *41*, 141–150.

30. Zhou, X.H.; Gao, S.L.; Gao, W.Z.; Li, N. Formation and distribution of marine-continental transitional lithologic reservoirs in Pingbei slope belt, Xihu sag, East China Sea Shelf Basin. *China Pet. Explor.* **2019**, *24*, 153–164.

 *energies*

MDPI

Article

Geological Characteristics and Development Techniques for Carbonate Gas Reservoir with Weathering Crust Formation in Ordos Basin, China

Haijun Yan [1,2], Ailin Jia [2], Jianlin Guo [2], Fankun Meng [3,*], Bo Ning [2] and Qinyu Xia [2]

[1] School of Energy Resource, China University of Geosciences (Beijing), Beijing 100083, China; yhj010@petrochina.com.cn
[2] Research Institute of Petroleum Exploration & Development, PetroChina, Beijing 100083, China; jal@petrochina.com.cn (A.J.); guojianl@petrochina.com.cn (J.G.); ningbo07@petrochina.com.cn (B.N.); xiaqy-ordos@petrochina.com.cn (Q.X.)
[3] School of Petroleum Engineering, Yangtze University, Wuhan 430100, China
* Correspondence: mengfk09021021@163.com

Abstract: The carbonate gas reservoir is one of the most important gas formation types; it comprises a large proportion of the global gas reserves and the annual gas production rate. However, a carbonate reservoir with weathering crust formation is rare, and it is of significant interest to illustrate the geological characteristics of this kind of formation and present the emerging problems and solution measures that have arisen during its exploitation. Therefore, in this research, a typical carbonate gas reservoir with weathering crust formation that is located in Ordos Basin, China, was comprehensively studied. In terms of formation geology, for this reservoir, the distribution area is broad and there are multiple gas-bearing layers with low abundance and strong heterogeneity, which have led to large differences in gas well production performance. Some areas in this reservoir are rich in water, which seriously affects gas well production. Regarding production dynamics, the main production areas in this gas reservoir have been stable on a scale of 5.5 billion cubic meters for more than a decade, and the peripheral area has been continually evaluated to improve production capacity. Nevertheless, after decades of exploration and development, the main areas of this reservoir are faced with several problems, including an unclear groove distribution, an unbalanced exploitation degree, low formation pressure, and increases in intermittent gas wells. To deal with these problems and maintain the stability of gas reservoir production, a series of technologies have been presented. In addition, several strategies have been proposed to solve issues that have emerged during the exploration and exploitation of peripheral reservoir areas, such as low-quality formation, unclear ancient land and complex formation-water distribution. These development measures employed in the carbonate gas reservoir with weathering crust formation in the Ordos Basin will surely provide some guidance for the efficient exploitation of similar reservoirs in other basins all over the world.

Keywords: Ordos Basin; carbonate gas reservoir; weathering crust formation; geological characteristics; development technologies

Citation: Yan, H.; Jia, A.; Guo, J.; Meng, F.; Ning, B.; Xia, Q. Geological Characteristics and Development Techniques for Carbonate Gas Reservoir with Weathering Crust Formation in Ordos Basin, China. *Energies* 2022, 15, 3461. https://doi.org/10.3390/en15093461

Academic Editor: Prabir Daripa

Received: 4 April 2022
Accepted: 5 May 2022
Published: 9 May 2022

Publisher's Note: MDPI stays neutral with regard to jurisdictional claims in published maps and institutional affiliations.

1. Introduction

Carbonate reservoirs play an important role in the global oil and gas industry. They comprise about 72% of global oil and gas reserves, and nearly 60% of global oil and gas production is produced from carbonate reservoirs [1–4]. In China, carbonate reservoirs comprise nearly 30% of natural gas reserves and 20% of gross gas production, and they have an important position in the supplement of natural gas. There are three basins in China that have large-scale carbonate reservoirs: the Tarim Basin (e.g., Tazhong and Lunnan), the Sichuan Basin (e.g., Gaoshiti, Moxi, Eastern Sichuan, Longgang, Yuanba, and Puguang), and the Ordos Basin (e.g., Jingbian and Gaoqiao). Globally, basins with giant carbonate gas reservoirs include the Persian Gulf Basin (e.g., North Field, Pars South, and Northwest Dome), the Pre-Caspian Basin (Astrakhan and Karachaganak), and the Zagros basin (Rag-E-Safid) [2]. These carbonate gas reservoirs can be divided into four categories depending on formation type: the fractured-vuggy reservoir, reef flat reservoir, bedded dolomite reservoir, and weathering crust reservoir [5–10]. The first three reservoirs are found all over the world and have been studied sufficiently, but the weathering crust carbonate reservoir is raw and has not been deeply investigated. Therefore, in this research, we studied the geological characteristics and optimal development strategies for a typical carbonate gas reservoir with weathering crust formation. This study can provide some insights into the exploration and development of similar carbonate gas reservoirs, such as the tight carbonate reservoirs in the north of Iraq, which possess Turonian–Campanian Kometan formations with low porosity and permeability [11].

Structures for weathering crust formation can be qualitatively separated according to color, mineral features, core properties, and other chemical indicators [12–16]. There are two main division schemes for this kind of formation: one identifies different layers, such as iron crust, fracture, lamination, and sandy weathering layers, and the other scheme identifies layers with different kinds of weathering zones, such as severely, weakly, and slightly weathered zones [13,16]. To date, many investigators have conducted studies regarding different kinds of formations with weathered crust features. Tian et al. [17] proposed a multi-layer artificial intelligence workflow to map the seismic attributes and represent the dissolution values of volcanic weathered crust formations. Zhu et al. [18] analyzed the vertical structure characteristics of granite weathering crusts for reservoirs in the western segment of the northern belt of Dongying Sag, Bohai Bay Basin, China, which possesses sandstone formations. Sidorova et al. [19] reported that the widespread weathering crust of crystalline basements can be used to study the mineral formation process of ancient weathering crust, though they did not study formation characteristics. However, these studies mainly focused on volcanic or sandstone reservoirs with weathering crust formation, and there has been a lack of the research regarding carbonate reservoirs.

There have been many geologic studies of the targeted gas reservoir, Jingbian gas field. Influenced by environmental variation at the end of the Middle Ordovician, the Ordos Basin was uplifted by the Caledonian movement and experienced 130–150 million years of weathering and erosion. In this period, the topography of the Jingbian platform was a large karst slope, in which the western area was higher than the eastern area and the surface water flowed from west to east. During this period, the effects of supergene leachate karsts were strong, which led to the formation of dendrite erosion grooves. In addition, the continuous replenishment of atmospheric and acidic aquifers caused the expansion of micro-cracks, solution pores, and intergranular solution pores, thus leading to the formation of large-area and layered area with porphyritic and honeycomb solution pores, as well as intensive weathering fractures and mechanical crusting fractures [20–22]. To determine the rock type, sedimentary characteristics, and environments of subsections 1 and 2 of member 5 in the Majiagou Formation, Xu et al. [23] distinguished sedimentary microfacies via geological laboratory analysis. The origins and gas sources of Ordovician paleo-weathering crust reservoirs can be determined with geochemical gas evidence, such as the carbon isotope reversal for the Ordos Basin [24,25]. To reconstruct the paleo-geomorphology of the weathering crust from the end of the Ordovician in the eastern part of the Ordos Basin, Wei et al. [26] studied paleo-geomorphic characteristics, the thickness of residual strata, and paleo-karsts. Li et.al [27] found that gases were accumulated in stratigraphic traps related to karst paleo-geomorphology and lithologic traps associated with the late diagenetic features of carbonate rocks. On the other hand, to study production dynamic in the Ordos Basin, Zhang et al. [28,29] used productivity testing in tandem with pressure build-up data and the "one point method" to estimate well productivity in the initial stages. Zhang et al. [30] and Yan et al. [31] analyzed the decline law of wellhead pressure under a constant production rate via geological modelling, numerical simulations, and gas reservoir engineering. Geologic investigations, well-logging, water production performance analysis, and the study of formation-water components and formation-water layer distribution have been used to determine the origins of the water produced in the Ordos Basin, and the produced water wells can be divided into four categories [32,33].

This comprehensive review of studies on the reservoir of interest (Jingbian gas field) clearly demonstrates that although many investigators have studied the geological characteristics and production performance of this reservoir, there is little understanding of its overall geologic features and development technologies. Most researchers have focused on one aspect, such as gas sources, sedimentary accumulation, gas well production behavior, and formation-water distribution. Therefore, the authors of this study describe the geological characteristics of this reservoir in general and then present the problems and solution strategies that have emerged during the exploration and development process. The innovations for this research mainly lie in the presentation of optimal exploitation technologies for the peripheral area based on comprehensive illustrations of the geology characteristics and encountered problems for this area, which have generally been ignored in previous studies. This study will be a significant reference for the efficient development of carbonate reservoirs in the Ordos Basin and similar gas reservoirs around the world.

2. Background of Gas Reservoir

The Jingbian gas field is a lower Paleozoic carbonate reservoir in the Ordos Basin, which is the first supergiant carbonate gas field in China. This reservoir is a typical weathering crust reservoir that is a part of the Changqing oil and gas field. The Ordos Basin is located in western North China Craton. Tectonic units are more stable in the central area than the margin areas. The Ordos Basin uplifts in the south and north margins and thrusts from west to east, which leads to the rise of the east margin. The whole basin can be subdivided into six tectonic units: Yimeng uplift, Weibei uplift, Jinxi flexing belt, western margin thrust belt, Yishan slop, and Tianhuan depression, as shown in Figure 1. Note that Yishan slop is the most significant tectonic unit for hydrocarbon accumulation. Lower Paleozoic carbonates in the Ordos Basin mainly were developed in the Cambrian and Ordovician periods. The Ordovician Majiagou Formation in the middle-east of the basin is the most important gas-bearing interval; it consists of six lithologic members, with member 1 at the bottom. Members 1, 3 and 5 of the Majiagou Formation are composed of dolomite with gypsum and salt rock. Members 2, 4, and 4 of the Majiagou Formation are composed of dolomite and limestone. Member 5 of the Majiagou Formation can be divided into 10 sub-members, starting from the top, among which sub-members 1, 2, and 4 are principal producing formations in the Jingbian weathering crust gas field and sub-members 5–10 comprise the gas reservoir formed by dolomite. The discovery of the Jingbian gas field suggested good development prospects for marine carbonate reservoir in the Ordos Basin and prompted the search for a large-scale gas reservoir in the basin [34]. developments of carbonate reservoirs under salt rock in the middle-east of basin and reef flat carbonate reservoirs in the western margin have achieved early success [34], thus reflecting the great potential for the exploitation of the lower Paleozoic carbonate gas reservoir in the Ordos Basin.

Regarding production dynamics, this gas field has experienced four development stages: the early comprehensive evaluation and testing production stage (1991–1996), the pre-production for well exploration stage (1997–1998), the large-scale development stage (1999–2003), and the stable production stage (from 2004 to present). The discovery of a lower Paleozoic gas field stimulated the large-scale development of natural gas in the Ordos Basin, and its successful exploitation has provided enormous support for successful implementation of a west–east major project for gas transmission in China. During the development of the lower Paleozoic gas field, a series of techniques have been proposed to develop large-scale carbonate reservoirs that can guarantee long-term stable gas production for the Jingbian gas field and provide great support for the 5000×10^4 t target in the Changqing oil and gas field. Moreover, stable gas supplementation can decrease the consumption of coal and other unclean resources in large urban areas, thus implicitly protecting the atmospheric environment.

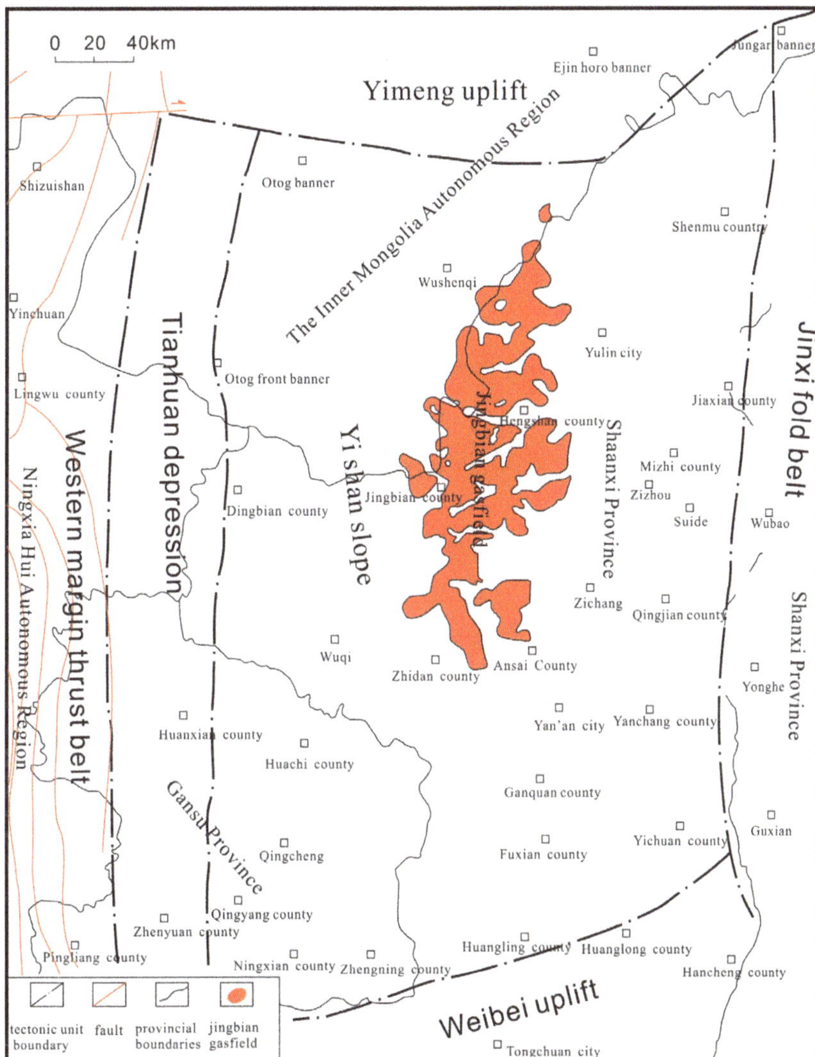

Figure 1. Division of tectonic units and location of the Jingbian gas field in the Ordos Basin.

3. Formation Geological Characteristics

The carbonate gas reservoir in the Ordos Basin, Jingbian gas field, is a weathering crust reservoir. It has unique features influenced by primary deposits, tectonic evolution, paleo-topography, ancient surface runoff, paleo-climatology, and gas source sufficiency. An analysis of carbonate rock composition showed that these rocks are composed of silty dolomite, dolomicrite, grained dolomite, gray dolomite, and cargneule, of which the silty dolomite is the primary mineral. For the formation-water, the main ions in the water are Ca+ and Cl⁻, which are formed in the closed environment.

(1) **Dominated by reservoir sedimentation, crustal uplift and water level lowering, the reservoir has stable horizons, widespread distribution area and multiple gas layers.**

The studied gas reservoir with weathering crust formation in the Ordos Basin is a combined-stratum lithologic subtle trap dominated by paleo-tectonics, lithofacies paleo-geography, and karst paleo-geomorphology. Morphologically deep and slope areas comprise the western Ordos Basin, and the central basin mostly comprises paleo-high tidal flat facies. The depression in the eastern basin is shallow and primarily composed of salt rocks. In this tectonic setting, the northeastern side of the central paleo-uplift is restricted to sea deposits that are composed of carbonates and evaporates. This kind of Paleozoic gas reservoir has stable horizons and large distribution areas that are influenced by a tidal flat environment. As discovered during drilling operations, anhydrite tubercles with dissolved pores are densely distributed in multiple layers, which causes the overlap of vadose and underflow zones. The rhythmical changes of shaliness during the original sedimentation, crustal uplifting, and water level lowering caused overlaps of the above-mentioned zones.

(2) **Controlled by reservoir lithology, karst development degree, direct runoff, and paleo-geomorphology, reservoirs are influenced by grooves of different grades that have caused the formation to have a thin effective thickness, low abundance, and strong heterogeneity.**

Although the distribution area for this kind of reservoir is large (affected by original sedimentation characteristics and lithology difference), its effective thickness is small. Statics analysis has shown that gas reserves per square kilometer are between 0.03×10^8 m^3 and 1.46×10^8 m^3, with an average value of only 0.54×10^8 m^3 in the main area, which means the abundance of gas reserves in the weathering crust gas reservoir is poor.

Weathering crust gas reservoirs have shown strong heterogeneity that is controlled by paleo-geomorphology, direct runoff, and karst development degree. The type and intensity of karstification in different geomorphic units have shown great differences that have led to the vertical partition of karst-rocks and reservoirs. Since the paleo-topography of the western Gaoqiao area is higher than that of the eastern area, the weather denudation is more intense in the west than in the east during Caledonian movement. Karst highlands and karst slopes were developed in karstic paleo-geomorphology from west to east (a karst basin was not developed), and these secondary paleo-geomorphologies can be divided into third-level paleo-geomorphologies. Direct runoff, karst development degree, and weathering crust depth are different in different areas. The Karst highland in the West Gaoqiao area is steeper, so the wells for direct runoff and karst water can reach the stratums beneath member 4 of the Majiagou Formation that penetrates mudstone interlayers. Gypsum rock mainly exists in sub-members 2, 3, and 4 of member 5 of the Majiagou Formation, with karstification in which vertical, deep, and dissolved fractures have intensely developed. Due to the high karst degree and the distribution of effective reservoirs in a paleo-hammock with well-preserved strata, sub-members 1 and 2 of member 5 of the Majiagou Formation are incomplete. Influenced by its relatively small slope angle, the central karst slope area does not have direct runoff. However, direct runoff has wells in grooves that are distributed between karst slope areas. Generally, an abrupt slope favors the rapid infiltration and lateral migration of surface water. The horizon outcropping gypsum rock gradually turns from west to east for the formation of sub-members 2, 3, and 4 of member 5 of the Majiagou Formation. The depth of the vertical leaching grows shallow, which shows that karst water flows slowly lengthwise and that karstification is weak. All of these factors have determined the strong heterogeneity of carbonate reservoirs with weathering crust formation.

(3) **Dynamic behaviors of gas wells show great differences influenced by serious formation heterogeneity.**

The statistical analysis of 677 producing wells in the main area of the Jingbian gas field showed that the cumulative gas production rates and gas reserves have great discrepancies among various gas wells. The average cumulative gas production for these wells is 1.06×10^8 m^3, and the maximum value is 7.6×10^8 m^3 for well G10-14. However, there are 284 producing wells (45% of total wells) for which the cumulative gas production is less than 0.5×10^8 m^3. The average gas reserve controlled by per well is 2.2×10^8 m^3, and the maximum value is 11.4×10^8 m^3 for well Longping1. Nevertheless, there are 298 producing wells (44% of total wells) for which the dominant gas reserve is less than 1×10^8 m^3.

(4) **Influenced by current formation structure, reservoir heterogeneity, paleo-geomorphology, tectonic reversal, and gas source abundance, gas reservoirs generally have no unified bottom and edge water, though they do have remaining interlayer water under sealing conditions, which has caused water-rich zones to be formed in some areas.**

Typically, the weathering crust gas reservoir showed no unified edge or bottom water, though interlayer water is retained in local areas and forms water-rich zones. The Gaoqiao area is located at the convergence of L-form water-rich areas in the Jingbian gas reservoir, which has a complex formation-water distribution pattern. The distribution of gas and water is controlled by regional tectonic setting, gas source abundance, low-amplitude structure, reservoir heterogeneity, and tectonic reversal. Regional tectonic setting is the basic condition that affects formation-water distribution, and insufficient gas sources are the primary reasons for the formation of water-rich zones. Low-amplitude structures and reservoir heterogeneity also play decisive roles in the distribution of local formation-water. The key factor that determines the complex distribution of gas and water is the distribution of ancient grooves and tectonic reversals.

(5) **The production performance of gas wells is seriously impacted by formation-water in some local areas.**

In the zone with retained interlayer water, the proportion of wells that produce water during production testing was shown to be significant. For these wells, gas and water are simultaneously produced, and water production rate variations are large. The water production rate and the probability of water breakthrough for horizontal wells is always larger than those of vertical wells. In the Gaoqiao area, for example, 38 of 150 vertical wells (25.3%) were found to produce water during production testing. The gas production rate was found to be between 0 and 7.97×10^4 m^3/d, with an average value of 1.43×10^4 m^3/d. The water production rate was found to be 0.5~33.5 m^3/d, with an average value of 6.7 m^3/d. In contrast, 10 of 18 total horizontal wells (55.5%) were found to produce water. The gas production rate for these wells was calculated as 0.86~20.41 $\times 10^4$ m^3/d, with an average value of 5.6×10^4 m^3/d. The water production rate was calculated as 8~90 m^3/d, with an average value of 31.0 m^3/d. An illustration of these statistical results is shown in Figure 2. These results can be attributed to the chances of encountering water-rich zones. The probability of drilling into water-rich areas during acid fracturing is low for the vertical wells but high for the horizontal wells. Furthermore, the improvement of reservoir stimulation during production testing has been shown to be able to increase the gas production rate and water production.

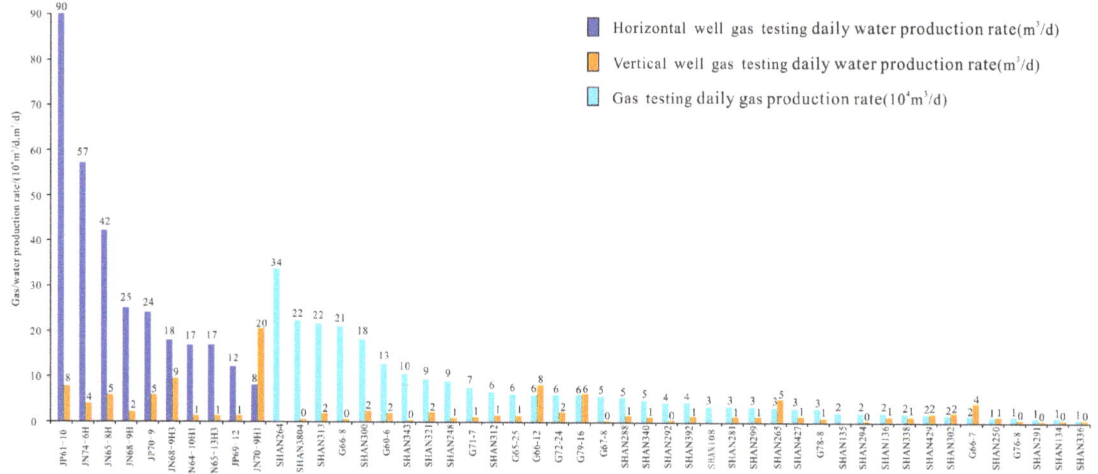

Figure 2. The distribution histogram of gas well testing production in Gaoqiao.

On the other hand, the production wells located in the area with retained interlayer water were found to be significantly influenced by formation-water. The Paleozoic gas reservoir in the Gaoqiao area was chosen as an example for the following reasons:

(1) Well-log interpretation showed that the reservoir is a gas–water formation, as illustrated in Figure 3.
(2) Production testing showed that water production is greater than 2 m3 and the water/gas ratio is larger than 1 m^3/104 m^3.
(3) The producing water–gas ratio is larger than 0.5 m^3/104 m^3.
(4) The content of Cl- is larger than 20,000 mg/L.
(5) The degree of mineralization is larger than 100 g/L.
(6) When the wellhead pressure is equal to 10 MPa, gas production is larger than 1.4 × 104 m^3.

Currently, there are 88 production wells in the Gaoqiao lower Paleozoic gas reservoir with similar characteristics, and 41 of these wells produce water (46.6%). The daily gas production rate for wells that produce water was found to be 0.13 × 10^4 m^3~5.29 × 10^4 m^3, and the daily water production rate was calculated as 0.2 m^3~18.59 m^3. The average daily gas and water production rate for these wells were shown to be 1.12 × 10^4 m^3 and 1.53 m^3, respectively, and the water–gas ratio was maintained at 0.19~14.2 m^3/10^4 m^3. It is obvious that the water and gas production rates for water-producing wells vary greatly, which indicates strong formation heterogeneity. Some gas wells, particular for horizontal wells, are seriously influenced by formation-water, which has led to the shut-down of some wells. For instance, Jingnan 57-9H2 comprises sub-members 1 and 2 of member 5 of the Majiagou Formation, and the dilled length ratio for effective formation is 62%. After the utilization of acid fracturing for five stages, gas well productivity was found to be 147.37 × 10^4 m^3, which demonstrates great production capacity. The gas production rate at the initial stage was calculated as 15~20 × 10^4 m^3/d and was shown to be influenced by formation-water; after 5 months of production, it rapidly decreased. This well only had been producing for 12 months and is now shut down. The production curve for this well is shown Figure 4.

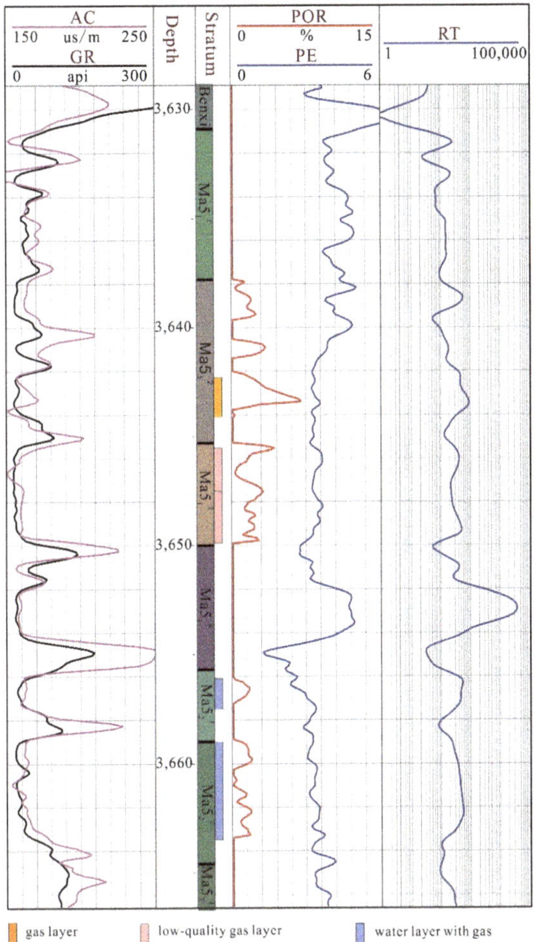

Figure 3. Typical well-log curve and its interpretation results. ($Ma5_1^1$, $Ma5_1^2$, $Ma5_1^3$, and $Ma5_1^4$ represents the first, second, third, and fourth layers, respectively, of sub-member 1 of member 5, Majiagou Formation; $Ma5_2^1$ and $Ma5_2^2$ represent the first and second layers, respectively, of sub-member 2 of member 5, Majiagou Formation).

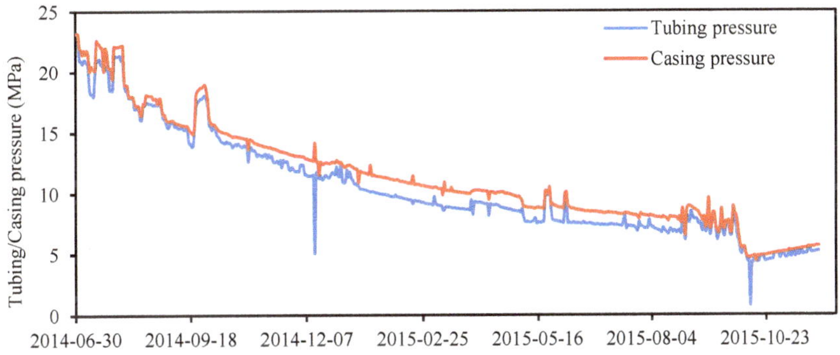

Figure 4. *Cont.*

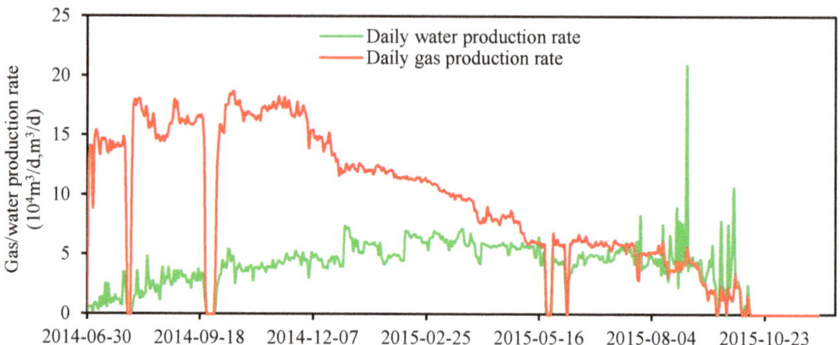

Figure 4. Production curves of well Jingnan 57-9H2.

4. Geology and Production Problems

At present, the development of the lower Paleozoic carbonate gas reservoir requires stable production in the main area and productivity construction in the peripheral area.

Problems for the stable production of the main area include:

(1) As the gas reservoir has entered into the middle or later stage of development, the fine characterization of secondary grooves in the main area is the key to the success of infill wells. Additionally, the basis of long-term stable production for this gas field is reserve distribution.

(2) The prominent contradiction for gas reservoir development is unbalanced exploitation, which has led to the unbalanced distribution of pressure (the pressure in the middle-high and low yield areas is low and high, respectively) and the unbalanced domination of reserves (in the middle-high yield area, the dynamic reserve degree is high, and it is less in the low yield area; in the vertical direction, the dynamic reserve degree for principal producing formation is high, and it is low for other layers). This unbalanced exploitation was found to strongly influence the stable production and regulation ability.

(3) Wellhead pressure is low. We found that the pressure distribution in middle zone and surrounding areas was low and high, respectively. In August 2009, the average reservoir pressure was 11.23 MPa and the wellhead pressure in middle-high production area was 9.37 MPa, which was close to the transport pressure.

(4) Belching wells, low production wells, and water production wells are increasing in number, and the management of gas reservoirs has thus become more difficult.

All these problems have hindered the improvement of gas reservoir recovery and long-term stable gas production in the main area.

The productivity problems of the peripheral area are as follows:

(1) Influenced by sedimentation, diagenism, and paleo-geomorphology, reservoir quality has become worse and the optimization of enrichment areas has become harder.

(2) The reconstruction of paleo-geomorphology in the peripheral area is difficult due to the scarce wells and tiny grooves.

(3) The distribution of formation-water in the western part of weathered crust is complicated.

These problems have increased the complexity of early evaluation and the risk of productivity construction.

5. Optimal Development Techniques and Results

To address the different problems that have emerged in the main and peripheral areas of the low Paleozoic gas reservoir in the Ordos Basin, a series of techniques that can help maintain stable production in the main area and enlarge the scale of productivity construction in the peripheral area have been established.

5.1. Stable Production Techniques and Results in the Main Area

5.1.1. Production Techniques

Five key techniques have been proposed to solve the problems in the main area and aid stable production in the Jingbian gas reservoir: the fine description of grooves with a comprehensive geological modeling technique, a gas reservoir dynamic analysis technique, a pressure-charged mining technique, an optimization technique for horizontal well locations in thin reservoir, and a fine management technique for the gas reservoir.

(1)   **Fine description of grooves and comprehensive geological modeling technique.**

Grooves are formation deficiencies that are developed in slope areas with surface runoff and caused by surface water erosion and chemical eluviation. The paleo-grooves for Ordovician weathered crust in the Jingbian gas field were mainly formed by surface runoff erosion and filling, which were controlled by paleo-tectonics, paleo-climate, paleo-hydrodynamic power, and formation lithology. The development of the Jingbian gas field has demonstrated that accurately recognizing the small grooves between wells can significantly influence the success of well drilling. Though restrained by well locations, secondary and tertiary level grooves can be recognized and tracked via the utilization of different groove surfaces in well-logging and seismic analysis—as well as the combination of static and dynamic data—to provide a fine description of the distribution of low Paleozoic grooves.

However, due to the complexity and heterogeneity of erosion grooves, for the geological modeling of weathering crust gas reservoirs, many investigators have proposed the concepts of groove and stratum facies under the consideration of carbonate reservoir peculiarities in the Jingbian gas field, which can be studied with controlled facies modeling [35]. Some researchers have divided the study area into groove facies, reservoir facies, and dry layer facies for geological modeling [36]. The distribution features of grooves should be characterized by facies-controlled modelling techniques, and then a property model can be developed with the aid of well testing, dynamic monition, and gas production data.

(2)   **Comprehensive gas reservoir dynamic analysis technique.**

In terms of the unbalanced exploitation of gas reservoirs, the formation pressure in the main area and dynamic reserves can be evaluated with pressure, well, and production testing data, and then comprehensive gas reservoir dynamic analysis technique can be used. Due to different percolation characteristics and dynamic features for gas wells, there are few formation pressure testing data and an unstable schedule for gas wells; accordingly, multi-method evaluation techniques based on pressure-drop and production rate transient analyses have been proposed for the estimation of gas reserves for low-permeability and heterogeneous reservoirs. These techniques can provide support for the evaluation of single-well dynamic reserves and their varied features in the Jingbian gas field. On the other hand, because formation permeability is low and the recovery of wellhead pressure after shutting down wells is slow, some pressure evaluation methods—such as corrections for wellhead pressure and the extension of deliverability equations—have been proposed to provide support for the fine evaluation of change laws and distribution features in gas field formation pressure. A diagram illustrating these methods is shown in Figure 5.

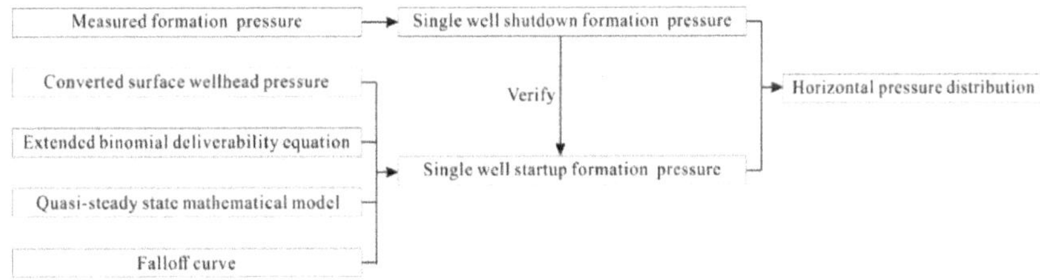

Figure 5. The evaluation technology for formation pressure measurements of shut-in gas wells.

(3) **Pressure-charged production technique.**

Following decades of development in the main area of the Jingbian gas field, the gas well pressure is continually decreasing. However, pressure-charged production can be used to improve gas reservoir recovery in the main area for long-term stable production. The surface pressure system in the main area of the Jingbian gas field was found to be 6.4 MPa. The abandonment pressure was found to be 9.7 MPa, with a depleted gas production rate of 1000 m^3/d. However, the abandonment pressure could be reduced to 6.1 MPa in the Jingbian gas field using the pressure-charged production technique, which has enormous potential due to the reservoir's large scale. However, the large area, many wells, and seriously unbalanced development also mean that the single-well pressure-charged technique has limitations. Table 1 compares the pros and cons of different pressure-charged techniques (Table 1).

Table 1. Comparison of different pressure-charged methods.

Pressurization Method	Advantages	Disadvantage	Prediction Gas Recovery after 30 Years (%) (Wellhead Pressure 2 MPa)	Total Investment (One Hundred Million Yuan)
Single-well pressurization	No need for pipe network reconstruction and the consideration of unbalanced gas field exploitation	Too many pressurized points, heavy workload, maintenance difficulty, and poor development benefits	56.3	23.7
Gas gathering station pressurization	Small workload and easy management	Too many stations, difficulties in compressor choice and management	56.2	15.3
Regional pressurization	Efficient reduction in station numbers and lowering of difficulties in production and management	Hard to divide compress cells due to pressure drawdown desynchrony between wells	56.4	17.6
Concentrated pressurization	Least required stations and little maintenance work	High pipe network reconstruction costs and high operation risks	56.3	25.1

The results show that the gas field development effects of different pressure-charged techniques are almost identical, and the selection of optimal pressure-charged techniques depends on economic and engineering factors [37].

(4) **Optimization technique for horizontal well locations in thin reservoir.**

Sub-members 1 and 2 of member 5 of the Ordovician Majiagou Formation in the Jingbian gas field comprise a trap that combines paleo-geomorphology and lithology and that has low porosity, low permeability, thin layers, and strong heterogeneity. The gas production rate and dynamic well gas reserves for conventional vertical wells are low, and there is significant unbalanced development. Fully developing geological reserves and improving single-well production and gas recovery are the keys to the long-term stable gas field production. For horizontal wells in weathering crust reservoirs that are full of erosion grooves and have thin primary formation, low-amplitude structure variation, and strong reservoir heterogeneity, optimization can be accomplished presented through the utilization of 3D seismic technology, geological formation evaluation, techno-economic analysis, and gas reservoir numerical simulation [38]. Firstly, an analysis of the relationship between formation permeability and the net present value (NPV) conducted with a comprehensive technical and economic evaluation method showed that the area that could be exploited with horizontal wells can be determined when permeability is larger than $0.1 \times 10^{-3} \mu m^3$. Secondly, according the distribution law of abundant natural gas resources and karst paleo-geomorphology analysis, geomorphic units of karst monadnock and gentle slope were chosen as the areas that can be exploited within horizontal wells. Finally, five principles for deployment with horizontal wells were determined as follows:

1. The results of reservoir evaluation demonstrated that the residual thickness of sub-members 1 and 2 of member 5 of the Majiagou Formation is larger than 20 m and that the horizontal distribution of formation is steady.
2. The thickness of sub-member 1 of member 5 of the Majiagou gas-bearing formation is greater than 2 m.
3. Reservoir physical properties is strong, and the formation is of class I or II.
4. The formation structure is relatively flat.
5. Production testing for adjacent vertical wells demonstrated that gas production is stable. The vertical distance between horizontal wells meets the requirement for production without interference.

(5) **Fine gas reservoir management.**

In recent years, development techniques and management system for the Jingbian gas field have been continually optimized, which has led to the development of a suitable management pattern. To propose a novel gas management pattern, which can be used to improve the accuracy of flowing unit evaluation and determine the necessary steps for gas reservoir developments, the authors considered the flowing element as the management object and conducted experiments that considered a combination of geology, engineering, and operator factors. According to the gas–water distribution, production dynamics, reservoir properties, monitored fluids properties, current formation pressure, and other static parameters, criteria for flowing unit classification were established. Additionally, various technical strategies for the stable production and enhancement of gas recovery were formulated for each unit. Table 2 shows the three primary classes.

Table 2. Classification criteria and results for developed units in Jingbian gas field.

Types	Class I	Class II	Class III
Dynamic reserve ratio/%	>30.0	15~30	
Average absolute open-flowing gas rate/(10^4 m^3/d)	>20.0	<20	
Average cumulative gas production for unit pressure drop (10^4 m^3/MPa)	>900	<900	Located in water-rich area
Recovery factor/%	>10	<10	
Average allocating gas production rate/(10^4 m^3/d)	>3.0	2.0~3.0	
Water–gas ratio (m^3/10^4 m^3)	<0.18	<0.20	>0.60
Division results	13	17	6

Class I: This kind of flowing unit is characterized by four large factors and one low factor that refer to a large ratio between dynamic and static reserves, a high recovery extent, a high gas well productivity, a large cumulative gas production rate with a unit pressure-drop, and a low gas well pressure, respectively. This kind of unit is primarily located in the main area of the gas field and should be addressed with the technical strategies of fine characterization, deep potential exploitation, and enhanced gas recovery. For areas that are not dominated by this well pattern, infill wells should be drilled to improve the area-dominated extent of gas reserves. The dominated degree in the vertical direction for this gas field can be improved by perforating the new gas-bearing layer and side-tracking.

Class II: This kind of flowing unit can be characterized by four low factors and one large factor: a low ratio between dynamic and static reserves, a low recovery factor, a low gas productivity, a low cumulative gas production rate with a unit pressure-drop, and a high formation pressure. This kind of unit is primarily located in the eastern part of the gas field and should be addressed with the technical strategies of block optimization, scale enlargement, and improvements in single-well gas production rate. Horizontal wells are preferable for use during development to improve individual well producing rates.

Class III: This kind of unit can be characterized by two large factors and one low factor: a high water production rate, a high water–gas ratio, and a low formation pressure. Influenced by water production, this kind of unit should be addressed with the technical strategies of evaluation enhancement, internal water drainage, and external water control; of these, drainage is the primary measure that can improve gas recovery. A pattern in which the flowing unit is the management object can improve the efficiency of gas management, which also can allow for fine gas reservoir evaluation.

5.1.2. Development Strategies with the Presented Techniques

The development plan and results of the use of the presented techniques in the main area are as follows:

(1)   Ten first-order grooves, seventy-two second-order grooves, three hundred and eighty-two third-order grooves, and some fourth-order grooves were characterized. Fine descriptions for different ranks of grooves have allowed for a more accurate characterization of low Paleozoic reservoir architecture, which has provided a good foundation for comprehensive reservoir geological research. Following four steps regarding facies-control, formation, physical properties, and gas content, a 3D geological model for the main area of gas reservoir could be developed. The new geological model could be used to recalculate the main area reserves in combination with production data to form the basis of stable production and gas recovery improvements for the main area.

(2)   The ratio between static and dynamic reserves was found to be 34.06% in the main area and 19.89% in the east of Qiantai, which indicates that the exploited gas reservoir extent is low for the whole reservoir and has great potential for gas recovery enhancement. According to the evaluated pressure distribution and fine gas reservoir description, the north Beier district, the Shan66 district, the Shan175 district, the south Naner district, and the Shan106 district are the best places for new well drilling, which can enhance the produced gas reserve degree in the main area.

(3)   Based on production dynamics and surface construction in gas field, the regional pressure-charged and gas collection pressure-charged techniques were determined to be main and auxiliary ways to enhance gas recovery. The principle of integrated planning and implementing by steps has also been employed, and the gas reservoir has been divided into thirty elements that can support the arrangement of the pressure-charged project. The pressure-charged experiments showed that the stable production period can be prolonged for 2–3 years, and the recovery degree of the gas reserves can be increased by 14.6% in the Jingbian gas field.

(4)   The gas well production rate can be greatly increased with the horizontal well development technique in thin gas formation, and development benefits also can be improved. In 2011, nine horizontal wells with an average length of 1145 m were drilled. The average absolute gas flow rate for five wells was found to be $10^8 \times 10^4$ m^3/d, which was nine times that of surrounding vertical wells. Horizontal length for the well Jingbian 012-6 was found to be 1161 m, and the drilled effective reservoir thickness was found to be 1048 m, which accounted for 90.3% of total drilled formation length. After acid fracturing for seven segments, the absolute gas flow rate was found to be 219.27×10^4 m^3/d.

(5)   The main area in the Jingbian gas field can be divided into 36 units, and the well-spacing density, recovery factor, and remaining gas reserves can be calculated for each unit to provide the basis for the adoption of development strategies in different flowing units.

5.2. Techniques for Improvement of Production Scale in Peripheral Area

Enrichment area optimization, paleo-geomorphic restoration, and formation-water distribution evaluation techniques have been established to solve problems and strongly support productivity construction in the peripheral area.

5.2.1. Production Techniques

(1) **Optimization technique for enrichment area.**

Each method for the evaluation of reservoir properties has disadvantages in carbonate reservoirs with strong heterogeneity. Therefore, during the optimization process for enrichment areas in the peripheral zones of the Jingbian gas field, a method that combines multiple factors was applied to screen for some quantitative and qualitative factors including micro-facies, reservoir/gas-bearing formation thickness, residual thickness for weathered crust, porosity, permeability, gas saturation, and shale content distribution. Then, a weight index was normalized and evaluated, and the weighted average process was conducted. Finally, reservoir properties could be comprehensively evaluated and the reserve for each layer could be determined. After the superposition and selection of these enrichment layers, the areas that are suitable for development could be determined.

(2) **Paleo-geomorphic restoration technique.**

Conventional paleo-geomorphic restoration techniques include the sedimentology restoration, impression, residual thickness, layer flattening, and high-resolution paleo-geomorphic sequential stratigraphy restoration methods. Each of these methods needs well drilling, well-logging, core, slice, and seismic data. Considering each method's advantages and disadvantages, the dual-interface paleo-geomorphic restoration method has been established as the most suitable method for the evaluation stage of gas reservoirs.

This method is based on two interfaces, namely a typical horizon for overlying strata and a basic horizon for underlying strata. The typical horizon for overlying strata is similar to the top reference surface in the impression method and the top flattening surface in the layer flattening method. The selection of the typical horizon for overlying strata is based on high-resolution sequence stratigraphy theory. The overlying isochronal stratigraphic framework was developed through the utilization of corrections for the high-resolution base level cycle. The transfer surface (sequence boundary or maximum flooding surface) of the base level cycle was chosen as overlying marker zone while considering sedimentary analysis results. The overlying marker surface is a horizon that lies on the flattened formation. This marker surface is the transfer surface in the base level cycle that can be easily correlated between wells and has good isochroneity. The formation's bottom morphology beneath the marker surface can reflect the original paleo-geomorphology features before formation deposition and after flat marker surface deposition. The basic underlying stratum is the starting horizon for the calculation of easily selected paleo-geomorphology values. The first well-developed formation under a weathering crust's unconformable surface should be determined first. Then, the maximum thickness (H_3, constant) from the overlying marker surface to the formation's bottom can be calculated. The underlying marker surface is a horizon that is lower than overlying marker surface for the value of H_3 (Figure 6). The paleo-geomorphology value of a drilled well should be defined after the recognition of two interfaces and a contour map for paleo-geomorphology, which should be drawn according to seismic data. Finally, a paleo-geomorphic unit can be divided by regional paleo-geomorphology features.

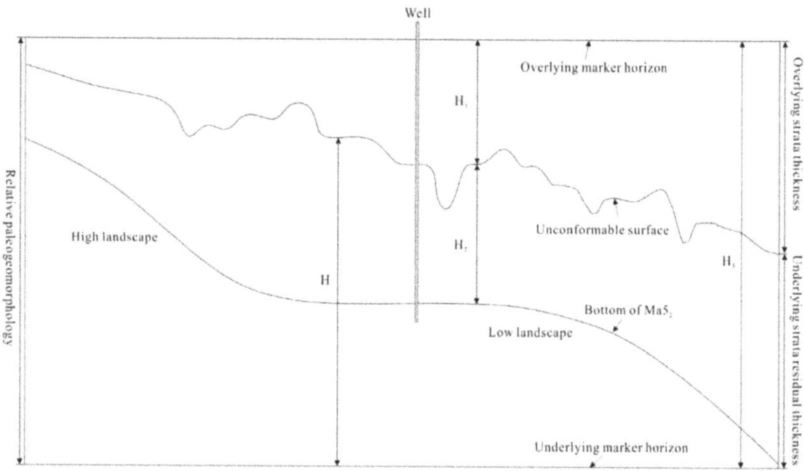

Figure 6. Geomorphology division principle for dual-interface method.

(3)   **Evaluation technique for formation-water distribution.**

The distribution of formation-water in the peripheral area of the Jingbian gas field is complicated. It has been difficult to find an effective way to avoid drilling formation-water during development. To solve this problem, we comprehensively utilized stationary and dynamic methods to characterize the distribution of gas and water; after identifying various types, a classification standard could be established for permeable formation bodies (Table 3). This standard subdivided the permeable reservoir bodies into six types: high-permeability formation without water, high-permeability formation with condensed water, high-permeability formation with sealed water, low-permeability formation with sealed water, low-permeability formation with condensed water, and low-permeability formation without water. The distribution of different types of permeable formation bodies was characterized with a combination of comprehensive evaluation results regarding the gas reservoir (Figure 7). Consequently, optimal well locations were determined according to the following order: high-permeability formations without water, low-permeability formations without water, high-permeability formations with condensed water, and low-permeability formations with condensed water.

Table 3. The dynamic and static classification standard of permeable formation bodies.

Permeable Formation Body Types	Static Criterion		Dynamic Criterion	
	Permeability (mD)	Saturability (%)	Gas Output (10^4 m^3/d)	Gas–Water Ratio (m^3/10^4 m^3)
High-permeability formation without water	≥0.6	≥75	≥1.2	≤0.2
High-permeability formation with condensed water		≥45–75		0.2–0.6
High-permeability formation with sealed water		<45		>0.6
Low-permeability formation with sealed water	<0.6	<50	<1.2	>0.6
Low-permeability formation with condensed water		≥50–80		0.2–0.6
Low-permeability formation without water		≥80%		≤0.2

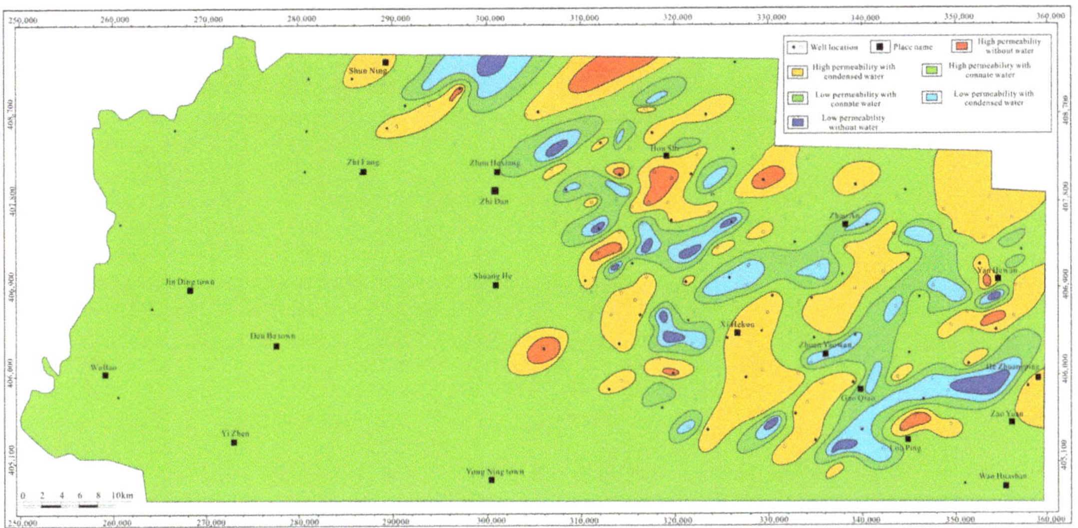

Figure 7. Distribution map of permeable formation bodies in Gaoqiao area, Ordos Basin.

The comprehensive evaluation of formation-water distribution with stationary and dynamic methods can guide the development of wells in areas rich in retained interlayer water in the peripheral area of the Jingbian gas field.

5.2.2. Development Strategies with the Presented Techniques

Through the utilization of the three presented production techniques for the peripheral area, the development plan for this area is as follows:

(1) With the optimization technique, the gas reserve for the enrichment area of the Shenmue gas field was found to be 44.2 billion cubic meters; this information can be used to guide the optimization of well placement and production facility construction.

(2) The paleo-geomorphology of the top Ordovician karst weathering crust in the Gaoqiao area was assessed using the dual-interface paleo-geomorphic restoration method. The first order paleo-geomorphic unit of weathering crust in the Gaoqiao area can be subdivided into two kinds of second order paleo-geomorphic units, namely karst highland and karst slope. Meanwhile, there are seven kinds of third order paleo-geomorphic unit, namely flat, platform, depression, monadnock, main monadnock, groove, and main groove. This classification scheme can guide the setting of well patterns in the Gaoqiao area.

(3) According to the evaluation results regarding the distribution of formation-water in the Gaoqiao area and data of the 21 horizontal wells in the lower Paleozoic, seven wells that can normally produce are located in an area with a low risk of water breakthrough and all other wells save one cannot produce normally and are located in zones with a high risk of water breakthrough (sealed aquifer).

Through the utilization of the presented optimal production techniques for the main and peripheral areas in the Ordos Basin, it is predicted that gas production can be maintained at 5.5 billion m^3 per year and be stable before 2025. It should be noted that the presented techniques have been conducted in practice. The techniques have been additionally used in geological exploration, gas reservoir development, and well operation, which has made it difficult to compare the results of different methods across the entire development process. Different comparative methods can be used to select optimal techniques. Therefore, a scheme that integrates all techniques will be optimal and lead to the most benefits.

6. Conclusions

The lower Paleozoic carbonate gas reservoir in the Ordos Basin is a typical karst weathering crust carbonate gas reservoir that has a large distribution area, low abundance, strong heterogeneity, and multiple gas layers. Currently, this reservoir faces many problems, including the complexity of groove distribution, a low degree of gas reservoir recovery, an unbalanced exploitation degree, and a low wellhead pressure. In addition, some local regions are rich in water, which significantly influences the production of gas wells.

To deal with the emerging problems of this carbonate gas reservoir, five key techniques have been presented to maintain stable gas production in the main area: (1) the fine description of grooves and comprehensive geological modeling, (2) the comprehensive dynamic analysis of the gas reservoir, (3) pressurized development, (4) the optimization of horizontal well locations in the thin reservoir, and (5) the fine management of the gas reservoir. For the peripheral area, which has faced worse reservoir quality, complicated formation-water distribution, and unclear paleo-geomorphological distribution, three key techniques have been developed: (1) the optimization of the enrichment area, (2) dual-interface paleo-geomorphic restoration, and (3) the evaluation of formation-water distribution, with the division of permeable formation bodies conducted via stationary and dynamic methods.

With the presented techniques, the gas production of the main area can be maintained at 5.5 billion cubic meters per year and be stable before 2025. The predicted natural gas reserve of the peripheral area was found to comprise 1 trillion cubic meters in the preliminary evaluation, and the potential gas production rate for this area could reach a value of 4.5 billion cubic meters per year, which would greatly contribute to the goal of one billion cubic meters of gas production per year for the lower Paleozoic carbonate gas reservoir in the Ordos Basin. Furthermore, considering the middle and lower gas layers in the east basin and the potential carbonate gas-bearing formations in the west basin. The Lower Paleozoic carbonate gas reservoir in the Ordos Basin has good prospects for exploration and development.

Author Contributions: Conceptualization, H.Y. and A.J.; methodology, J.G.; formal analysis, H.Y. and B.N.; investigation, A.J.; resources, J.G.; data curation, Q.X.; writing—original draft preparation, Q.X.; writing—review and editing, F.M.; visualization, H.Y.; supervision, A.J.; project administration, B.N.; funding acquisition, A.J. All authors have read and agreed to the published version of the manuscript.

Funding: This work was supported by the national science and technology major project (No. 2016ZX05015-003) and the CNPC Science and Technology Major Project (Grant No. 2021DJ1504).

Institutional Review Board Statement: Not applicable.

Informed Consent Statement: Not applicable.

Data Availability Statement: All data used in this research are easily accessible by downloading the various documents appropriately cited in the paper.

Acknowledgments: The authors greatly appreciate the Exploration and Development Research Institute of Changqing Oilfield Company for providing data access and for permission to publish the results.

Conflicts of Interest: The authors declare no conflict of interest.

References

1. Fan, J.S. Characteristics of carbonate reservoirs for oil and gas fields in the world and essential controlling factors for their formation. *Earth Sci.* **2005**, *12*, 23–30.
2. Jia, A.L.; Yan, H.J.; Guo, J.L.; He, D.B.; Wei, T.J. Characteristics and experiences of the development of various giant gas fields all over the world. *Nat. Gas Ind.* **2014**, *34*, 33–46.
3. Bai, G.P. Distribution patterns of giant carbonate fields in the world. *J. Paleogeography* **2006**, *8*, 241–250.
4. Wang, J.P.; Shen, A.J.; Cai, X.Y.; Luo, Y.C.; Li, Y. A review of the Ordovician carbonate reservoirs in the world. *J. Stratigr.* **2008**, *32*, 363–373.
5. Jia, A.L.; Yan, H.J.; Guo, J.L.; He, D.B.; Cheng, L.H.; Jia, C.Y. Development characteristics for different types of carbonate gas reservoirs. *Acta Pet. Sin.* **2013**, *34*, 914–923.

6.   Jia, A.L.; Yan, H.J. Problems and countermeasures for various types of typical carbonate gas reservoirs development. *Acta Pet. Sin.* **2014**, *35*, 519–527.

7.   Jia, A.L.; Fu, N.H.; Cheng, L.H.; Guo, J.L.; Yan, H.J. The evaluatioin and recoverability analysis of low-quality reserves in Jingbian gas field. *Acta Pet. Sin.* **2012**, *33*, 160–165.

8.   Yan, H.J.; Jia, A.L.; He, D.B.; Guo, J.L.; Yang, X.F.; Zhu, Z.M. Development problems and strategies of reef-shola carbonate gas reservoir. *Nat. Gas Geosci.* **2014**, *25*, 414–422.

9.   Zhao, W.Z.; Wang, Z.C.; Hu, S.Y.; Pan, W.Q.; Yang, Y.; Wang, H.J. Large-scale hydrocarbon accumulation factors and characteristics of marine carbonate reservoirs in three large onshore cratonic basins in China. *Acta Pet. Sin.* **2012**, *33*, 1–10.

10.   Wang, Z.C.; Zhao, W.Z.; Hu, S.Y.; Jiang, H.; Pan, W.Q.; Yang, Y.; Bao, H.P. Reservoir types and distribution characteristics of large marine carbonate oil and gas fields in China. *Oil Gas Geol.* **2013**, *34*, 153–160.

11.   Rashid, F.; Glover, P.W.J.; Lorinczi, P.; Collier, R.; Lawrence, J. Porosity and permeability of tight carbonate reservoir rocks in the north of Iraq. *J. Pet. Sci. Eng.* **2015**, *133*, 147–161. [CrossRef]

12.   Howat, M.D. Completely weathered granite- soil orrock? *Q. J. Eng. Geol.* **1986**, *19*, 433–438.

13.   Irfan, T.Y. Mineralogy, fabric properties and classi-fication of weathered granites in Hong Kong. *Q. J. Eng. Geol.* **1996**, *29*, 5–35. [CrossRef]

14.   Hill, S.E.; Rosenbaum, M.S. Assessing the significant factors in a rock weathering system. *Q. J. Eng. Geol.* **1998**, *3*, 85–94. [CrossRef]

15.   Price, D.G. A suggested method for the classification of rock mass weathering by a ratings system. *Q. J. Eng. Geol.* **1993**, *26*, 69–76. [CrossRef]

16.   Dewande, B.; Lachassagne, P.; Wyns, R. A generalized 3-D geological and hydrogeological conceptual mode of granite aquifers controlled by single or multiphase weathering. *J. Hydrol.* **2006**, *330*, 260–284. [CrossRef]

17.   Tian, M.; Xu, H.M.; Cai, J.; Wang, J.; Wang, Z.Z. Artificial neural network assisted prediction of dissolution spatial distribution in the volcanic weathered crust: A case study from Chepaizi Bulge of Junggar Basin, northwestern China. *Mar. Pet. Geol.* **2019**, *110*, 928–940. [CrossRef]

18.   Zhu, M.L.; Liu, Z.; Liu, H.M.; Li, X.K.; Liang, S.Z.; Gong, J.Q.; Zhang, P.F. Structural division of granite weathering crusts and effective reservoir evaluation in the western segment of the northern belt of Dongying Sag, Bohai Bay Basin, NE China. *Mar. Pet. Geol.* **2020**, *121*, 104612. [CrossRef]

19.   Sidorova, E.; Sitdikova, L.; Izotov, V. The major types of the weathering crust of the eastern Russian plates and its mineralogical and geochemical features. *Procedia Earth Planet. Sci.* **2015**, *15*, 573–578.

20.   Yan, H.J.; Jia, A.L.; Ji, G.; Guo, J.L.; Xu, W.Z.; Meng, D.W.; Xia, Q.Y.; Huang, H.J. Gas-water distribution characteristic of the karst weathering crust type water-bearing gas reservoirs and its development counter measures: Case study of Lower Paleozoic gas reservoir in Gaoqiao, Ordos Basin. *Nat. Gas Geosci.* **2017**, *28*, 801–811.

21.   Wang, H.; Zhou, Q.; Zhou, W.; Zhang, Y.; He, J. Carbonate platform reef-shoal reservoir architecture study and characteristic evaluation: A case of S field in Turkmenistan. *Energies* **2022**, *15*, 226. [CrossRef]

22.   He, J.; Feng, C.Q.; Ma, L.; Qiao, L.; Wang, Y. Diagenesis and diagenetic facies of crust-weathered ancient karst carbonate reservoirs. *Pet. Geol. Exp.* **2015**, *37*, 8–16.

23.   Liu, Q.Y.; Chen, M.J.; Liu, W.H.; Li, J.; Han, P.L.; Guo, Y.R. Origin of natural gas from the Ordovician paleo-weathering crust and gas-filling model in Jingbian gas field, Ordos basin, China. *J. Asian Earth Sci.* **2009**, *35*, 74–88. [CrossRef]

24.   Wang, K.; Pang, X.Q.; Zhao, Z.F.; Wang, S.; Hu, T.; Zhang, K.; Zheng, T.Y. Geochemical characteristics and origin of natural gas in southern Jingbian gas field, Ordos Basin, China. *J. Nat. Gas Sci. Eng.* **2017**, *46*, 515–525. [CrossRef]

25.   Xu, X.; Feng, Q.H.; Wei, Q.S.; Yang, S.G.; Zhang, J.C.; Pang, Q.; Zhu, Y.S. Sedimentary characteristics and reservoir origin of the mound and shoal microfacies of the Ma51+2 submember of the Majiagou Formation in the Jingbian area. *J. Pet. Sci. Eng.* **2021**, *196*, 108041. [CrossRef]

26.   Wei, X.S.; Ren, J.F.; Zhang, J.X.; Zhang, D.F.; Luo, S.S.; Wei, L.B.; Chen, J.P. Paleogeomorphy evolution of the Ordovician weathering crust and its implication for reservoir development, eastern Ordos Basin. *Pet. Res.* **2018**, *3*, 77–89. [CrossRef]

27.   Li, J.; Zhang, W.Z.; Luo, X.; Hu, G.Y. Paleokarst reservoirs and gas accumulation in the Jingbian field. *Ordos Basin. Mar. Pet. Geol.* **2008**, *25*, 401–415. [CrossRef]

28.   Zhang, Z.L.; Zhao, Z.J.; Zhang, Q.; Sa, Q.F.; Tang, T.Z. Jing Bian gas field well productivity verification & proper production proration. *Nat. Gas Ind.* **2006**, *26*, 106–108.

29.   Zhang, Z.L.; Hu, J.G. A method of determining deliverability with build-updata of gas Wells—An example from the Jingbian gas field. *Oil Gas Geol.* **2009**, *30*, 250–254.

30.   Zhang, Z.L.; Wu, Z.; Zhang, Q.; Sa, Q.F.; Wang, Z.J. Constant-rate well test and pressure decline analysis of gas wells in Jingbian. *Nat. Gas Ind.* **2007**, *27*, 100–101.

31.   Yan, N.P.; Wang, X.; Lv, H.; Huang, W.K.; Huang, G. Deliverability decline law of heterogeneous Lower Paleozoic gas reservoirs in the Jingbian Gas Field, Ordos Basin. *Nat. Gas Ind.* **2013**, *33*, 43–47.

32.   Guo, C.H.; Zhou, W.; Kang, Y.L.; Yu, Y. A Comprehensive estimation method on the origin of water production in gas wells of Jingbian gas field. *Nat. Gas Ind.* **2007**, *27*, 97–99.

33.   Zhou, W.; Luo, G.B.; Chen, Q.; Zhang, Z.L. Characteristics and Recognition of the Water Formation in M51 Gas Reservoir of Jingbian Gas Field. *Nat. Gas Ind.* **2008**, *28*, 61–63.

34. Yao, J.L.; Bao, H.P.; Ren, J.F.; Sun, L.Y.; Ma, Z.R. Exploration of Ordovician subsalt natural gas reservoirs in Ordos basin. *China Pet. Explor.* **2015**, *20*, 1–12.
35. Lan, Y.F.; Wang, D.X.; Fan, Y.H. Application of facies-controlling modeling in the erosion carbonate reservoir. *Nat. Gas Ind.* **2007**, *27*, 52–53.
36. Zhang, H.Y.; He, S.L.; Men, C.Q.; Guo, D.H. The countermeasures and application of geological modeling in complex carbonate reservoir: Example from the Jingbian gas field. *Nat. Gas Geosci.* **2012**, *23*, 1155–1162.
37. Zhang, J.G.; Wang, D.X.; Lan, Y.F.; Ai, Q.L. Study on the optimization of boosting stimulation in Jingbian gas field. *Drill. Prod. Technol.* **2013**, *36*, 31–32.
38. Liu, H.F.; Wang, D.X.; Xia, Y.; He, L.; Xu, Y.L.; Zhang, B.G. Development geology key technology for horizontal wells of carbonate gas reservoir with low permeability and thin layer. *Nat. Gas Geosci.* **2013**, *24*, 1037–1041.

Article

High-Frequency Sea-Level Cycle Reconstruction and Vertical Distribution of Carbonate Ramp Shoal Facies Dolomite Reservoir in Gucheng Area, East Tarim Basin

Tong Lin [1,2,3], Kedan Zhu [4,5,*], You Zhang [4,5,*], Zihui Feng [2,3], Xingping Zheng [4,5], Bin Li [6] and Qifan Yi [7]

[1] College of Energy, Chengdu University of Technology, Chengdu 610059, China; ltong0118@163.com
[2] Exploration and Development Research Institute, Daqing Oilfield Company Ltd., Daqing 163712, China; fengzihui@petrochina.com.cn
[3] Research Branch of Daqing Oilfield, Key Laboratory of Carbonate Reservoirs, CNPC, Daqing 163712, China
[4] Petro China Hangzhou Research Institute of Geology, Hangzhou 310023, China; zhengxp_hz@petrochina.com.cn
[5] Key Laboratory of Carbonate Reservoirs, CNPC, Hangzhou 310023, China
[6] College of Geosciences, China University of Petroleum (Beijing), Beijing 102249, China; 2020215026@student.cup.edu.cn
[7] School of Earth Sciences, Northeast Petroleum University, Daqing 163318, China; yiqifan@163.com
* Correspondence: redcloudszkd@163.com (K.Z.); zhangshiyouda@126.com (Y.Z.)

Citation: Lin, T.; Zhu, K.; Zhang, Y.; Feng, Z.; Zheng, X.; Li, B.; Yi, Q. High-Frequency Sea-Level Cycle Reconstruction and Vertical Distribution of Carbonate Ramp Shoal Facies Dolomite Reservoir in Gucheng Area, East Tarim Basin. *Energies* **2022**, *15*, 4287. https://doi.org/10.3390/en15124287

Academic Editors: Bo Zhang and Yuming Liu

Received: 8 April 2022
Accepted: 6 June 2022
Published: 11 June 2022

Publisher's Note: MDPI stays neutral with regard to jurisdictional claims in published maps and institutional affiliations.

Abstract: During the sedimentary period of the Ordovician Yingshan Formation, the carbonate platform of the Gucheng area in the Tarim basin was characterized by a distally steepened ramp. Relative sea-level changes exerted a strong influence on the shoal facie dolomite reservoirs of the 3rd Member of the Ordovician Yingshan Formation (the Ying 3 member), sedimented in the context of a shallow water environment on the carbonate ramp. However, previous studies that lacked high-frequency sea-level changes in the Gucheng area prevent further dolomite reservoir characterization. The current work carries out systematic sampling based on the continuous core from the upper and middle parts of the Ying 3 member in two newly drilled exploration wells (GC17 and GC601) and a series of geochemistry analyses, such as C-O isotope, Sr isotope, and rare earth elements (REE), which helps to investigate the features of the shoal facies dolomite reservoir development against high-frequency sea-level changes. With the help of Fischer plots of these two wells, high-density $\delta13C$ data (sample interval is about 0.272 m) were merged to construct a comprehensive curve, contributing to characterizing the high-frequency sea-level changes of the upper and middle parts of the Ying 3 member in the Gucheng area and validating the relationship between the pore-vug vertical distribution and high-frequency sea-level changes. Results revealed that the porosity of dolomite reservoirs increased when the high-frequency sea-level fell and decreased when it rose. Furthermore, the karst surface can be found at the top of the upward-shallowing cycle during the high-frequency sea-level falling; the pore-vug reservoirs are concentrated below the karst exposure surface, and porous spaces are more developed closer to the top of the cycle. The high frequency sea-level curve built in this study can be used as a standard for further research of regional sea-levels in the Gucheng area, and this understanding is highly practical in the prediction of shoal facies carbonate reservoir in carbonate ramp.

Keywords: high-frequency sea-level cycle; shoal facies dolomite reservoir; carbonate ramp; Ordovician Ying 3 member; Tarim basin

1. Introduction

Studies on paleo-relative sea-level changes are widely applied in many fields, such as carbonate sedimentology, paleoenvironment reconstruction, and hydrocarbon exploration [1–3]. In general, relative sea-level changes dominate the carbonate platform margin type and the structure of carbonate sediments [4]; these sediments thus record that information in their sedimentation, allowing researchers to study paleo-relative sea-level changes despite a lack of direct access to its measurement [1]. The most extensively used methods are as follows: sequence stratigraphy and sediment structure oriented [5–8], geochemistry oriented [9,10], natural gamma-ray spectral logging oriented [11–14], and Fischer plot oriented [6,15,16]. It is clear that the cycle order of relative sea-level changes varies since each method allows for individual scale, data density, and time-space resolution. Moreover, the conception of higher sea-level cycle orders requires more continuous data with better time-space resolution. In other words, characterizing the 4th and above order, especially to high-frequency sea-level cycles [17] in a region, must be based on obtaining high-frequency, high-continuity data. The methods in the previous studies mentioned above all have certain drawbacks when used alone. For example, although the seismic data used in the study of sequence stratigraphy have good lateral continuity, the vertical resolution is insufficient. Geochemical analysis requires a large amount of continuous data. It is relatively easy to sample outcrops in the field [18], but in underground hydrocarbon exploration, a continuous core is very precious and the cost is very high. Although the vertical resolution of the logging curve is high, it is necessary to analyze the sensitivity to reduce ambiguity. Thus, a combination of multiple methods seems sensible to pursue a better outcome.

With the deepening of China's marine carbonate hydrocarbon exploration, breakthroughs in shoal facies dolomite reservoirs inspired researchers, including the Deyang-Anyue area in the Sichuan basin and the Gucheng area in the Tarim basin [19–23]. These exploration examples demonstrate a strong link between the shoal facies dolomite reservoirs and the distribution of shoal facies carbonate sediments under the control of sea-level changes [22]. Relative sea-level changes influenced the productivity and type of carbonate by controlling sedimentary patterns and water energy changes and further influenced the sequence structures of carbonate rocks [24]. For the carbonate ramp, shoal facies dolomite reservoirs cover a broader range with greater gross thickness [25], owing to their long-distance migration along the shoreline and the swing of the high hydrodynamic energy zone; however, high-frequency relative sea-level changes produce more frequent facies migration, because of their gentle declivity and absence of barriers [5,26], which makes the stacking of thin shoal facies dolomite reservoirs more complex [27]. It can even show a mosaic-like distribution pattern lacking clear and regular trends in facies-to-facies transitions [18].

The 3rd member of the Ordovician Yingshan Formation (the Ying 3 member) in the Gucheng area of the Tarim basin witnesses considerable cycled thin intervals in pore-vug beds of its shoal facies dolomite reservoirs, and karst exposure surfaces as well, which indicates a strong relationship with high frequent sea-level changes. However, the restriction of incomplete coring formation and low sampling density in the existing exploration wells leaves studies on paleo-relative sea-level changes crude, let alone high-frequent sea-level changes, preventing further fine correlation and description. Therefore, the current work carries out a systematic sampling, based on the continuous core from the upper and middle parts of the Ying 3 member in two newly drilled exploration wells (GC17 and GC601), and a series of geochemistry analyses, such as the C-O isotope, Sr isotope, and rare earth elements (REE), which helps to investigate the features of the shoal facies dolomite reservoir development in the context of high-frequency sea-level changes. With the help of Fischer plots of these two wells, high-density $\delta^{13}C$ data contribute to characterizing high-frequency sea-level changes of the upper and middle parts of the Ying 3 member in the Gucheng area and validating the relationship between vertical pore-vug distribution and high-frequency sea-level changes. The current work fills the blank of the high-frequency sea-level cycle in the Ying 3 member of the Gucheng area. The high frequency sea-level curve built in this study can be used as a standard for further research of regional sea-levels in this area, which is also highly practical in the prediction of shoal facies carbonate reservoir in carbonate ramps.

2. Geological Background

The Gucheng area is situated in the mid-south of the North Depression in the Tarim Basin, on the slope of the Tazhong uplift [28]. It is a second-order structural unit located on the southeast margin of the West Tarim platform [29], adjacent to the Manxi low bulge in the north, and to the Tadong uplift to the east [30,31]. The top surface of the Ordovician carbonate rocks in the target area is structurally a large-wide nose-like uplift inclining toward the northwest, cut by NE faults into several fault blocks with grabens alternating with horsts [28] (Figure 1a).

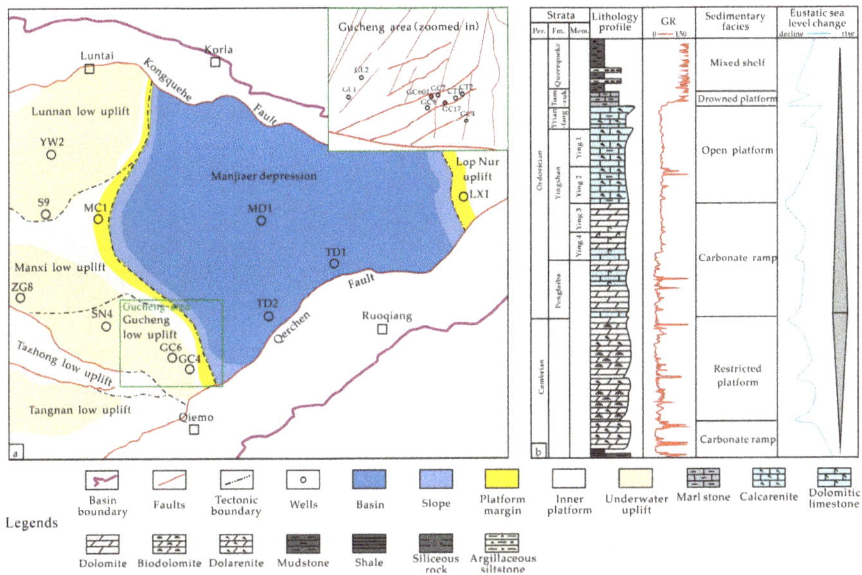

Figure 1. (**a**) Ordovician lithofacies paleogeography of Tarim basin and location of the Gucheng area; (**b**) Strata column of the Gucheng area.

In the weak extensional tectonic setting of the Cambrian–Early Ordovician period [32], carbonate platforms developed widely in the Gucheng area with a general rise in sea-level. The seismic response showed that the platform style was a distally steepened ramp in the period of the Yingshan Formation [33,34]. Additionally, practice experience of the shoal facies reservoirs shows that the Ordovician Yingshan Formation in this area developed an extensive inner platform shoal and marginal shoal composed of dolomites. Among them, the lower Yingshan Formation (including the Ying 3 member) was dominated by fine- to medium-crystal dolomite with residual granular structures, which were considered favorable reservoir rocks, while the upper Yingshan Formation by calcarenite and dolomite grainstones [28] is a lower quality reservoir than the lower Yingshan Formation (Figure 1b).

After the extensive analysis of lithofacies, several distinct microfacies were found, namely sand shoals (Figure 2a,b), dolomitized shoals (Figure 2c,d), inter-shoal marine (Figure 2e,f), and tidal flat (Figure 2g,h). Moreover, sedimentary structures, such as cross bedding (Figure 2c,d), karstic mosaic (Figure 2i), and seepage silt (Figure 2j), are common in dolomitized shoals, indicating the hydrodynamic environment, during the depositional period, generally featured relatively high-energy turbulence. When it comes to cathodo-luminescence analysis, the fine-medium crystalline dolomite of dolomitized shoals is generally the most common in dim, or tan light, and brown light under the cathode rays (Figure 2k,l). This indicates that shoal facie dolomite is mainly affected by the burial process and may be partially transformed by late hydrothermal fluids.

Figure 2. Macroscopic and microscopic scale images of typical facies and sedimentary phenomenon in the Ying 3 member. (**a**) Calcarenite, core sample, GC601 6065.25 m, sand shoals; (**b**) Calcarenite, plane-polarized light, GC601 6044 m, sand shoals; (**c**) Dolomite, core sample, cross bedding, GC601 6046.5 m, dolomitized shoals; (**d**) Dolomite, plane-polarized light, GC601 6061.87 m, dolomitized shoals; (**e**) Micrite, core sample, pelitic strip, GC601 6162.42 m, inter-shoal marine; (**f**) Micrite, plane-polarized light, pelitic strip and dolomite crystal, GC601 6047 m, inter-shoal marine; (**g**) Crystal powder dolomite, core sample, GC601 6067.05 m, tidal flat; (**h**) Crystal powder dolomite, plane-polarized light, GC601 6067.05 m, tidal flat; (**i**) Dolomite with karstic mosaic, core sample, GC601 6131.52 m; (**j**) Crystal dolomite with seepage silt, plane-polarized light, GC601 6066.64 m; (**k**) Dolomite, plane-polarized light, GC601 6111.36 m, dolomitized shoals; (**l**) CL image in the same field of vision with (**k**).

3. Materials and Methods

3.1. Experimental Materials and Methods

GC17 and GC601, two exploration wells, were drilled in Guchengarea within the last 5 years, which have collected over 150 m of continuous core in the Ying 3 member, lower Ordovician, as well as wire logging data, such as GR logging, natural gamma-ray spectral logging, and resistivity logging. In this study, oxygen and carbon stable isotope analysis was performed on 547 dolomite and limestone wall rock samples taken from core and sidewall coring at an average sampling interval of about 0.272 m, with the average data density reaching about 3.68 per meter. Among these samples, 52 were tested by hole rock element analysis and rear earth element (REE) analysis, and 85 were analyzed by the strontium isotope. Two batches of these geochemistry experiments were carried out in different laboratories: the Key Laboratory of Carbonate Reservoirs, CNPC, and the Experiment Center of Exploration and Development Research Institute of Daqing Oilfield. At the same time, the porosity test data of 679 carbonate rock samples in the Ying 3 Member of these two wells were collected.

The oxygen and carbon stable isotope values adopted the Vienna Peedee Belemnite standard (VPDB). Carbonate powders were reacted with 100% phosphoric acid for 4 h at 25 °C for calcite and at 50 °C for dolomite, and the resultant CO_2 was measured to determine its oxygen and carbon isotopic ratios using a Delta V advantage + Gasbench mass spectrometer. The reproducibility values of the isotopic measurement for both carbon and oxygen isotopes were better than ±0.01%.

The REE analysis method and process follow the general method of inductively coupled plasma mass spectrometry (ICP-MS) with an Element XR inductively coupled plasma mass spectrometry instrument, whose limit of detection is 10^{-12}, and the measurement error can be regulated to within 5%, meeting the measurement accuracy standards. Selected samples are then crushed to 75 μm in the agate mortar and put into paper sample bags for later use. Acetic acid and nitric acid are used to dissolve and purify, and scaling solution is added to fix the volume to 10 mL. Then, the sample well is shaken for the final instrument test.

The $^{87}Sr/^{86}Sr$ isotope ratios are tested for selected matrix dolomite and dolomite cements using a Neptune Plus mass spectrometer (MC-ICP-MS). The test temperature was 20 °C, the humidity was 40% RH, the single band was Ta band, and the ionization temperature was 1450 °C The analytical precision of the individual runs is determined to be 0.00005 (2σ). The $^{87}Sr/^{86}Sr$ of the instrument test standard NBS987 is 0.710244 ± 0.000004. The mean standard error of the mass spectrometer performance was ±0.00003, which conforms to the Chinese national standard GB/T 17672-1999.

3.2. High-Frequency Sea-Level Curve Reconstruction Methods

In the study of paleo-relative sea-level changes, the oxygen and carbon stable isotopes set a solid foundation for capturing information on sea-level fluctuation [1,9,35,36], with $\delta^{13}C$ having been widely applied [36,37], whose principle lies in the fact that the increase of $\delta^{13}C$ in carbonate rocks usually relates to the grown biological productivity or the deepened burial of organic matter when the relative sea-level rises. Therefore, there is a positive correlation between $\delta^{13}C$ and the relative sea-level in geological history—specifically, $\delta^{13}C$ increases when the sea-level climbs up and declines when it drops [38–41]. Given this, the carbon isotope data of Wells GC601 and GC17 were used to characterize the high-frequency sea-level cycles of the Ying 3 Member in the target region. In order to exploit the collected data of Wells GC17 and GC601 as fully as possible, and to develop a more complete pattern of sea-level changes, we managed to merge the data vertically and get a comprehensive curve on the following conditions:

First, the consistent features that are correlative between Wells GC17 and GC601 must be recognized, which helps to identify the depth correlation between these two wells that can be mapped to a vertical axis. However, it is hardly possible to directly apply lithologic characteristics and well-logging features to finish the stratigraphic correlation. The Fischer plot is thus introduced in this study to assist the task. It is a semi-quantitative graphic of sea-level changes drawn by cumulative departure from mean cycle thickness (CMDT) proposed by Fischer in 1964 [42,43]. After further improvement by Sadler et al. [44] and Day [45], this method has been widely used to research relative sea-level changes based on data from outcrop sections and subsurface cores. Generally, the vertical axis of the plot is the CDMT, while the horizontal axis is the cycle number in the time domain (which can be transformed into a depth domain). In addition, the average cycle thickness is chosen as the subsidence correction factor, and the difference between the thickness of a given cycle and the average thickness brings about the net variation (growth/reduction) of the accommodation, so the overall accommodation variation tendency (Figure 3) can be illustrated by the continuous connection of the net variation with polygonal lines. In the carbonate sedimentary environment of shoal facies, for instance, the accommodation change has a close relation with the change of relative sea-levels, where the plots can lead to a visually quantitative determination of the relative sea-level changes trend. The cycle thickness for Fischer plots can be obtained either from cores or outcrops but also from cycle statistics in indirect data (e.g., well logs), which makes this method highly applicable.

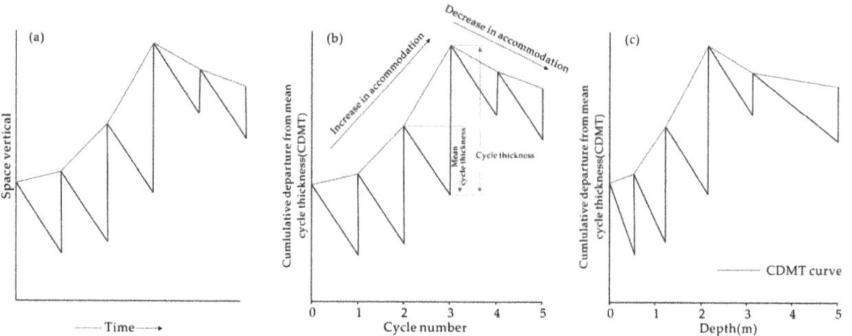

Figure 3. (**a**) Original concept of Fischer plots introduced by Fischer (1964); (**b**) Schematic form of Fischer plot in time domain; (**c**) Schematic form of Fischer plots in depth domain. Cited from Yang et al., 2021 [2].

Secondly, errors and anomalies are inevitable, so it is neither realistic nor desirable to blindly pursue a restoration of the original variation tendency of carbon isotopes with depth. Instead, it is necessary to refine the curve smoothing method to obtain a better reflection of the high-frequency variation tendency of carbon isotopes, with a simultaneous reduction of the interference of random errors and anomalies of data. In this study, the moving average smoothing method was used to process the original carbon isotope data, which is used to successively calculate the arithmetic mean of a given set of values by dividing data into several sets according to a specified time span to reflect the long-term trend. The moving average method can eliminate the influence of periodic and random fluctuations in the time series and capture the development orientation and trend of data changes [46], which is more practical for highlighting the trend of continuous time-dependent sea-level changes during geological history. The moving average can be expressed as follows:

$$T_i = \frac{1}{2m+1} \sum_{k=-m}^{m} T_{i+k} \tag{1}$$

T_i: The data point to be smoothed.

m: The data interval centered at point T_i.

Finally, we need to verify whether the established carbon isotope ratio variation with depth can reflect high-frequency sea-level cycles. In this study, the anomalies of the REE cerium (δCe) and strontium isotope ratios were used to prove the pattern of sea-level changes. As the result of the ionic Ce in the oxidation state, the anomaly of the REE Cerium (Ce) was first proposed by Elderfield and Greaves (1982), and its value is usually presented by δCe [47].

$$\delta Ce = \log [3Ce_n/(2La_n + Nd_n)] \tag{2}$$

where Ce_n, La_n and Nd_n are normalized values to the North American shale composite (NASC; Gromet et al., 1984) [48].

Wilde et al. pointed out that δCe is a potential indicator of paleo-relative sea-level changes that can serve as supplemental evidence in the study [49].The negative δCe deviation of the bulk rock indicates more of a reducing environment or sea-level rising, while the positive deviation suggests a more oxidizing environment or sea-level falling [50]. As a symbol of the level of oxygen deficiency and eustatic changes, independent of sedimentological or seismic factors, the bulk rock δCe has been applied in research on sea-level changes and marine environments from the early Paleozoic to Precambrian worldwide [49,51]. However, the $^{87}Sr/^{86}Sr$ isotope ratio of sedimentary carbonate rocks are mainly dominated by two factors: the mantle-derived strontium with lower initial values, and crust-derived strontium with higher initial values from weathering of ancient aluminosilicate rocks in the continental crust. Without extensive submarine volcanic activities, the input of crust-derived strontium is the main controlling factor of the marine strontium isotope ratio. Additionally, the intensity of weathering has a primary influence on sea-level changes—the rise in sea-levels leads to a decrease in the weathering rate, a decrease in the input of the crust-derived strontium, and ultimately a drop in the marine strontium isotope ratio; on the contrary, the fall of sea-levels results in the growth of the marine strontium isotope ratio [35].

According to the methods shown above, 4 steps and a workflow were designed to realize this study (Figure 4).

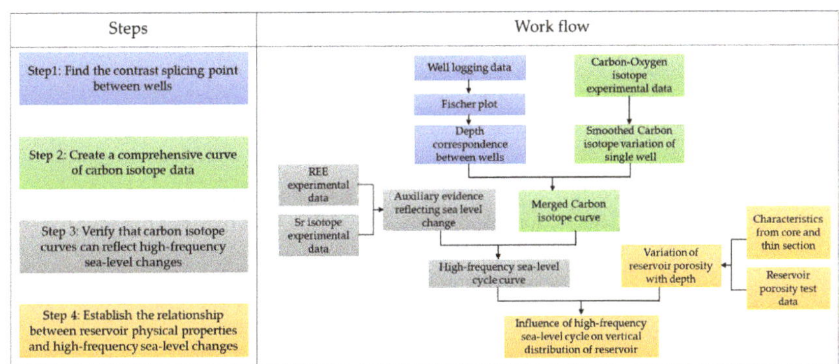

Figure 4. Steps and workflow of this study.

4. Experimental Results

The results of the C-O isotope PDB values, Sr isotope ratios, and calculated δCe values from the Wells GC601 and GC17 samples are presented in Figure 5 and Table 1, with all original data mapped in depth domain. The δ^{18}O values of the samples ranged from -14% to -2%, and the δ^{13}C values of the dolomites ranged from -3.9% to 0.5%. Moreover, a fluctuation pattern can be drawn from the C-O isotope values. However, δ^{13}C values display more convergence and fluctuation than δ^{18}O values. The Sr isotope ratios of over three-quarters of the samples are within the range of the Latest Cambrian to Middle Ordovician seawater (0.7079–0.7092) [52]. In addition, δCe values show some fluctuation pattern as well, despite their sparse distribution.

Table 1. An overview of experimental data statistics.

Well No.	Analysis	Amount	Result Range
GC 601	C-O isotope	396	δ^{13}C: $-2.802\%\sim-0.537\%$ δ^{18}O: $-13.917\%\sim-4.085\%$
	^{87}Sr/^{86}Sr	73	0.708896~0.712001
	δCe	40	$-0.0746\sim0.0287$
GC 17	C-O isotope	151	δ^{13}C: $-2.927\%\sim-0.740\%$ δ^{18}O: $-10.190\%\sim-5.909\%$
	^{87}Sr/^{86}Sr	12	0.708927~0.710174
	δCe	12	$-0.0456\sim-0.0031$

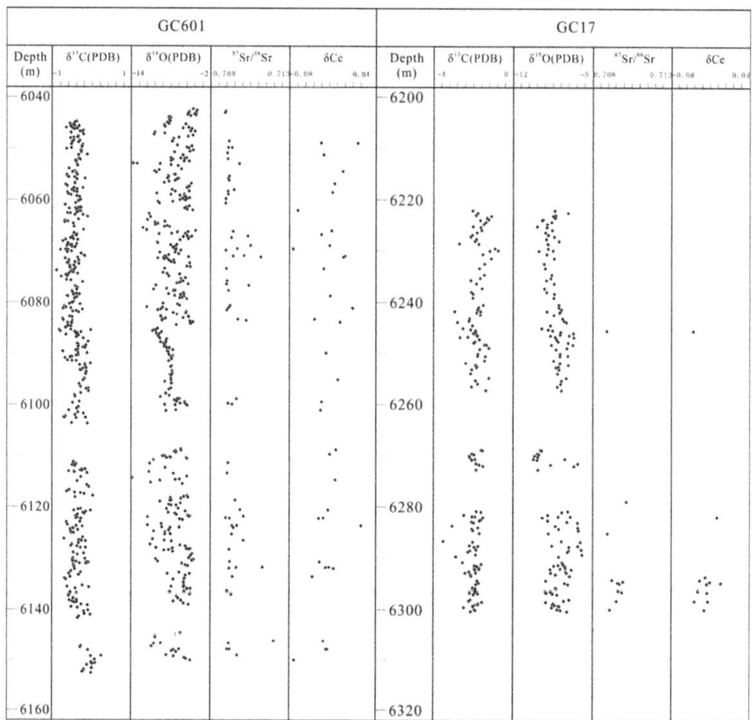

Figure 5. Original testing data of samples from well GC601 and GC 17.

5. Discussion

5.1. Correlation by Fischer Plot

In the current work, the Python program coded by Yang [2] was employed for automatic plotting based on GR curves, overcoming the subjectivity caused by manual identification of cycles. The Fischer plots (Figure 6) of the two wells, highly correlated with one another, demonstrate that sea-levels of the middle and upper parts of the Ying 3 member fluctuate and the variation is characterized by falling, then rising, subsequently maintaining, and finally dropping again. This trend is consistent with the second-order sea-level fluctuation of the Tarim Basin [10], and the third-order sea-level change curve of the Ying 3 member in the southeast Tarim [3], which provides good support for the application of such a method. A comparison of the Fischer plots of the middle and upper parts of the Ying 3 Member of these two wells shows that there are locally minor differences in high-frequency cycles, with overall consistency, which may result from multiple interpretations of GR logs or local differences in sedimentary palaeogeomorphology. GR logs work with the key inflection points in the Fischer plots together to fulfill the depth correlation between the two wells—5995 m in Well GC601 corresponds to 6223 m in Well GC17 (Figure 6).

5.2. Reconstruction of Carbon Isotope Comprehensive Curve

Geochemical features of ancient carbonate rocks are prone to changes triggered by later diagenesis, resulting in partial or complete loss of geochemical information of seawater in the original sedimentary period [38,52]. It is necessary to ensure that the carbon and oxygen isotope data are reliable to reflect the characteristics of the original seawater. A commonly used method for reliability evaluation is to check whether the carbon and oxygen isotope values of the samples are correlated [54]. Because of a more sensitivity of the oxygen isotopes, rather than the carbon isotopes, to the diagenetic alteration, higher correlations between carbon and oxygen isotope values indicate that the carbon and oxygen isotopes are subjected to coordinated changes during diagenesis, and the data are far from reliable; on the contrary, when the correlation is low, samples are less affected by the diagenetic alteration, of which isotopic values can reflect the isotope composition of the original seawater [55]. Therefore, when the carbon and oxygen isotope values of Wells GC601 and GC17 were cross plotted, it was found that $\delta^{13}C$ and $\delta^{18}O$ of the two wells are highly scattered with no distinct linear correlation (Figure 7), and the data are thus considered reliable.

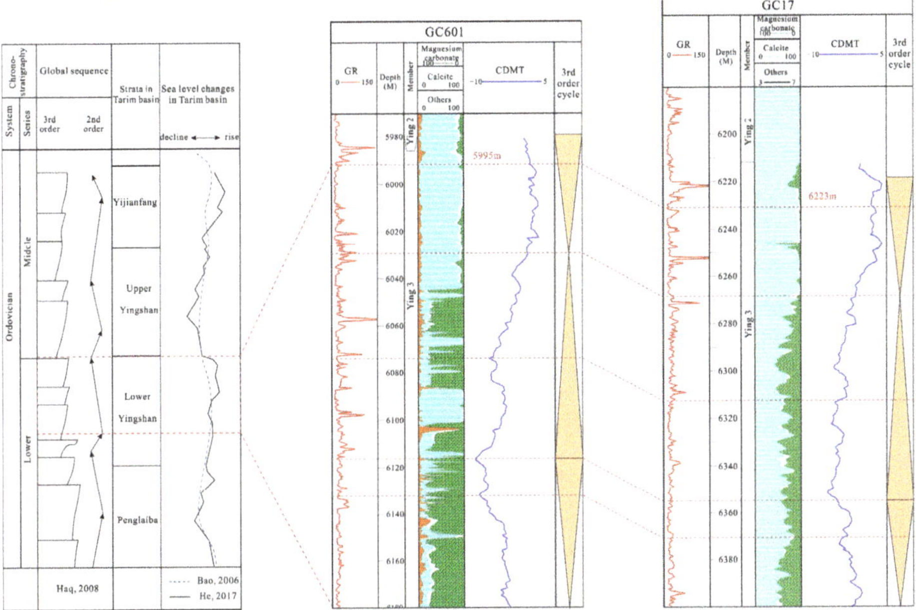

Figure 6. Fischer plots of well GC601 and GC 17 and their comparison with the previous third-order sea-levels. The global sequence is cited from Reference [53], the sea-level fluctuation of the Tarim Basin is cited from References [3,10].

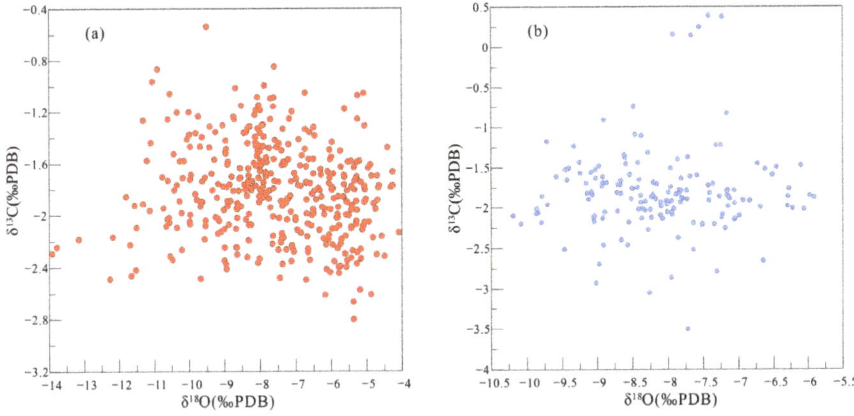

Figure 7. Cross plot of carbon and oxygen isotope values. (**a**) GC601. (**b**) GC17.

When the $\delta^{13}C$ data of Wells GC601 and GC17 were mapped by depth (Figure 5) and displayed with a proper vertical scale and identical horizontal scale range, a fluctuating pattern can then be identified, which is consistent with the periodic rise and fall of sea-levels. Due to the periodic and random fluctuations, the original $\delta^{13}C$ data fluctuated fiercely; it is not easy to make a direct comparison or to identify high-frequency cycles. However, by comparing the envelopes of the data points, it can still be found that the $\delta^{13}C$ fluctuation at 6267–6300 m of Well GC17 and 6043–6076 m of Well GC601 are in line with each other (Figure 8), and the corresponding depth correlation complies with that drawn from the Fischer plots (Figure 6), indicating that this correlation makes the carbon isotope curve mergence feasible. After the implementation of the 5-point moving average smoothing, the data points become more convergent, and the fluctuation trend becomes more obvious (Figure 9, smoothed $\delta^{13}C$ column). The smoothed $\delta^{13}C$ data can be merged by the depth correlation drawn from the Fischer plots, and then the high-frequency carbon isotope fluctuation curve of the whole middle and upper parts of the Ying 3 member in the Gucheng area can be obtained (Figure 9).

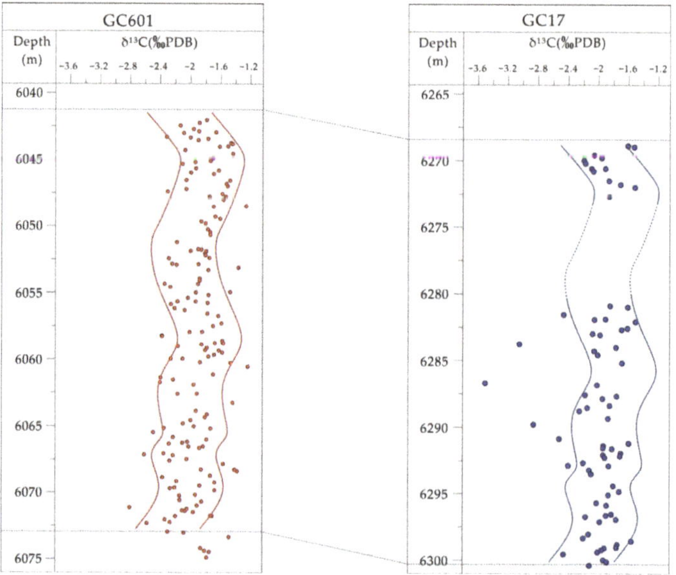

Figure 8. Comparison of carbon isotope variation with depth between the two wells.

5.3. Verification of Carbon Isotope Comprehensive Curve

In this study, bulk rock REE analysis was carried out in 52 core samples from Wells GC17 and GC601, with δCe calculated. The REE analysis and δCe calculation results were then mapped by depth, revealing that the interval with a positive carbon isotope value anomaly is accompanied by the negative deviation of δCe; on the contrary, the positive deviation of δCe tends to occur in the case of negative deviation of the carbon isotope value. In addition, the strontium isotope ratio of the interval with the negative carbon isotope deviation is often higher than that of the Ordovician seawater, implying the influences of the meteoric freshwater. Therefore, it is concluded that the high-frequency carbon isotope value fluctuation can work as the high-frequency sea-level changes, and four fourth-order cycles, and 21 high-frequency cycles were identified by the curve (Figure 10).

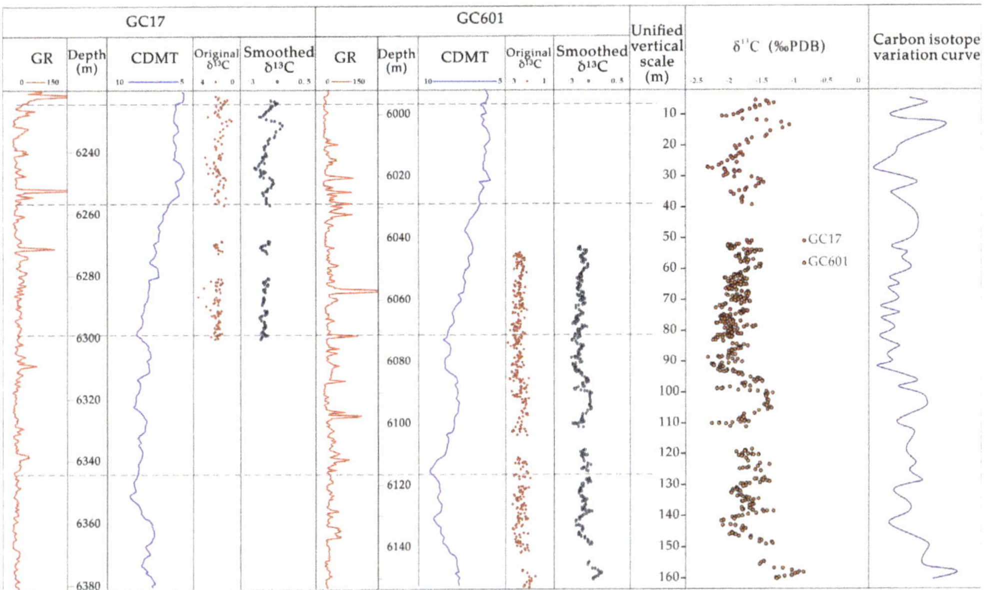

Figure 9. The developed comprehensive curve of carbon isotope variation with depth.

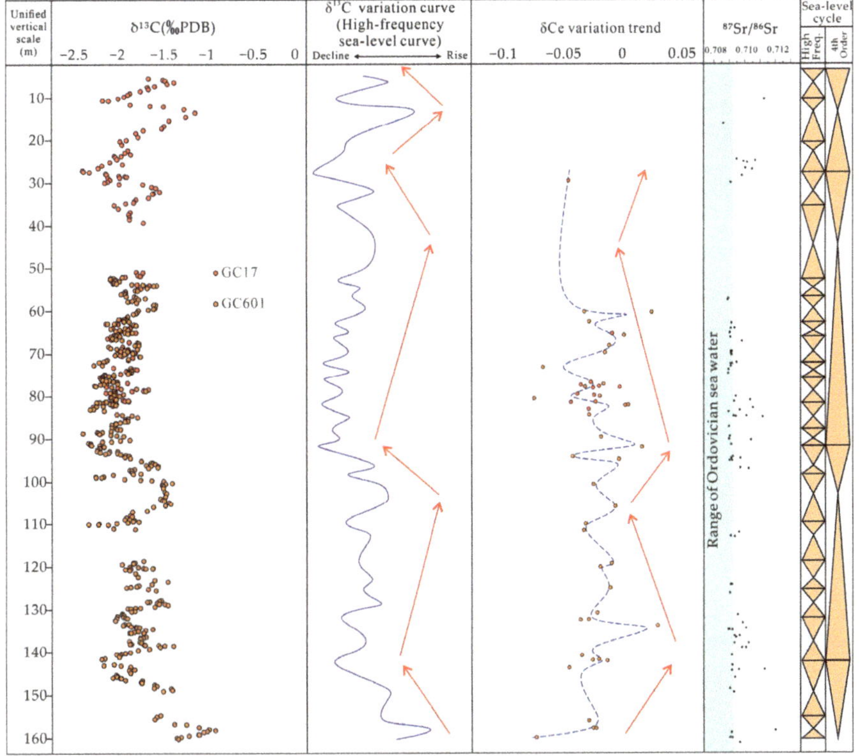

Figure 10. Verification of the effectiveness of the merged carbon isotope variation curve with depths to represent sea-level changes. Red arrows showed the trend of $\delta^{13}C$ and δCe.

5.4. Relation between High-Frequency Sea-Level Changes and Reservoir Characteristics

The current work carried out some analyses to explore the way high-frequency sea-level changes affect the carbonate ramp shoal facies dolomite reservoir, namely macroscopic core observation, thin section microscopic analysis, reservoir physical property testing, and geochemical testing. More than 20 cyclic karst surfaces as a result of the penecontemporaneous exposure dissolution were identified based on the systematic observation of the dolomite cores of the Ying 3 member in Wells GC17 and GC601 (Figure 11). These karst surfaces are commonly filled with sparry calcuate, due to the irregular space created by karstification, and macroscopically feature chaotic and mottling and a common geopetal texture. The microscopy also reveals typical karst features, such as the geopetal fabric and the vadose zone of silts (Figure 2i,j and Figure 12). Additionally, the $\delta^{13}C$ has seen considerable negative deviations near karst surfaces.

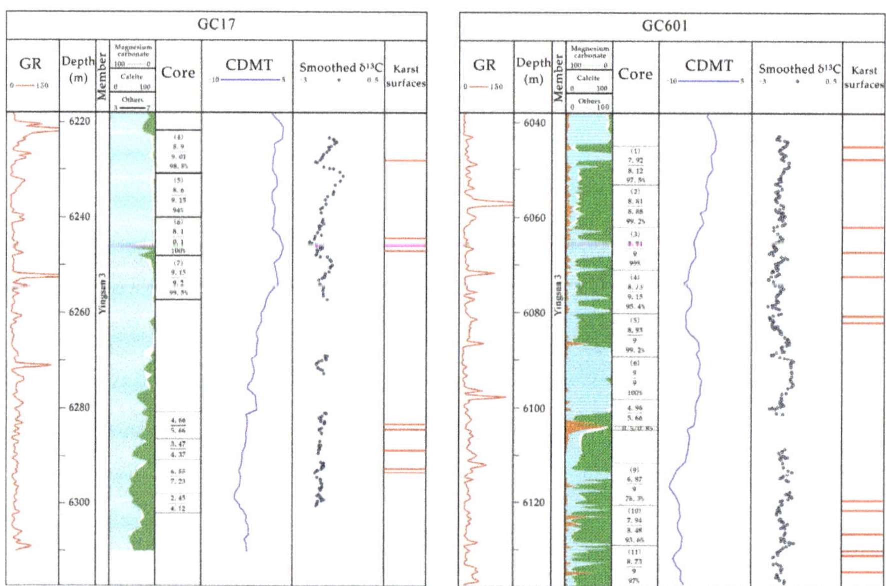

Figure 11. Cyclic karst exposure surfaces in Wells GC17 and GC601.

The measured porosity of the shoal facies dolomite cores from 6280–6300 m of Well GC17 is mapped by depth (trend is marked in red line) and compared with the high-frequency sea-level change curve of the interval at the same depth of the Ying 3 member issued from the carbon isotope value (Figure 13). In general, there is an obvious negative correlation between both, with dolomite porosity increasing when the high-frequency sea-level falls and decreasing when it rises. Furthermore, at 6295–6300 m, the Sr isotope ratios of the high-porosity samples are much higher than those of the Ordovician seawater (ranging 0.7079–0.7092) [52]. As shown in the high-frequency sea-level change curve, the interval mentioned above responds to one maximum sea regression; however, the distinct positive deviation of the Sr isotope ratio indicates that the exposure dissolution intensifies when the sea-level falls, and thus, a mixture with the terrigenous Sr occurs due to the meteoric freshwater invasion contributing to the development of pores in this interval.

Similarly, when it comes to mapping the porosity of the continuously-cored shoal facies dolomite ranging from 6040–6150 m (nearly 90 m, trend is marked in red line) of Well GC601 by depth (Figure 14), we find that the correlation between the porosity and the high-frequency sea-level change is negative as well, even over an extensive interval, which is consistent with the pattern in Well GC17. In addition, the Sr isotope ratio is often higher in the intervals with high porosity than that of the seawater of the same period. The

REE pattern of samples from the karst surface showed a negative anomaly of element Eu (Figure 15), which means they experienced modification by low temperature and oxidizing fluid in the penecontemporaneous period. This is also evidence of exposure.

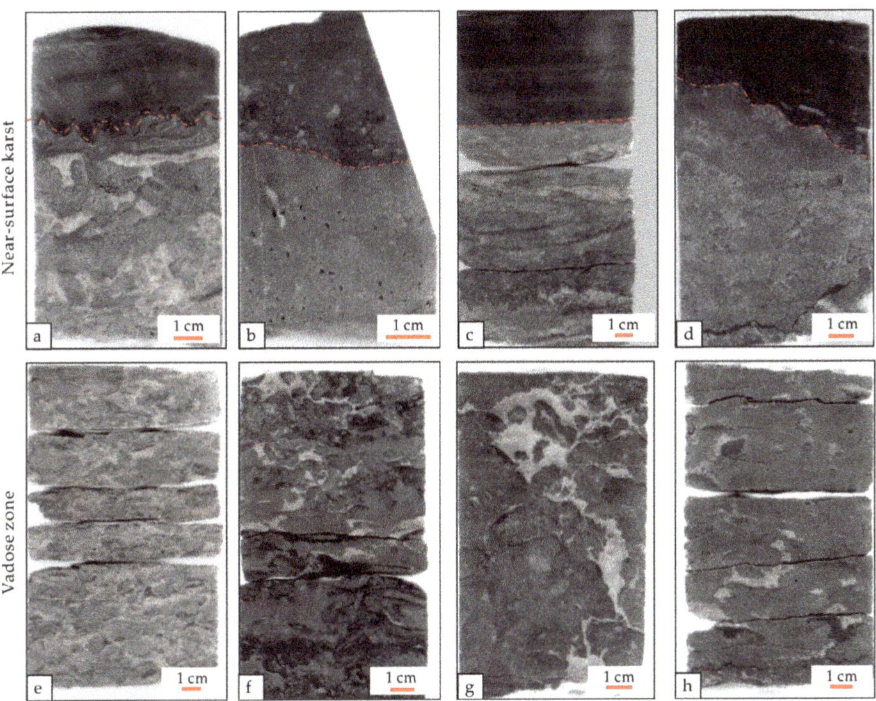

Figure 12. Macroscopic characteristics of the karst exposure surfaces in Well GC601. Karst surfaces are marked in dashed lines. (**a**) 6167.52 m. (**b**) 6125.37 m. (**c**) 6109.81 m. (**d**) 6067.15 m. (**e**) 6167.7 m. (**f**) 6122.12 m. (**g**) 6079.12 m. (**h**) 6057.84 m.

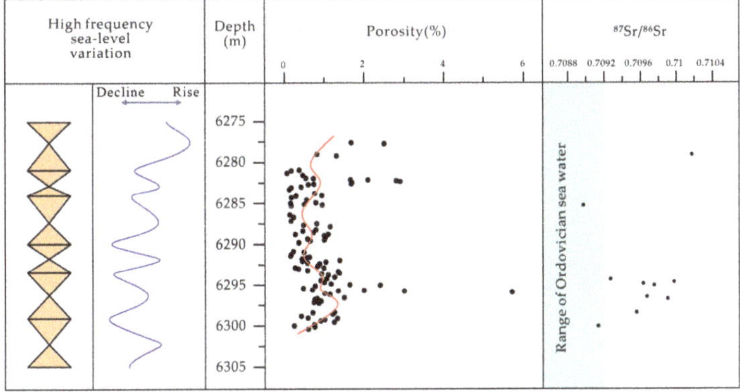

Figure 13. Correlation between reservoir porosity and high-frequency sea-levels and Sr isotope ratio characteristics in GC17. The red line showed the trend of porosity.

Core analysis of Well GC601 shows that such penecontemporaneous exposure dissolution occurs frequently during the sea-level falling, with the corresponding cycle thickness even at the meter scale. For example, the interval ranging from 6080–6089 m responds to an obvious high-frequency sea-level fall (Figure 14, marked by dashed box). When the vertical scale is enlarged (Figure 16), we can recognize several obvious meter-scale cycles in the carbon isotope value variation, which is highly consistent with the core records, showing that the high-frequency sea-level falling is associated with the frequent exposure and modification of the shallow-water ramp belt. Furthermore, pore-vug intervals are primarily found near the karst exposure surface. Such intervals, recognized either by cores or thin sections, are generally located under the karst exposure surface (Figure 13). The tight dolomite with less porosity of the grain shoal facies occurs in an alternating manner with the porous dolomite of the grain shoal facies, resulting in many upward-shallowing cycles. Pores tend to develop closer to the top cycle, which demonstrates that the physical properties of the dolomitized reservoirs of the shoal facies are dependent on the periodic penecontemporaneous exposure dissolution.

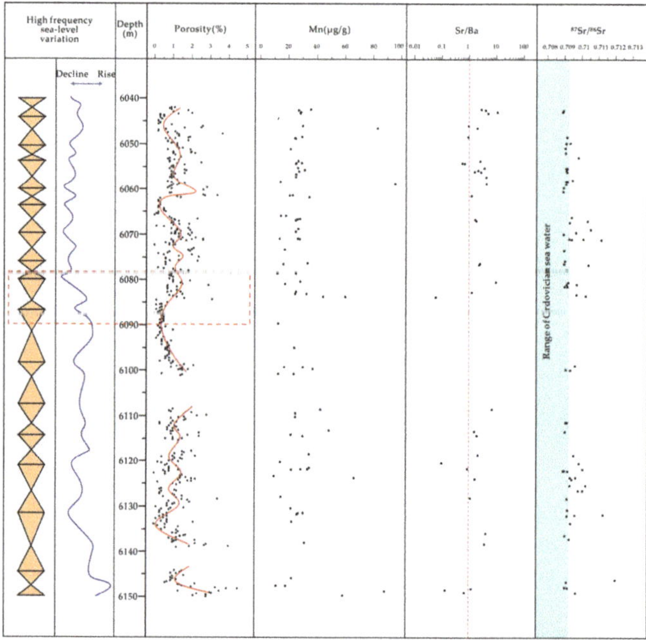

Figure 14. Correlation between reservoir porosity and high-frequency sea-levels and other geochemical evidence in GC601. The red line showed the trend of porosity. The part in dashed box will be showed in Figure 16.

5.5. Limitations and Future Work

In the current work, a relatively reliable high-frequency sea-level change curve of the Ying 3 member has been built. With the help of this curve, the relation between the shoal facies dolomite reservoir and the high-frequency sea-level cycle was established and examined in Wells GC17 and GC601. Moreover, this study proves that the Fischer plot can be applied in bridging data from different wells. However, there are still several limitations that need to be overcome in future work. For example, in the current work, the GR curve was used to build the Fischer plot, which is the key step to correlate these two wells. However, when it comes to some other wells in this area, they may show a less comparable result. This may have resulted in the loss of original sea water environment information during the burial process and diagenesis. To solve this problem, it is necessary to further optimize the logging curve and to do more work on sample selection. If this can be overcome in future work, more data can be applied in the study to build a more accurate sea-level curve. With more wells involved, it may be possible to build a general model of the shoal facies dolomite reservoir.

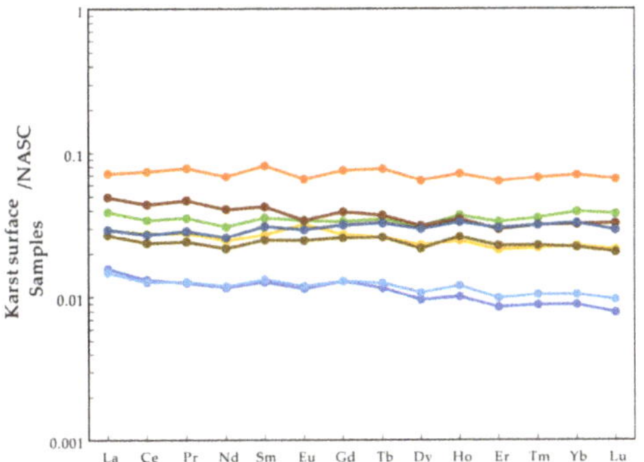

Figure 15. The REE pattern of samples from the karst surface. Each line with colors represents one sample from karst surface.

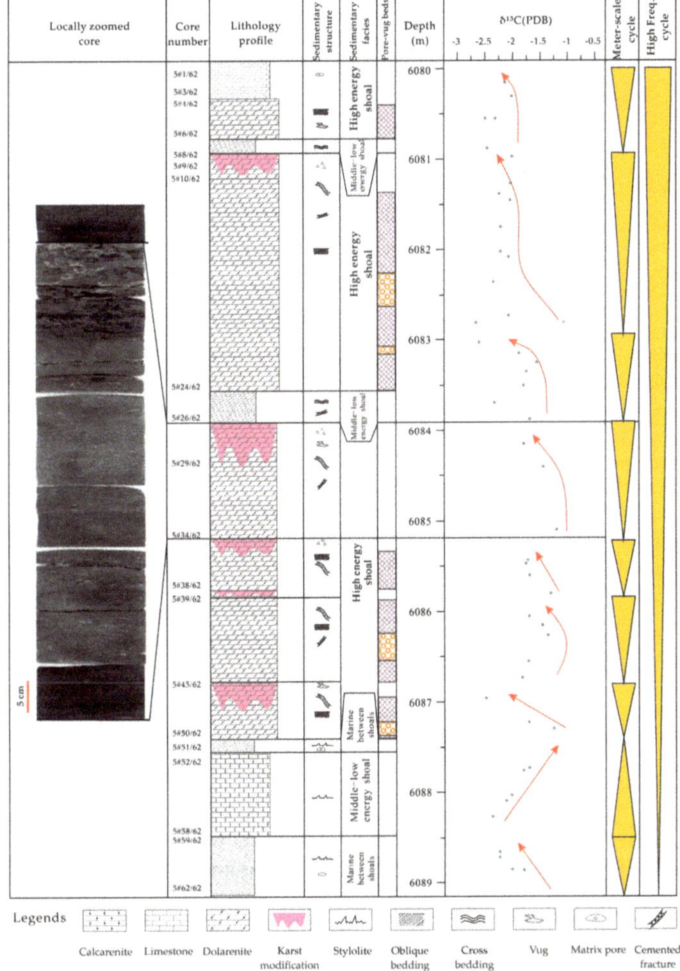

Figure 16. Variation pattern of the meter-scale cycles in Well GC601. Red arrows showed the trend of δ^{13}C.

6. Conclusions

With the Fischer plots used as the basis for the inter-well merging, the carbon isotope data of Wells GC17 and GC601 with high sampling density (the sampling interval is about 0.272 m) were integrated to develop a relatively reliable high-frequency sea-level change curve for the Gucheng area of the Tarim Basin, where four fourth-order cycles and 21 high-frequency cycles have been identified. Moreover, the merged curve shows that the vertical distribution of the porosity of the carbonate ramp shoal dolomite reservoir is closely related to the high-frequency sea-level change. Specifically, the porosity of the dolomite increases with the high-frequency sea-level falling and decreases when it rises. Meanwhile, the karst exposure surface is developed at the top of the upward-shallowing cycle due to the frequent exposure and modification of the shallow-water ramp belt when the high-frequency sea-level falls. The pore-vug reservoirs are concentrated below the karst exposure surface, and pores are more developed closer to the top of the cycle, indicating that the properties of the dolomitized reservoirs of the shoal facies are dominated by the penecontemporaneous periodic exposure dissolution.

Author Contributions: Conceptualization, T.L. and K.Z.; methodology, K.Z. and T.L.; software, Y.Z.; validation, Y.Z., K.Z. and X.Z.; formal analysis, K.Z. and T.L.; investigation, T.L.; resources, T.L. and Y.Z.; data curation, K.Z.; writing—original draft preparation, T.L. and K.Z.; writing—review and editing, Y.Z., X.Z., Q.Y. and B.L.; visualization, K.Z., T.L. and B.L.; supervision, Z.F. and Y.Z.; project administration, Y.Z.; funding acquisition, Y.Z. All authors have read and agreed to the published version of the manuscript.

Funding: This work is supported by the Scientific Research and Technology Development Project of CNPC (grant number 2021DJ0501).

Institutional Review Board Statement: Not applicable.

Informed Consent Statement: Not applicable.

Data Availability Statement: Not applicable.

Acknowledgments: The authors are grateful to the reviewers, whose comments greatly improved the original manuscript.

Conflicts of Interest: The authors declare no conflict of interest.

References

1. Xiao, J.; Wang, L.; Chen, M.; Chen, Z.; Zhou, J.; Chen, P. Multiple scale fluctuations of the Early Triassic sea level and its influence on reservoirs in the Sichuan Basin. *Pet. Geol. Exp.* **2017**, *39*, 618–624.

2. Yang, D.; Huang, Y.; Chen, Z.; Huang, Q.; Ren, Y.; Wang, C. A Python Code for Automatic Construction of Fischer Plots Using Proxy Data. *Sci. Rep.* **2021**, *11*, 10518. [CrossRef] [PubMed]

3. He, F.; Lin, C.; Liu, J.; Zhang, Z.; Zhang, J.; Yan, B.; Qu, T. Migration of the Cambrian and Middle-Lower Ordovician carbonate platform margin and its relation to relative sea level changes in southeastern Tarim Basin. *Oil Gas Geol.* **2017**, *38*, 711–721.

4. Read, J.F. Carbonate Platforms of Passive (Extensional) Continental Margins: Types, Characteristics and Evolution. *Tectonophysics* **1982**, *81*, 195–212. [CrossRef]

5. Mei, M. From vertical stacking pattern of cycles to discerning and division of sequences: The third advance in sequence stratigraphy. *J. Palaeogeogr.* **2011**, *13*, 37–54.

6. Wang, Q.; Han, J.; Li, H.; Sun, Y.; He, H.; Ren, S. Carbonate sequence architecture, sedimentary evolution and sea level fluctuation of the Middle and Lower Ordovician on outcrops at the northwestern margin of Tarim Basin. *Oil Gas Geol.* **2019**, *40*, 835–850, 916.

7. Zhang, Y.; Chen, J.; Zhou, J.; Yuan, Y. Sedimentological Sequence and Depositional Evolutionary Model of Lower Triassic Carbonate Rocks in the South Yellow Sea Basin. *China Geol.* **2019**, *2*, 301–314. [CrossRef]

8. Jamaludin, S.N.F.; Pubellier, M.; Menier, D. Structural Restoration of Carbonate Platform in the Southern Part of Central Luconia, Malaysia. *J. Earth Sci.* **2018**, *29*, 155–168. [CrossRef]

9. Li, W.; Zhang, J.; Hao, Y.; Ni, C.; Tian, H.; Zeng, Y.; Yao, Q.; Shan, S.; Cao, J.; Zou, Q. Characteristics of carbon and oxygen isotopic, paleoceanographic environment and their relationship with reservoirs of the Xixiangchi Formation, southeastern Sichuan Basin. *Acta Geol. Sin.* **2019**, *93*, 487–500.

10. Bao, Z.; Jin, Z.; Sun, L.; Wang, Z.; Wang, Q.; Zhang, Q.; Shi, X.; Li, W.; Wu, M.; Gu, Q.; et al. Sea-Level Fluctuation of the Tarim Area in the Early Paleozoic: Respondence from Geochemistry and Karst. *Acta Geol. Sin.* **2006**, *80*, 366–373.

11. Gao, D.; Lin, C.; Hu, M.; Huang, L. Using Spectral Gamma Ray Log to Recognize High-frequency Sequences in Carbonate Strata: A case study from the Lianglitage Formation from Well T1 in Tazhong area, Tarim Basin. *Acta Sedimentol. Sin.* **2016**, *34*, 707–715.

12. Zong, Y.; Shen, Y.; Qin, Y.; Jin, J.; Liu, J.; Tong, G.; Zheng, J.; Zhang, Y. High Frequency Cyclic Sequence Based on the Milankovitch Cycles in Upper Permian Coal Measures in Panxian, Western Guizhou Province. *Geol. J. China Univ.* **2019**, *25*, 598–609.

13. Wang, G.; Deng, Q.; Tang, W. The application of spectral analysis of logs in depositional cycle studies. *Pet. Explor. Dev.* **2002**, *29*, 93–95.

14. Zhang, Z.; Zhang, C.; He, Z. Recognition of Stratigraphic High-frequency Cyclical Properties with Sliding Window Spectrum of Logs. *J. Oil Gas Technol.* **2003**, *25*, 56–58.

15. Yi, H. Application of well log cycle analysis in studies of sequence stratigraphy of carbonate rocks. *J. Palaeogeogr.* **2011**, *13*, 456–466.

16. Shao, C.; Fan, T.; Sun, Y. A case study on Yaojia formation of Changyuan district: Fischer plot analysis based on natural gamma data. *Resour. Ind.* **2013**, *15*, 64–70.

17. Liu, Y.; Meng, X. The sea-level change forcing cycies of oolitic carbonate and cycioc-hrological applications. *Chin. J. Geol.* **1999**, *4*, 442–450.

18. Amour, F.; Mutti, M.; Christ, N.; Immenhauser, A.; Benson, G.S.; Agar, S.M.; Tomás, S.; Kabiri, L. Outcrop Analog for an Oolitic Carbonate Ramp Reservoir: A Scale-Dependent Geologic Modeling Approach Based on Stratigraphic Hierarchy. *AAPG Bull.* **2013**, *97*, 845–871. [CrossRef]

19. Ma, X.; Yang, Y.; Wen, L.; Luo, B. Distribution and exploration direction of medium- and large-sized marine carbonate gas fields in Sichuan Basin, SW China. *Pet. Explor. Dev.* **2019**, *46*, 1–13. [CrossRef]

20. Chen, Y.; Zhang, J.; Li, W.; Pan, L.; She, M. Lithofacies paleogeography, reservoir origin and distribution of the Cambrian Longwangmiao Formation in Sichuan Basin. *Mar. Orig. Pet. Geol.* **2020**, *25*, 171–180.

21. Wang, Z.; Yang, H.; Qi, Y.; Chen, Y.; Xu, Y. Ordovician gas exploration breakthrough in the Gucheng lower uplift of the Tarim Basin and its enlightenment. *Nat. Gas Ind.* **2014**, *34*, 1–9.

22. Zhang, Y.; Li, Q.; Zheng, X.; Li, Y.; Shen, A.; Zhu, M.; Xiong, R.; Zhu, K.; Wang, X.; Qi, J.; et al. Types, evolution and favorable reservoir facies belts in the Cambrian-Ordovician platform in Gucheng-Xiaotang area, eastern Tarim Basin. *Acta Pet. Sin.* **2021**, *42*, 447–465.

23. Liu, Y.; Hou, J.; Li, Y.; Dong, Y.; Ma, X.; Wang, X. Characterization of Architectural Elements of Ordovician Fractured-cavernous Carbonate Reservoirs, Tahe Oilfield, China. *J. Geol. Soc. India* **2018**, *91*, 315–322. [CrossRef]

24. Chen, H.; Zhong, Y.; Hou, M.; Lin, L.; Dong, G.; Liu, J. Sequence Styles and Hydrocarbon Accumulation Effects of Carbonate Rock Platform in the Changxing-Feixianguan Formations in the Northeastern Sichuan Basin. *Oil Gas Geol.* **2009**, *30*, 539–547.

25. Zhu, Y.; Ni, X.; Liu, L.; Qiao, Z.; Chen, Y.; Zheng, J. Depositional Differentiation and Reservoir Potential and Distribution of Ramp Systems during Post-rift Period: An example from the Lower Cambrian Xiaoerbulake Formation in the Tarim Basin, NW China. *Acta Sedimentol. Sin.* **2019**, *37*, 1044–1057.

26. Huang, X.; Fu, M.; Zhao, L.; Zhou, W.; Wang, Y. Identification and sgnificance of meter-scale cycle of carbonate rocks in Mishrif Formation, HF Oilfield, Irag. *Mar. Orig. Pet. Geol.* **2019**, *24*, 44–50.

27. Handford, C.R.; Loucks, R.G. Carbonate Depositional Sequences and Systems Tracts–Responses of Carbonate Platforms to Relative Sea-Level Changes: Chapter 1. 1993. Available online: http://archives.datapages.com/data/specpubs/seismic2/data/a168/a168/0001/0000/0003.htm (accessed on 9 May 2021).

28. Feng, J.; Zhang, Y.; Zhang, Z.; Fu, X.; Wang, H.; Wang, Y.; Liu, Y.; Zhang, J.; Li, Q.; Feng, Z. Characteristics and main control factors of Ordovician shoal dolomite gas reservoir in Gucheng area, Tarim Basin, NW China. *Pet. Explor. Dev.* **2022**, *49*, 45–55. [CrossRef]

29. Zhang, J.; Hu, M.; Feng, Z.; Li, Q.; He, X.; Zhang, B.; Yan, B.; Wei, G.; Zhu, G.; Zhang, Y. Types of the Cambrian platform margin mound-shoal complexes and their relationship with paleogeomorphology in Gucheng area, Tarim Basin, NW. *Pet. Explor. Dev.* **2021**, *48*, 94–105. [CrossRef]

30. Cao, Y.; Wang, S.; Zhang, Y.; Yang, M.; Yan, L.; Zhao, Y.; Zhang, J.; Wang, X.; Zhou, X.; Wang, H. Petroleum geological conditions and exploration potential of Lower Paleozoic carbonate rocks in Gucheng Area, Tarim Basin, China. *Pet. Explor. Dev.* **2019**, *46*, 1099–1114. [CrossRef]

31. Zhang, J.; Feng, Z.; Li, Q.; Zhang, B. Evolution of Cambrian mound-beach gas reservoirs in Gucheng platform margin zone, Tarim Basin. *Pet. Geol. Exp.* **2018**, *40*, 655–661.

32. Shen, A.; Fu, X.; Zhang, Y.; Zheng, X.; Liu, W.; Shao, G.; Cao, Y. A study of source rocks & carbonate reservoirs and its implication on exploration plays from Sinian to Lower Paleozoic in the east of Tarim Basin, northwest China. *Nat. Gas Geosci.* **2018**, *29*, 1–16.

33. Ren, Y.; Zhang, J.; Qi, J.; Zhang, Y.; Zhang, B.; Liu, Y. Sedimentary characteristics and evolution laws of Cambrian-Ordovician carbonate rocks in tadong region. *Pet. Geol. Oilfield Dev. Daqing* **2014**, *33*, 103–110.

34. Zhang, Y.; Gao, Z.; Li, J.; Zhang, B.; Gu, Q.; Lu, Y. Identification and distribution of marine hydrocarbon source rocks in the Ordovician and Cambrian of the Tarim Basin. *Pet. Explor. Dev.* **2012**, *39*, 285–294. [CrossRef]

35. Jiang, M.; Zhu, J. Carbon and strontium isotopic characteristics of Ordovician carbonate rocks in the Tarim Basin and their responses to sea level changes. *Sci. Sin.* **2002**, *32*, 36–42.

36. Zhao, G. Middle-Late Ordovician Sea-Level Changes in the Bachu Area, Tarim Basin, Xinjiang: Carbon, Oxygen and Strontium Isotope Records. Ph.D. Thesis, Jilin University, Changchun, China, 2013.

37. Kaufman, A.J.; Knoll, A.H. Neoproterozoic Variations in the C-Isotopic Composition of Seawater: Stratigraphic and Biogeochemical Implications. *Precambrian Res.* **1995**, *73*, 27–49. [CrossRef]
38. Hoefs, J. *Stable Isotope Geochemistry*, 6th ed.; Springer: Berlin, Germany, 2009.
39. Liu, C.; Zhang, Y.; Li, H.; Cao, Y.; Zhao, Y.; Yang, M.; Zhou, B. Sequence stratigraphy classification and its geologic implications of Ordovician Yingshan formation in Gucheng area, Tarim basin. *J. Northeast. Pet. Univ.* **2017**, *41*, 82–96.
40. Wenzel, B.; Joachimski, M.M. Carbon and Oxygen Isotopic Composition of Silurian Brachiopods (Gotland/Sweden): Palaeoceanographic Implications. *Palaeogeogr. Palaeoclimatol. Palaeoecol.* **1996**, *122*, 143–166. [CrossRef]
41. Cramer, B.D.; Saltzman, M.R. Sequestration of 12C in the Deep Ocean during the Early Wenlock (Silurian) Positive Carbon Isotope Excursion. *Palaeogeogr. Palaeoclimatol. Palaeoecol.* **2005**, *219*, 333–349. [CrossRef]
42. Fischer, A.G. The Lofer Cyclothem of the Alpine Triassic. In Symposium on cyclic sedimentation. *Kans. State Geol. Surv. Bull.* **1964**, *169*, 107–149.
43. Read, J.F.; Goldhammer, R.K. Use of Fischer Plots to Define Third-Order Sea-Level Curves in Ordovician Peritidal Cyclic Carbonates, Appalachians. *Geology* **1988**, *16*, 895–899. [CrossRef]
44. Sadler, P.M.; Osleger, D.A.; Montanez, I.P. On the Labeling, Length, and Objective Basis of Fischer Plots. *J. Sediment. Res.* **1993**, *63*, 360–368.
45. Day, P.I. The Fischer Diagram in the Depth Domain: A Tool for Sequence Stratigraphy. *J. Sediment. Res.* **1997**, *67*, 982–984. [CrossRef]
46. Guo, Y.; Wang, F.; Gan, F.; Yan, B. Forecasting of Spring Flow based on Moving Average Model and Exponential Smoothing Model. *J. Hebei GEO Univ.* **2020**, *43*, 19–25.
47. Elderfield, H.; Greaves, M.J. The Rare Earth Elements in Seawater. *Nature* **1982**, *296*, 214–219. [CrossRef]
48. Gromet, L.P.; Haskin, L.A.; Korotev, R.L.; Dymek, R.F. The "North American Shale Composite": Its Compilation, Major and Trace Element Characteristics. *Geochim. Cosmochim. Acta* **1984**, *48*, 2469–2482. [CrossRef]
49. Wilde, P.; Quinby-Hunt, M.S.; Erdtmann, B.D. The Whole-Rock Cerium Anomaly: A Potential Indicator of Eustatic Sea-Level Changes in Shales of the Anoxic Facies. *Sediment. Geol.* **1996**, *101*, 43–53. [CrossRef]
50. Li, G. The Carbon Isotope Fluctuations and Its Paleoenvironmental Significance of the Upper Jurassic Bulk Carbonate from Amdo Area, Tibet. Ph.D. Thesis, Chengdu University of Technology, Chengdu, China, 2020.
51. Yang, J.; Sun, W.; Wang, Z. Variations in Sr and C Isotopes and Ce Anomalies in Successions from China: Evidence for the Oxygenation of Neoproterozoic Seawater? *Precambrian Res.* **1999**, *93*, 215–233.
52. Veizer, J.; Ala, D.; Azmy, K.; Bruckschen, P.; Buhl, D.; Bruhn, F.; Carden, G.A.; Diener, A.; Ebneth, S.; Godderis, Y. 87Sr/86Sr, Δ13C and Δ18O Evolution of Phanerozoic Seawater. *Chem. Geol.* **1999**, *161*, 59–88. [CrossRef]
53. Haq, B.U.; Schutter, S.R. A Chronology of Paleozoic Sea-Level Changes. *Science* **2008**, *322*, 64–68. [CrossRef]
54. Horacek, M.; Brandner, R.; Abart, R. Carbon Isotope Record of the P/T Boundary and the Lower Triassic in the Southern Alps: Evidence for Rapid Changes in Storage of Organic Carbon. *Palaeogeogr. Palaeoclimatol. Palaeoecol.* **2007**, *252*, 347–354. [CrossRef]
55. Mazzullo, S.J.; Harris, P.M. An Overview of Dissolution Porosity Development in the Deep-Burial Environment, with Examples from Carbonate Reservoirs in the Permian Basin. *West Tex. Geol. Soc. Midl. TX* **1991**, 89–91.

Article

Non-Matrix Quick Pass: A Rapid Evaluation Method for Natural Fractures and Karst Features in Core

Paul J. Moore [1,*] and Fermin Fernández-Ibáñez [2]

[1] ExxonMobil Research Qatar, Doha, Qatar
[2] ExxonMobil Upstream Research Company, Houston, TX 77098, USA;
     fermin.fernandez.ibanez@exxonmobil.com
*   Correspondence: pj.moore@exxonmobil.com

Abstract: Mechanical and chemical processes experienced by carbonate rocks result in a complex network of natural fractures and dissolution features that have direct implications on porosity, permeability, and connectivity in reservoirs. Characterization of natural fractures is best done in core; however, it can be time-consuming due to the large amounts of individual features present and the long list of attributes typically collected for each feature. Additionally, karst features in core, such as vugs and small cavities, are seldom characterized in a quantitative way or are overlooked. We introduce a new methodology, called the non-matrix quick pass (NMQP), which allows for the collection of non-matrix features in a rapid yet quantitative fashion at a rate of 12 to 20 m of core per hour. The NMQP methodology offers enough vertical resolution so that observations can be integrated with other wellbore data types (e.g., wireline logs, well tests, and production logs). This method also yields estimates of density and porosity that are rigorous enough to provide the technical basis to build first-generation dual-porosity models describing the non-matrix component of a carbonate reservoir and its potential impact on field performance.

Keywords: dual porosity; core logging; carbonate reservoir characterization; non-matrix characterization; karst; fractures

Citation: Moore, P.J.; Fernández-Ibáñez, F. Non-Matrix Quick Pass: A Rapid Evaluation Method for Natural Fractures and Karst Features in Core. *Energies* **2022**, *15*, 4347. https://doi.org/10.3390/en15124347

Academic Editors: Yuming Liu and Bo Zhang

Received: 22 March 2022
Accepted: 9 June 2022
Published: 14 June 2022

Publisher's Note: MDPI stays neutral with regard to jurisdictional claims in published maps and institutional affiliations.

1. Introduction

Conventional core is a key dataset used to understand the geologic processes occurring within a reservoir. The wealth of information gained comes from observations made on the core and subsequent analyses done on core samples. Geologic descriptions from core focus on understanding how the processes of sedimentology, stratigraphy, and chemical and mechanical diagenesis have evolved over time [1,2]. Such observations typically evaluate changes in rock fabric, grain size and sorting, stacking patterns, porosity, and structural discontinuities. This information is critical in deciphering the geologic history of a reservoir, such as developing a sequence stratigraphic framework or modeling the intensity and distribution of natural fractures [3,4].

In carbonate reservoirs and aquifers, geologic core has been instrumental in addressing the complexity of the carbonate pore system, which can be separated into two main categories of matrix and non-matrix porosity. Matrix porosity is represented by pore types that are equal to or smaller than the surrounding host grains. The matrix pore system has been studied using core by a number of methods, including by classifying pore types by genetic origin [5], differentiating pore types by connectivity of the pore space [6,7], relating pore space to petrophysical rock properties [8–11], characterizing pore throat size and distributions [12,13], quantifying pore types and geometries [14–17], and characterizing and defining microporosity [18–24].

In contrast to matrix porosity, non-matrix porosity is represented by pore types that are larger than the surrounding host grains and can extend several orders of magnitude larger in size compared to matrix pores. The two processes responsible for non-matrix porosity are through the development of natural fractures and karstification, which results in features including joints, faults, touching vugs, and caves. Although the multiscale nature of the non-matrix pore system often requires multiple datasets for proper characterization [25,26], there is a wealth of knowledge to be gained from utilizing geologic core for characterizing fractures and karst because of the ability to address the processes responsible for their development [25–34]. For example, Tinker et al. [25] utilized more than 8500 m of core and complimentary wireline logs to quantify the contribution of karst porosity within the Permian Yates field in southwest Texas, USA. Ibrayev et al. [34] used a combination of static (e.g., core, wireline, and image) and dynamic (e.g., drilling data and well tests) datasets to develop a genetic-based understanding of fracture development in the Carboniferous Kashagan field in the North Caspian offshore, Kazakhstan. Ahdyar et al. [26] incorporated static (core, wireline, image, and 3D seismic) and dynamic (drilling data, production logs, and pressure transient analysis) datasets to characterize the distribution and pore volume associated with non-matrix processes within the Oligo-Miocene Banyu Urip field onshore, Indonesia.

Non-matrix features are often observed and described in core [28,30,34]; however, typically, only the fracture portion of the non-matrix pore system is cataloged in a systematic way that addresses fracture geometries and spacing [33,35]. Although karst features are commonly recognized in core, such observations are rarely cataloged along with and in addition to the fracture observations to understand the impact that these combined features may have on the non-matrix pore system. Instead, observations of karst features are often used to support the building of a sequence stratigraphic framework or to understand the diagenetic processes that likely occurred in the reservoir [30,36,37]. Nonetheless, the non-matrix pore system is a combination of chemical and mechanical processes that often occur in similar locations, e.g., along the margin of carbonate platforms. Consequently, in order to be predict the magnitude and distribution of all non-matrix features and their associated pore volume within a given field, systematic cataloging of these different non-matrix features is warranted.

In this paper, we present a methodology that arose from the need to collect an integrated karst and fracture dataset from a large amount of core in multiple wells, i.e., total lengths in excess of 5000 m per study. We were tasked with collecting quantitative data from non-matrix features observed in cores that could be used for reservoir characterization and development of initial geologic concepts. The description of such large amounts of core had to be achieved using a rapid yet adequate quantitative data collection strategy so that business deadlines could be met while ensuring the technical integrity of our work. Therefore, detailed data collections strategies such as those employed in traditional fracture characterization [2] were not an option, as they cover less than about 20 m of core per day [38,39]. The data collected should also be quickly processed and suitable for integration with other wellbore-based datasets (e.g., drilling data, wellbore images, wireline logs, and seismic and production data) while providing some basic inputs for geostatistical reservoir models. The fast and quantitative yet simple data collection approach allows a team of two to cover more than 100 m of core in less than two days of work. The final methodology described herein has been successfully implemented as standard practice for the characterization of non-matrix in cores collected in several carbonate reservoirs that we have worked. Obviously, non-matrix observations collected using this method are not free of sampling bias inherent to core. For example, bias due to fracture spacing and the likelihood of intersecting a fracture must be considered [40,41].

2. Methodology and Workflow

Non-matrix quick pass (NMQP) is based on a set of rules and methodologies that ensures consistency within and between cored wells. This workflow was developed based on a balance between time spent on the core and ensuring that sufficient observations were made and recorded to capture the variability in non-matrix features observed in core, which can be used to compliment additional datasets. This method does not replace detailed data collection. Instead, it provides a means for describing large amounts of core that can identify specific areas for further detailed work. The methodology is designed to be run in teams of two. Although each individual reports on the observable non-matrix features, one person is tasked with measuring (i.e., quantifying) each feature, and the other person inputs the data directly in a computer.

2.1. Ground Rules

In order to ensure consistency, the first step of the data collection process is to establish some ground rules and clearly define how the data will be collected, as well as the amount of detail required to meet the objectives of the study. The list below highlights some key rules that we have found useful in collecting the appropriate amount of detail while optimizing the time spent on each core interval described with the NMQP methodology:

1. Always utilize the whole core whenever possible, i.e., use both sides of a slabbed core to ensure maximum core coverage. Refrain from using core photos. Most core descriptions are performed on a 1/3 slabbed portion. Although this portion of the core is sufficient for making observations on sedimentology, NMQP works best if non-matrix observations are made on the 2/3 portion of the core. The logic is simple. The larger sample size allows for the recognition of non-matrix features that are often under-sampled in the 1/3 portion. We have also found that many non-matrix features are often found on the backside of the core due to the core being slabbed to avoid intersecting non-matrix features that can compromise the integrity of the 1/3 slabbed portion. In the case of fractures, the spacing, orientation, and angle of the fracture with respect to the core can lead to significant sampling bias.

2. Define a collection interval. For simplicity, we use the length of core that fits in the box used to store the core. This length is usually one meter and provides enough resolution to identify trends and intervals of interest for further analyses or data collection. Alternatively, the user can define custom intervals that could be driven by lithologic or mechanically distinctive units or simply a smaller regular interval for higher-resolution results.

3. Evaluate the integrity and layout of the core. This step is meant to account for core handling and previous sampling efforts, which can create issues with core quality, incorrect orientation (e.g., upside-down core), mislabeled core boxes, and missing core. Core integrity is also a flag for reliability or confidence with respect to the collected data, analysis, and derived results. In our experience, non-matrix features are easily overlooked when core integrity is low. Low core integrity can result from the presence of non-matrix features but also from poor core handling or drilling operations.

4. Define a target maximum amount of time to be spent on each interval (or core box). We have determined that no more than five minutes per core box (assuming one meter) is a target sufficient to capture the proper amount of information. Of course, there might be core intervals with a large amount of features and complex relationships that might require more than the maximum specified time, and this is appropriate as long as the majority of the core intervals are completed within five minutes. If the analysts spends more than an average of five minutes per interval, too much detail is being collected and perhaps the ground rules need to be revisited. Figure 1 shows the distribution of times spent collecting data in more than 300 boxes of core with different levels of non-matrix abundance and complexity. On average, we spent just under three minutes to collect all the information we deemed necessary in a typical interval. Boxes that took longer than five minutes usually corresponded to a high density of non-matrix features or low core integrity, which required some degree of core reconstruction. The time per interval reported in Figure 1 includes data collected by different analysts with varying degrees of experience with the NMQP methodology.

5. Define a minimum size and amount of features to be characterized. We typically log only fractures that are >2 cm in length with apertures ≥0.05 mm and more than 10 cm cumulative length per box. In the case of karst features, we only log features that are

>1 cm^2 with evidence of dissolution, i.e., molds and other depositional pores, such as fenestrae, are excluded. For simplicity in the NMQP, we refer to any dissolution-enhanced pore as a vug. A touching vug implies that multiple dissolution-enhanced vugs are connected in the core [42]. The lower limits of recorded observations, e.g., vug >1 cm^2 and fractures >2 cm in length, were set to increase efficiency while on the core. The main idea here is that smaller features, regardless of abundance, are not expected to significantly contribute to the non-matrix pore system flow.

6. Agree on the threshold between fracture and karst feature. Although this might be an overstatement, we have run into many situations where fractures are enhanced so much by dissolution that they develop vug-like aspect ratios (e.g., Figure 4C in [43]). Therefore, it is critical to make appropriate observations on cross-cutting relationships, such as by establishing whether a vug was intersected by a fracture or developed along a fracture. In any case, do not capture the same observation twice, i.e., classify as both fracture and karst.

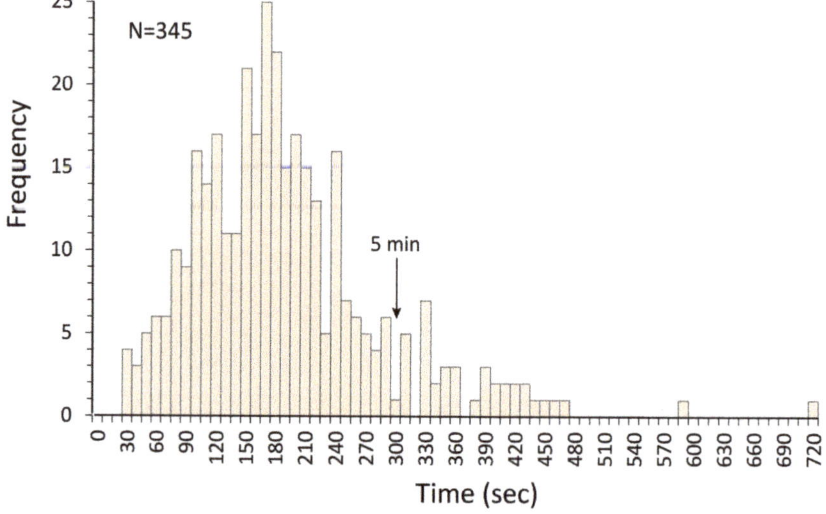

Figure 1. Histogram showing the distribution of time spent collecting non-matrix observations on more than 345 boxes of core using the NMQP method.

2.2. Data Collection

Measuring tape, a small ruler or a comparator [43], a hand lens, a computer, and a clean core surface are the minimum requirements to get started. Figure 2 illustrates the spreadsheet format we use to collect the information that is reported during an NMQP. Data are easily recorded using an alphanumerical system that can easily be translated into other software packages optimized for well log analysis. Table 1 shows an example key that can be used to record information in the spreadsheet. With a bit of practice, the alphanumeric key can be easily remembered by analysts, which reduces the time spent on each interval. Each row in the spreadsheet represents a single core box. For each core box, the top and base of the interval are recorded, along with the core dimensions, which include the core diameter and cut, such as the 1/3 or 2/3 portion of the slabbed core. Core integrity is typically input on this first pass of the core, together with the top and bottom of each interval (or core box). The core integrity scale ranges from 1 to 5 (Figure 3). Once the depth, dimensions, and quality of the core have been captured, each core box is evaluated for the presence of fractures and karst features.

Box	Depth		Core Dimensions		Core Integrity	Fractures								Karst						
	Top	Base	Width	Cut		Type	Count	$\Sigma_{lengths}$	Width		Distribution	Openness	Fill	Type	Count	Size		Distribution	Openness	Fill
	(m)	(m)	(cm)	(slab)			(min)	(cm)	Avg (mm)	Max (mm)						Avg (cm^2)	Max (cm^2)			
1	1910.00	1910.91	10	2/3	5	-	-	-	-	-	-	-	-	-	-	-	-	-	-	-
2	1910.92	1911.79	10	2/3	5	-	-	-	-	-	-	-	-	-	-	-	-	-	-	-
3	1911.79	1912.70	10	2/3	5	2	10	40	0.05	0.05	1,2,3	0	1	4	2	3	4	3	3	2
4	1912.70	1913.56	10	2/3	5	2	3	30	0.05	0.2	2	0	1	4	1	5	5	1	4	2,7
5	1913.56	1914.46	10	2/3	5	-	-	-	-	-		-	-	3	2	6	6	1	1	1
6	1914.46	1915.37	10	2/3	5	2	5	25	0.1	1	3	0	2	-	-	-	-	-	-	-
7	1915.37	1916.27	10	2/3	5	2	5	70	0.1	0.1	1	0	1,2	-	-	-	-	-	-	-
8	1916.27	1917.18	10	2/3	2	2	3	75	0.1	0.1	2	0	1,6	7	4	18	24	1,2,3	1	2
9	1917.18	1918.12	10	2/3	3	-	-	-	-	-		-	-	-	-	-	-	-	-	-
10	1918.12	1918.98	10	2/3	3	2	5	110	0.05	0.1	1,2,3	1	1,6	-	-	-	-	-	-	-
11	1918.98	1919.89	10	2/3	4	2	1	10	0.1	0.2	1	0	1,6	-	-	-	-	-	-	-
12	1919.89	1920.81	10	2/3	5	2	8	160	0.05	0.1	1,2,3	0	1,2	-	-	-	-	-	-	-
13	1920.81	1921.72	10	2/3	3	-	-	-	-	-		-	-	4	2	18	32	2	3	1
14	1921.72	1922.63	10	2/3	3	-	-	-	-	-		-	-	-	-	-	-	-	-	-
15	1922.63	1923.48	10	2/3	5	2	2	50	0.05	0.05	1,2,3	0	2	5	1	30	30	1	3	2,7
16	1923.48	1924.31	10	2/3	5	1	1	12	1	6	1	0	4	-	-	-	-	-	-	-
17	1924.31	1925.19	10	2/3	5	-	-	-	-	-	-	-	-	-	-	-	-	-	-	-
18	1925.19	1926.10	10	2/3	3	-	-	-	-	-	-	-	-	-	-	-	-	-	-	-
19	1926.10	1927.00	10	2/3	4	2	4	75	0.05	1	3	1	1	5	1	40	40	2	0	1,2
20	1927.00	1927.90	10	2/3	4	-	-	-	-	-	-	-	-	4	1	42	42	1	2	2

Figure 2. Example of a spreadsheet format used in the NMQP methodology. A detailed explanation of the non-matrix metrics that are recorded is provided in the text.

Table 1. Example key of the alphanumerical system used in the NMQP methodology.

Core Integrity	Non-Matrix Types	Distribution	Openness	Fill
1–Rubble	1–1st Gen Fractures	1–Upper 1/3	0–Closed (<10%)	1–Bitumen
2–Rubble w/intact sections	2–2nd Gen Fractures	2–Middle 1/3	1–Slightly Open (10–30%)	2–Calcite cement
3–Partially intact *	3–Breccia	3–Lower 1/3	2–Partly Open (30–70%)	3–Clay
4–Mostly intact *	4–Vug		3–Mostly Open (70–90%)	4–Debris
5–Completely intact [†]	5–Touching Vugs		4–Open (>90%)	5–Anhydrite
	6–Vugs on Fracture			6–Stylolite residue
	7–Vug in breccia			7–Breccia clasts

* Can have missing pieces; [†] may be broken.

Figure 3. Examples showing variations in core integrity. (**A**) rubble, (**B**) rubble with intact sections, (**C**) partially intact, (**D**) mostly intact, (**E**) completely intact. Core intervals with missing pieces or being broken does not impact the core integrity call.

2.2.1. Fracture Metrics

In an ideal situation, some sort of fracture paragenesis work should have been done prior to collecting these data. This can be achieved by using geochemical techniques that analyze cement types and paragenesis or more simply by observing the type of cement fills and cross-cutting relationships. When a fracture paragenesis exists, the first step is to identify the genetic fracture types present in the analyzed interval. Genetic types refer to the relative timing and processes responsible for fracture development, which also has implications for the type of fill that may be present. For example, early fractures that form contemporaneously with sediment deposition are commonly open at the surface [34]. Consequently, such fractures are often filled with carbonate debris and soil-derived clays. Conversely, late fractures that develop during burial often crosscut the early fractures and are filled with calcite cement. Conducting NMQP within a fracture-paragenetic sequence allows the user to distribute properties and fracture types in a model using a genetic-based approach. This effectively means being able to define areas with a higher probability of finding certain types of fractures and assigning different properties of each fracture generation. NMQP can also be performed in the absence of fracture paragenesis work; however, there will be uncertainty on understanding what controls fracture distribution within the reservoir. Regardless, NMQP can still be used to identify general distribution trends and sweet spots or make inferences about mechanical stratigraphy. The following fracture metrics should be collected independently for each fracture generation (or type):

- *Count*: number of fractures of a given genetic type that meet the conditions defined by the ground rules. When counting fractures, one should avoid counting the same fracture twice at the intersection with the slabbed face and the back of the core. The user should not forget that the number of fractures sampled by the core is highly dependent of fracture spacing and orientation; therefore, the core is only a partial representation of fractures in the subsurface.
- *Cumulative Length*: summation of all fracture trace lengths of the same fracture generation.
- *Characteristic width*: a rough estimate of an average or most representative width of each fracture type observed in the core interval. Width is defined as the distance from wall to wall of a given fracture, regardless of whether the fracture is filled or open. A simple measuring scale or comparator [44] can be used.
- *Maximum width*: the maximum observed width of a given fracture type. In reservoirs with evidence of dissolution, maximum widths correspond with vugs that developed along fractures.
- *Openness*: the amount of visible open space in a fracture under the naked eye. We use a Likert-type scale [45] with 5 classes (Table 1) that covers non-uniform ranges in an attempt to account for human bias and the inability to visually quantify the proportion of the fracture that is open. The classes range from 0 (completely filled with cement) to 4 (more than 90% of the fracture is open).
- *Fill type*: describes the different types of cements that can be observed with the assistance of a hand lens. In the case of multiple cement generations, the order in which they are recorded in the spreadsheet represents the relative timing of the cements, from late to early. For example, in box 7 (interval 1915.17–1916.27 m), the numeric codes 1,2 depict two cement generations observed, where the fractures are coated by calcite cement, followed by a lining of bitumen (Figure 2). As this is a general description over a meter-long interval, the fill sequence described here would be the one that is most commonly observed.
- *Distribution*: defines in which third(s) of the core the described features occur. This attribute is an attempt to further refine the vertical distribution of features within a box. For instance, a 1,3 distribution means that the fractures occur in the upper and lower thirds of the interval under consideration; a 1,2,3 distribution implies that fractures occur throughout the entire box.

One advantage of the NMQP approach is that it is highly adaptable. For example, in our experience, we typically have wireline image logs that cover the cored intervals. In such cases, we use the image logs to determine dip and orientation of fractures as a way to optimize our time on the core. Conversely, if image logs do not exist over the cored intervals, then recording orientation and dip information can easily be added to the NMQP workflow; however, time allocation per interval will need to be considered.

2.2.2. Karst Metrics

For this workflow approach, we define karst as either an early epigenic process associated with subaerial exposure and meteoric water or a late hypogenic process during burial that results from acids and fluids decoupled from the overlying surface [46]. Although there is difficultly in differentiating between early meteoric karst features and late burial-related karst features solely from core, there are some key observations that can be made to distinguish the two karst types. For example, meteoric karst features are associated with an exposure horizon in core. Additional lines of evidence for meteoric karst includes the development of touching vugs (i.e., multiple vugs that are connected [42]), the presence of cave deposits, such as speleothems or collapse breccia associated with early fill, and rubble zones associated with karst voids. Evidence of burial karst features can be more challenging to determine in core. Nonetheless, some observations that point to porosity generation during burial include vugs that develop along stylolites or burial fractures, as well as further enlargement of syndepositional fractures and early meteoric karst features.

Because of the time consumption that can happen with detailed differentiation of early and burial karst processes during the NMQP process, we focus efforts on the quantitative geometric description of these features that arise from such processes. Qualitative observations of early versus burial features are recorded as comments. Deciphering of the relative abundance of early versus burial karst comes after the NMQP and requires integration with additional datasets, including a sequence stratigraphic framework, optical petrography, and diagenetic studies [31,37]. The following karst metrics should be collected for each karst type:

- *Count*: number of karst features of a given genetic type that meet the conditions defined by the ground rules. The most common example of karst at the core scale is a combination of isolated dissolution-enhanced voids and touching vugs, reflecting the earliest stages of coastal karst development [47,48], which are often associated with carbonate reservoirs [26–28,37,41,49].
- *Average size*: an estimate of the mean cross-sectional area of the same karst type, which is calculated by multiplying the approximate length of the short and long axes that define each feature.
- *Maximum size*: the maximum measured cross-sectional area of a certain type of karst feature.
- *Openness*: amount of visible open space in a karst features. We use the same Likert-type scale [45] used for fractures.
- *Fill type*: describes the different types of cements that can be observed with the assistance of a hand lens. Rules for fill type reporting and data collection are similar to those described in the case of fractures.
- *Distribution*: similar to the fracture metrics, *distribution* indicates which third(s) of the core contain(s) the features being described.

2.3. Data Processing

The data collection strategy described in the prior section is geared towards this step of the workflow, where information is processed to provide reasonable estimates of non-matrix properties and their vertical variations. The main objective of the data processing step is to deliver a quantitative interpretation of density and porosity associated with the non-matrix pore system. Although there might be a certain degree of uncertainty with respect to the absolute values, this methodology provides early insights on non-matrix variations along the core and in between wells.

Fracture density. The fracture density derived from this workflow yields a P_{21} (as defined by [50]), which is calculated as the summation of all fracture trace lengths (of the same generation) divided by the core surface area:

$$P_{21} = \sum_1^n L / A,$$ (1)

where L is fracture trace length, and A is the core surface area.

Fracture porosity. Fracture porosity (P_{22}, [4]) is calculated as the product of the cumulative fracture length multiplied by the characteristic fracture width, minus the amount of pore space occupied by any filling cement:

$$P_{22} = \sum_1^n L \times \alpha \times [1 - f],\tag{2}$$

where L is fracture length, α is characteristic width, and f is the proportion of the fracture filled with cement (defined as the midpoint of the reported interval range). Potential fracture porosity (P_{22pot}) is defined as the porosity that would result from removing all fracture-filling cements (i.e., $f = 0$ in Equation (2)).

Karst porosity. Karst porosity ($\varnothing_k$) is derived from the average vug size multiplied by the number of vugs over a given area:

$$\varnothing_k = \frac{\overline{K_A} \times K_n}{A}(1 - f),\tag{3}$$

where $\overline{K_A}$ is the average vug size, K_n is the total count of vug features, and A is the core surface area. The area should include the slab face area and the area of the core outer surface to account for the vugs in the back of the core. Potential karst porosity ($\varnothing_{kpot}$) is defined as the porosity that would result from removing all filling cements (i.e., $f = 0$ in Equation (3)). A different approach to estimate karst porosity involves calculating an equivalent radius (based on average vug cross-sectional area) and the volume of the vugs, assuming that they approximate spheres, and therefore taking into account the volume of the core interval. We found that this approach, which is an oversimplification of the vug pore geometry, tends to produce lower porosity numbers than the method in Equation (3). Nonetheless, incorporating both estimations provides a range of karst porosity that can be useful for addressing uncertainty.

3. Results and Discussion

The following section is dedicated to showcasing three examples where NMQP was used to describe core. These cases are not related to each other. They are intended to show how NMQP can be applied in different scenarios. The first case is an example of core from a single well with a variety of non-matrix features that includes both fractures and karst features. The second case is an example of multiple cores wherein epigenic karst is the main diagenetic process, resulting in a high density of vug features with variable vertical distribution. The third case illustrates how to compare NMQP with additional datasets, including borehole acoustic image log and previous fracture interpretation, as well as the beginning of an interval in the well where total losses occurred. We will use these examples as an opportunity to discuss some of the details and direct implications of this methodology for aquifer/reservoir characterization purposes.

3.1. Case 1: Single-Well NMQP

Case 1 represents a detailed example of the NMQP workflow. Figure 4 illustrates the type of dataset that captures the variability expected within carbonate reservoirs. In our experience, carbonate reservoirs can exhibit a range of non-matrix features, whereby some reservoirs are mostly karst-dominated [26], fracture-dominated [51], or a mix of both karst features and fractures [41,52]. This dataset reflects what one can expected to find in core from a carbonate reservoir that has experienced both karst and fracture processes. Figure 4 demonstrates how the NMQP methodology provides a holistic view on the magnitude and distribution of non-matrix features observed in core, which can be compared with numerous datasets, including sedimentary core descriptions, wireline and image logs, and dynamic data, such as production logging tools (PLTs) and well tests. Together, the integration of these datasets provides key insight on the controls driving the total carbonate pore system within a given reservoir [24,26,41]. The following subsections step through the types of observations and classifications that can be made on the core, as well as what considerations should be made to understand the potential impact of non-matrix features within the reservoir.

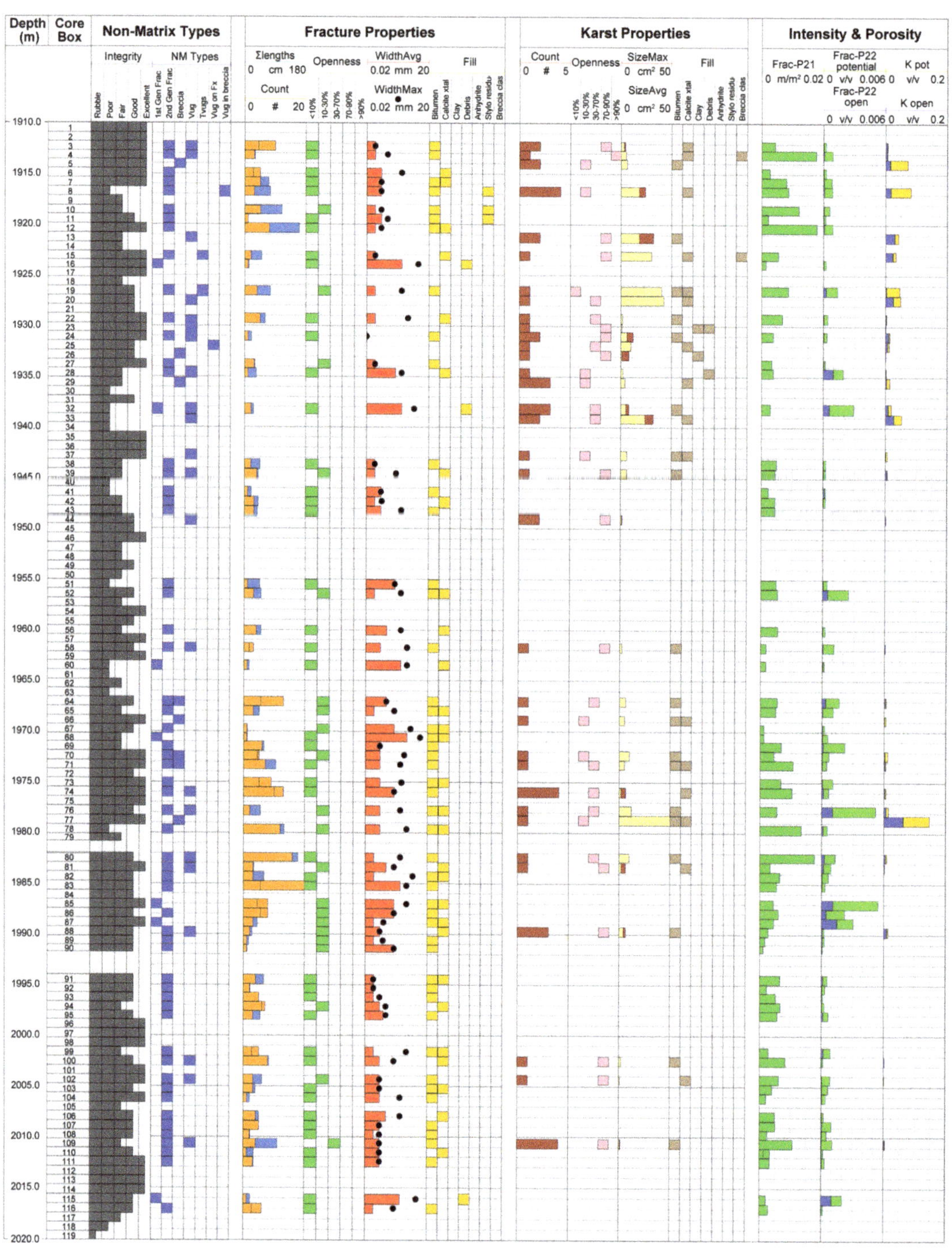

Figure 4. Single well showing the vertical distribution of non-matrix observations obtained with the NMQP methodology.

3.1.1. Core Quality

Results from NMQP show that the overall core quality is good with 88% ranging from fair (22%) to good (31%) to excellent (35%). Only 12% of the core is of poor quality. In reservoirs where non-matrix features are known to exist, missing core and rubble intervals need to be evaluated to determine the cause of little-to-no core recovery. Although drilling operations and wellbore conditions can commonly lead to poor core retrieval, these intervals may reflect non-matrix features, including fracture zones or karst. An evaluation of the core integrity demonstrates 97% core recovery, with two intervals that account for 3.5 m of missing rock from 1981.00 m to 1982.5 m and from 1992.0 m to 1994.0 m. These two missing intervals may indicate non-matrix features, which needs to be corroborated with additional data, such as caliper, image log, or a PLT run showing flow. Our experience has shown that numerous rubble zones and missing intervals are indications of non-matrix features that were previously misinterpreted as the result of coring operations or wellbore integrity (e.g., breakout) during drilling.

3.1.2. Non-Matrix Types

Non-matrix types within this example include a combination of fractures and small-scale dissolution features that can be diagnostic of incipient karst processes. Fractures account for 68% of the observed non-matrix features. We classify fractures based on their relative timing, e.g., syndepositional, early burial, and late burial [34]. Similarly to the proposal of Ibrayev et al. [34], first-generation fractures correspond to fractures that developed contemporaneously with sediment deposition. These fractures often exhibit large apertures (≥ 1 cm) and are filled with carbonate debris derived from the platform top and early marine calcite cements. Second-generation fractures develop during early burial due to loading and compaction. Such fractures typically have apertures of less than 1 mm and are often associated with stylolites. Late fractures can often be differentiated by cross-cutting relationships with earlier fractures due to continued burial or tectonics.

Breccia accounts for 6% of the observed non-matrix types. Karst-related breccia observed in core typically results from cave collapse or exposure, i.e., epikarst. Dissolved vugs and touching vugs account for 20% and 3% of the observed non-matrix features, respectively. Although such features may not be explicitly defined as karst, e.g., small-scale vugs that have experienced varying degrees of dissolution, their presence provides insight into areas within the reservoir where dissolution may have been favorable for karst development under the right flow conditions. This type of understanding can be extremely useful when predicting where karst processes may impact reservoir properties, such as connected pore volume and high-permeability zones. Consequently, we also flag any vugs that are present within the breccia and along fractures, as such observations may indicate possible flow paths within these features.

3.1.3. Fracture Properties

The fracture properties collected during NMQP include a total count and summation of the total fracture length per box, as well as the degree of open pore volume, average and maximum aperture, and fill material. Per Rule 5, the minimum total sum of fractures > 2 cm in length with apertures ≥ 0.05 mm must be at least 10 cm per box.

At 1920 m, core box 12 has eight fractures with a combined total of 160 cm in length (Figure 4). Another core box at 1982 m (core box 80) also has a total fracture length of 160 cm with a fracture count of sixteen. This core box is also next to an interval with no core recovery, whereas there is a box just above the missing core interval at 1979 m with twelve fractures for a total of 120 cm (core box 78). Given the high values of total fracture lengths above and below the missing core interval, additional data should be evaluated to determine whether the interval has any non-matrix features. One straightforward approach would be to examine drilling data to determine whether any losses were recorded during coring at this depth. If so, the volume versus rate (VvR) plot described by Fernández-Ibáñez et al. [52] would be of use, as it utilizes information collected within lost circulation zones to determine whether the mud losses are due to fractures or karst (see also Case 3 below). If no losses occurred, then an evaluation of image logs would help to determine whether this missing interval is the result of non-matrix features [33].

Second-generation fractures account for 91% of all fractures observed, with 70% of these fractures with an average width in the range of 0.05 to 0.1 mm. The remaining 20% of second -generation fractures have an average width ranging from 0.15 to 0.5 mm. Maximum width is useful to understand the variability of the fracture sets within one box. For example, core box 3 has 40 cm of total fracture length with both an average and maximum width of 0.5 mm. Conversely, core box 32 also has 40 cm of total fracture length, but the average and maximum width are 0.1 mm and 1 cm, respectively, and the number of fractures is less than that observed in core box 3 (Figure 4). Recognizing such variation and distribution of fracture density and width in core can provide insight that can help to address the impact that these features have on variations in non-matrix flow [53].

First-generation fractures account for 9% to the total observed fractures. This type of distribution between first- and second-generation fractures is common in many carbonate reservoirs that we have studied [34]. Although first-generation fractures are often fewer in number compared to second-generation fractures, their commonly wider apertures and vertical extent can have a significant impact on large-scale flow in certain portions of a reservoir [54]. An evaluation of fracture fill material shows that all but one box (core box 109) have less than 30% open pore volume per fracture. Most of the fracture fill is in the form of bitumen and calcite crystals.

3.1.4. Karst Properties

The karst properties collected during NMQP include a count of individual karst features per box, as well as the degree of open pore volume, average and maximum size of each feature, and fill material. According to Rule 5, only vugs that are greater than 1 cm^2 and show evidence of dissolution are recorded, i.e., molds and other isolated pores are excluded. The idea is that vugs that have experienced dissolution are in the earliest stages of karst development. All large karst features, such as caves, start out as small-scale pre-solution openings that include bedding-plane partings, fractures, and/or pores associated with the matrix host rock. Depending on the hydrogeologic conditions under which caves form, the result may be angular or curvilinear conduits, maze-like networks, or meter-scale isolated voids [46]. Although not all vugs that have experienced dissolution are sensu stricto karst, their presence indicates specific locations within the reservoir where karst processes may have occurred beyond what can be explicitly viewed in core. Touching vugs, on the other hand, demonstrate areas where dissolution has progressed enough to generate a well-connected pore system at a local scale, i.e., core.

There are 74 individual karst features observed in the core, with a maximum count of five features in a box. The average and maximum size is about 12 cm^2 and 130 cm^2, respectively. The 130 cm^2 feature is in core box 77 and is a karst breccia that still has an open pore volume estimated at 30%. In core boxes 19 and 20, the observed features are one touching vug and one isolated vug, respectively. Although there is only one feature per box, their proximity and size (about 50 cm^2 each) suggest that sufficient dissolution and fluid flow could have been favorable for karst development within this portion of the reservoir. The degree of open pore volume within the karst features ranges from less than 10% to 90%, with 65% of all features ranging between 25% and 70% open. In our reservoir studies, we have used such observations of open pore volume to guide estimates of field-wide karst porosity [41].

3.1.5. Density and Porosity

Estimates of non-matrix properties and their vertical variations are evaluated through a quantitative interpretation of density and porosity (Equations (1)–(3)). Such values provide an understanding of the potential impact that non-matrix features may have on flow and storativity at the core scale. Fracture density (P_{21}) was calculated using Equation (1).

Fracture density over the entire cored interval has an average of 0.005 fractures/m, with a minimum and maximum density of 0.001 and 0.018 fractures/m, respectively. There are two intervals where density is elevated above the average range: from 1911 m to 1920 m and from 1973 m to 1983 m. Within the interval from 1973 m to 1983 m, the fracture density reaches a maximum above and below a section with missing core (Figure 4), suggesting that the missing section may have resulted from poor core recovery due to the presence of non-matrix features.

Fracture porosity, as estimated here, represents an attempt to capture the range of pore volume associated with fractures as observed in core using Equation (2). The range of pore volume associated with fractures is estimated by observing the amount of open pore volume seen on core (P_{22open}), i.e., accounting for fill within the fractures, and estimating the maximum amount of fracture pore volume, assuming no fill (P_{22pot}). The average values of fracture porosity is a P_{22open} of 0.02% and a P_{22pot} of 0.1%, suggesting that 80% of the fractures are filled mostly with a combination of bitumen and calcite (Figure 4). Maximum values for P_{22open} and P_{22pot} are 0.2% and 0.5%, respectively. Similar to fracture porosity, karst porosity evaluates the range of actual pore volume ($\varnothing_k$) versus maximum possible pore volume ($\varnothing_{kpot}$) associated with karst features in core using Equation (3). The average values for $\varnothing_k$ and $\varnothing_{kpot}$ are 0.9% and 2%, respectively, indicating that only 56% of karst is filled. Maximum values for $\varnothing_k$ and $\varnothing_{kpot}$ are 4% and 15%, respectively. Combined, these average and maximum values of non-matrix porosity provide insight into uncertainty related to pore volume estimates as observed at the core scale.

3.2. Case 2: Using NMQP to Correlate across Wells

NMQP results can also be used to evaluate lateral and vertical trends between wells. In the Case 2 example, NMQP was used to quantify the distribution and porosity associated with the non-matrix features observed in core from a shallow carbonate aquifer. In this case, non-matrix features are in the form of dissolution-enhanced vugs, i.e., karst. Wells in this aquifer are drilled to an average depth of ~25 m, and core is collected in every well with core recovery >95%. Lateral distance between wells ranges from about 0.5 to 1 km.

Figure 5 shows the karst porosity distribution from core collected in seven wells using a common depth scale reference to mean sea level. Vug size and count were used to estimate $\varnothing_k$ and $\varnothing_{kpot}$ for each well. The NMQP results highlight a preferential concentration of vugs around 12 m below mean sea level (m bmsl). There is also an increase in vuggy features below ~20 m bmsl. Whereas the vugs at ~12 m bmsl in most of the cases are open, the features located below ~20 m show a varying degree of openness, which appears to reflect a regional trend. For example, the cluster of wells located in the northern part of the field that run west to east (to the left of Figure 5) show a mostly open, vuggy pore system. Conversely, the cluster of wells in the south part of the field run north to south and shows a vug system that is mostly occluded with cement (Figure 6). We interpret the concentration of vug porosity at two distinct levels as the result of freshwater lens diagenesis. The observations and regional trend derived from NMQP in these wells provide a general framework for describing non-matrix distribution in the aquifer, as well as to impose a decreasing aquifer porosity and permeability trend towards the south.

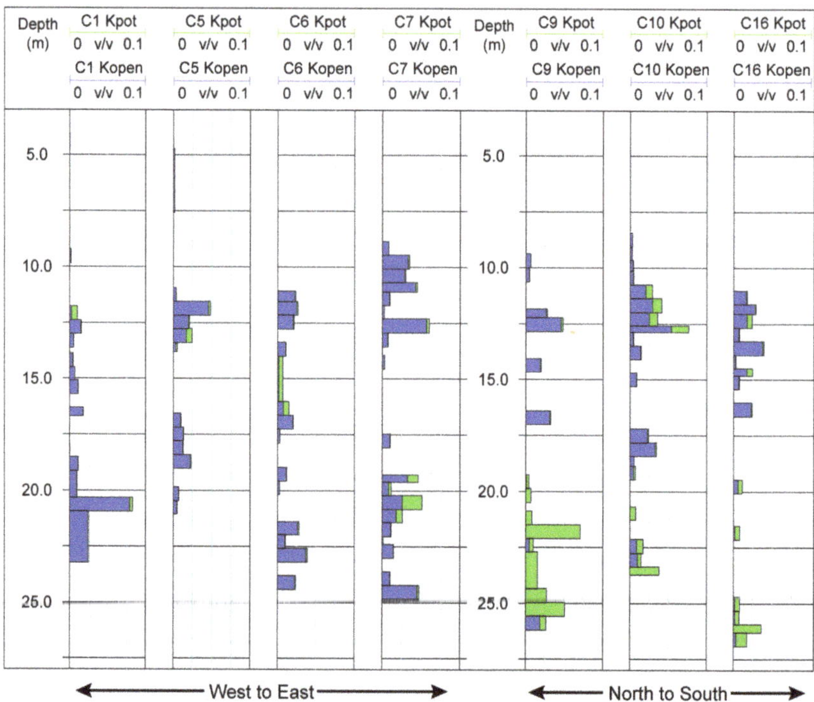

Figure 5. Cross section of multiple wells showing the lateral and vertical distribution of dissolution-enhanced vugs (karst) observed using the NMQP methodology. Attributes displayed correspond to φ22pot and φ22 (Kpot and Kopen, respectively, in log headers), with the well name preceding each of them (e.g., C1 Kpot). The blue fill represents the proportion of vug porosity that is open, and the green fill corresponds to the proportion of vug porosity filled with cements.

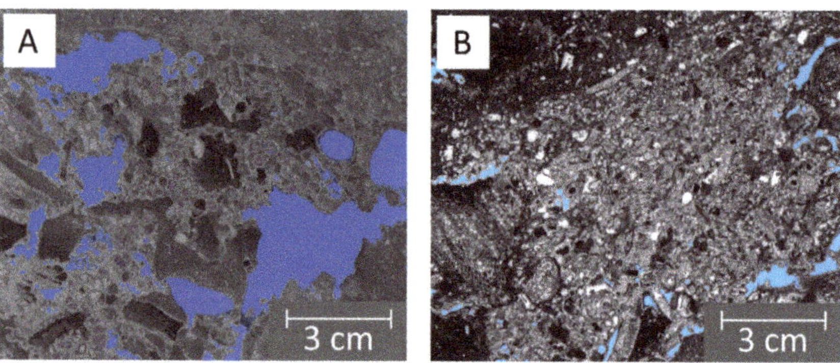

Figure 6. Images showing increase in cementation and decrease in vug size moving from north to south, as shown in Figure 5. Both images are from thin sections collected at 25 m depth from each well. Blue represents porosity. (**A**) Well C9; (**B**) Well C16.

3.3. Case 3: Combining NMQP with Additional Datasets

The integration of NMQP results with additional datasets can provide key insights with respect to how non-matrix features enhance dynamic performance, including mud losses during drilling or flow potential during production. For example, Case 3 shows the results of NMQP over 33 m of core from a carbonate reservoir, which also has a borehole acoustic image log, associated fracture interpretation from image log, and lost circulation zone (LCZ), where severe losses during drilling started at 5075 m (Figure 7). The fracture interpretation suggests that fractures start where losses initiated, i.e., 5075 m, and continue within the LCZ. Conversely, results from NMQP show the presence of vugs throughout the core, with an increased concentration of vugs starting at 5068 m, where minor losses were recorded, and just above the LCZ, where total losses occurred (Figure 7). Equations (1)–(3) were used to estimate the fracture density and porosity of both fractures and karst. The increase in observed non-matrix features in the core just above and at the start of losses suggests the presence of a well-connected non-matrix pore network that promotes high flow rates. Interpretations using only image logs indicate that fractures alone are the main feature responsible for the losses; however, integration with NMQP clearly demonstrates the occurrence of dissolution-enhanced vugs that also play a role.

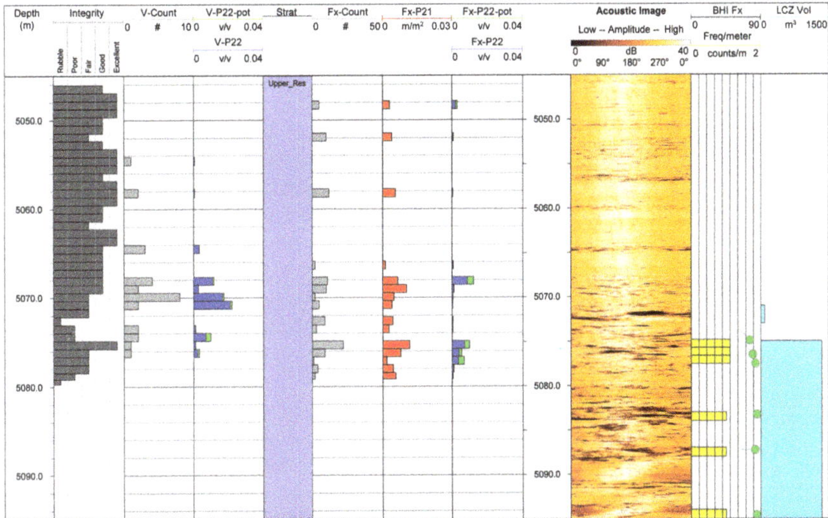

Figure 7. Single well showing the vertical distribution of non-matrix observations using the NMQP methodology compared with additional data, including borehole acoustic image log, fracture interpretation from image, and lost circulation zone (LCZ). From left to right: depth, core integrity, vug count, vug porosity (V-P22) and vug potential porosity (V-P22pot), stratigraphic unit, fracture count (Fx-count), fracture density (Fx-P21), fracture porosity (Fx-P22) and fracture potential porosity (Fx-P22pot), acoustic image log (amplitude), tad poles to image log fracture interpretation and fracture frequency, and top and base of lost circulation zone with volume lost (LCZ). In the vug and fracture potential porosity tracks, the blue fill represents the actual open porosity, and the green represents the proportion of space filled by cements.

Recognizing the difference between fracture and karst development is critical when characterizing reservoirs and developing conceptual models. For example, predicting the magnitude and distribution of fractures away from well control requires an understanding of local and regional paleostress regimes that could be responsible for their development. Conversely, the development of dissolution-enhanced vugs and other karst features require processes including controls on fluid flow and dissolution reactions, which are often tied to exposure horizons and freshwater lens diagenesis. Although karst processes are commonly superimposed on fractures, e.g., dissolution-enhanced fractures, the development of karst pore volume does not require a pre-existing fracture network [47]. Consequently, the integration of multiple datasets, as illustrated in Figure 7, provides insight into the processes responsible for the development of non-matrix features and how to accurately model their presence in the subsurface.

4. Further Applications

The NMQP method provides a fast-pass, semi-quantitative characterization of non-matrix features in core. It requires minimal prior expertise and delivers an integrated approach to extracting basic metrics of both karst features and fractures. NMQP is especially useful in fields with long lengths of core distributed between many wells, as it offers a glance at the types and distribution of non-matrix features. In our experience, this is one of the first tasks to be completed when core is available. The following subsections highlight some benefits and additional applications of the NMQP approach.

4.1. Time Management and Efficiency

Other non-matrix data collection methods, especially those designed for detailed fracture characterization, are typically tedious and time-consuming. Having a NMQP log available first has helped us to target further detailed fracture data collection efforts. Detailed fracture work includes recording properties for individual fractures of all types, lengths, spacing, width, openness, and roughness; as well as exhaustive information on the lithology of the host rock. Although results from such efforts deliver world-class databases used to generate robust statistical inputs to fracture models, the pace at which data is collected is extremely slow, i.e., between 0.5 and 5 m of core described per hour, depending on the complexity of the non-matrix network. When these types of meticulous core descriptions are performed, it is often difficult to decide when a sufficient amount of data has been collected or whether the selected wells are appropriate. Such efforts often result in scope creep and overspending of time and resources.

In our experience, having an available NMQP log provides an opportunity to develop a detailed data collection strategy that allows for targeting of specific intervals in the most representative cored wells. NMQP can help optimize time and resources in situations where business needs typically set a limited time for acquiring core data. For example, consider a case in which we have only one week to characterize 120 m of core. With the standard times per meter reported in Figure 1, one should be able to become familiarized with the main non-matrix core types and complete NMQP within the first few days. This information can then be used then to target detailed data collection in certain intervals of interest (e.g., areas of increased fracture density or areas with a certain type of non-matrix feature). By doing so, we ensure a balanced dataset that contains enough detailed measurements to develop suitable statistical distributions of properties while describing, at a more general level, the non-matrix properties throughout the available core.

4.2. Rapid Data Acquisition and Testing of Concepts

NMQP allows for rapid acquisition of data that can facilitate the testing of concepts in real time, such as by determining whether the presence of increased fracture density represents clusters, corridors, or fault zones. Another example of rapid integration of the NMQP approach is to compare fracture width populations between multiple wells as a proxy for strain. Figure 8 shows an example of how NMQP results can be quickly evaluated to observe differences in fracture widths between wells. In this example not related to any case discussed above, a box plot of fracture widths from four different wells shows that wells B and D have wider fractures, on average, compared to wells A and C (Figure 8). Such differences could be interpreted as variations in the amount of strain experienced by the rocks and thus provide insight into how fracture properties vary spatially within a given reservoir or aquifer.

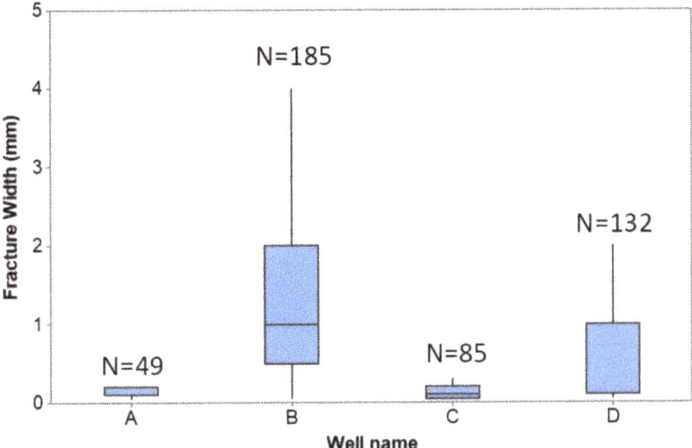

Figure 8. Box plot of fracture widths for four different wells using NMQP data.

NMQP results can also be used test concepts of mechanical stratigraphy. In this situation, the mechanical units can be smaller than the selected interval (core box), in which case the interval could be further subdivided using the *distribution* attribute and compared with sedimentological core descriptions to observe changes in lithology. Additional uses of NMQP results include areal controls on non-matrix density or types, impact of proximity to platform margin, or relations to paleotopography and stratigraphy.

4.3. Input for Numerical Modeling

Data collected with the NMQP approach can also be used to develop early realizations of discrete fracture network (DFN) models, which can later be upgraded with more detailed fracture data collection or calibrated to a well test. Figure 9A shows a cumulative frequency distribution of fracture width constructed using NMQP characteristic width data from 185 core boxes. A similar approach can be employed to model vug distribution (Figure 9B). The collection of width values fits a log-normal distribution (dashed line) that can be use as input for a DFN model (Figure 9).

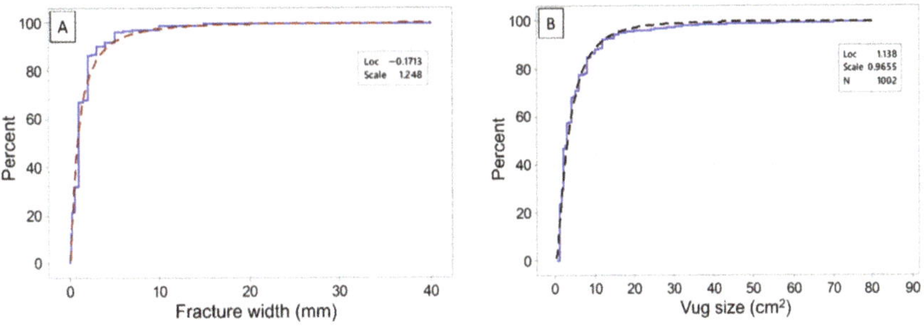

Figure 9. Cumulative frequency distributions of fractures (**A**) and vugs (**B**) based on NMQP of 185 core boxes. Blue lines represent width data collected on core. Dashed lines represent log-normal distributions of widths.

Additionally, carbonate reservoirs with abundant non-matrix features and high-permeability contrast typically require dual-porosity, dual-permeability formulations for more accurate forecast during reservoir simulation and performance prediction. Dual-permeability simulations require a non-matrix porosity input to initialize the models.

NMQP is a reliable source for quantifying non-matrix porosity, not only for the fracture component but for the karst component of the total non-matrix systems.

5. Summary

- NMQP is a comprehensive method for collecting non-matrix (karst and natural fractures) quantitative information in a rapid yet adequate fashion that allows a group of two researchers to describe 12–20 m of core per hour.
- NMQP provides a first-pass approach to understanding non-matrix types and distribution that can be used for reservoir characterization purposes. It also provides useful context for designing and targeting intervals of interest for further detailed data collection.
- NMQP offers enough vertical resolution to define trends and integrate observations with other wellbore data types, such as borehole image logs, losses during drilling, mechanical stratigraphy, or upscaled log properties.
- NMQP-based porosity and density logs provide reasonable values that can be used as input to discrete fracture network models and to initialize dual-porosity, dual-permeability reservoir simulations.

Author Contributions: Initial design and implementation of methodology were jointly performed by P.J.M. and F.F.-I. F.F.-I. developed the data processing step of the methodology. P.J.M. wrote the manuscript with support from F.F.-I. All authors have read and agreed to the published version of the manuscript.

Funding: ExxonMobil Upstream Research Company.

Acknowledgments: The authors thank ExxonMobil Upstream Research Company and ExxonMobil Research, Qatar, for permission to publish this work. We also thank the three anonymous reviewers, who provided constructive feedback, which improved the quality of this manuscript.

Conflicts of Interest: The authors declare that this study received funding from ExxonMobil Upstream Company. The funder was not involved in the study design, collection, analysis, interpretation of data, the writing of this article or the decision to submit it for publication.

References

1. Folk, R.L. *Petrology of Sedimentary Rocks*; Hemphill Publishers: Cedar Hill, TX, USA, 1974; p. 159.
2. Kulander, B.R.; Dean, S.L.; Ward, B.J. *Fractured Core Analysis: Interpretation, Logging, and Use of Natural and Induced Fractures in Core*; American Association of Petroleum Geologists: Tulsa, OK, USA, 1990; Volume 8.
3. Kerans, C.; Tinker, S.W. *Sequence Stratigraphy and Characterization of Carbonate Reservoirs*; SEPM: Broken Arrow, OK, USA, 1997; Volume 40, p. 130.
4. Dershowitz, W.S.; Herda, H.H. Interpretation of Fracture Spacing and Intensity. In Proceedings of the 33rd US Symposium on Rock Mechanics (USRMS), Santa Fe, NM, USA, 3–5 June 1992.
5. Choquette, P.W.; Pray, L.C. Geologic nomenclature and classification of porosity in sedimentary carbonates. *AAPG Bull.* **1970**, *54*, 207–250.
6. Lucia, F.J. Petrophysical parameters estimated from visual descriptions of carbonate rocks: A field classification of carbonate pore space. *J. Pet. Technol.* **1983**, *35*, 629–637. [CrossRef]
7. Lucia, F.J. Rock-fabric/petrophysical classification of carbonate pore space for reservoir characterization. *AAPG Bull.* **1995**, *79*, 1275–1300.
8. Archie, G.E. Classification of carbonate reservoir rocks and petrophysical considerations. *AAPG Bull.* **1952**, *36*, 278–298.
9. Lønøy, A. Making sense of carbonate pore systems. *AAPG Bull.* **2006**, *90*, 1381–1405. [CrossRef]
10. Weger, R.J.; Eberli, G.P.; Baechle, G.T.; Massaferro, J.L.; Sun, Y.F. Quantification of pore structure and its effect on sonic velocity and permeability in carbonates. *AAPG Bull.* **2009**, *93*, 1297–1317. [CrossRef]
11. Skalinski, M.; Kenter, J.A. Carbonate petrophysical rock typing: Integrating geological attributes and petrophysical properties while linking with dynamic behaviour. *Geol. Soc. Lond. Spec. Publ.* **2015**, *406*, 229–259. [CrossRef]
12. Clerke, E.A.; Mueller, H.W., III; Phillips, E.C.; Eyvazzadeh, R.Y.; Jones, D.H.; Ramamoorthy, R.; Srivastava, A. Application of Thomeer Hyperbolas to decode the pore systems, facies and reservoir properties of the Upper Jurassic Arab D Limestone, Ghawar field, Saudi Arabia: A "Rosetta Stone" approach. *GeoArabia* **2008**, *13*, 113–160. [CrossRef]
13. Clerke, E.A. Permeability, relative permeability, microscopic displacement efficiency and pore geometry of M_1 bimodal pore systems in Arab-D limestone. *SPE J.* **2009**, *14*, 524–531. [CrossRef]

14. Anselmetti, F.S.; Luthi, S.; Eberli, G.P. Quantitative characterization of carbonate pore systems by digital image analysis. *AAPG Bull.* **1998**, *82*, 1815–1836.
15. Hollis, C.; Vahrenkamp, V.; Tull, S.; Mookerjee, A.; Taberner, C.; Huang, Y. Pore system characterisation in heterogeneous carbonates: An alternative approach to widely-used rock-typing methodologies. *Mar. Pet. Geol.* **2010**, *27*, 772–793. [CrossRef]
16. Fullmer, S.M.; Guidry, S.A.; Gournay, J.; Bowlin, E.; Ottinger, G.; Al Neyadi, A.M.; Edwards, E. Microporosity: Characterization, distribution, and influence on oil recovery. In Proceedings of the International Petroleum Technology Conference, Doha, Qatar, 19–22 January 2014.
17. Buono, A.; Peterson, K.; Luck, K.; Fullmer, S.; Moore, P.J. Quantitative Digital Petrography: A Novel Approach to Reservoir Characterization. *J. Sediment. Res.* **2019**, *18*, 285–293.
18. Pittman, E.D. Microporosity in carbonate rocks. *AAPG Bull.* **1971**, *55*, 1873–1878.
19. Ahr, W.M. Early diagenetic microporosity in the Cotton Valley Limestone of east Texas. *Sediment. Geol.* **1989**, *63*, 275–292. [CrossRef]
20. Budd, D.A. Micro-rhombic calcite and microporosity in limestones: A geochemical study of the Lower Cretaceous Thamama Group, UAE. *Sediment. Geol.* **1989**, *63*, 293–311. [CrossRef]
21. Loucks, R.G.; Lucia, F.J.; Waite, L.E. Origin and description of the micropore network within the lower Cretaceous Stuart City Trend tight-gas limestone reservoir in Pawnee Field in South Texas. *Gulf Coast Assoc. Geol. Soc.* **2013**, *2*, 29–41.
22. Lucia, F.J.; Loucks, R.G. Micropores in carbonate mud: Early development and petrophysics. *Gulf Coast Assoc. Geol. Soc.* **2013**, *2*, 1–10.
23. Kaczmarek, S.E.; Fullmer, S.M.; Hasiuk, F.J. A universal classification scheme for the microcrystals that host limestone microporosity. *J. Sediment. Res.* **2015**, *85*, 1197–1212. [CrossRef]
24. Fullmer, S.; Al Qassab, H.; Buono, A.; Gao, B.; Kelley, B.; Moore, P.J. Carbonate pore-system influence on hydrocarbon displacement and potential recovery. *J. Sediment. Res.* **2019**, *18*, 268–284.
25. Tinker, S.W.; Ehrets, J.R.; Brondos, M.D. Multiple karst events related to stratigraphic cyclicity: San Andres Formation, Yates field, west Texas. In *Unconformities and Porosity in Carbonate Strata*; Budd, D.A., Saller, A.H., Harris, P.M., Eds.; AAPG Memoir: Tulsa, OK, USA, 1995; Volume 63, pp. 213–238.
26. Ahdyar, L.O.; Sekti, R.P.; Mohammad, S.R.; Fernandez-Ibanez, F.; Moore, P.J. *Integrated Carbonate Non-Matrix Characterization in Banyu Urip Field*; Indonesian Petroleum Association: South Jakarta, Indonesia, 2019.
27. Craig, D.H. Caves and other features of Permian karst in San Andres dolomite, Yates field reservoir, west Texas. In *Paleokarst*; James, N.P., Choquette, P.W., Eds.; Springer: Berlin/Heidelberg, Germany, 1988; pp. 342–363.
28. Kerans, C. Karst-controlled reservoir heterogeneity in Ellenburger Group carbonates of west Texas. *AAPG Bull.* **1988**, *72*, 1160–1183.
29. Lorenz, J.; Hill, R. Measurement and analysis of fractures in core. In *Geological Studies Relevant to Horizontal Drilling: Examples from Western North America*; Schmoker, J.W., Coalson, E.B., Brown, C.A., Eds.; Association of Geologists: Denver, CO, USA, 1992; pp. 47–57.
30. Loucks, R.G. Paleocave carbonate reservoirs: Origins, burial-depth modifications, spatial complexity, and reservoir implications. *AAPG Bull.* **1999**, *83*, 1795–1834.
31. Heward, A.P.; Chuenbunchom, S.; Mäkel, G.; Marsland, D.; Spring, L. Nang Nuan oil field, B6/27, Gulf of Thailand: Karst reservoirs of meteoric or deep-burial origin? *Pet. Geosci.* **2000**, *6*, 15–27. [CrossRef]
32. Lorenz, J.C.; Cooper, S.P. *Atlas of Natural Fractures and Coring-Induced Structures in Core*; John Wiley & Sons: Hoboken, NJ, USA, 2017; p. 320.
33. Fernández-Ibáñez, F.; DeGraff, J.M.; Ibrayev, F. Integrating borehole image logs with core: A method to enhance subsurface fracture characterization. *AAPG Bull.* **2008**, *102*, 1067–1090. [CrossRef]
34. Ibrayev, F.; Fernandez-Ibanez, F.; DeGraff, J.M. Using a genetic-based approach to enhance natural fracture characterization in a giant carbonate field. In Proceedings of the SPE Annual Caspian Technical Conference & Exhibition, Astana, Kazakhstan, 1–3 November 2016.
35. Tonkins, M.C.; Coggan, J.S. Characterization of Rock Fracturing for Vertical Boreability. *Procedia Eng.* **2017**, *191*, 112–118. [CrossRef]
36. Kerans, C.; Donaldson, J.A. Proterozoic paleokarst profile, Dismal Lakes Group, NWT, Canada. In *Paleokarst*; Springer: Berlin/Heidelberg, Germany, 1988; pp. 167–182.
37. Ronchi, P.; Ortenzi, A.; Borromeo, O.; Claps, M.; Zempolich, W.G. Depositional setting and diagenetic processes and their impact on the reservoir quality in the late Visean–Bashkirian Kashagan carbonate platform (Pre-Caspian Basin, Kazakhstan). *AAPG Bull.* **2010**, *94*, 1313–1348. [CrossRef]
38. Lockman, D.F.; George, R.P.; Hayes, M.J. *A Systematic Technique for Describing and Quantifying Fractures in Cores*. MP-44; The Pacific Section American Association of Petroleum Geologists: Bakersfield, CA, USA, 1997; pp. 1–33.
39. DeGraff, J.M.; (ExxonMobil (retired), Spring, TX, USA). Personal communication, 2018.
40. Watkins, H.; Bond, C.E.; Healy, D.; Butler, R.W. Appraisal of fracture sampling methods and a new workflow to characterise heterogeneous fracture networks at outcrop. *J. Struct. Geol.* **2015**, *72*, 67–82. [CrossRef]
41. Fernández-Ibáñez, F.; Moore, P.J.; Jones, G.D. Quantitative assessment of karst pore volume in carbonate reservoirs. *AAPG Bull.* **2019**, *103*, 1111–1131. [CrossRef]

42. Lucia, F.J. Touching Vug Reservoirs. In *Carbonate Reservoir Characterization, An Integrated Approach*; Lucia, F.J., Ed.; Springer: Berlin/Heidelberg, Germany, 2007; pp. 301–331.
43. Fernández-Ibáñez, F.; Jones, G.D.; Mimoun, J.G.; Bowen, M.G.; Simo, J.A.; Marcon, V.; Esch, W.L. Excess permeability in the Brazil pre-Salt: Nonmatrix types, concepts, diagnostic indicators, and reservoir implications. *AAPG Bull.* **2022**, *106*, 701–738. [CrossRef]
44. Ortega, O.J.; Marrett, R.A.; Laubach, S.E. A scale-independent approach to fracture intensity and average spacing measurement. *AAPG Bull.* **2006**, *90*, 193–208. [CrossRef]
45. Likert, R. A technique for the measurement of attitudes. *Arch. Psychol.* **1932**, *22*, 1–55.
46. Palmer, A.N. Origin and morphology of limestone caves. *Geol. Soc. Am. Bull.* **1991**, *103*, 1–21. [CrossRef]
47. Vacher, H.L.; Mylroie, J.E. Eogenetic karst from the perspective of an equivalent porous medium. *Carbonates Evaporites* **2002**, *17*, 182–196. [CrossRef]
48. Breithaupt, C.I.; Gulley, J.D.; Moore, P.J.; Fullmer, S.M.; Kerans, C.; Mejia, J.Z. Flank margin caves can connect to regionally extensive touching vug networks before burial: Implications for cave formation and fluid flow. *Earth Surf. Processes Landf.* **2021**, *46*, 1458–1481. [CrossRef]
49. Kosters, M.; Hague, P.F.; Hofmann, R.A.; Hughes, B.L. Integrated modeling of karstification of a central Luconia Field, Sarawak. In *IPTC 2008: International Petroleum Technology Conference 2008*; European Association of Geoscientists & Engineers: Houten, The Netherlands, 2008; p. cp-148.
50. Van Dijk, J. Analysis and Modeling of Fractured Reservoirs. In Proceedings of the European Petroleum Conference, SPE-50570-MS, The Hague, The Netherlands, 20–22 October 1998. [CrossRef]
51. Belfield, W.C. Characterization of a Naturally Fractured Carbonate Reservoir: Lisburne Field, Prudhoe Bay, Alaska. In Proceedings of the SPE Annual Technical Conference and Exhibition, Houston, TX, USA, 2–5 October 1988. [CrossRef]
52. Fernandez-Ibanez, F.; DeGraff, J.M.; Moore, P.J.; Ahdyar, L.; Nolting, A. Characterization of non-matrix type and flow potential using lost circulation information. *J. Pet. Sci. Eng.* **2019**, *180*, 89–95. [CrossRef]
53. Bisdom, K.; Bertotti, G.; Nick, H.M. The impact of different aperture distribution models and critical stress criteria on equivalent permeability in fractured rocks. *J. Geophys. Res. Solid Earth* **2016**, *121*, 4045–4063. [CrossRef]
54. Odling, N.E.; Gillespie, P.; Bourgine, B.; Castaing, C.; Chiles, J.P.; Christensen, N.P.; Watterson, J. Variations in fracture system geometry and their implications for fluid flow in fractures hydrocarbon reservoirs. *Pet. Geosci.* **1999**, *5*, 373–384. [CrossRef]

Article

Application of Machine Learning for Lithofacies Prediction and Cluster Analysis Approach to Identify Rock Type

Mazahir Hussain [1], Shuang Liu [1,*], Umar Ashraf [2], Muhammad Ali [1], Wakeel Hussain [3], Nafees Ali [4,5] and Aqsa Anees [2]

[1] Institute of Geophysics and Geomatics, China University of Geosciences, Wuhan 430074, China; mazahir@cug.edu.cn (M.H.); muhammad_ali@cug.edu.cn (M.A.)
[2] Institute for Ecological Research and Pollution Control of Plateau Lakes, School of Ecology and Environmental Science, Yunnan University, Kunming 650500, China; umarashraf@ynu.edu.cn (U.A.); aqsaanees@ynu.edu.cn (A.A.)
[3] Department of Geological Resources and Engineering, Faculty of Earth Resources, China University of Geosciences, Wuhan 430074, China; wakeelhussain90@cug.edu.cn
[4] State Key Laboratory of Geomechanics and Geotechnical Engineering, Institute of Rock and Soil Mechanics, Chinese Academy of Sciences, Wuhan 430071, China; nafeesali@mails.ucas.ac.cn
[5] University of Chinese Academy of Sciences, Beijing 100049, China
* Correspondence: lius@cug.edu.cn

Abstract: Nowadays, there are significant issues in the classification of lithofacies and the identification of rock types in particular. Zamzama gas field demonstrates the complex nature of lithofacies due to the heterogeneous nature of the reservoir formation, while it is quite challenging to identify the lithofacies. Using our machine learning approach and cluster analysis, we can not only resolve these difficulties, but also minimize their time-consuming aspects and provide an accurate result even when the user is inexperienced. To constrain accurate reservoir models, rock type identification is a critical step in reservoir characterization. Many empirical and statistical methodologies have been established based on the effect of rock type on reservoir performance. Only well-logged data are provided, and no cores are sampled. Given these circumstances, and the fact that traditional methods such as regression are intractable, we have chosen to apply three strategies: (1) using a self-organizing map (SOM) to arrange depth intervals with similar facies into clusters; (2) clustering to split various facies into specific zones; and (3) the cluster analysis technique is used to identify rock type. In the Zamzama gas field, SOM and cluster analysis techniques discovered four group of facies, each of which was internally comparable in petrophysical properties but distinct from the others. Gamma Ray (GR), Effective Porosity(eff), Permeability (Perm) and Water Saturation (Sw) are used to generate these results. The findings and behavior of four facies shows that facies-01 and facies-02 have good characteristics for acting as gas-bearing sediments, whereas facies-03 and facies-04 are non-reservoir sediments. The outcomes of this study stated that facies-01 is an excellent rock-type zone in the reservoir of the Zamzama gas field.

Keywords: self-organizing map; cluster analysis; lithofacies; Zamzama gas field; rock type

Citation: Hussain, M.; Liu, S.; Ashraf, U.; Ali, M.; Hussain, W.; Ali, N.; Anees, A. Application of Machine Learning for Lithofacies Prediction and Cluster Analysis Approach to Identify Rock Type. *Energies* **2022**, *15*, 4501. https://doi.org/10.3390/en15124501

Academic Editor: Reza Rezaee

Received: 11 May 2022
Accepted: 18 June 2022
Published: 20 June 2022

Publisher's Note: MDPI stays neutral with regard to jurisdictional claims in published maps and institutional affiliations.

1. Introduction

Machine learning emerged as a subfield of artificial intelligence (AI) in the second decade of the twentieth century, using self-learning algorithms that gathered information from data to make predictions [1–4]. Machine learning offers a more efficient option to capture the information in data to gradually improve the performance of prediction models and make data-driven decisions [5–8], rather than needing humans to manually create rules and build models from analyzing massive volumes of data [9–11]. Machine learning is divided into three categories: supervised learning, unsupervised learning, and reinforcement learning [12,13]. Each type has its application and algorithm; however, because of the lack of outcome information in our case study, we primarily focused on unsupervised learning. Furthermore, unsupervised learning takes into account the fact that it may automatically extract hidden patterns without human instruction, making it more similar to machine learning than other varieties [14]. We used a machine learning model to categorize the facies for Zamzama gas field and tested the findings against real facies data in this study. Our model likewise uses data from this field, but we used a novel model called the self-organizing map (SOM) to tackle the problem [15]. In the situation of a lack of facies data or geologically inexperienced users, our model would be the best fit [16]. The principal component analysis (PCA) is our model's first unsupervised learning approach [17]. This is a linear mathematical strategy for condensing a big set of variables (seismic characteristics) into a smaller set that retains the majority of the independent information variation found in the larger data set [18,19]. One can distinguish sedimentary units with similar log characteristics by gathering data from several good logs [20–24]. In the literature, sedimentary units established on this basis and characterized from wireline logs were referred to as electrofacies or logfacies [17,25–27]. One of the most accurate and impactful procedures in oil-bearing clastic reservoirs is multivariate cluster analysis (referred to as the best method of data grouping in the literature) [8,16].

The aim of this research is to classify gamma-ray, porosity, permeability and water saturation into logfacies and rock types. Our study compares and evaluates lithofacies and various rock type identifications, utilizing SOM and cluster analysis, via hierarchical and non-hierarchical approaches to calibrate the appropriate model for researching lithofacies and rock-type identification. The rock type classification is performed using the cluster analysis method, which aims to discover groups of well-log data with similar characteristics. This classification is based on the unique properties of well-log measurements, which reflect lithofacies within the recorded interval, and does not require any artificial segmentation of the data population [28].

A cluster analysis can be performed using a variety of methods. Furthermore, unsupervised learning is more similar to machine learning than other forms since it can automatically extract hidden patterns without human assistance [29–31]. For lithofacies classification, there is an unsupervised learning model called support vector machine (SVM). The SVM is a useful approach for higher-dimensional datasets that is also versatile, as alternative kernels can be specified according to the user's needs. The SOM is the next step in the process. There are massive data analysis challenges, particularly in the classification of lithofacies and the identification of rock types, both of which generate large amounts of data, as well as the fact that humans are unable to fully appreciate the link between seismic properties [32]. Using the advanced machine learning approach and cluster analysis, we can not only solve these problems, but also reduce their time-consuming nature, and deliver an accurate result even when the user is unskilled. Finally, clustering is used to classify subgroups (facies) based on their dissimilarity. In this research, we want to systematize the essential background of the SOM and then apply this workflow to facies classification in two real examples. Based on the final results, which are compared, several discussions are presented of lithofacies identification.

General Geology and Stratigraphy of the Study Area

The Zamzama gas field is located on the eastern edge of the Kirthar Foldbelt and is a broad, trust-related anticline northeast of the fields of Bhit and Badhra, and south of the fields of Mehar, Sofiya, and Mazarani (Figure 1a). In the frontal folds on the Kirthar foredeep, these field are situated along the western edge of the Lower Indus Basin [33,34]. The Zamzama gas area is situated along the Kirthar folds and thrust belts of Pakistan's Southern Indus Basin. The southern Indus Basin is limited by the Indian Shield to the east and the Indian plate's marginal zone to the west, as well as the Sukkur rift from the north to the offshore Indus in the south [35,36]. The Kirthar Foldbelt in Pakistan is part of the lateral mountain belt linking Makran's accretionary wedge to the Himalayan orogeny. Parallel to the regional plate motion vector, the area is undergoing oblique deformation (Figure 1b).

In 1998, with estimated wet gas in place at discovery, the Zamzama Field was discovered and has provided condensate cumulatively to date [25,26].

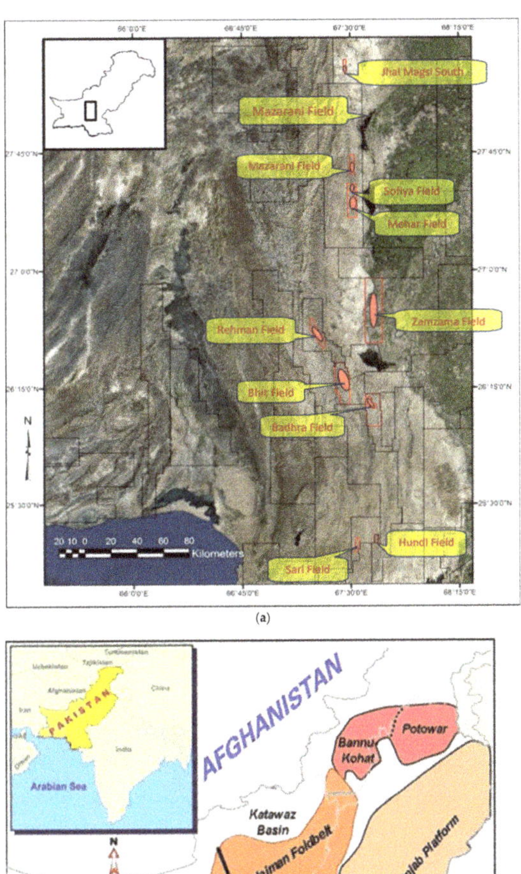

Figure 1. (**a**) Satellite map showing the locations of the major oil and gas filed within the Middle and Southern Indus Basin. Zamzama gas field is present in the middle towards the eastern side of the map. (**b**) Regional tectonic map of Pakistan, showing the major basins and tectonic regions.

Most of the production comes from late Cretaceous Pab Formation fluvial and shallow marine sandstones, but the Zamzama area also produces sandstones from the estuarine Palaeocene Khadro formation, which are from the Pab formation in stratigraphic pressure isolation. In the Zamzama region, the Sembar's Cretaceous shales and the Goru formations are regarded as the principal source rock [37–40]. Through a majority of the Southern Indus Basin, the Sembar Formation was deposited in marine settings [41–43]. The Lower Goru Formation was deposited over the whole basin of the Southern Indus [33,37]. The early Cretaceous Goru Formation, which is divided into two sections (the Lower Goru Formation (LGF) and Upper Goru Formation (UGF)), superimposes the Sembar Formation [44,45]. The Goru Formation was accumulated in a shallow marine environment such as a shoreface to the fluvial-based proximal delta-front depositional framework [1,41]. Quite coarse to fine, porous, and permeable sediments are preserved in fluvial networks and create reservoirs in fluvial-based depositional systems [9,37,39]. The Lower Goru Formation, which includes quite coarse to fine sediments, is the largest reservoir rock in the Lower Indus Basin [33,34]. However, in our study area, Goru formation is acting as a source rock, while Pab sandstone is the main reservoir rock within the Zamzama gas field (Figure 2). The main producing reservoir in the Zamzama gas field is the Maastrichtian Pab Formation, which shows the deposition of the sand-rich fluvio-deltaic coastal plain and shoreface depositional system that passes westwards into deep marine turbidites. An alternative target is sandstone reservoirs within the underlying Palaeocene Khadro Formation, which are separated by varied thicknesses of coastal plain shales and mudstones, across the top Pab Formation unconformity. The Khadro Formation sandstones are made up of estuary, intertidal, and shoreface deposits, with the shoreface units cut by tidal channels, and are hence very discontinuous and variable in distribution. The Palaeocene Girdo (Ranikot) Formation marine shales serve as the top seal for the Khadro Formation reservoir sand, which is present throughout the field and offers a durable continuous pressure barrier, even when cut by thrusts. Because the basal Khadro Formation shales form a good seal from the underlying Pab Formation reservoir, Khadro Formation sandstones are anticipated to be closer to virgin field pressures, unless depleted by production from Khadro Formation producing wells [22–25].

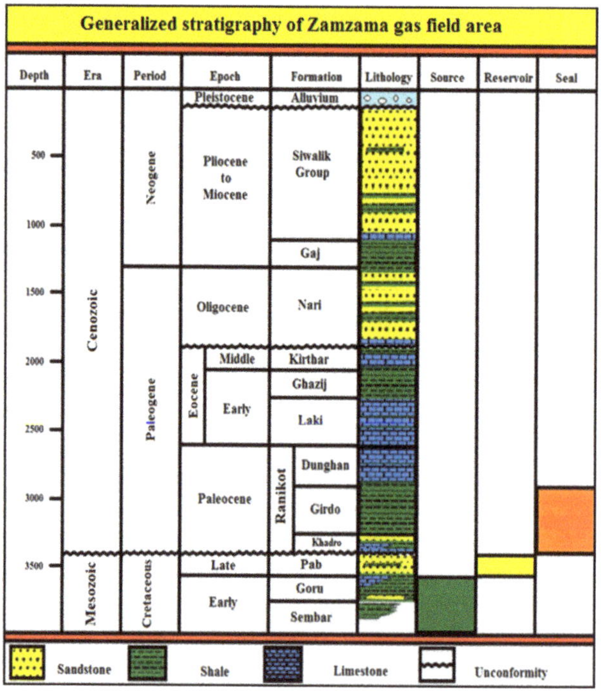

Figure 2. Generalized lithostratigraphy of the Lower Indus Basin [34].

2. Data and Methods

2.1. Dataset

For the interest, a dataset of four wells (well-2, -3, -4, and -5) were used in the current study. The model was trained in well-3, and the experiment models were verified in the other wells. We have focused on the gas-bearing zone in the Zamzama gas field, which is present within the Lower Goru formation of Cretaceous age. The primary goal of the study was to fix the machine learning model's configuration settings to attain the maximum classification accuracy feasible in the dataset. Many unsupervised learning algorithms were performed for classifying lithofacies and rock type identification in the Zamzama gas field, as evidenced by the majority of the evaluation's findings being validated on well dataset samples. The study compares and evaluates lithofacies and identification of various rock types, using methods such as SOM and cluster analysis to calibrate the appropriate model for researching lithofacies and rock type identification. Lithofacies distributed the reservoir interval by combining sedimentological explanations and reacting to gamma-ray log response. We have used the commercial "Interactive Petrophysics (IP) Software" and coding for machine learning throughout the whole study. The IP software is used to incorporate all the well-logged data for computing and then evaluating the inputs of several petrophysical properties for accurate and adequate assessment of the formation's lithofacies.

2.2. Methods

2.2.1. SOM

The SOM is a mathematical technique for organizing data into groups to build a map. It is a neuro-computational clustering approach that uses supervised and unsupervised learning processes to uncover new and valuable knowledge hidden in massive datasets [11]. Geoscientists can use the SOM to analyze rock characteristics and reservoir fluids since it delivers high-quality data. SOM's capacity to learn and organize data without requiring associated dependent output values for the input pattern is one of its most attractive features [15]. The topology of SOM is determined by several nodes (varying from a few dozens to thousands) linked to surrounding nodes and organized on a regular low-dimensional grid. The nodes for the entire dataset are created by a training method in SOM.

Electro facies assessment is an important stage in determining the accuracy of reservoir rock evaluation. To decrease the uncertainties and evaluate the electro facies, a type of artificial neural network (ANN) called SOM was utilized in this study. It is a model of unsupervised learning. The SOM is followed as

$$W_v 1 = W_v(s) + \theta(u, v, s).a(s).[D(t) - W_v(s)] \tag{1}$$

where s represents the current iteration, t represents the index of the target input data vector in the input dataset, $D(t)$ represents the vector of target input data, v represents the node index in the map, $W_v(s)$ represents the current weight vector of node v, u represents the index on the map for best matching units (BMUs) (SOM node having the shortest aggregate distance to one of the input vectors), $\theta(u, v, s)$ represents even though due to the distance from BMU, commonly referred to as the neighborhood and α represents, based on iteration development, the learning restriction.

To begin the training process, the weights in each node are assigned to a random value. After the map has been initialized, the input data is sent to it. For each level of input data, the BMU is determined, which is the node that most closely represents the input data. The Euclidean distance represents the weight vectors of each node and the provided input vector is calculated as follows:

$$\text{Distance} = \sqrt{\sum_{i=0}^{i=n}(V_i - W_i)^2} \tag{2}$$

Here, V is the present input vector and W is the weight vector of the node. The BMU is the node where such a distance evaluates to the smallest value. The following equation is used to change the weight vectors of the successful node such that they are closer to the input vector:

$$W_{t+1} = W_t + L_t \, (V_t - W_t) \tag{3}$$

where 't' is the current training pass (or time-step), 'W' is the weight vector, 'V' is the input vector and 'L' is a variable called the learning rate:

$$0 < L < 1 \tag{4}$$

The learning rate lowers over time (per training pass) and declines with the following equation for every repetition of the training pass.

$$L_t = L_0 \, \exp\left(-\frac{t}{\lambda}\right) \tag{5}$$

where L_0 represents the initial learning rate before training, 't' represents the current training pass repetition, and 'λ' represents a time constant determined by the equation:

$$\lambda = \frac{t}{\log_{\sigma_0}} \tag{6}$$

where σ_0 is the initial radius of the neighborhood of effect, as discussed below.

The node with the least Geometric difference between the input vector and all nodes is picked, and its neighboring nodes within a specific radius are slightly altered to match the input vector. The neighborhood radius is set to half of the map grid width at the start. However, as time passes, the radius of the neighborhood reduces, and at the end of the training, the radius is reduced to a single node. With training passes, the neighborhood radius decreases as follows:

$$\sigma_0 \, \exp\left(-\frac{t}{\lambda}\right) \tag{7}$$

where the radius of the neighborhood is denoted by 'σ'.

$$W_{t+1} = W_t + \theta_t L_t (V_t - W_t) \tag{8}$$

Here, 'θ' is the impact of a node's distance from the BMU on its weighting correction, as calculated by the equation.

$$\theta_t = \exp\left(-\frac{\mathrm{dist}^2}{2\sigma_t^2}\right) \tag{9}$$

Here, 'dist' is the distance between the node and the BMU, as measured by Pythagoras' theorem. $\sigma(t)$ is the radius of the neighborhood function, which controls how far neighbor nodes are checked. It becomes smaller and smaller over time.

The technique described above is carried out for the specified number of training iterations. The weights of the input dataset are optimized at each iteration step until the best and most reliable set of weights for the network is found. The above exercise is ended to guarantee that a minimum error criterion is met. It is worth noting that geological heterogeneities influenced the number of clusters; the more heterogeneous the geology, the more clusters; hence, process levels use SOM weight planes and local geological information at the same time. SOM can be used to analyze financial stability in addition to facies evaluation for oil and gas exploration.

2.2.2. Clustering Procedure

Due to various factors that affect the logs, similar facies may have distinct log responses. Because statistical methods and processes are required, data are clustered with a minimal distance and maximum homogeneity in the clustering procedure. It is self-evident that different geological factors can be linked to a set of data known as logfacies, which geologists can utilize for further interpretation. All log readings are treated as "observations" in this calculation, and the user logs are treated as "values of the observations." The lowest distances are joined together to form a pair in cluster analysis. Because the number of logfacies is usually smaller than the number of readings, pairs of vectors are linked to form a cluster (logfacies). To create higher rank kinds, lower-rank clusters are joined together. This process is repeated until a single cluster (representing all of the data) is formed. There are several methods for connecting two clusters. To link the cluster components in some of them, the least distance between them is used. Using IP software, the clustering module was completed in two stages: To begin, the data (gamma-ray log, porosity, and water saturation) are separated into easily understandable data clusters. The number of clusters should be sufficient to cover all of the data ranges seen in the logs. For most data sets, fifteen to twenty clusters appear to be an acceptable quantity. The second, more labor-intensive phase is to organize these 15 to 20 clusters into a reasonable number of geological facies. This could mean condensing the data into five or six groups. The K-mean statistical technique is used in the first stage of "Facies Clustering" to cluster the data into a known number of clusters. To make this work, an estimate of the mean value of each cluster for each input log must be made first. The starting assumption can have an impact on the findings; therefore, make sure the beginning values cover the entire range of the logs. Each input data point is assigned to a cluster in K-mean clustering. The method tries to reduce the sums of squares of the difference between the data point and the cluster mean value inside each cluster. The method works by computing the sum of squares difference between a data point and each cluster mean, then allocating the data point to the cluster with the smallest difference. The new mean values in each cluster are determined when all of the data points have been assigned to the clusters. The programs begin with reassigning the data to the clusters using the updated mean values. This loop is repeated until the mean values between loops do not change. Before starting, all input log data are adjusted so that each input log has the same dynamic range. The mean and standard deviation of the log are calculated, and the data are then normalized by subtracting the mean and dividing by the standard deviation.

Stage-2 Cluster Consolidation

Cluster consolidation can be carried out entirely by hand, utilizing the cross plot and log plot output to group data, or using a hierarchical cluster approach to group data. Hierarchical clustering works by calculating the distances between all clusters and then combining the two clusters that are the most closely related. After that, the new cluster distance to all other clusters is recalculated, and the two closest clusters are combined once more. This technique is repeated until only one cluster remains. A dendrogram can be made from the results. The dendrogram depicts how and in what order the clusters were fused. The merging sequence is shown by the numbers at the top of each branch. The original K-mean clustering findings are presented at the bottom of the plot. There are five main clustering strategies in IP software that determine how the clusters are combined. The outcomes of the various strategies will be vastly different. The distance calculation is updated differently in each of the five approaches after two clusters have been connected. Assume that clusters "A" and "B" have recently been linked to from cluster "Z," and that we need to compute the distance between "Z" and another cluster, designated "C", in the diagram below.

The computations for the various techniques are as follows: (1) the minimum distance between all clustered objects—the distance between Z and C is the shortest of the distances (A to C and B to C). (2) Maximum distance between all clustered objects—the distance between Z and C is the greatest of the distances (A to C and B to C). (3) Average distance between merged clusters—the distance from Z to C is the average distance between all objects in the cluster generated by merging clusters and C. (4) Average distance between all objects in clusters—the distance between clusters Z and C is the average distance between cluster Z and cluster C (Figure 3).

2.2.3. Non-Hierarchical or K-means Clustering Methods

In these methods, the required number of clusters is specified in advance, and the best solution is selected. When working with big data sets, non-hierarchical cluster analysis is frequently utilized since it allows individuals to shift from one cluster to another, which is not possible with hierarchical cluster analysis [11,34,46]. The k-mean cluster analysis has

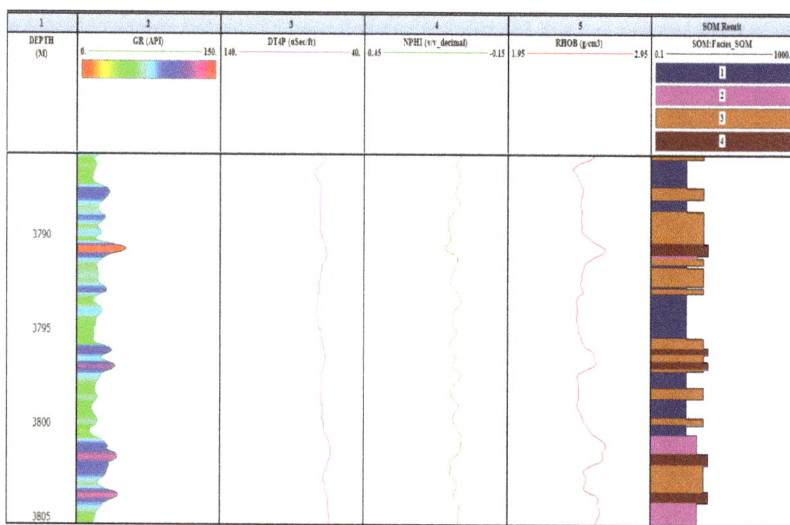

Figure 6. SOM vertical distribution model showing the interpreted lithofacies for the four input parameters after calibration.

Table 1. The results of "cluster means" plus other statistics of user data for each cluster.

			GR	eff	Perm	Sw
		K-Mean Cluster Results				
Facies	Points	Rock Typing	Mean	Mean	Mean	Mean
1	13	Excellent-quality rock type	19.44	0.12	32.49	0.16
2	50	Good-quality rock type	20.32	0.10	8.94	0.26
3	62	Moderate-quality rock type	22.59	0.05	0.37	0.77
4	35	Poor-quality rock type	33.34	0.02	0.01	34.69

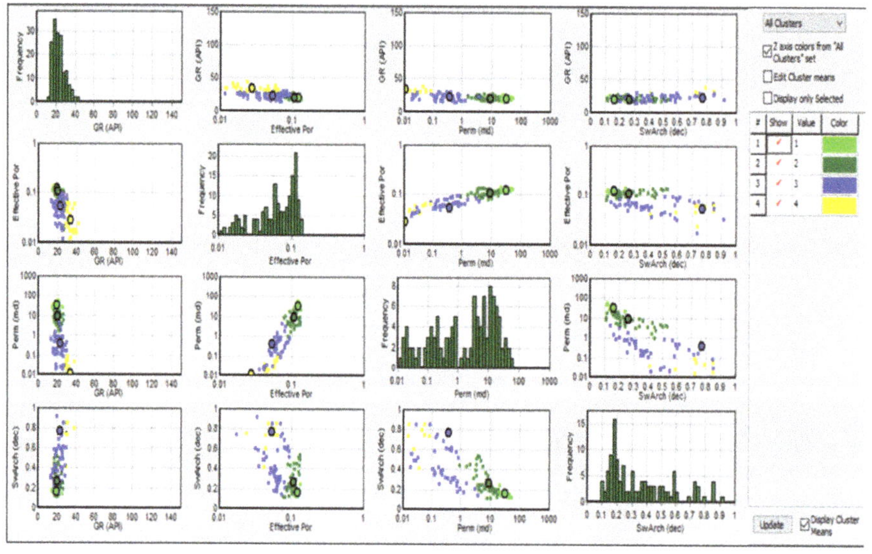

Figure 7. The final graphical result of clustering analysis.

Table 2. The properties for groups of Facies.

S. No	Rock Typing	GR	eff	Perm	Sw
Facies-01	Excellent-quality rock type	Very low	Good to excellent	Good to excellent	Very low
Facies-02	Good-quality rock type	low	Good	Good	low
Facies-03	Moderate-quality rock type	Medium	Fair to Good	Fair to Good	Medium
Facies-04	Poor-quality rock type	High	Low	Low	Very high

3.3. Hierarchical and Non-Hierarchical

After calculating the distance between objects in the dataset, connecting distance data can be used to identify how objects in the dataset should be grouped into clusters. The objects with the shortest distance between them were joined together to form new clusters. These newly generated clusters link to one other and to add items to form larger clusters, eventually linking all of the objects in the original dataset in a hierarchical tree. In general, "minimum distance between all things in clusters" produces long, thin clusters, whereas "maximum distance between all objects in clusters" produces larger spherical clusters. The "minimize the within-cluster sum of squares distance" and "average distance between all objects in clusters" are likely to produce clusters that are comparable to those created with "average distance between all objects in clusters." The clusters (electro-facies) were then constructed based on the data cluster tree or dendrogram (Figure 7). A dendrogram is a hierarchical tree with many U-shaped lines connecting things. The distance between two objects being connected is represented by the height of each U. Two objects with the shortest distance connect in the cluster tree to form a new, larger cluster. This sequence would repeat itself until just one cluster remained. For the dataset from all available wells, the procedure described above was used. As seen in Figure 8 of the dendrogram, the default approach "minimize the within-cluster sum of squares distance" produces good results for splitting the distinct log lithologies into different clusters. Stopping the grouping at a specific cutoff level makes it simple to divide the clusters into a defined number of groups. It is feasible to examine the groupings to determine whether adding another cluster adds more information or merely adds noise at which level. This information can be found in the "Cluster Randomness Plot." The "Cluster Randomness Plot" that determines the perceived randomness of the data for each cluster group is shown in (Figure 9). The greater the score, the less random the clusters are, indicating that the data are more structured. The average number of depth levels per cluster, for example, the average thickness of a cluster layer, is used to determine unpredictability. This is carried out on the original log data. The theoretical average thickness is then determined, assuming that the clusters are assigned at each depth level at random. The ratio of the two is randomness. A value of 1 would be completely random, whereas higher values would be less so.

$$\text{average thickness} = \text{number of depth levels}/\text{number of cluster layers}$$

$$\text{random thickness} = \sum \frac{p_i}{(1 - p_i)}$$

where p_i is the proportion of depth levels assigned to the i_{th} cluster.

Randomness index = average thickness = random thickness. The plot is interpreted by picking the number of least random clusters (highest peaks).

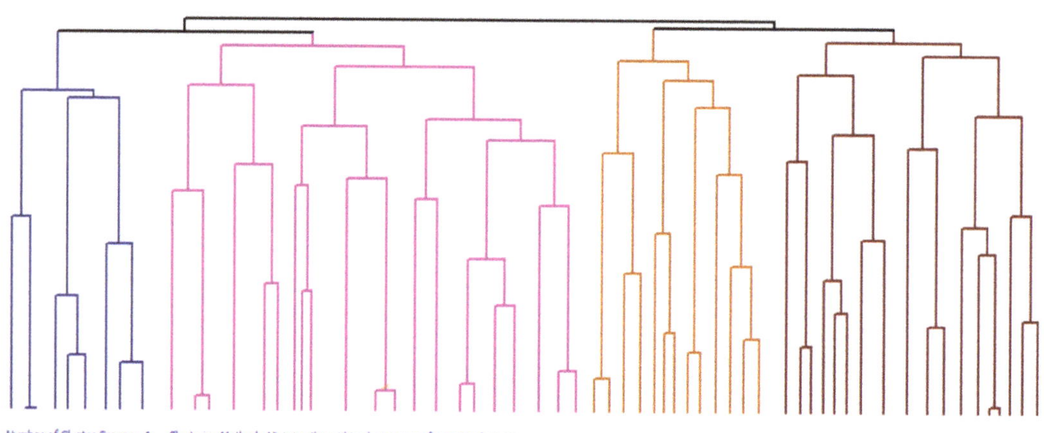

Figure 8. Cluster grouping dendrogram of the Zamzama Gas Field.

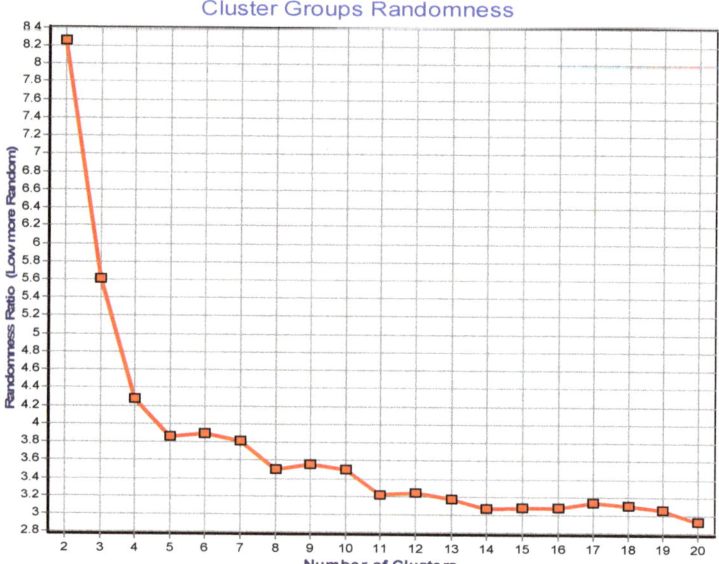

Figure 9. Cluster groups randomness of Zamzama Gas Field.

4. Discussion

The Zamzama Gas field in Pakistan was investigated using unsupervised learning and cluster analysis to categorize lithofacies and identify rock types. As both classifiers, hierarchical and non-hierarchical and SOFM produced more trustworthy results in lithofacies classification and rock type identification. These classifiers were able to accurately predict shaly sandstone and precision sandstone. However, when compared to other facies, the PR-area-under-curve score for shale was below average, indicating machine learning misclassification in predicting one facies to other facies despite the usually good accuracy. In comparison to the other facies, the existence of a mixture of shaly sand and sandstone has similar rock physical properties. Sandstone and shaly sandstone can be accurately categorized using the results of log-facies classification. On the other hand, shale distribution has been inconsistent. Furthermore, despite differences in rock physical properties between the different depths regions due to different rates of compaction and diagenetic processes, the machine learning model was able to properly predict lithofacies and rock type in both sections [27]. Due to the post-depositional processes that each formation underwent, the sensitivity of such a result may vary at deeper intervals, but despite this, the machine learning model has an overall high result and has effectively supported the geological investigation in a much shorter period [29,32]. The main contributions of this work were a simple approach for lithofacies classification and rock type identification in the Zamzama Gas field in Pakistan using SOM and cluster analysis, as well as hierarchical and non-hierarchical approaches, evaluation of each multiclass of unsupervised learning methods, high accuracy results despite some misclassification, log-facies classification analysis, and rock physics analysis based on unsupervised learning [33,34]. Most importantly, this study has demonstrated that the effectiveness of hierarchical and non-hierarchical SOFM can be evaluated from both a machine learning and geological perspective [29,35,36].

5. Conclusions

We have systematized the fundamental background of the SOM in this work. The U-matrix can also be used to view it, and BMUs can be seen directly. In addition, certain new changes have been made, such as normalization to standardize the input data and eliminating the scale value gap between curves. These contribute to a more accurate and realistic outcome. We also offered certain mathematical calculations, such as SOM, to help illustrate the process. Furthermore, attributes of the facies log are shown, such as the shape, measurement, and depth values. The input of the clustering process is also detailed in terms of SOM building. Based on log data, cluster analysis is a straightforward approach for determining the rock type for a reservoir. As illustrated in the roundness figure, cluster analysis of log data for wells that penetrated the reservoir were classified into four groups. Based on the cluster analysis, four facies have been identified; the findings of each facies are shown in (Table 1) and the behavior of each facies indicated in (Table 2). The results of these facies are shown in (Table 1) and (Table 2). Gamma Ray (GR), Effective Porosity(eff), Permeability (Perm) and Water Saturation (Sw) are used to generate these results. The Facies-01 zone in the reservoir for the Zamzama gas field is the most productive in the reservoir, as shown by plotting rock type in the continuous form in the well.

Author Contributions: M.H. and M.A. conceived this study. S.L. supervised this project. N.A. and A.A. undertook the responsibility of arranging the data for this project. M.H. and U.A. wrote the manuscript. W.H., A.A. and S.L. reviewed the manuscript and provided the necessary funding. All authors have read and agreed to the published version of the manuscript.

Funding: This research was supported by the National Natural Science Foundation of China (Grant Nos. 41874122 and 42122030).

Acknowledgments: The authors would like to thank the Directorate General of Petroleum Concessions (DGPC), Pakistan, for the release of well data. I am grateful to my supervisor for providing the necessary data, guidance, support, software, and technical help to accomplish this research. I am also thankful for my lab mates at the China University of Geosciences.

Conflicts of Interest: The authors declare no conflict of interest.

Nomenclature

SOM	Self-Organizing Map
SOFM	Self-Organizing Feature Map
PCA	Principal Component Analysis
LGF	Lower Goru Formation
UGF	Upper Goru Formation
BMU	Best Matching Unit

References

1. Ashraf, U.; Zhang, H.; Anees, A.; Mangi, H.N.; Ali, M.; Zhang, X.; Imraz, M.; Abbasi, S.S.; Abbas, A.; Ullah, Z. A Core Logging, Machine Learning and Geostatistical Modeling Interactive Approach for Subsurface Imaging of Lenticular Geobodies in a Clastic Depositional System, SE Pakistan. *Nat. Resour. Res.* **2021**, *30*, 2807–2830. [CrossRef]
2. Ali, M.; Jiang, R.; Huolin, M.; Pan, H.; Abbas, K.; Ashraf, U.; Ullah, J. Machine Learning-A Novel Approach of Well Logs Similarity Based on Synchronization Measures to Predict Shear Sonic Logs. *J. Pet. Sci. Eng.* **2021**, *203*, 108602. [CrossRef]
3. Bressan, T.S.; Kehl de Souza, M.; Girelli, T.J.; Junior, F.C. Evaluation of Machine Learning Methods for Lithology Classification Using Geophysical Data. *Comput. Geosci.* **2020**, *139*, 104475. [CrossRef]
4. Ashraf, U.; Zhang, H.; Anees, A.; Mangi, H.N.; Ali, M.; Ullah, Z.; Zhang, X. Application of Unconventional Seismic Attributes and Unsupervised Machine Learning for the Identification of Fault and Fracture Network. *Appl. Sci.* **2020**, *10*, 3864. [CrossRef]
5. Safaei-Farouji, M.; Vo Thanh, H.; Sheini Dashtgoli, D.; Yasin, Q.; Radwan, A.E.; Ashraf, U.; Lee, K.K. Application of Robust Intelligent Schemes for Accurate Modelling Interfacial Tension of CO_2 Brine Systems: Implications for Structural CO_2 Trapping. *Fuel* **2022**, *319*, 123821. [CrossRef]
6. Vo-Thanh, H.; Amar, M.N.; Lee, K.K. Robust Machine Learning Models of Carbon Dioxide Trapping Indexes at Geological Storage Sites. *Fuel* **2022**, *316*, 123391. [CrossRef]
7. Vo Thanh, H.; Lee, K.K. Application of Machine Learning to Predict CO_2 Trapping Performance in Deep Saline Aquifers. *Energy* **2022**, *239*, 122457. [CrossRef]
8. Thanh, H.V.; Van Binh, D.; Kantoush, S.A.; Nourani, V.; Saber, M.; Lee, K.; Sumi, T.; Sciences, E.; Korea, S.; Resources, W.; et al. Reconstructing Daily Discharge in a Megadelta Using Machine Learning Techniques. *Water Resour. Res.* **2022**, *58*. [CrossRef]
9. Klose, C.D. Self-Organizing Maps for Geoscientific Data Analysis: Geological Interpretation of Multidimensional Geophysical Data. *Comput. Geosci.* **2006**, *10*, 265–277. [CrossRef]
10. Al-Baldawi, B.A. Applying the Cluster Analysis Technique in Logfacies Determination for Mishrif Formation, Amara Oil Field, South Eastern Iraq. *Arab. J. Geosci.* **2015**, *8*, 3767–3776. [CrossRef]
11. Nguyen, T.T.; Kawamura, A.; Tong, T.N.; Nakagawa, N.; Amaguchi, H.; Gilbuena, R. Clustering Spatio-Seasonal Hydrogeochemical Data Using Self-Organizing Maps for Groundwater Quality Assessment in the Red River Delta, Vietnam. *J. Hydrol.* **2015**, *522*, 661–673. [CrossRef]
12. Sfidari, E.; Kadkhodaie-Ilkhchi, A.; Rahimpour-Bbonab, H.; Soltani, B. A Hybrid Approach for Litho-Facies Characterization in the Framework of Sequence Stratigraphy: A Case Study from the South Pars Gas Field, the Persian Gulf Basin. *J. Pet. Sci. Eng.* **2014**, *121*, 87–102. [CrossRef]
13. Garća, H.L.; González, I.M. Self-Organizing Map and Clustering for Wastewater Treatment Monitoring. *Eng. Appl. Artif. Intell.* **2004**, *17*, 215–225. [CrossRef]
14. Unglert, K.; Radić, V.; Jellinek, A.M. Principal Component Analysis vs. Self-Organizing Maps Combined with Hierarchical Clustering for Pattern Recognition in Volcano Seismic Spectra. *J. Volcanol. Geotherm. Res.* **2016**, *320*, 58–74. [CrossRef]
15. Hsieh, B.Z.; Lewis, C.; Lin, Z.S. Lithology Identification of Aquifers from Geophysical Well Logs and Fuzzy Logic Analysis: Shui-Lin Area, Taiwan. *Comput. Geosci.* **2005**, *31*, 263–275. [CrossRef]
16. Al-Anazi, A.; Gates, I.D. A Support Vector Machine Algorithm to Classify Lithofacies and Model Permeability in Heterogeneous Reservoirs. *Eng. Geol.* **2010**, *114*, 267–277. [CrossRef]
17. Imamverdiyev, Y.; Sukhostat, L. Lithological Facies Classification Using Deep Convolutional Neural Network. *J. Pet. Sci. Eng.* **2019**, *174*, 216–228. [CrossRef]
18. Male, F.; Duncan, I.J. Lessons for Machine Learning from the Analysis of Porosity-Permeability Transforms for Carbonate Reservoirs. *J. Pet. Sci. Eng.* **2020**, *187*, 106825. [CrossRef]
19. Deng, Z.; Zhu, X.; Cheng, D.; Zong, M.; Zhang, S. Efficient KNN Classification Algorithm for Big Data. *Neurocomputing* **2016**, *195*, 143–148. [CrossRef]
20. Vo Thanh, H.; Sugai, Y.; Nguele, R.; Sasaki, K. A New Petrophysical Modeling Workflow for Fractured Granite Basement Reservoir in Cuu Long Basin, Offshore Vietnam. In Proceedings of the 81st EAGE Conference and Exhibition, London, UK, 3–6 June 2019; European Association of Geoscientists & Engineers: London, UK, 2019; pp. 1–5.
21. Anees, A.; Zhang, H.; Ashraf, U.; Wang, R.; Liu, K.; Abbas, A.; Ullah, Z.; Zhang, X.; Duan, L.; Liu, F.; et al. Sedimentary Facies Controls for Reservoir Quality Prediction of Lower Shihezi Member-1 of the Hangjinqi Area, Ordos Basin. *Minerals* **2022**, *12*, 126. [CrossRef]
22. Anees, A.; Zhang, H.; Ashraf, U.; Wang, R.; Liu, K.; Mangi, H.N.; Jiang, R.; Zhang, X.; Liu, Q.; Tan, S.; et al. Identification of Favorable Zones of Gas Accumulation via Fault Distribution and Sedimentary Facies: Insights from Hangjinqi Area, Northern Ordos Basin. *Front. Earth Sci.* **2022**, *9*, 822670. [CrossRef]
23. Jiang, R.; Zhao, L.; Xu, A.; Ashraf, U.; Yin, J.; Song, H.; Su, N.; Du, B.; Anees, A. Sweet Spots Prediction through Fracture Genesis Using Multi-Scale Geological and Geophysical Data in the Karst Reservoirs of Cambrian Longwangmiao Carbonate Formation, Moxi-Gaoshiti Area in Sichuan Basin, South China. *J. Pet. Explor. Prod. Technol.* **2021**, *12*, 1313–1328. [CrossRef]
24. Ullah, J.; Luo, M.; Ashraf, U.; Pan, H.; Anees, A.; Li, D.; Ali, M.; Ali, J. Evaluation of the Geothermal Parameters to Decipher the Thermal Structure of the Upper Crust of the Longmenshan Fault Zone Derived from Borehole Data. *Geothermics* **2022**, *98*, 102268. [CrossRef]

25. Abbas, A.; Zhu, H.; Anees, A.; Ashraf, U.; Akhtar, N. Integrated Seismic Interpretation, 2d Modeling along with Petrophysical and Seismic Atribute Analysis to Decipher the Hydrocarbon Potential of Missakeswal Area. *Pakistan. J. Geol. Geophys.* **2019**, *7*, 1–12.

26. Anees, A.; Zhong, S.W.; Ashraf, U.; Abbas, A. Development of a Computer Program for Zoeppritz Energy Partition Equations and Their Various Approximations to Affirm Presence of Hydrocarbon in Missakeswal Area. *Geosciences* **2017**, *7*, 55–67. [CrossRef]

27. Shehata, A.A.; Osman, O.A.; Nabawy, B.S. Journal of Natural Gas Science and Engineering Neural Network Application to Petrophysical and Lithofacies Analysis Based on Multi-Scale Data: An Integrated Study Using Conventional Well Log, Core and Borehole Image Data. *J. Nat. Gas Sci. Eng.* **2021**, *93*, 104015. [CrossRef]

28. Wang, G.; Carr, T.R.; Ju, Y.; Li, C. Identifying Organic-Rich Marcellus Shale Lithofacies by Support Vector Machine Classifier in the Appalachian Basin. *Comput. Geosci.* **2014**, *64*, 52–60. [CrossRef]

29. Freund, Y. Boosting a Weak Learning Algorithm by Majority. *Inf. Comput.* **1995**, *121*, 256–285. [CrossRef]

30. Al Kattan, W.; Jawad, S.N.A.L.; Jomaah, H.A. Cluster Analysis Approach to Identify Rock Type in Tertiary Reservoir of Khabaz Oil Field Case Study. *Iraqi J. Chem. Pet. Eng.* **2018**, *19*, 9–13.

31. Mandal, P.P.; Rezaee, R. Facies Classification with Different Machine Learning Algorithm–An Efficient Artificial Intelligence Technique for Improved Classification. *ASEG Ext. Abstr.* **2019**, *2019*, 1–6. [CrossRef]

32. Shahid, A.R.; Khan, S.; Yan, H. Human Expression Recognition Using Facial Shape Based Fourier Descriptors Fusion. In Proceedings of the Twelfth International Conference on Machine Vision (ICMV 2019), Amsterdam, The Netherlands, 16–18 November 2019; Volume 11433, p. 48. [CrossRef]

33. Qureshi, M.A.; Ghazi, S.; Riaz, M.; Ahmad, S. Geo-Seismic Model for Petroleum Plays an Assessment of the Zamzama Area, Southern Indus Basin, Pakistan. *J. Pet. Explor. Prod. Technol.* **2020**, 1–12. [CrossRef]

34. Ahmed Abbasi, S.; Asim, S.; Solangi, S.H.; Khan, F. Study of Fault Configuration Related Mysteries through Multi Seismic Attribute Analysis Technique in Zamzama Gas Field Area, Southern Indus Basin, Pakistan. *Geod. Geodyn.* **2016**, *7*, 132–142. [CrossRef]

35. Mangi, H.N.; Chi, R.; DeTian, Y.; Sindhu, L.; Lijin; He, D.; Ashraf, U.; Fu, H.; Zixuan, L.; Zhou, W.; et al. The Ungrind and Grinded Effects on the Pore Geometry and Adsorption Mechanism of the Coal Particles. *J. Nat. Gas Sci. Eng.* **2022**, *100*, 104463. [CrossRef]

36. Mangi, H.N.; Detian, Y.; Hameed, N.; Ashraf, U.; Rajper, R.H. Pore Structure Characteristics and Fractal Dimension Analysis of Low Rank Coal in the Lower Indus Basin, SE Pakistan. *J. Nat. Gas Sci. Eng.* **2020**, *77*, 103231. [CrossRef]

37. Ehsan, M.; Gu, H.; Akhtar, M.M.; Abbasi, S.S.; Ehsan, U. A Geological Study of Reservoir Formations and Exploratory Well Depths Statistical Analysis in Sindh Province, Southern Lower Indus Basin, Pakistan. *Kuwait J. Sci.* **2018**, *45*, 84–93.

38. Foredeep, K. A radical seismic interpretation re-think resolves the structural complexities of the zamzama field, kirthar foredeep, pakistan. In Proceedings of the SPE Annual Technical Conference, Islamabad, Pakistan, 10–12 December 2018; pp. 1–17.

39. Asim, S.; Qureshi, S.N.; Asif, S.K.; Abbasi, S.A.; Solangi, S.; Mirza, M.Q. Structural and Stratigraphical Correlation of Seismic Profiles between Drigri Anticline and Bahawalpur High in Central Indus Basin of Pakistan. *Int. J. Geosci.* **2014**, *5*, 1231–1240. [CrossRef]

40. Sirimangkhala, K.; Pimpunchat, B.; Amornsamankul, S.; Triampo, W. Modelling Greenhouse Gas Generation for Landfill. *Int. J. Simul. Syst. Sci. Technol.* **2018**, *19*, 16.1–16.7. [CrossRef]

41. Ashraf, U.; Zhu, P.; Yasin, Q.; Anees, A.; Imraz, M.; Mangi, H.N.; Shakeel, S. Classification of Reservoir Facies Using Well Log and 3D Seismic Attributes for Prospect Evaluation and Field Development: A Case Study of Sawan Gas Field, Pakistan. *J. Pet. Sci. Eng.* **2019**, *175*, 338–351. [CrossRef]

42. Ashraf, U.; Zhang, H.; Anees, A.; Ali, M.; Zhang, X.; Shakeel Abbasi, S.; Nasir Mangi, H. Controls on Reservoir Heterogeneity of a Shallow-Marine Reservoir in Sawan Gas Field, SE Pakistan: Implications for Reservoir Quality Prediction Using Acoustic Impedance Inversion. *Water* **2020**, *12*, 2972. [CrossRef]

43. Ali, M.; Ma, H.; Pan, H.; Ashraf, U.; Jiang, R. Building a Rock Physics Model for the Formation Evaluation of the Lower Goru Sand Reservoir of the Southern Indus Basin in Pakistan. *J. Pet. Sci. Eng.* **2020**, *194*, 107461. [CrossRef]

44. Dar, Q.U.Z.; Pu, R.; Baiyegunhi, C.; Shabeer, G.; Ali, R.I.; Ashraf, U.; Sajid, Z.; Mehmood, M. The Impact of Diagenesis on the Reservoir Quality of the Early Cretaceous Lower Goru Sandstones in the Lower Indus Basin, Pakistan. *J. Pet. Explor. Prod. Technol.* **2021**, *12*, 1437–1452. [CrossRef]

45. Ali, N.; Chen, J.; Fu, X.; Hussain, W.; Ali, M.; Hussain, M.; Anees, A.; Rashid, M.; Thanh, H.V. Prediction of Cretaceous Reservoir Zone through Petrophysical Modeling: Insights from Kadanwari Gas Field, Middle Indus Basin. *Geosystems Geoenviron.* **2022**, *1*, 100058. [CrossRef]

46. Chon, T.S. Self-Organizing Maps Applied to Ecological Sciences. *Ecol. Inform.* **2011**, *6*, 50–61. [CrossRef]

Article

Spatial Fractional Darcy's Law on the Diffusion Equation with a Fractional Time Derivative in Single-Porosity Naturally Fractured Reservoirs

Fernando Alcántara-López [1], Carlos Fuentes [2,*], Rodolfo G. Camacho-Velázquez [3], Fernando Brambila-Paz [1] and Carlos Chávez [4,*]

[1] Department of Mathematics, Faculty of Science, National Autonomous University of Mexico, Circuito Exterior S/N, Mexico City 04510, Mexico; alcantaralopez@comunidad.unam.mx (F.A.-L.); fernandobrambila@gmail.com (F.B.-P.)

[2] Mexican Institute of Water Technology, Paseo Cuauhnáhuac Num. 8532, Jiutepec 62550, Mexico

[3] Engineering Faculty, National Autonomous University of Mexico, Circuito Exterior, Ciudad Universitaria, Mexico City 04510, Mexico; camachovrodolfo@gmail.com

[4] Water Research Center, Department of Irrigation and Drainage Engineering, Autonomous University of Querétaro, Cerro de las Campanas S/N, Col. Las Campanas, Querétaro 76010, Mexico

* Correspondence: cbfuentesr@gmail.com (C.F.); chagcarlos@uaq.mx (C.C.)

Abstract: Due to the complexity imposed by all the attributes of the fracture network of many naturally fractured reservoirs, it has been observed that fluid flow does not necessarily represent a normal diffusion, i.e., Darcy's law. Thus, to capture the sub-diffusion process, various tools have been implemented, from fractal geometry to characterize the structure of the porous medium to fractional calculus to include the memory effect in the fluid flow. Considering infinite naturally fractured reservoirs (Type I system of Nelson), a spatial fractional Darcy's law is proposed, where the spatial derivative is replaced by the Weyl fractional derivative, and the resulting flow model also considers Caputo's fractional derivative in time. The proposed model maintains its dimensional balance and is solved numerically. The results of analyzing the effect of the spatial fractional Darcy's law on the pressure drop and its Bourdet derivative are shown, proving that two definitions of fractional derivatives are compatible. Finally, the results of the proposed model are compared with models that consider fractal geometry showing a good agreement. It is shown that modified Darcy's law, which considers the dependency of the fluid flow path, includes the intrinsic geometry of the porous medium, thus recovering the heterogeneity at the phenomenological level.

Keywords: Weyl fractional derivative; Caputo fractional derivative; fractal porous media; naturally fractured reservoir

Citation: Alcántara-López, F.; Fuentes, C.; Camacho-Velázquez, R.G.; Brambila-Paz, F.; Chávez, C. Spatial Fractional Darcy's Law on the Diffusion Equation with a Fractional Time Derivative in Single-Porosity Naturally Fractured Reservoirs. *Energies* **2022**, *15*, 4837. https://doi.org/10.3390/en15134837

Academic Editors: Yuming Liu and Bo Zhang

Received: 12 June 2022
Accepted: 28 June 2022
Published: 1 July 2022

Publisher's Note: MDPI stays neutral with regard to jurisdictional claims in published maps and institutional affiliations.

1. Introduction

The modeling of physical phenomena is always a complex task in which hypotheses and idealizations are used in order to implement already studied laws or similar models. In particular, modeling the transient pressure behavior in naturally fractured reservoirs is a task that has been studied and worked on by various researchers over decades, where the heterogeneity of the porous medium and anomalous fluid flow has been a challenge for which more complex models have been generated and implemented, obtaining results that constantly improve the understanding of the phenomenon.

Considering the complexity of the porous medium, Chang and Yortsos [1] are the pioneers that considered the porous medium as a fractal system, allowing them to describe reservoirs with spatial disorder and, therefore, a complex fluid flow path.

In recent years, fractional calculus has become a useful tool that, applied to diffusion-type problems, explains the behavior of anomalous flow (where the movement of the fluid does not have a Brownian-type behavior) by demonstrating that the fluid has memory [2].

The study of Metzler et al. [3] has become an important reference when studying the phenomenon of anomalous diffusion in a medium with a fractal structure, allowing endless researchers to explore the implications considering different approaches [4–7].

Although all these models have allowed a more precise understanding of the phenomenon in non-homogeneous reservoirs; the task of understanding this complex phenomenon has also included modifications to Darcy's law. When considering non-Darcy flows, i.e., a sub-diffusive process, to include the memory effect, the fractional derivative approach is used.

To include the memory effect in Darcy's law, Caputo's fractional time derivative is used in Darcy's law modeling, to account for the effect of a decrease in permeability with time [8]. Raghavan [9] shows that using the properties of the fractional derivative, one can translate the fractional time derivative into the continuity equation.

Several variants of the temporal fractional derivative have been applied to Darcy's law. Chang et al. [10] replace the spatial derivative in Darcy's law with the Riemann–Liouville fractional derivative to quantify the spatial path dependence of a fluid flow. El-Amin [11] constructs the fractional mass equation and combines it with variants of Darcy's law that include adding the fractional time derivative and replacing the space derivative with Caputo's fractional derivative. Chang and Sun [12] replace the time derivative with Caputo's fractional time derivative to describe the heavy tail decay at long times and the space derivative with the Riemann–Liouville fractional time derivative to describe the dependence of non-local concentration change on a wide range of spatial areas.

Likewise, Caputo and Plastino [13], considering infinite media, add a term proportional to the spatial fractional derivative of the Caputo type to the classic Darcy's law, which represents the effect of spatial memory, i.e., the pressure gradient investigated from the initial point to the measured point.

Obembe et al. [14], in their excellent review work, show a model of Darcy's law, among others, where the temporal fractional derivative is applied to a complex combination of a spatial fractional derivative and complementary temporal fractional derivative; that reflects the presence of flow hindrances, while the spatial fractional derivatives consider flow buffers.

Finally, in order to consider the intrinsic geometry of the porous medium, Cloot and Botha [15] replace the spatial derivative in Darcy's law with Weyl's fractional derivative; they consider that the fluid flow at a given point is governed not only by the properties of the pressure field at a specific position but also depends on the global spatial distribution of that field.

Therefore, in order to model the transient pressure behavior in a naturally fractured reservoir, including spatial and temporal memory, a new model will be proposed that incorporates both the fractional time derivative in the Caputo sense, in the continuity equation, and the fractional Weyl derivative, in Darcy's law. This work is organized as follows: Section 2 describes the essential mathematical tools used; Section 3 shows the development of the model, explaining the modified Darcy's law used and the spatial fractional diffusion model developed; Section 4 shows the results of solving this model and the effect of every parameter of the model, Section 5 compares the proposed model with models that consider the porous medium as fractal; in the end, Section 6 describes the main conclusions reached.

2. Mathematical Background

This section presents the mathematical tools that will be applied throughout this work. For more details, see [16–18].

Definition 1. *Let $t_0 < \infty$. The Riemann–Liouville fractional integral $^{RL}I_{t_0+}^{\alpha}y$ of order $\alpha \in \mathbb{R}$ is defined by*

$$\left(^{RL}I_{t_0+}^{\alpha}y\right)(t) = \frac{1}{\Gamma(\alpha)} \int_{t_0}^{t} (t-\tau)^{\alpha-1} y(\tau) d\tau, \qquad t > t_0 \tag{1}$$

where $\Gamma(\cdot)$ is Euler's gamma function.

The Caputo fractional derivative, expressed from the fractional integral, is defined as follows:

Definition 2. *Let $t_0 < \infty$. For $\alpha \in \mathbb{R}$, the Caputo fractional derivative, $^{C}D_{t_0+}^{\alpha}y$ is defined by*

$$\left(^{C}D_{t_0+}^{\alpha}y\right)(t) = \frac{1}{\Gamma(n-\alpha)} \int_{t_0}^{t} (t-\tau)^{n-\alpha-1} y^{(n)}(\tau)d\tau = \left(^{RL}I_{t_0+}^{n-\alpha} D^{n}y\right)(t), \qquad t > t_0 \quad (2)$$

where $D = d/dt$ and $n \in \mathbb{N}$ with $n - 1 < \alpha \leq n$.

Proposition 1. *The Caputo fractional derivative provides an inverse operator to the Riemann–Liouvile fractional integral, that is*

$$\left(^{C}D_{t_0+}^{\alpha}\,^{RL}I_{t_0+}^{\alpha}y\right)(t) = y(t); \tag{3}$$

$$\left(^{RL}I_{t_0+}^{\alpha}\,^{C}D_{t_0+}^{\alpha}y\right)(t) = y(t) - \sum_{k=0}^{n-1} \frac{y^{(k)}(t_0)}{k!}(t-t_0)^{k}. \tag{4}$$

The above definitions consider $t_0 < \infty$; however, similar definitions exist on the whole axis $\mathbb{R}$.

Definition 3. *The Weyl fractional integral of order $\beta \in \mathbb{R}$ is defined by*

$$\left(W^{-\beta}y\right)(t) = \frac{1}{\Gamma(\beta)} \int_{t}^{\infty} (\iota - t)^{\beta-1} y(\tau)d\tau, \qquad t > 0. \tag{5}$$

Indeed, the Weyl fractional integral is interesting; however, care should be taken when using this definition because it may not be applied to all functions.

Partcularly, if $y(t)$ is integrable on any finite subinterval of $J = [0, \infty)$, and, if $y(t)$ behaves like $t^{-\mu}$ for t large, then the Weyl fractional integral of y of order β will exist if $0 < \beta < \mu$.

Definition 4. *If $y(t)$ is a function for which $W^{-\beta}y(t)$ exists and has n continous derivatives; then, the Weyl fractional derivative of y of order $\nu \in \mathbb{R}$ is defined by*

$$W^{\beta}y(t) = (-1)^{n}D^{n}\left[W^{-(n-\beta)}y(t)\right] \tag{6}$$

where $D = d/dt$ and $n \in \mathbb{N}$ with $n - 1 < \beta \leq n$.

Before finishing this long list of definitions, some interesting properties of fractional operators will be shown.

Proposition 2. *Let $\mu, \beta \in \mathbb{R}$ and $n \in \mathbb{N}$ with $n - 1 < \beta \leq n$. Both, the fractional Weyl integral and the fractional Weyl derivative satisfy the following:*
- $D^{n}\left[W^{-\beta}y(t)\right] = W^{-\beta}[D^{n}y(t)]$.
- $W^{-\mu}\left[W^{-\beta}y(t)\right] = W^{-\beta}[W^{-\mu}y(t)] = W^{-(\mu+\beta)}y(t)$.
- $W^{-\beta}\left[W^{\beta}y(t)\right] = y(t) = W^{\beta}\left[W^{-\beta}y(t)\right]$.
- $W^{\mu}\left[W^{\beta}y(t)\right] = W^{\beta}[W^{\mu}y(t)] = W^{\mu+\beta}y(t)$.

3. Model Development

In this section, the proposed Darcy's law will be presented, which integrates the fractional derivative of Weyl, and the flow equation with spatial fractional Darcy flow will be constructed.

We consider a fully penetrated well in an infinite porous media with a single porosity system, i.e., a Type I fractured system of Nelson [19], constant initial pressure, permeability, density and viscosity. Furthermore, we consider that the fluid flow at a given point is governed not only by the properties of the pressure field at that specific position but also depends on the global spatial distribution of that field, and lastly, we consider that the radial symmetry is valid. The model in dimensionless variables with a fractional time derivative for the fluid transfer equation, resulting from the combination of the continuity equation and Darcy's law, in radial coordinates and considering a Euclidean porous medium, is given by

$$\tau_D^{\alpha-1}\frac{\partial^\alpha p_D}{\partial t_D^\alpha} = -\frac{1}{r_D}\frac{\partial}{\partial r_D}(r_D q_D), \qquad q_D = -\frac{\partial p_D}{\partial r_D}; \tag{7}$$

whereas, when considering a fractal porous medium, that is, a fractal reservoir, the model is

$$\frac{\partial^\alpha p_D}{\partial t_D^\alpha} = -\frac{1}{r_D^{d_{mf}-1}}\frac{\partial}{\partial r_D}(r_D^\gamma q_D), \qquad q_D = -\frac{\partial p_D}{\partial r_D}, \tag{8}$$

where $\gamma = d_{mf} - \theta - 1$. Additionally, it is defined $\nu = \frac{1-\gamma}{\theta+2}$ [20].

In Equations (7) and (8), the fractional time derivative is the Caputo fractional derivative, $\frac{\partial^\alpha p_D}{\partial t_D^\alpha} \equiv {}^C D_{0+}^\alpha p_D$, with $0 < \alpha \le 1$. For both models, the following initial and boundary conditions are considered.

- Initial condition

$$p_D(r_D, t_D = 0) = 0. \tag{9}$$

- Inner boundary condition

$$\lim_{r_D \to 1} r_D q_D(r_D, t_D) = -1. \tag{10}$$

- Outer boundary condition

$$\lim_{r_D \to \infty} p_D(r_D, t_D) = 0. \tag{11}$$

The dimensionless variables are defined as follows:

$$r_D = \frac{r}{r_w}; \quad p_D = \frac{2\pi h\kappa}{Q_0 B_0 \mu}(P_i - p); \quad q_D = \frac{2\pi h r_w}{Q_0 B_0}q;$$

$$t_D = \frac{\kappa}{\phi c r_w^2 \mu}t; \quad \tau_D = \frac{\kappa}{\phi c r_w^2 \mu}\tau;$$

where τ is a constant introduced for the purpose of maintaining dimensional balance.

The proposed model considers a slightly compressible liquid; that is, all the complexities due to the multiphase gas–oil interactions are not taken into consideration [21]. Furthermore, the model presented in this paper does not consider petrophysical properties dependent on the stress state, which even for a Nelson Type I reservoir, imposes a challenge [22].

3.1. Spatial Fractional Darcy's Law

Figure 1 shows the solution of Equations (7) and (9)–(11), i.e., the behavior of the pressure drop, p_D, throughout space for different times and values of the fractional time derivative order, α.

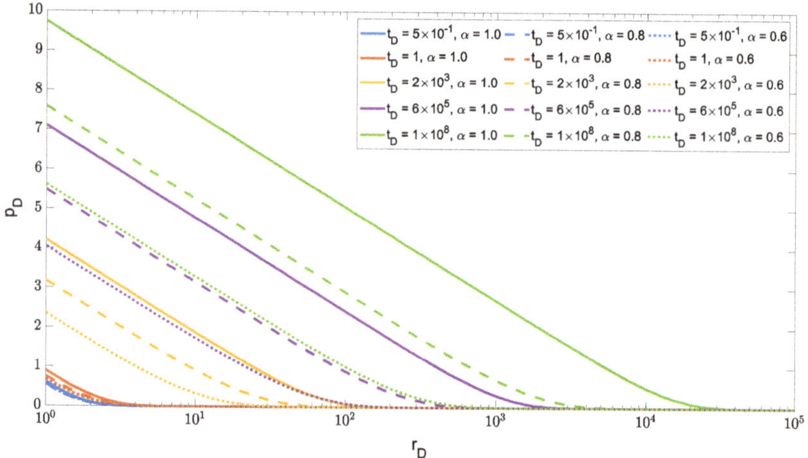

Figure 1. Spatial pressure behavior at different times, considering the Euclidean porous medium.

Figure 1 shows the decreasing behavior of the pressure drop in space and, therefore, shows that it is valid to apply the Weyl fractional derivative to p_D.

For that reason, it is proposed to modify Darcy's law by substituting the spatial derivative for the Weyl fractional derivative, obtaining the following spatial fractional Darcy's law:

$$q_{\beta,D} = -\delta_{\beta,D}\frac{\partial^\beta}{\partial r_D^\beta}p_D, \qquad \text{with } \frac{\partial^\beta}{\partial r_D^\beta}p_D := -W^\beta p_D \quad \text{and } \delta_{\beta,D} = r_w^{1-\beta}\delta^{\beta-1}; \tag{12}$$

where $0 < \beta < 2$ and δ is a constant term included to maintain dimensional balance in the flow equation. Note that setting $\beta = 1$, $\frac{\partial^\beta}{\partial r_D^\beta}p_D = \frac{\partial}{\partial r_D}p_D$ and, therefore, the classical Darcy's law is recovered.

3.2. Spatial Fractional Diffusion Model

Given the continuity equation, the first equation in (7), the traditional Darcy's law, q_D, is replaced by the spatial fractional Darcy's law, Equation (12), where the spatial fractional diffusion model is obtained, namely

$$\tau_D^{\alpha-1}\frac{\partial^\alpha p_D}{\partial t_D^\alpha} = \frac{1}{r_D}\frac{\partial}{\partial r_D}\left(r_D\delta_{\beta,D}\frac{\partial^\beta}{\partial r_D^\beta}p_D\right). \tag{13}$$

As a consequence of the spatial fractional Darcy's law, the inner boundary condition is also modified, namely

$$\lim_{r_D\to 1} r_D q_{\beta,D}(r_D, t_D) = -1. \tag{14}$$

Therefore, the spatial fractional diffusion model to be solved is the one constituted by Equations (9), (11), (13) and (14).

With these modifications, it is intended that the fluid flow at a given point in the porous medium be governed by the global spatial distribution of the pressure field and not only by the behavior in the direct neighborhood of the pressure around that point.

4. Results

In this section, the solution of the spatial fractional diffusion model will be shown, as well as its behavior when its parameters vary, and such behavior will be compared with the model that considers a fractal reservoir.

The spatial fractional diffusion model was solved using the finite difference method, considering an implicit scheme in time and Crank–Nicholson in space.

Figure 2 shows the numerical solution of solving the spatial fractional diffusion model considering the classical time derivative, $\alpha = 1$, and varying the order in the fractional Darcy's law, β. The solution in the well, $p_{wD} = p_D(r_D = 1, t_D)$, and its Bourdet semi-logarithmic derivative, p'_{wD} [23], are shown.

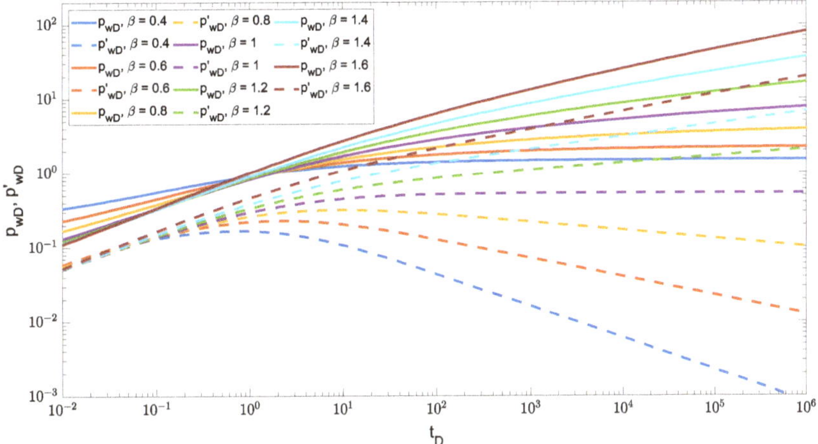

Figure 2. Effect of spatial fractional Darcy's law with Weyl's fractional derivative, considering the classical time derivative $\alpha = 1$ with $\delta_{\beta,D} = 1$. The solid lines denote the pressure drop in the well, p_{wD}, and the dashed lines their respective Bourdet derivative, p'_{wD}.

In Figure 2, two important features can be described when modifying β in the spatial fractional Darcy's law.

At short times, the pressure drop is lower than the pressure drop of the classical case, $\beta = 1$, which corresponds to a higher order in the spatial fractional Darcy's law; while if β is lower than one (i.e., classical case), a higher pressure drop will be obtained. However, in the Bourdet derivative, β in the spatial fractional Darcy's law does not seem to have a remarkable impact, except by keeping a power-law behavior with a slope of 0.5.

At long times, when $\beta < 1$, the pressure drop is lower than that in the classic case and its respective Bourdet derivative shows a decrease following a power-law behaviour, showing a greater connectivity of pores that provide preferential flow paths; while for $\beta > 1$, the pressure drop is greater than in the classic case following a power-law behaviour, and the same happens with the pressure derivative. Both power-law behaviors are parallel, evidencing the creation of flow buffers in the porous medium.

Figure 3 shows the combined effect of the fractional temporal derivative and the spatial fractional Darcy's law, that is, for different values of β in the space fractional Darcy's law, on the pressure drop and its Bourdet derivative for different values of the order of the fractional time derivative, α.

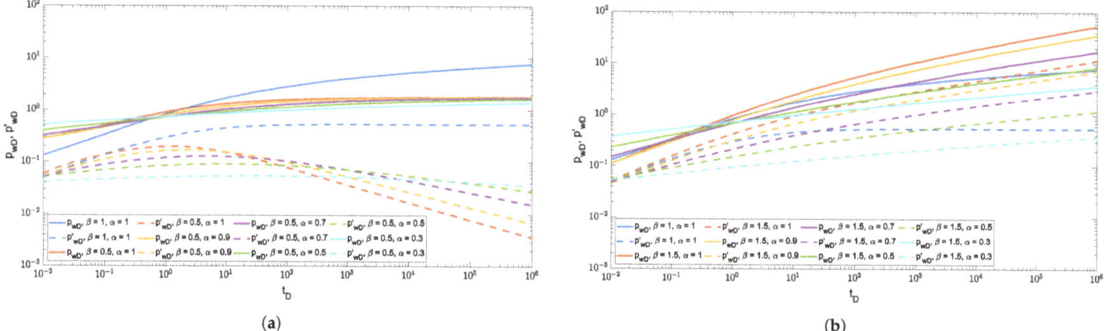

Figure 3. Effect on the pressure drop and its Bourdet derivative of the combined effect of the fractional temporal derivative and the spatial fractional Darcy's law. (**a**) For $\beta < 1$, the behavior when varying the order of the fractional time derivative. (**b**) For $\beta > 1$, the behavior when varying the order of the fractional time derivative. In both cases, $\delta_{\beta,D} = 1$.

It is necessary to highlight that the results shown in Figure 3 demonstrate the compatibility of two different definitions of fractional derivative; that is, Caputo's fractional derivative for the temporal derivative and Weyl's fractional derivative for the spatial derivative in Darcy's law, which allows better modeling of a complex phenomenon by making each fractional derivative capture different characteristics of this phenomenon.

Figure 3a shows for $\beta < 1$ the effect of varying the order of the fractional time derivative, α. In the case with the integer time derivative, $\alpha = 1$, the effect of spatial fractional Darcy's law creates a larger short-time pressure drop that quickly loses its effect, making the long-time pressure drop smaller the smaller α is; this is a consequence of a greater connectivity of pores that provide preferential flow paths. In the case of $\beta = 0.5$, Figure 3a, we can observe that the influence of α is not as big as that observed in the pressure derivative, where we observe power-law behaviors at late times, which could be an indication of flow restrictions.

Figure 3b shows for $\beta > 1$ the effect of varying the order of the fractional time derivative, α, on the pressure deficit and its Bourdet derivative. For the case with an integer time derivative, $\alpha = 1$; at short times, the spatial fractional Darcy's law causes the fluid flow velocity to have a behavior similar to the case with the classical Darcy's law, i.e., a linear flow behavior; however, the pressure drop is smaller. Subsequently, for long times, the pressure drop is lower than in the classical case, the lower α is; this is caused by the displacement of dead pores that creates preferential flow paths, which is reflected in the increase in flow velocity and, therefore, in the decrease of the pressure drop. By incorporating the memory effect of the fluid flow, the dependence of the previous pressure on the subsequent pressures is observed; that is, the flow of the fluid receives an increase in the phenomenon that is reflected in the decrease of the pressure drop and its Bourdet derivative.

It is important to point out that in all cases with $\beta = 1.5$ and α less or equal to one, Figure 3b, at late times, there is parallel power-law behavior for both the pressure drop and semilog pressure derivative. This behavior is similar to the fractal behavior, and both cases are expressions of anomalous diffusion; in this case, there is a super-diffusion because of the memory effect, and in the fractal case, there is sub-diffusion.

Figure 4 shows the effect of $\delta_{\beta,D}$, the dimensionless variable introduced in the spatial fractional Darcy's law to maintain the dimensional balance in the fractional flow equation, for the cases $\beta < 1$ and $\beta > 1$ with $\alpha = 1$.

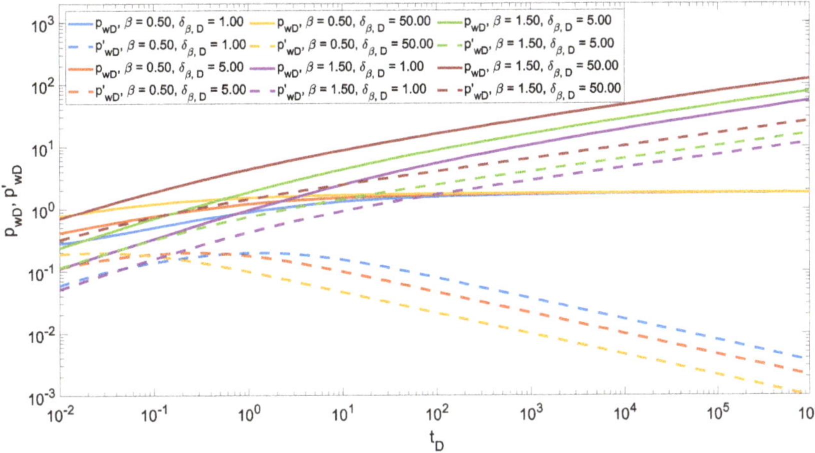

Figure 4. Effect of $\delta_{\beta,D}$ in pressure drop and Bourdet derivative of the spatial fractional diffusion model with $\alpha = 1.0$.

The effect of $\delta_{\beta,D}$ is remarkable throughout the whole phenomenon. For $\beta < 1$ and in short times, it is observed that the greater the value of $\delta_{\beta,D}$, the greater the value of p_{wD}, also affecting the transition phase, presenting itself at different times for each value of $\delta_{\beta,D}$ and finally reaching the same value of p_{wD} in the stability phase. Bourdet's derivative shows that although in short times, the greater the value of $\delta_{\beta,D}$ the greater p'_{wD} will be and the faster the decrement in p'_{wD} to such a degree that upon reaching the stability phase, it will be shown that the higher the value of $\delta_{\beta,D}$, the lower the value of p'_{wD}, also showing a parallel behavior for all the values of $\delta_{\beta,D}$.

On the other hand, for $\beta > 1$ at short times, it is observed that in the same way as in the previous case, the greater $\delta_{\beta,D}$, the greater the value of p_{wD}, changing the growth rate in the transient phase and subsequently having a constant growth rate for each value of $\delta_{\beta,D}$. This behavior is maintained in the Bourdet derivative, where, in short times, the higher the value for p'_{wD}, the higher the $\delta_{\beta,D}$, followed by a parallel growth for the different values of $\delta_{\beta,D}$, changing the growth rate during the transient phase until reaching a constant and parallel growth for the different values of $\delta_{\beta,D}$.

5. Discussion

In this section, the spatial fractional diffusion model, derived from the equation of continuity with a fractional time derivative and the proposed spatial fractional Darcy's law, and the results from the fractal model are compared.

Figure 5 shows, on the one hand, the model of Razminia et al. [4], which considers a fractal reservoir, Equations (8)–(11); and on the other hand, the spatial fractional diffusion model, Equations (9), (11), (13) and (14).

The purpose is to compare, through the results shown, the approaches considered by both models. The spatial fractional diffusion model shows a great agreement with the data of the model that considers a fractal reservoir, both for the pressure drop and its Bourdet derivative; thereby, it can be considered that by including the spatial fractional Darcy's law, the heterogeneity of the porous medium is recovered through the use of the fractional approach.

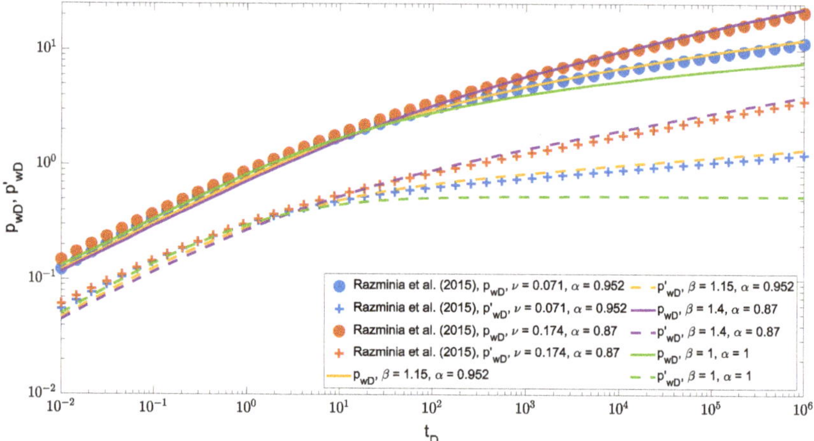

Figure 5. Effect of fractal spatial variables versus the spatial fractional Darcy's law with $\delta_{\beta,D} = 1$.

Furthermore, the above makes sense considering that the spatial fractional Darcy's law takes into account the geometry of the porous medium in two ways; on the one hand, the geometry of the porous medium is implicitly included through the parameter in the Weyl fractional derivative, β; while, on the other hand, it is explicitly included through the δ parameter, which also helps with dimensional balance.

With the above, it is shown that the spatial fractional Darcy's law, by including the dependence on the fluid flow path, i.e., spatial memory, also takes into account the intrinsic geometry of the porous medium.

The proposed model can be part of a robust deep learning model, such as the one suggested by [24], for the automated analysis of pressure tests.

6. Conclusions

Considering the hypotheses that the Type I naturally fractured reservoir of Nelson is embedded in an infinite porous medium, with a slightly compressible fluid and stress-independent petrophysical properties, we have the following conclusions:

1. Fluid flow at a given point is governed not only by the properties of the pressure field at that specific position but also depends on the global spatial distribution of that field; it is proposed to modify Darcy's law by replacing the spatial derivative with the Weyl fractional derivative.
2. The proposed model, with the Caputo fractional time derivative and Weyl fractional derivative on the gradient, proves that two different definitions of the fractional derivative can be compatible.
3. The results show that the spatial fractional Darcy's law can, on the one hand, reflect the preferential flow paths, and on the other hand, the creation of flow buffers during the phenomenon, whereas the time-fractional derivative incorporates the memory effect of the fluid flow.
4. The proposed model not only resembles the results obtained in models that incorporate fractal geometry, but it also incorporates the intrinsic geometry of the porous medium and, therefore, allows recovering the heterogeneity of the porous medium. This is important because it is confirmed for the first time that both the fractal and fractional approaches represent two possible ways of capturing anomalous diffusion.

Author Contributions: F.A.-L., C.F., R.G.C.-V., F.B.-P. and C.C. contributed equally to this work. All authors have read and agreed to the published version of the manuscript.

Funding: This research received no external funding.

Institutional Review Board Statement: Not applicable.

Informed Consent Statement: Not applicable.

Data Availability Statement: Not applicable.

Acknowledgments: The first author is grateful to CONACYT for the scholarship grant, scholarship number 548429. Additionally, the UNAM PAPIIT program is thanked for its support.

Conflicts of Interest: The authors declare no conflict of interest.

Nomenclature

The following abbreviations are used in this manuscript:

p_D	Dimensionless pressure of the porous medium
p_{wD}	Dimensionless pressure at the well
t_D	Dimensionless time
r_D	Dimensionless distance
q_D	Dimensionless flow rate of fluid of the porous medium
α	Order of the fractional time derivative
τ_D	Auxiliary constant to get dimensional balance in the dimensionless fractional differential equation
d_{mf}	Mass fractal dimension
θ	Conductivity index
P_i	Constant initial pressure (Pa)
p	Pressure in the porous media (Pa)
ϕ	Porosity of the porous media (m^3/m^3)
c	Compressibility of the porous medium (Pa^{-1})
κ	Permeability of the porous medium (m^2)
μ	Dynamic viscosity of the fluid (Pa·s)
t	Time (s)
r	Distance variable (m)
r_w	Well radius (m)
h	Reservoir height (m^2)
Q_0	Flow rate of extracted fluid ($m^3{\cdot}s^{-1}$)
B_0	Oil formation volume factor, RB/STB
q	Flow rate of fluid per unit area of the porous medium (m/s)
τ	Auxiliary constant to get dimensional balance in the fractional differential equation (s)
$q_{\beta,D}$	Dimensionless spatial fractional Darcy's law
$\delta_{\beta,D}$	Auxiliary constant to get dimensional balance in the dimensionless spatial fractional Darcy's law
δ	Auxiliary constant to get dimensional balance in the spatial fractional Darcy's law (m)

References

1. Chang, J.; Yortsos, Y.C. Pressure-Transient Analysis of Fractal Reservoirs. *SPE Form. Eval.* **1990**, *5*, 31–38. [CrossRef]
2. Alcántara-López, F.; Fuentes, C.; Brambila-Paz, F.; López-Estrada, J. Quasi-Analytical Model of the Transient Behavior Pressure in an Oil Reservoir Made Up of Three Porous Media Considering the Fractional Time Derivative. *Math. Comput. Appl.* **2020**, *25*, 74. [CrossRef]
3. Metzler, R.; Glöckle, W.G.; Nonnenmacher, T.F. Fractional model equation for anomalous diffusion. *Phys. A Stat. Mech. Its Appl.* **1994**, *211*, 13–24. [CrossRef]
4. Razminia, K.; Razminia, A.; Torres, D.F. Pressure responses of a vertically hydraulic fractured well in a reservoir with fractal structure. *Appl. Math. Comput.* **2015**, *257*, 374–380. [CrossRef]
5. Raghavan, R. Fractional derivatives: Application to transient flow. *J. Pet. Sci. Eng.* **2011**, *80*, 7–13. [CrossRef]
6. Camacho-Velázquez, R.; Fuentes-Cruz, G.; Vásquez-Cruz, M. Decline-Curve Analysis of Fractured Reservoirs with Fractal Geometry. *SPE Reserv. Eval. Eng.* **2008**, *11*, 606–619. [CrossRef]
7. Tian, J.; ke Tong, D. The Flow Analysis of Fiuids in Fractal Reservoir with the Fractional Derivative. *J. Hydrodyn.* **2006**, *18*, 287–293. [CrossRef]
8. Caputo, M. Diffusion of fluids in porous media with memory. *Geothermics* **1999**, *28*, 113–130. [CrossRef]
9. Raghavan, R. Fractional diffusion: Performance of fractured wells. *J. Pet. Sci. Eng.* **2012**, *92–93*, 167–173. [CrossRef]

10. Chang, A.; Sun, H.; Zhang, Y.; Zheng, C.; Min, F. Spatial fractional Darcy's law to quantify fluid flow in natural reservoirs. *Phys. A Stat. Mech. Its Appl.* **2019**, *519*, 119–126. [CrossRef]

11. El-Amin, M.F. Derivation of fractional-derivative models of multiphase fluid flows in porous media. *J. King Saud Univ.—Sci.* **2021**, *33*, 101346. [CrossRef]

12. Chang, A.; Sun, H. Time-space fractional derivative models for CO2 transport in heterogeneous media. *Fract. Calc. Appl. Anal.* **2018**, *21*, 151–173. [CrossRef]

13. Caputo, M.; Plastino, W. Diffusion with Space Memory. In *Geodesy—The Challenge of the 3rd Millennium*; Springer: Berlin/Heidelberg, Germany, 2003; pp. 429–435. [CrossRef]

14. Obembe, A.D.; Al-Yousef, H.Y.; Hossain, M.E.; Abu-Khamsin, S.A. Fractional derivatives and their applications in reservoir engineering problems: A review. *J. Pet. Sci. Eng.* **2017**, *157*, 312–327. [CrossRef]

15. Cloot, A.; Botha, J. A generalised groundwater flow equation using the concept of non-integer order derivatives. *Water SA* **2007**, *32*, 1–7. [CrossRef]

16. Baleanu, D.; Diethelm, K.; Scalas, E.; Trujillo, J.J. *Fractional Calculus: Models and Numerical Methods*; World Scientific: Singapore, 2012; Volume 3.

17. Samko, S.G.; Kilbas, A.A.; Marichev, O.I. *Fractional Integrals and Derivatives*; Gordon and Breach Science Publishers: Yverdon Yverdon-les-Bains, Switzerland, 1993; Volume 1.

18. Miller, K.S.; Ross, B. *An Introduction to the Fractional Calculus and Fractional Differential Equations*; Wiley: Hoboken, NJ, USA, 1993.

19. Nelson, R. *Geologic Analysis of Naturally Fractured Reservoirs*; Elsevier: Amsterdam, The Netherlands, 2001. [CrossRef]

20. Flamenco-López, F.; Camacho-Velázquez, R. Determination of Fractal Parameters of Fracture Networks Using Pressure-Transient Data. *SPE Reserv. Eval. Eng.* **2003**, *6*, 39–47. [CrossRef]

21. Yassin, M.R.; Alinejad, A.; Asl, T.S.; Dehghanpour, H. Unconventional well shut-in and reopening: Multiphase gas-oil interactions and their consequences on well performance. *J. Pet. Sci. Eng.* **2022**, *215*, 110613. [CrossRef]

22. Martyushev, D.A.; Galkin, S.V.; Shelepov, V.V. The Influence of the Rock Stress State on Matrix and Fracture Permeability under Conditions of Various Lithofacial Zones of the Tournaisian–Fammenian Oil Fields in the Upper Kama Region. *Mosc. Univ. Geol. Bull.* **2019**, *74*, 573–581. [CrossRef]

23. Bourdet, D.; Ayoub, J.A.; Pirard, Y.M. Use of Pressure Derivative in Well-Test Interpretation. *SPE Form. Eval.* **1989**, *4*, 293–302. [CrossRef]

24. Pandey, R.K.; Kumar, A.; Mandal, A. A robust deep structured prediction model for petroleum reservoir characterization using pressure transient test data. *Pet. Res.* **2021**. [CrossRef]

Article

Insights into Heterogeneity and Representative Elementary Volume of Vuggy Dolostones

Yufang Xue [1,2], Zhongxian Cai [1,2,*], Heng Zhang [1,2], Qingbing Liu [2], Lanpu Chen [1,2], Jiyuan Gao [1,2] and Fangjie Hu [3]

1   Key Laboratory of Tectonics and Petroleum Resources, Ministry of Education, China University of Geosciences, Lumo Road No.388, Wuhan 430074, China
2   School of Earth Resources, China University of Geosciences, Wuhan 430074, China
3   Research Institute of Exploration and Development, PetroChina Tarim Oilfield Company, Korla 841000, China
*   Correspondence: zxcai@cug.edu.cn; Tel.: +86-02767847611

Abstract: Carbonate reservoirs commonly have significant heterogeneity and complex pore systems due to the multi-scale characteristic. Therefore, it is quite challenging to predict the petrophysical properties of such reservoirs based on restricted experimental data. In order to study the heterogeneity and size of the representative elementary volume (REV) of vuggy dolostones, a total of 26 samples with pore sizes ranging from micrometers to centimeters were collected from the Cambrian Xiaoerbulake Formation at the Kalping uplift in the Tarim Basin of northwestern China. In terms of the distribution of pore size and contribution of pores to porosity obtained by medical computed tomography testing, four types of pore systems (Types I–IV) were identified. The heterogeneity of carbonate reservoirs was further quantitatively evaluated by calculating the parameters of pore structure, heterogeneity, and porosity cyclicity. The results indicate that different pore systems yield variable porosities, pore structures, and heterogeneity. The porosity is relatively higher in Type-II and Type-IV samples compared to those of Type-I and Type-III. It is caused by well-developed large vugs in the former two types of samples, which increase porosity and reduce heterogeneity. Furthermore, the REV was calculated by deriving the coefficient of variation. Nine of the twenty-six samples reach the REV within the volume of traditional core plugs, which indicates that the REV sizes of vuggy dolostones are commonly much larger than the volume of traditional core plugs. Finally, this study indicates that REV sizes are affected by diverse factors. It can be effectively predicted by a new model established based on the relationship between REV sizes and quantitative parameters. The correlated coefficient of this model reaches 0.9320. The results of this study give more insights into accurately evaluating the petrophysical properties of vuggy carbonate reservoirs.

Keywords: medical-CT; vuggy carbonate; representative elementary volume; quantitative characterization of heterogeneity; Xiaoerbulake Formation

Citation: Xue, Y.; Cai, Z.; Zhang, H.; Liu, Q.; Chen, L.; Gao, J.; Hu, F. Insights into Heterogeneity and Representative Elementary Volume of Vuggy Dolostones. *Energies* **2022**, *15*, 5817. https://doi.org/10.3390/en15165817

Academic Editors: Yuming Liu, Bo Zhang and Reza Rezaee

Received: 7 June 2022
Accepted: 8 August 2022
Published: 10 August 2022

Publisher's Note: MDPI stays neutral with regard to jurisdictional claims in published maps and institutional affiliations.

1. Introduction

Carbonate reservoirs are widely investigated because they host more than 60% of oil and gas worldwide [1–3]. It is well known that carbonate reservoirs are generally heterogeneous [4–6]. Diverse depositional environments and complex diagenetic alterations commonly lead to the development of pore systems with sizes ranging from micrometers to centimeters [7–10]. The heterogeneity of carbonate reservoirs brings a series of challenges in evaluating their petrophysical properties. Therefore, a quantitative evaluation of heterogeneity is extremely significant. Heterogeneity, defined as an inherent, ubiquitous, and critical attribute of ecological systems, has been extensively studied in previous studies [11–13]. It is significantly affected by the spatio-temporal scale of observation and measurement methods [14,15]. For example, carbonate samples may be homogeneous on a macro-scale, but heterogeneous at a micro-scale in terms of pore structures [13,16,17].

A representative elementary volume (REV) is proposed to quantitatively characterize heterogeneity and could provide a link between the pore scale and continuum scale. The REV is defined as the minimum volume of a porous medium that is large enough to represent the macroscopic property of a heterogeneous rock [18–21]. The schematic diagram of the REV is shown in Figure 1, where V is defined as the volume of porous medium and n is defined as the property of the rocks, e.g., porosity [22–25]. It is noteworthy that REV is various when different physical properties are investigated, even for the same porous medium [26–28]. Porosity is the most common property to determine the REV [28,29]. The other properties include permeability [26,30–32], tortuosity [23,33], coordination number [34], specific surface area [32], moisture saturation [35], local void ration [36], and fractal dimension [25,37].

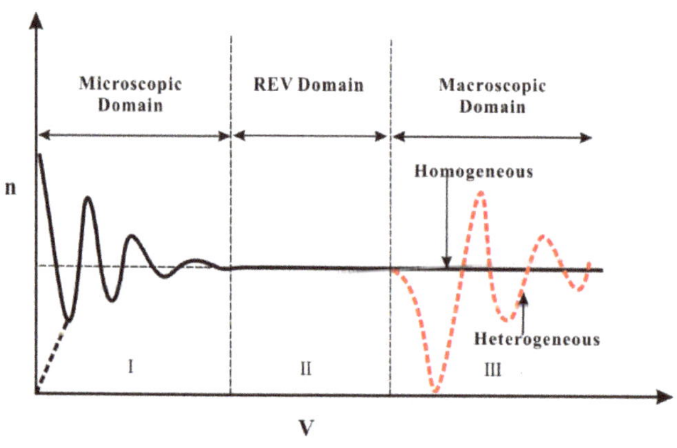

Figure 1. Schematic graph of the representative elementary volume (REV) (modified from [18,30]).

Porosity generally varies with the sample's volume in carbonate rocks. During measurements, the testing data of traditional core plugs (a cylinder with a diameter of 2.5 cm and a height of 5 cm) have been commonly used to represent the data of whole cores. However, the porosity of whole cores is usually higher than that of traditional core plugs [38]. When the volume of the analyzed sample is smaller than the REV, there will be a large discrepancy between the measured and real porosity of the sample. As such, precisely determining the REV sizes is very crucial for accurately calculating the petrophysical property and evaluating the reservoir. Extensive papers have been published on determining REV sizes [22,39–41]. However, they have mainly focused on determining REV sizes in microscopic domain microstructures of rocks by semi-quantitative to quantitative methods. Only a few studies were conducted on the REV sizes of vuggy carbonate rocks with sizes ranging from micrometers to centimeters [26].

Recently, X-ray computed tomography (CT), as a non-destructive technique, has been widely used to characterize the pore structure of a porous medium [42–44]. High-resolution CT such as micro-CT and nano-CT can capture the microstructures of the pore system. However, the trade-off between the sample size and spatial resolution makes the targeted field of high-resolution CT relatively microscopic. Therefore, it is very challenging to display the macroscopic pore space [45–48]. For vuggy carbonate rocks, medical-CT is an effective technique to characterize the pore system as it is both more efficient and cost-effective compared with micro-CT and nano-CT [43,49]. More importantly, the image resolution of medical-CT is more suitable for studying the pore structure and REV of vuggy carbonate reservoirs with pore sizes ranging from micrometers to centimeters. The aims of this study are to quantitatively characterize the heterogeneity and precisely determine the REV size of dolostones with well-developed vugs. This study can provide a foundation for accurately calculating physical properties and precisely evaluating the reservoirs.

In this study, a total of 26 samples were collected from the Shihuiyao section at the Kalping uplift in the northwestern Tarim Basin to study the heterogeneity of vuggy dolostones. The samples were firstly prepared as cylinders with diameters of 5 cm and heights of 49–100 cm. Then CT was used to characterize the types of pore systems and calculate the parameters of pore structure and heterogeneity. Subsequently, the REV size was determined by deriving the coefficient of variation. Finally, a prediction model of REV was established based on the parameters of pore geometrical and heterogeneity. The potential of this model to predict the REV sizes of vuggy carbonates is also discussed.

2. Geological Setting

The Tarim Basin is a foreland basin with a total area of 5.6×10^5 km^2 in the Xinjiang Province in northwestern China (Figure 2a). It is bordered by the Kunlun-Altyn Mountains to the south and the Tianshan Mountains to the north [50–52]. The Tarim Basin has suffered a complex tectonic evolution and consists of one low uplift, four uplifts, and six depressions, namely, the Shuntuo Low Uplift, Tabei Uplift, Bachu Uplift, Tazhong Uplift, Tadong Uplift, Kuqa Depression, Awati Depression, Manjiaer Depression, Southwest Depression, Tanggu Depression, and Southeast Depression (Figure 2b) [53,54]. The Kalping Uplift is located in the northwest, along the edge of the Bachu Uplift (Figure 2b,c).

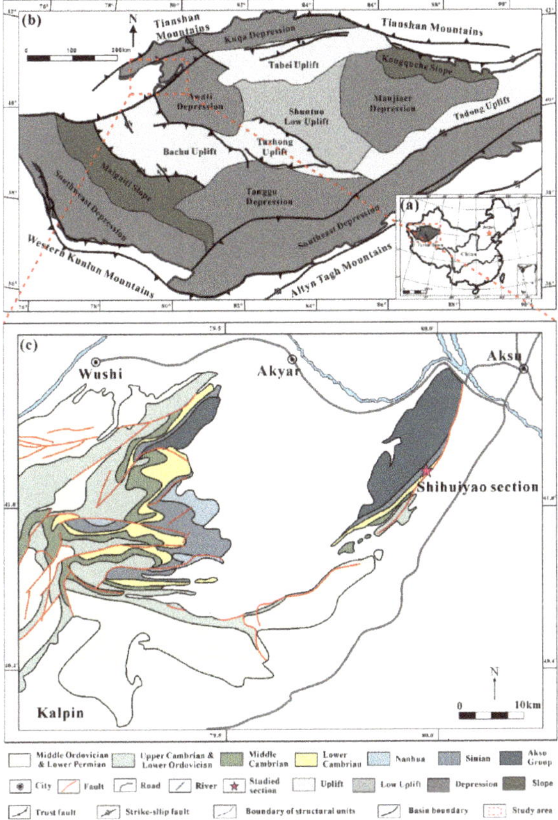

Figure 2. (**a**) Location of the Tarim Basin (modified from [50]). (**b**) Structural units of the Tarim Basin (modified from [51]). (**c**) Geological setting of outcrop section in the northwestern Tarim Basin (modified from [52]).

The Shihuiyao section, located at the northeastern edge of the Kalping Uplift (Figure 2c), is exposed with consecutive Cambrian strata [53,55,56]. The Lower Cambrian strata, from bottom to top, comprises the Yuertusi, Xiaoerbulake, and Wusongeger Formations. The Xiaoerbulake Formation at the Shihuiyao section is 138.5 m thick and composed of two sequences: the upper Xiaoerbulake and the lower Xiaoerbulake intervals (Figure 3). Algal lamina silty-micritic crystalline dolostone, silty-micritic crystalline dolostone, rubble dolostone, algal clot dolostone, and fabric-obliterated fine-to-medium-crystalline dolostone are well developed in the lower Xiaoerbulake interval, whereas foam spongy texture dolostone, stromatolithic dolostone, algal arene dolostone, and oncolite dolostone are present in the upper Xiaoerbulake interval.

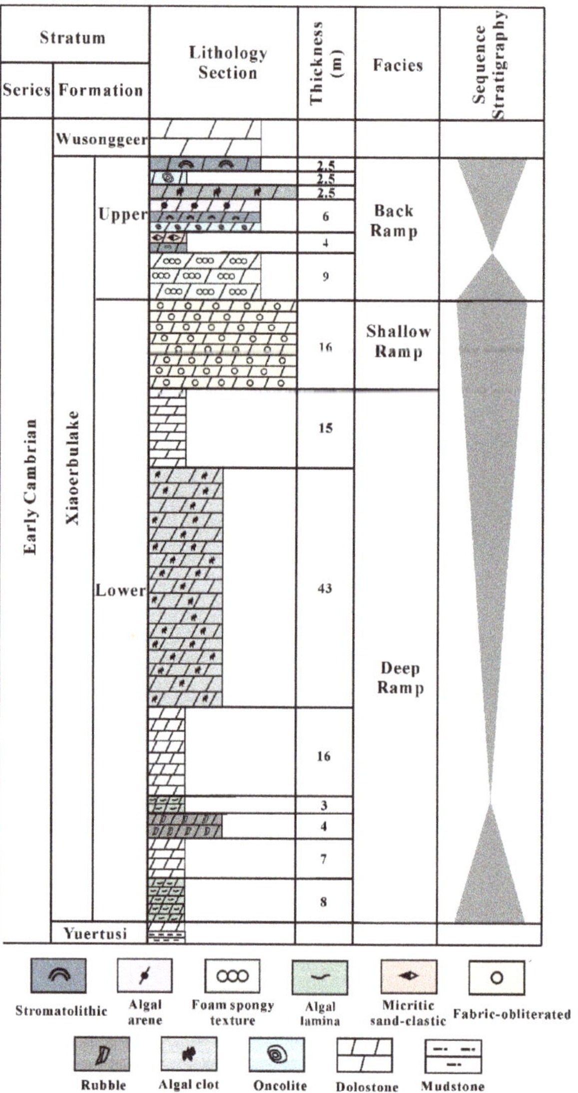

Figure 3. Simplified stratigraphic framework of the Early Cambrian from the Shihuiyao section.

3. Samples and Methods

A total of 26 dolostones with well-developed vugs were collected from the Xiaoerbulake Formation in the Shihuiyao section (Figure 4). Rock types of these samples were determined through detailed field and microscopic investigation (Figure 5).

Figure 4. Image of the vuggy dolostones with millimeter-to-centimeter-scale vugs in the Xiaoerbulake Formation from the Shihuiyao section. (**a**) Image of cylinders of vuggy dolostones. (**b**) Image of sample SHY002, flat vugs are developed with a diameter of 2–6 mm. (**c**) Image of sample SHY007, the sample develops vugs and fissures and the vugs are half-filled by secondary carbonates. (**d**) Image of sample SHY016, flat vugs are developed and the distribution of vugs is scattered.

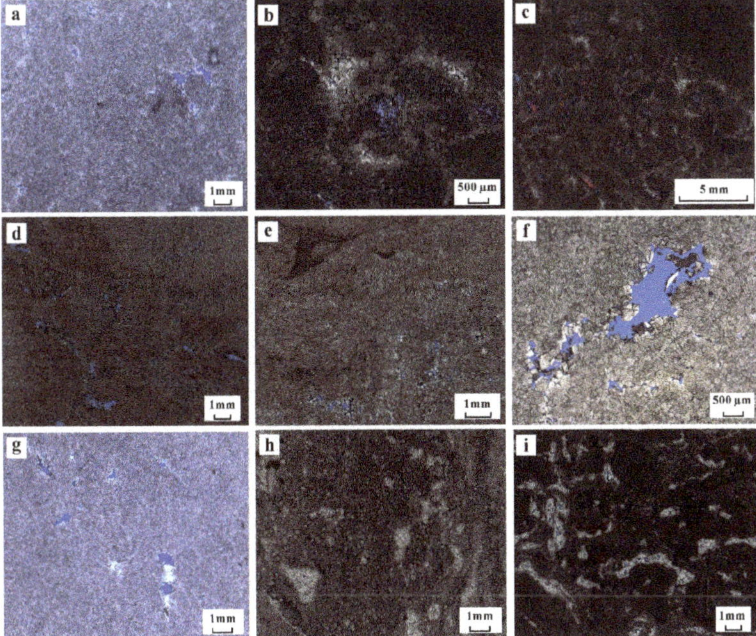

Figure 5. Plane polarized light microscopic images of porous dolostones in the Xiaoerbulake Formation from the Shihuiyao section. (**a**) Rock Type i, fabric-obliterated fine-crystalline dolostone with grain ghosts, intercrystalline pores, and intercrystalline pores. (**b**) Rock Type ii, rubble dolostone, intercrystalline pores. (**c**) Rock Type ii, rubble dolostone, intercrystalline pores, some intercrystalline pores filled with calcite. (**d**) Rock Type iii, the thrombolite dolostone with fabric obliterated by dolomitization or recrystallization, dissolution enlarged pores and fissures. (**e**) Rock Type iii, the thrombolite dolostone with fabric obliterated by dolomitization or recrystallization, intercrystalline pores. (**f**) Rock Type iv, crystalline dolostone, dissolution enlarged pores and intercrystalline dissolution pores. (**g**) Rock Type iv, crystalline dolostone, intercrystalline pores, and intercrystalline pores. (**h**) Rock Type v, foam spongy texture silty crystalline dolostone with fabric partly preserved. (**i**) Rock Type vi, oncolite dolostone, intercrystalline pores, and intercrystalline pores.

The 26 samples were prepared as cylinders with diameters of 5 cm and heights of 49–100 cm, which were subsequently scanned using Philips Brilliance iCT at a resolution with a voxel size of $97.66 \times 97.66 \times 335.00 \ \mu m^3$. An operating voltage of 140 kV and a filament current of 188 mA were applied. To obtain high-quality images, three preprocessing steps were used to reduce the drawbacks. The first step was filtering, which was beneficial in reducing high frequencies and noise. The second step was ring-artifact removal, where rings were removed from the images by comparing and adjusting all the voxel values of each ring. The third step was a beam-hardening correction, which reduced the homogeneous effect of the cylindrical samples. Finally, solid and void phases were distinguished by binarization and threshold segmentation.

Wavelet transform has been widely used in different disciplines of earth science, especially in the quantitative research of sequence stratigraphy [57,58], which hence is used in this study to quantitatively characterize the cyclicity of porosity.

The REV size of the vuggy dolostones was determined by a conventional statistical approach in which porosity was measured using medical-CT images. A total of 10 cylindrical subsamples were randomly selected from the individually large cylinders (Figure 6). The volume of these subsamples was gradually increased from a diameter of 1 mm and height of 1 mm until it was close to the volume of the largest cylinders. As shown in Figure 1, the REV size can be determined based on the covariation of property and volume. In region I, the fluctuation of n reduces with an increasing V, where the fluctuation is dominated by microscopic properties. In region II, the value of n is essentially consistent, which means that the observed property is not affected by the increase in the sample volume. Therefore, the boundary of regions I and II is defined as the REV. In region III, the value of n may remain stable or fluctuate with an increasing V as it depends on whether the porous medium is homogeneous or not. The fluctuations are dominated by macroscopic properties in this region.

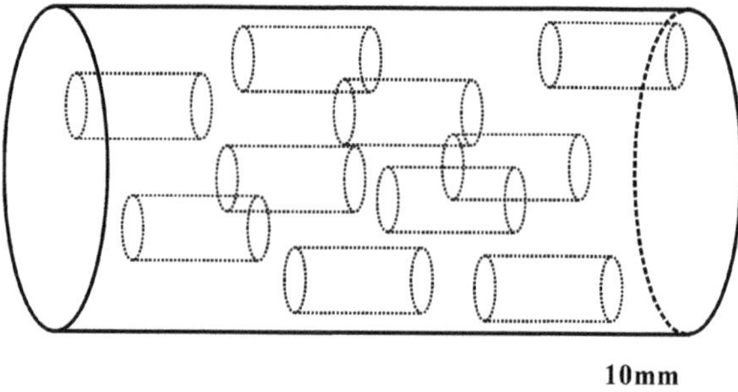

Figure 6. Schematic of subsamples selection.

4. Quantitative Calculation of Parameters

4.1. Pore Geometrical Parameters

The following parameters are calculated to better evaluate the influences of pore volume and morphology on heterogeneity.

(1)   The dominant pore volume

The pore volume fraction on the cumulative curve at 50% is defined as the dominant pore volume (V_{50}). In other words, a half of sample's porosity is composed of pores with volumes larger than V_{50}. This parameter indicates the dominant size of the well-developed vugs in the sample.

(2)   The number of large vugs

When the cumulative pore volume fraction is greater than 50%, the total number of vugs is defined as the number of large vugs (N_{LV}). This value is then adopted to characterize the heterogeneity of pore distribution.

(3)   Average shape factor

The average shape factor (Ave_SF) is used to quantitatively characterize pore shapes. The parameters equal to 1 indicate regular spherical pores, whereas the larger the parameters are, the more irregular the pore shapes are. The average shape factor is defined as

$$\text{Ave\_SF} = \frac{1}{n} * \sum_{i=1}^{n} \frac{S_i^3}{36 * \pi * V_i^2},$$ (1)

where S_i is the area of the *i*th pore, V_i is the volume of the *i*th pore, and *n* is the total number of pores.

4.2. Heterogeneity Parameters

(1)   Coefficient of variation

The coefficient of variation (Cv) is calculated to show the variability relative to the mean value [15]. Since the Cv value of a homogeneous sample is zero, positive Cv values thus indicate heterogeneous samples. In this paper, the Cv values based on CT images of subsamples (Cv_sub) and along the slice direction (Cv_sli) were calculated.

$$Cv = \frac{\sqrt{Var(x)}}{\bar{x}},$$ (2)

where Cv is the coefficient of variation, $\sqrt{Var(x)}$ is the standard deviation, and $\bar{x}$ is the arithmetic mean value.

(2)   Heterogeneous factor

The heterogeneous factor (H) can magnify the effect in the large vugs of carbonate rocks on heterogeneity [59].

$$V = V_{Bi} + V_{Bo},$$ (3)

$$V_{Bo} = V_{Bm} + V_{Bp},$$ (4)

where V is the bulk volume of the sample, V_{Bi} is the bulk volume of the inner large vugs of the sample (the volume of each vug is greater than or equal to V_{50}), V_{Bo} is the volume of the sample excluding V_{Bi}, V_{Bm} is the volume of the rock matrix, and V_{Bp} is the pore volume excluding V_{Bi} (Figure 7).

$$\varphi_i = \frac{V_{Bi}}{V_{Bi}} = 1,$$ (5)

$$\varphi = \frac{V_p}{V} = \frac{V_{Bi}\varphi_i + V_{Bo}\varphi_o}{V_{Bi} + V_{Bo}} = \frac{V_{Bi} + V_{Bo}\varphi_o}{V_{Bi} + V_{Bo}},$$ (6)

where φ is the porosity of the sample, V_p is the pore volume, φ_i is the porosity of the inner large vugs, and φ_o is the porosity of the sample excluding the inner large vugs.

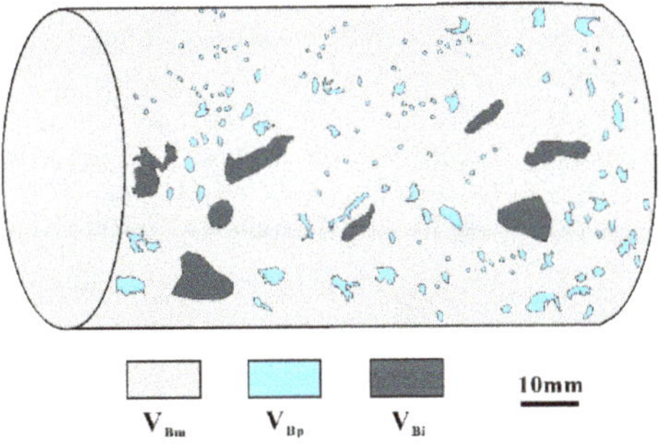

Figure 7. Schematic of volume introduction of each component of the sample.

If the ratio of the bulk volume of the inner large vugs compared with the bulk volume of the sample is defined as F, then

$$F = \frac{V_{Bi}}{V} \text{ and } (1 - F) = \frac{V_{Bo}}{V}.$$

(7)

Substituting Equation (7) into Equation (6),

$$\varphi = \varphi_o + F(1 - \varphi_o)$$

(8)

If the ratio of φ_i to φ_o is defined as R, then

$$R = \frac{\varphi_i}{\varphi_o} = \frac{1}{\varphi_o}.$$

(9)

If the heterogeneous factor (H) is defined as $(\varphi - \varphi_o)/\varphi_o$, then

$$H = \frac{(\varphi - \varphi_o)}{\varphi_o} = F(R - 1).$$

(10)

4.3. Cyclicity of Porosity

REV can be clearly defined only in two situations: (i) materials displaying periodic geometry and (ii) a sample volume containing a very large set of microscale elements of statistically homogeneous and ergodic properties [60,61]. The periodic distribution of sample porosity is analyzed based on a wavelet transform.

$$C(\alpha, \beta, f(t), \Psi(t)) = \int_{-\infty}^{\infty} f(t) \cdot \frac{1}{\sqrt{\alpha}} \Psi * \left(\frac{t - \beta}{\alpha} \right) dt,$$

(11)

where α is the scale parameter ($\alpha > 0$), β is the position parameter, $f(t)$ is the signal, and $\Psi(t)$ is the analyzing wavelet. The wavelet used here is a complex wavelet.

For a better comparison, the number of periodicities is calculated per unit length due to the different slice numbers of CT. It is defined as N_P

$$N_P = \frac{P}{L} * 100,$$

(12)

where P is the total number of periodicities, and L is the number of CT slices.

5. Results

5.1. Rock Type

The Lower Cambrian Xiaoerbulake Formation has been deeply buried, during which it has suffered intense and prolonged diagenetic modifications, resulting in complex rock types. Based on detailed field and microscope investigation, we have found that most of the dolostone shows, originally, depositional and microbial fabrics. In addition, some dolostones have been recrystallized with their original fabrics having vanished. In view of the rock fabrics and genesis, for the samples investigated, classification divided the 26 samples into three major categories, namely, microbial dolostone, hydrodynamic dolostone, and crystalline dolostone. In addition, the microbial dolostones can be further subdivided by their textures into thrombolite dolostone, foam spongy texture silty crystalline dolostone, and oncolite dolostone. The hydrodynamic dolostones were further subdivided into rubble dolostone and fabric-obliterated fine-crystalline dolostone. The twenty-six samples were in classified into six rock types (Figure 5).

5.2. Pore System Classification

To study the influence of variation in pore size on heterogeneity and REV in the samples, four types of pore systems (Types I–IV) were classified based on the distribution of pore size and the contribution of pores to porosity (Figure 8, Table 1).

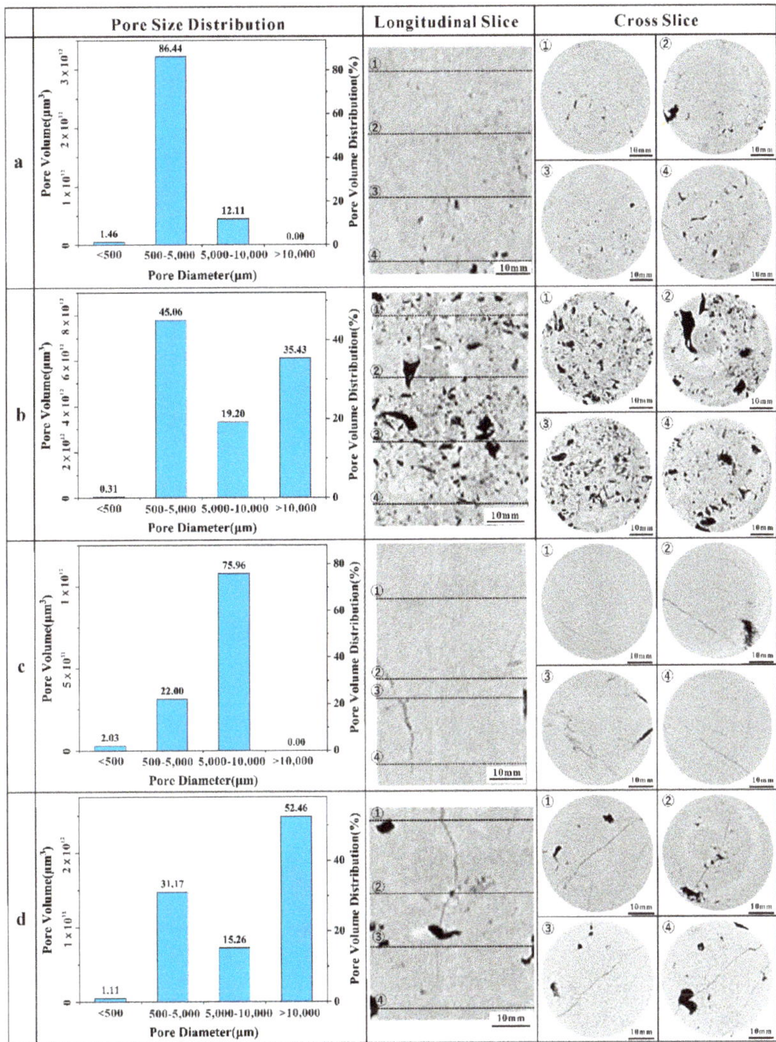

Figure 8. Examples of carbonate types of pore system classification from CT images. (**a**) Type-I, sample SHY026, mainly developed small vugs with a diameter of 1–2 mm, locally developed, and large vugs with a diameter of 7–8 mm. (**b**) Type-II, sample SHY002, flat vugs with a diameter of 2–6 mm. (**c**) Type-III, sample SHY006, cracks. (**d**) Type-IV, sample SHY009, a multitude of flat vugs with a diameter of 3–4 mm.

Table 1. The calculating parameters and determining the REV.

Sample Name	Pore Type	Porosity (%)	V_{50}	N_{LV}	Ave_SF	Cv	H	CREV	DREV
SHY001	I	1.11	1.14×10^9	139	1.19	0.92	1.00	3.30×10^{13}	2.27×10^{13}
SHY002	II	13.90	1.43×10^{11}	13	1.84	0.17	0.87	1.04×10^{13}	1.55×10^{13}
SHY003	II	8.43	3.31×10^{10}	43	1.68	0.31	0.92	6.64×10^{13}	4.54×10^{13}
SHY004	II	7.05	5.14×10^9	299	1.53	0.17	0.93	1.19×10^{13}	1.82×10^{13}
SHY005	II	9.13	2.00×10^{10}	56	1.54	0.34	0.91	1.78×10^{13}	2.23×10^{13}
SHY006	III	1.11	1.72×10^{11}	2	1.54	0.54	1.38	3.27×10^{13}	3.77×10^{13}
SHY007	III	1.14	9.47×10^{10}	3	2.54	0.41	1.03	5.98×10^{13}	5.98×10^{13}
SHY008	I	3.12	1.71×10^{10}	22	1.90	0.76	0.99	2.27×10^{13}	2.44×10^{13}
SHY009	IV	4.84	1.34×10^{11}	3	1.50	0.50	1.10	1.42×10^{13}	1.88×10^{13}
SHY010	I	2.82	2.42×10^{10}	30	1.61	0.80	0.99	2.41×10^{13}	2.61×10^{13}
SHY011	II	8.22	4.07×10^{10}	20	1.90	0.31	0.94	1.82×10^{13}	1.98×10^{13}
SHY012	IV	5.62	4.20×10^{11}	6	2.07	0.74	0.98	6.87×10^{13}	2.18×10^{13}
SHY013	I	0.69	2.98×10^9	50	1.28	0.19	1.00	4.00×10^{13}	$4.00 \times 10_{13}$
SHY014	I	0.15	1.14×10^9	32	1.14	0.65	1.03	6.58×10^{13}	4.61×10^{13}
SHY015	I	1.12	1.13×10^{10}	36	1.50	0.55	0.99	5.39×10^{13}	5.03×10^{13}
SHY016	I	2.81	1.16×10^{10}	72	1.54	0.75	0.98	3.42×10^{13}	3.69×10^{13}
SHY017	I	1.85	8.36×10^9	47	1.37	0.43	0.98	3.42×10^{13}	2.75×10^{13}
SHY018	I	1.40	1.76×10^{10}	35	1.55	0.54	0.99	5.44×10^{13}	4.02×10^{13}
SHY019	I	0.67	2.78×10^9	58	1.29	0.44	1.01	2.71×10^{13}	4.23×10^{13}
SHY020	I	1.28	2.81×10^9	32	1.08	0.40	1.00	2.88×10^{13}	2.70×10^{13}
SHY021	I	0.20	8.21×10^8	60	1.08	0.74	1.01	2.37×10^{13}	4.83×10^{13}
SHY022	I	2.83	1.03×10^9	952	1.08	0.73	0.97	1.71×10^{13}	3.41×10^{13}
SHY023	I	3.96	2.19×10^{10}	39	1.41	0.29	0.97	1.53×10^{13}	1.67×10^{13}
SHY024	I	2.10	1.47×10^9	409	1.07	1.11	0.98	3.24×10^{13}	3.97×10^{13}
SHY025	I	1.12	2.81×10^9	112	1.26	0.36	0.99	-	8.45×10^{13}
SHY026	I	3.23	5.89×10^9	86	1.43	0.60	0.97	5.81×10^{13}	6.15×10^{13}

The Type-I pore system is present in all rock types except for Rock Type ii (Figure 9), and mainly comprises small vugs with diameters of 500–5000 μm (Figure 8a). These small vugs contribute to more than 60% of the total porosity. No large vugs with diameters larger than 10,000 μm were observed in the Type-I pore system. The porosity of this type is lower than 4% (Figure 9).

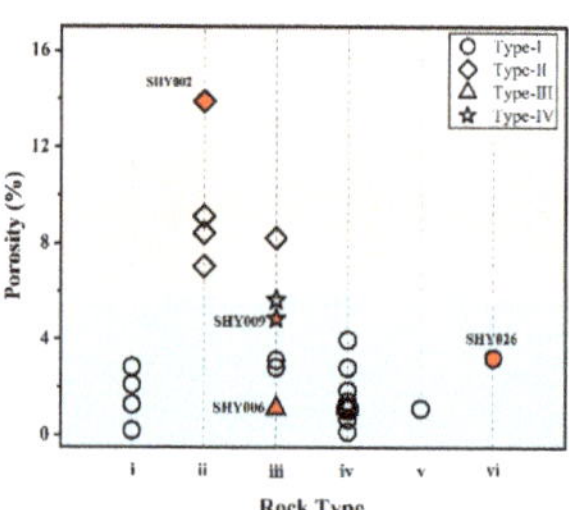

Figure 9. Porosity plotted against rock type. Marker color represents pore type.

The Type-II pore system comprises vugs with diameters mainly of 500–5000 μm (Figure 8b). However, large vugs with diameters greater than 10,000 μm are also observed. The porosity of this type of pore system is the highest among the four types, which ranges from 7.05% to 13.90% (Figure 9). This type of pore system is developed in Rock Type ii and Rock Type iii.

The Type-III pore system, developed in Rock Type iii and Rock Type iv (Figure 9), consists of vugs with diameters mostly between 5000 and 10,000 μm. Larger vugs with diameters greater than 10,000 μm are not developed (Figure 8c). The porosity of this type of pore system is lower than 4%.

The Type-IV pore system is mainly composed of large vugs with diameters greater than 10,000 μm (Figure 8d). The samples characterized by this type of pore system have a porosity which is generally greater than 4% (Figure 9). This type of pore system is predominantly developed in Rock Type iii.

5.3. Quantified Pore Geometrical Parameters

The 50% threshold on the curve of the cumulative pore volume fraction can be used to calculate parameters of V_{50} and N_{LV}. The calculated parameters are summarized in Table 1 and illustrated on Figure 10a,b. Figure 10a shows the lowest N_{LV} value and a relatively high V_{50} value in sample SHY006, whereas Figure 10b shows the highest N_{LV} value and a relatively low V_{50} value in sample SHY022. The parameters of V_{50} and N_{LV} thus reveal distinct pore structures of the studied samples. A smaller V_{50} and a larger N_{LV} indicate more uniform pore sizes of the sample and vice versa. On the plot of porosity versus $\log_{10} V_{50}$ (Figure 10c), different pore system types can be well distinguished. In detail, Type-I samples have low V_{50} values and high N_{LV} values, with the V_{50} values ranging from 8.21×10^8 to 2.42×10^{10} μm^3 and N_{LV} values greater than 22 (Table 1). The V_{50} values for Type-II samples, however, are in a relatively large range with N_{LV} being greater than 13. Type-III and Type-IV samples have relatively higher V_{50} and lower N_{LV} values, indicating that the pore size distribution in these samples is more uneven than those of Type-I and Type-II samples. The parameter of Ave_SF, depicting pore circularity, is broadly positively correlated with porosity (Figure 10d). Samples with low porosity have low Ave_SF values, demonstrating that pore structures are relatively uniform in these samples (Table 1).

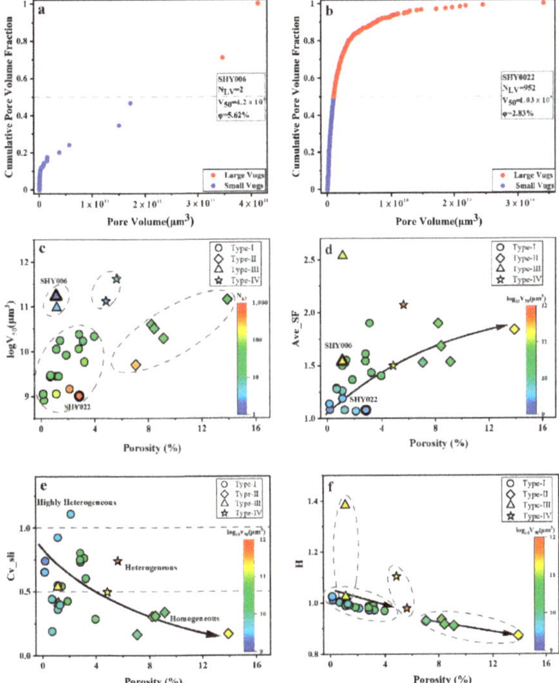

Figure 10. Cross plots of porosity and pore geometrical parameters and heterogeneity parameters. Marker shape represents pore type. (**a**) Pore volume cumulative curve of sample SHY006. Long axial vugs are developed along the cracks, with a diameter of 3–4 mm. (**b**) Pore volume cumulative curve of sample SHY022. A multitude of flat vugs are developed with a diameter of 1–2 mm and distributed in a laminar structure. (**c**) Cross-plot of porosity and dominant pore volume ($\log_{10} V_{50}$) with the number of large vugs superimposed in color. (**d**) Cross-plot of porosity and average shape factor (Ave_SF) with dominant pore volume ($\log_{10} V_{50}$) superimposed in color. (**e**) Cross-plot of porosity and the coefficient of variation of CT slices with dominant pore volume ($\log_{10} V_{50}$) superimposed in color. (**f**) Cross-plot of porosity and heterogeneous factor with dominant pore volume ($\log_{10} V_{50}$) superimposed in color.

5.4. Quantification of Heterogeneity

Based on the values of Cv, the heterogeneity of the samples can be divided into three grades, namely highly heterogeneous (Cv > 1.0), heterogeneous (0.5 < Cv < 1.0), and homogeneous (0.0 < Cv < 0.5). To quantitatively analyze the heterogeneity of carbonates, the coefficient of variation is calculated using CT slices along with the whole sample (Table 1). The values of Cv_sli show an evidently decreasing trend with the increase of porosity (Figure 10e). The Cv_sli values range from 0.19 to 1.11 in Type-I samples. For a given low porosity, these samples yield variable Cv_sli values from high heterogeneity to homogeneity. The Type-II samples have lower Cv_sli values and a higher porosity compared with the other pore system types. The values of Cv_sli of Type-II samples are below 0.5, suggesting that they are homogeneous. The Cv_sli values range from 0.41 to 0.74 in Type-III and Type-IV samples. Meanwhile, they yield relatively higher $\log_{10}V_{50}$ values than the other pore system types.

The values of heterogeneous factor H decrease with increasing porosity within individual types of pore systems (Figure 10f). However, the slope varies among the different types of pore system. The H values range from 0.97 to 1.03 in Type-I samples that have low $\log_{10}V_{50}$ values (Table 1). Samples of Type-II have the lowest H values among the four types of pore system, which are less than 0.94. Type-III and Type-IV samples generally have high H and $\log_{10}V_{50}$ values (Figure 10f).

5.5. Cycle Analysis of Porosity

As shown in Figure 11, porosity periodically fluctuates along with the slices, ranging from 8.36% to 19.35%. The cycles of porosity were quantitatively calculated using the complex wavelet transform. The rightmost part of Figure 11 is the number of calculated cycles, which shows three cycles in this sample.

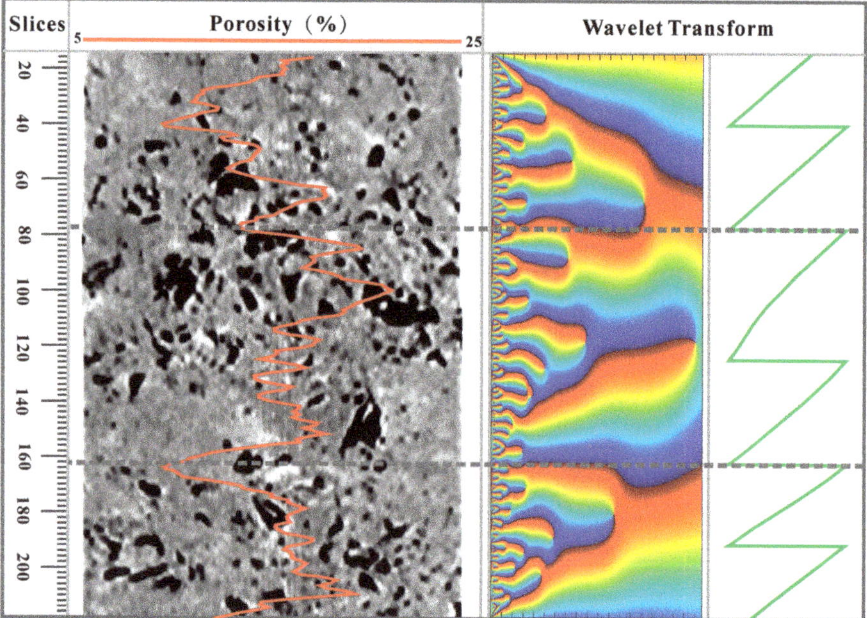

Figure 11. Identification of the porosity cyclicity based on complex wavelet transform. The red line is porosity and the green line is the cycles of porosity.

In the plot of porosity versus N_p (Figure 12), the data of Type-I samples are dispersive with N_p values ranging from 0.92 to 1.98, while those of Type-II samples range from 1.46 to

1.95 (Table 1). The N_p values are greater than 1.6 in Type-III samples, but less than 1.6 for Type-IV samples. Clearly, N_p decreases with increasing porosity when the porosity is less than 6%.

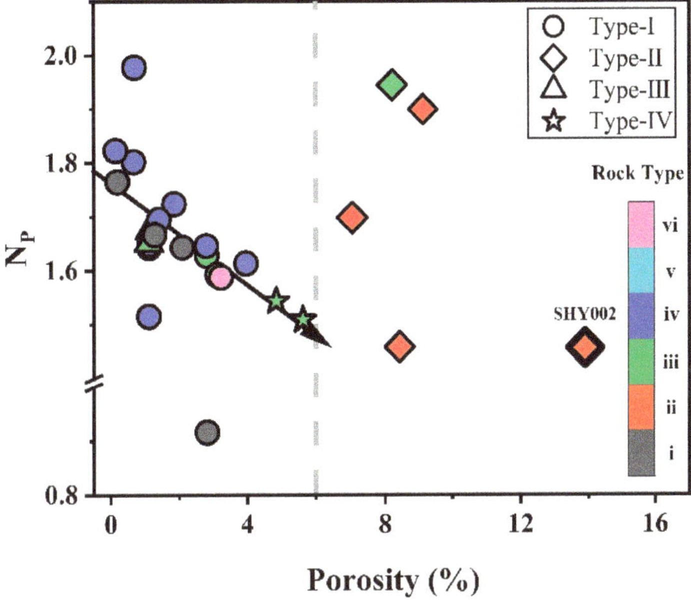

Figure 12. Cross-plot of porosity and porosity cyclicity with rock type superimposed in color. Marker shape represents pore type.

6. Discussion

6.1. Evaluation on Heterogeneity of Pore Systems

The classification schemes for carbonate pore systems have been proposed in view of pore geometry, rock fabric, genesis, flow properties, and pore-scale modeling [62–64]. In addition, definitions and classifications for vug or vuggy porosity have been discussed in previous works [63,65–67]. Choquette and Pray [63] defined a pore to be a vug if it is (1) approximately equant and not remarkably elongated; (2) large enough to be visible with the naked eye (pore diameter exceeding 1/16 mm); and (3) not fabric selective. Lucia stated that vugs that are within crystals or grains or that are markedly larger than crystals or grains are considered pore spaces [65–67]. The authors further subdivided vuggy pore spaces into separate vugs and touching vugs based on whether the vugs are interconnected or not. Luo and Machel [68] proposed a new pore size classification for complex carbonate reservoirs: microporosity (diameter < 1 μm), mesoporosity (diameter 1–1000 μm), macroporosity (diameter 1–256 mm), and megaporosity (diameter > 256 mm). The vugs of that study were grouped into mesoporosity. Li et al. [69] quantitatively divided the carbonate rock into matrix, fractured, and vuggy based on a new function of the carbonate rock index. Based on the definition of vugs by Lucia [65–67], the classification scheme in this paper is proposed by taking into account the pore size distribution and contributions of pores to porosity. Four types of pore systems (Types I–IV) were classified in this study (Figure 8). The scheme is rewarding for studying the influence of pore size distribution on heterogeneity and REV.

Large vugs with diameters greater than 10,000 μm are both developed in Type-II and Type-IV samples. The difference between these two types is their distinct pore size distributions that contribute to the total porosity (Figure 8b,d). The vugs' diameters in Type-II samples are mainly in the range of 500–5000 μm (Figure 8b), whereas those in Type-IV samples are greater than 10,000 μm (Figure 8d). The small vugs, with diameters less than

500 μm, contribute less to the rock porosity in all types (Figure 8), suggesting that large vugs are the main source for the reservoir space of vuggy dolostones. The relationships between rock types and types of pore systems are complex due to the dual control of the deposition environment and diagenesis. Individual types of pore systems could be developed in multiple rock types (Figure 9). Moreover, different degrees of dissolution in the same rock type could result in variation in the pore systems. The Rock Type ii samples have a relatively high porosity (Figure 9), indicating that the development degree of vugs is the highest in these samples.

Quantifying reservoir heterogeneity is an important but difficult process. Many diagenetic parameters and pore structure parameters could result in variations of reservoir properties. The values of $\log_{10}V_{50}$ and Ave_SF in Type-I samples are relatively low, indicating a more uniform pore structure. Nevertheless, heterogeneity could be high for a given low porosity in these samples (Figure 10c–f). Type-II samples are characterized by a relatively low heterogeneity but high porosity, indicating that the intensive development of vugs leads to low heterogeneity in carbonate reservoirs. However, the heterogeneity of the samples will increase only if a small number of large vugs is developed at a low porosity, such as Type-III samples (Table 1). The heterogeneity of different pore systems is highly variable in the vuggy carbonate reservoirs (Figure 10e,f). As such, it is a challenging task to select suitable samples for analysis. Therefore, precisely determining the REV sizes is very crucial for accurately calculating the petrophysical properties and evaluating the reservoir quality.

6.2. REV Analysis

This paper applies two methods to determine the REV size. One method is to determine the size of the representative elementary volume based on the cutoff value of Cv_sub when it is less than 0.1 (CREV), then the REV is obtained. Another method is to take the derivative of Cv_sub to determine the size of the representative elementary volume (DREV), where the REV is considered to be reached when the derivative of Cv_sub is between -9×10^{-15} and 9×10^{-15} (Figure 13). As shown in Figure 13a,c and e, the values of the DREV are greater than the CREV. There will be slight fluctuations when the value of Cv_sub is less than 0.1, resulting in a certain change in the derivative of Cv_sub. However, the DREV values could be equal to (Figure 13b) or less than (Figure 13d) the CREV, indicating that the value of Cv_sub is relatively stable when it is equal to or larger than 0.1, respectively. Figure 13f shows that the REV can be determined by the derivative of Cv_sub, but the size of the REV cannot be determined based on the cutoff value of Cv_sub. There are nine samples with values of the DREV which are lower than the CREV (the REV size of sample SHY025 cannot be obtained based on the cutoff value of Cv_sub.), two samples with a DREV equal to the CREV, and fifteen samples with a DREV greater than the CREV (Figure 14, Table 1). The REV determined based on the derivative of Cv_sub (DREV) is more accurate compared to the cutoff value of the Cv_sub (CREV), so the DREV is used for the following REV analysis.

Factors affecting the REV size have been studied in previous works based on rock samples and numerical models at different scales. Gitman et al. [61] stated that the REV sizes depended on the investigated petrophysical properties, the contrast of properties, volume fractions of the microstructure, required relative precision, and the number of realizations of the microstructure. Tavakoli [17] demonstrated that primary depositional settings and diagenesis controlled the REV sizes of the reservoirs. Moreover, the REV size also depends on the scale of observation [17]. Compared to previous studies, we mainly focused on the influence of measurement methods, porosity, type of pore systems, and heterogeneity of pore structures on the REV sizes, with an aim to reveal the effect of the heterogeneity of the developed vugs on REV sizes.

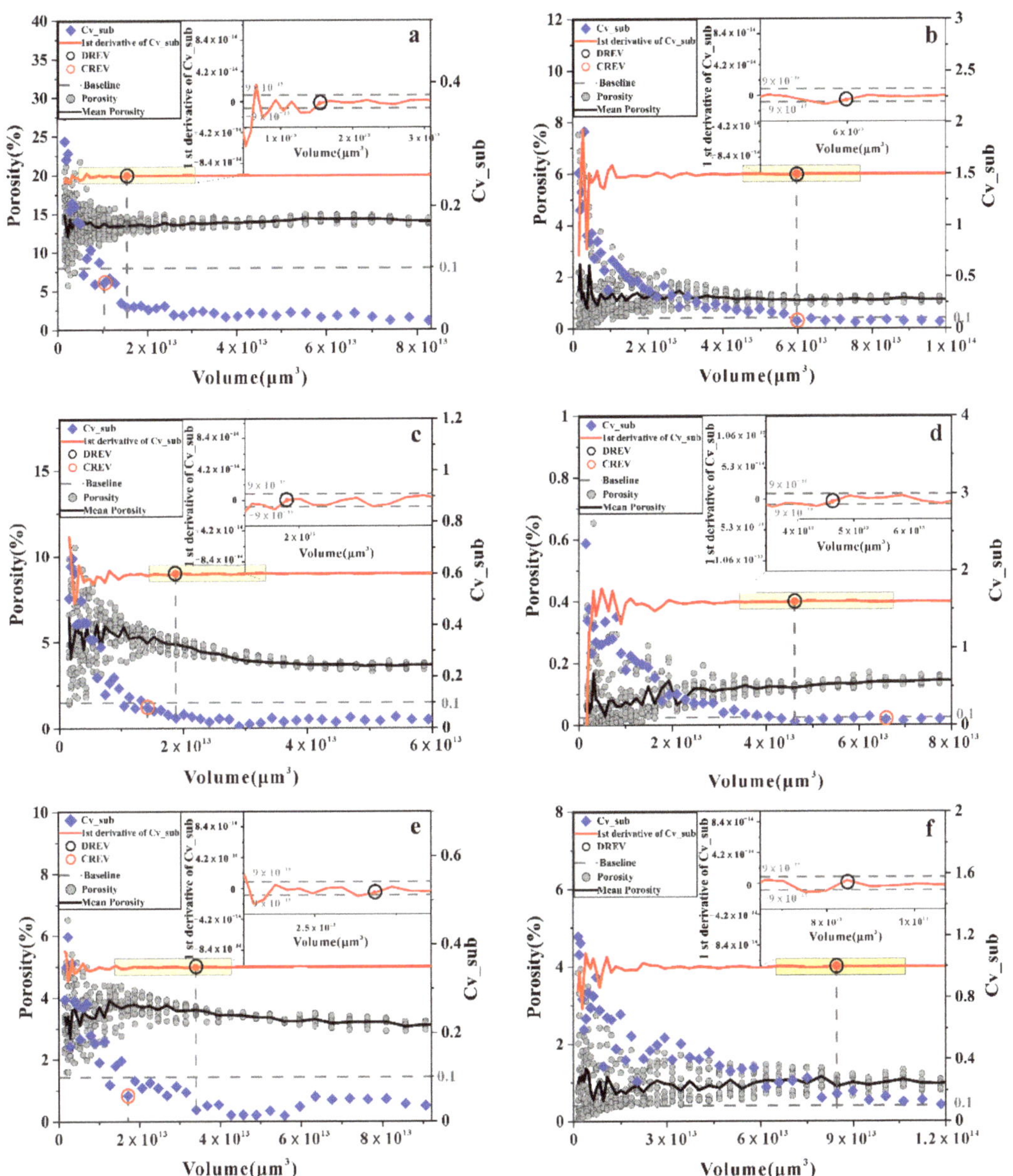

Figure 13. Determination of REV for porosity. (**a**) Sample SHY002 developed flat vugs with a diameter of 2–6 mm. (**b**) Sample SHY007 developed vugs and cracks, and vugs half-filled by secondary carbonates. (**c**) Sample SHY009 developed a multitude of flat vugs developed with a diameter of 3–4 mm. (**d**) Sample SHY014 developed small amount of vugs with a diameter of 0.5–1 mm. (**e**) Sample SHY022 developed a multitude of flat vugs with a diameter of 1–2 mm. (**f**) Sample SHY025 developed flat vugs with a diameter of 2–6 mm and partially filled by secondary carbonates. The insets in the six images are magnifications of light yellow rectangle indicated in individual images.

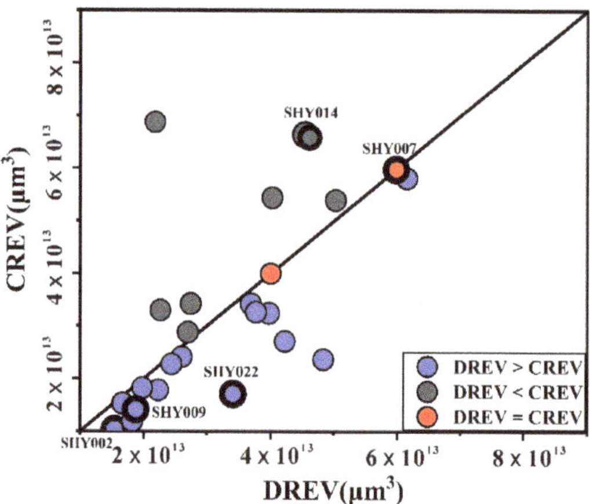

Figure 14. Cross-plot of DREV versus CREV.

The REV sizes are variable when using different methods, and the results obtained by the derivative of Cv_sub (DREV) are relatively more accurate (Figure 14). In addition, porosity is one of the important factors that influences the REV size. Clearly, the values of the DREV decrease as porosity increases (Figure 15). Nine of the twenty-six samples investigated have DREV values of less than 2.45×10^{13} μm^3 (the volume of the cylinder with a diameter of 2.5 cm and a height of 5 cm) and a porosity greater than 3.5%, except for one outlier (Figure 15). Type-II and Type-IV samples required smaller volumes to attain REV compared to those of Type-I and Type-III samples, indicating that the development of large vugs has negative impacts on the REV size. In addition, the REV size was affected by the parameters of heterogeneity and pore structure. However, compared with porosity, the correlation between these parameters and the DREV was not high, as revealed by the weak correlations in Figure 16. As such, the REV size is very likely the result of the interplay of multiple factors. It is necessary to comprehensively consider the multiple parameters of a sample to accurately determine the REV size.

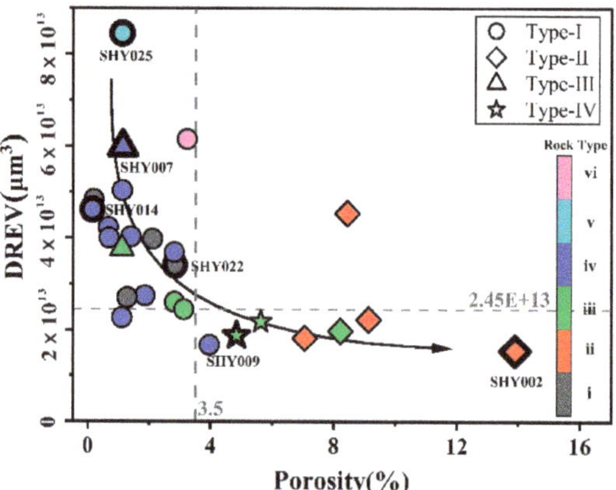

Figure 15. Cross-plot of porosity and DREV with rock type superimposed in color. Marker shape represents pore type.

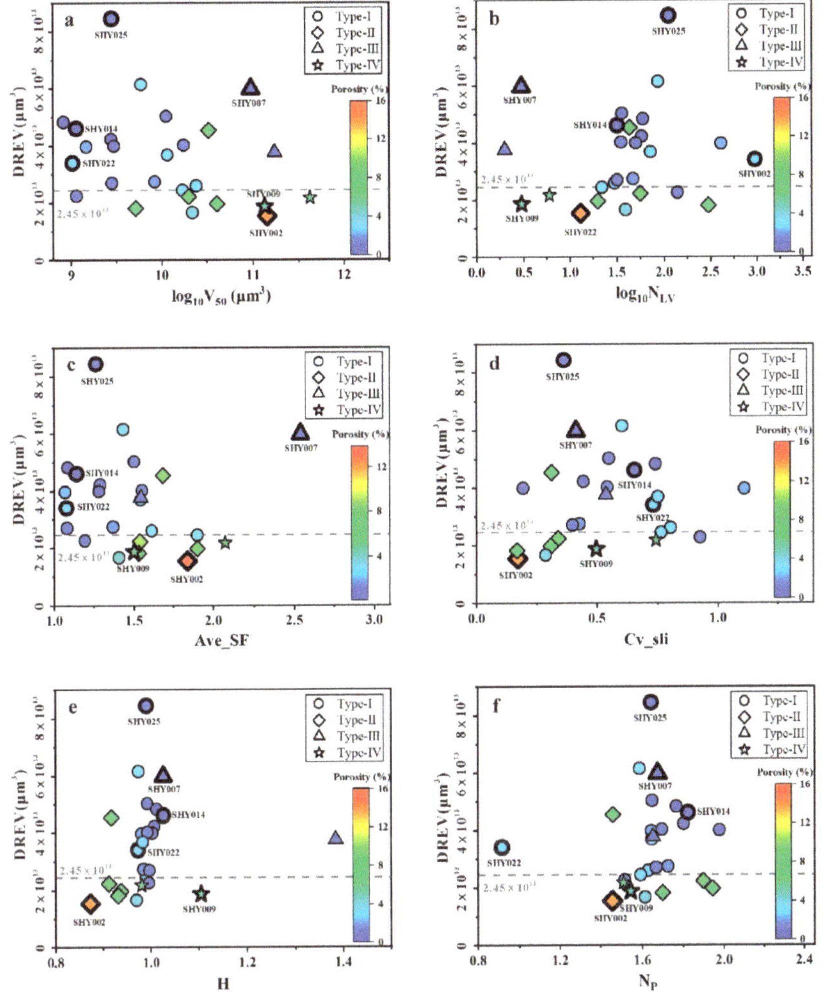

Figure 16. Cross-plots of DREV and five parameters with porosity superimposed in color. Marker shape represents pore type. (**a**) DREV versus $\log_{10}V_{50}$. (**b**) DREV versus $\log_{10}N_{LV}$. (**c**) DREV versus Ave_SF. (**d**) DREV versus Cv_sli. (**e**) DREV versus H. (**f**) DREV versus N_P.

6.3. The REV Prediction Model

Several researchers have attempted to determine the REV sizes of porous media. Clausnitzer et al. [39] showed that the side length of the REV for porosity was approximately 5.5 times of the diameter of a random pack of uniform glass beads. Bažant [40] found that the REV size could be calculated by $V = \updownarrow^{n_d}$, where $\updownarrow$ is the characteristic length, and n_d is the number of spatial dimensions. The characteristic length is approximately 2.7 times the maximum inclusion size [41]. Razavi et al. [34] suggested a systematic method to quantify the calculation of the REV, in which the REV radius of spherical glass beads was approximately two to three times the identified average diameters. The REV radius of silica sand was between 5 and 11 times the d_{50} (the median particle size). The REV radius of Ottawa sandstone was between 9 and 16 times the d_{50} (the median particle size). Vik et al. [26] showed that the REV sizes were close to the average values of all porosity measured when the bulk sample volume was greater than 1300 cm^3, which was expected to be above the REV. Clearly, the REV sizes determined by the above studies were mainly based on the characteristic length or diameter. For significantly heterogeneous carbonate rocks, any single parameter is inadequate to properly evaluate it, as addressed above (Figure 16).

Multiple statistical regressions (MLRs) are a useful method to analyze the effect of multiple independent variables on the dependent variable [70,71]. The model is defined by Equation (13).

$$P_m = \theta + \theta_0 P_e + \theta_1 X_1 + \theta_2 X_2 + \ldots \theta_n X_n, \tag{13}$$

where P_m is the determined property, P_e is the predicted property, X_n represents independent variables, and θ_n are coefficients determined by the regression. The adjusted coefficient of determination ($\overline{R}^2$) is used to determine how this statistical model fits to the data of the determined property.

$$\overline{R}^2 = 1 - \left(1 - R^2\right) \frac{m - 1}{m - p - 1}, \tag{14}$$

where m is the sample size, p represents the total number of independent variables in the linear model, and R^2 is the determination coefficient. $\overline{R}^2$ is different to R^2 because it considers the degree of freedom of the data set. $\overline{R}^2$ will increase when a new independent variable is included, with an exception that it improves R^2 more than expected by chance.

To reduce the uncertainties of the estimated REV, MLRs was used to evaluate the best statistical fit between the DREV and the calculated multi-factors using the adjusted coefficient of determination ($\overline{R}^2$). The independent variables were added one by one to the equation (Table 2). The independent variable was removed if the added independent variables did not contribute much to the fitted equation (e.g., parameters of Cv_sli and N_P) (Figure 17b,d). It is worth noting that parameter N_P is not used in the predicting model, but this parameter is the premise of whether the sample can obtain the REV or not (see Section 4.3). Combinations of various factors have been tested (Figure 17, Table 2). Of which, the best model consists of parameters of $\log\varphi$, H, Ave_SF, and $\log N_{LV}$ that yield the maximum $\overline{R}^2$ of 0.9320 with a p-value below 0.05 and D-W greater than 1.5. This final model passes the tests of the significance and sample independence. The REV sizes can be obtained by a non-destructive CT technique in this method, meaning there are intact sample aspects for later petrophysical tests. More importantly, the REV sizes obtained for multiple parameters are more accurate than those obtained for a single parameter.

Table 2. Adjusted Coefficient of Determination between Determined and Estimated REV.

Dependent Variable	Independent Variables	R^2	$\overline{R}^2$	p-Value	D-W
$\log(DREV/V_{50})$	(a) φ	0.3559	0.3290	0.0013	1.5702
	(b) $\varphi + Cv\_sli$	0.3570	0.3012	0.0062	1.5708
	(c) $\varphi + H$	0.5853	0.5492	0.0000	1.6838
	(d) $\varphi + H + N_P$	0.5874	0.5311	0.0002	1.6674
	(e) $\varphi + H + Ave\_SF$	0.8195	0.7949	0.0000	1.3571
	(f) $\varphi + H + Ave\_SF + N_{LV}$	0.8366	0.8054	0.0000	1.2550
	(g) $\log \varphi + H + Ave\_SF + N_{LV}$	0.8713	0.8468	0.0000	1.4798
	(h) $\log \varphi + H + Ave\_SF + \log N_{LV}$	0.9429	0.9320	0.0000	1.5530

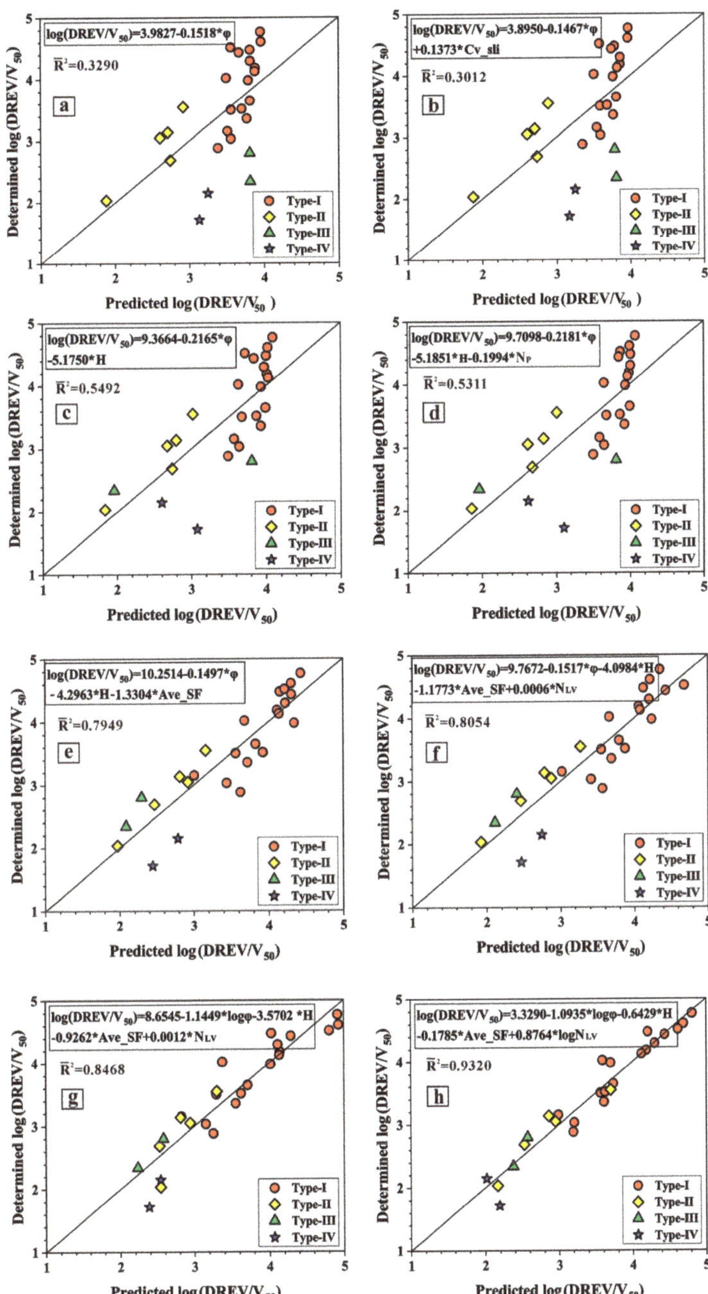

Figure 17. The fitting prediction model of DREV. (**a**) Fitting model established for φ. (**b**) Fitting model established for φ and Cv_sli. (**c**) Fitting model established for φ and H. (**d**) Fitting model established for φ, H and N_P. (**e**) Fitting model established for φ, H and Ave_SF. (**f**) Fitting model established for φ, H, Ave_SF and N_{LV}. (**g**) Fitting model established for $\log \varphi$, H, Ave_SF and N_{LV} (**h**) Fitting model established for $\log \varphi$, H, Ave_SF and $\log N_{LV}$.

The link between the REV sizes and porosity, pore structure parameters, and heterogeneity parameter is thus established based on the prediction model. In other words,

REV sizes can be obtained if these parameters are known. This potentially makes it a very effective model to predict the REV sizes of vuggy dolostones. This study provides a foundation for accurately calculating petrophysical properties and precisely evaluating the reservoir's quality.

7. Conclusions

(1) A total of 26 vuggy dolostones collected from the Cambrian Xiaoerbulake Formation at the Kalping uplift are classified into four types of pore systems based on the pore size distribution and contribution of pores to porosity.

(2) The different degrees of dissolution in different types of pore systems yield variation in porosity, pore structure parameters, and heterogeneity. The development of numerous vugs increases porosity and reduces heterogeneity, while the development of a small amounts of large vugs increases the sample's heterogeneity.

(3) The REV determined by the derivative of Cv_sub is more accurate than that determined by the cutoff value of Cv_sub. Only nine out of twenty-six samples have a DREV less than the volume of traditional core plugs (2.45×10^{13} μm^3), hence the traditional core plugs are unrepresentative for most vuggy carbonate rocks.

(4) The REV sizes are influenced by various factors. Any individual parameter only is inadequate to properly evaluate the REV sizes, so that multi-factors should be considered. A prediction model has been established based on the relationship between the REV sizes and the quantitative parameters of V_{50}, N_{LV}, Ave_SF, φ, and H, with the correlation coefficient reaching 0.9320. Thus, our model could be very effective for predicting REV sizes of vuggy dolostones.

Author Contributions: Conceptualization, Y.X. and Z.C.; methodology, Y.X.; software, Y.X., Q.L. and J.G.; validation. Y.X., Z.C. and H.Z.; formal analysis, Y.X. and L.C.; investigation, Y.X.; resources, Y.X. and F.H.; data curation, Y.X.; writing—original draft preparation, Y.X.; writing—review and editing, Y.X. and H.Z.; project administration, Z.C.; funding acquisition, Z.C. All authors have read and agreed to the published version of the manuscript.

Funding: This work is financially supported by the National Key Research and Development Program of China (Grant No. 2019YFC0605502), the Strategic Priority Research Program of the Chinese Academy of Sciences (Grant No. XDA14010302), and the National Natural Science Foundation of China (Grant No. 42002170).

Institutional Review Board Statement: Not applicable.

Informed Consent Statement: Not applicable.

Data Availability Statement: Not applicable.

Acknowledgments: We appreciate Jie Li and Peng Tang for their collaboration and enthusiastic support of fieldwork. The authors especially thank Yan Xiong for her help during the medical-CT experiment.

Conflicts of Interest: The authors declare no conflict of interest.

Nomenclature

REV	the representative elementary volume
V_{50}	the dominant pore volume
N_{LV}	the pore number on the cumulative curve at greater than 50%
Ave_SF	the average of the shape factor
S_i	the area of the ith pore
V_i	the volume of the ith pore
n	the total number of pores
Cv	the coefficient of variation
$\sqrt{Var(x)}$	the standard deviation
$\overline{x}$	the arithmetic mean value

Cv_sli	the coefficient of variation of the medical-CT along the slice direction
Cv_sub	the coefficient of variation when determining the REV.
H	the heterogeneous factor
V	the bulk volume of the sample
V_{Bi}	the bulk volume of the inner large vugs of the sample (the volume of each vug is greater than or equal to V_{50})
V_{Bo}	the volume of the sample, excluding V_{Bi}
V_{Bm}	the volume of the rock matrix
V_{Bp}	the volume of pores, excluding V_{Bi}
φ	the porosity of the sample
V_p	the volume of the pores
φ_i	the porosity of the inner large vugs
φ_o	the porosity of the sample, excluding the inner large vugs
F	the ratio of the bulk volume of the inner large vugs to the bulk volume of the sample
R	the ratio of the porosity of the inner large vugs to the porosity of a sample excluding the inner large vugs
α	the scale parameter ($\alpha > 0$)
β	the position parameter
$f(t)$	the signal
$\Psi(t)$	the analyzing wavelet (the wavelet used is a complex wavelet)
N_P	the number of periodicities per unit length
P	the total number of periodicities
L	the number of CT slices
CREV	the determining REV based on the cutoff value of Cv_sli
DREV	the determining REV based on the derivative of Cv_sub
$\updownarrow$	the characteristic length
n_d	the number of spatial dimensions
d_{50}	the median particle size
P_m	the determined property
P_e	the predicted property
X_n	independent variables
θ_n	coefficients determined by the regression
$\overline{R}^2$	the adjusted coefficient of determination
m	the sample size
p	the total number of independent variables
R^2	the determination coefficient

References

1. Ahr, W.M. *Geology of Carbonate Reservoirs: The Identification, Description, and Characterization of Hydrocarbon Reservoirs in Carbonate Rocks*; Wiley: Hoboken, NJ, USA, 2008. [CrossRef]
2. Jia, C. *Characteristics of Chinese Petroleum Geology: Geological Features and Exploration Cases of Stratigraphic, Foreland and Deep Formation Traps*; Springer: Berlin/Heidelberg, Germany, 2012. [CrossRef]
3. Bagrintseva, K.I. *Carbonate Reservoir Rocks*; John Wiley & Sons: Hoboken, NJ, USA, 2015.
4. Apolinarska, K. *Book Reviews. Origin of Carbonate Sedimentary Rocks*; De Gruyter Open: Berlin, Germany, 2017. [CrossRef]
5. Issoufou Aboubacar, M.S.; Cai, Z. A Quadruple-Porosity Model for Consistent Petrophysical Evaluation of Naturally Fractured Vuggy Reservoirs. *SPE J.* **2020**, *25*, 2678–2693. [CrossRef]
6. Zhang, H.; Ait Abderrahmane, H.; Arif, M.; Al Kobaisi, M.; Sassi, M. Influence of Heterogeneity on Carbonate Permeability Upscaling: A Renormalization Approach Coupled with the Pore Network Model. *Energy Fuels* **2022**, *36*, 3003–3015. [CrossRef]
7. Gundogar, A.S.; Ross, C.M.; Akin, S.; Kovscek, A.R. Multiscale pore structure characterization of middle east carbonates. *J. Pet. Sci. Eng.* **2016**, *146*, 570–583. [CrossRef]
8. Sadeghnejad, S.; Gostick, J. Multiscale Reconstruction of Vuggy Carbonates by Pore-Network Modeling and Image-Based Technique. *SPE J.* **2020**, *25*, 253–267. [CrossRef]
9. Radwan, A.E.; Trippetta, F.; Kassem, A.A.; Kania, M. Multi-scale characterization of unconventional tight carbonate reservoir: Insights from October oil filed, Gulf of Suez rift basin, Egypt. *J. Pet. Sci. Eng.* **2021**, *197*, 107968. [CrossRef]
10. Li, J.; Zhang, H.; Cai, Z.; Zou, H.; Hao, F.; Wang, G.; Li, P.; Zhang, Y.; He, J.; Fei, W. Making sense of pore systems and the diagenetic impacts in the Lower Triassic porous dolostones, northeast Sichuan Basin. *J. Pet. Sci. Eng.* **2021**, *197*, 107949. [CrossRef]

11. Chen, J.; Yang, S.; Mei, Q.; Chen, J.; Chen, H.; Zou, C.; Li, J.; Yang, S. Influence of Pore Structure on Gas Flow and Recovery in Ultradeep Carbonate Gas Reservoirs at Multiple Scales. *Energy Fuels* **2021**, *35*, 3951–3971. [CrossRef]
12. Jiang, Z.; van Dijke, M.I.J.; Sorbie, K.S.; Couples, G.D. Representation of multiscale heterogeneity via multiscale pore networks. *Water Resour. Res.* **2013**, *49*, 5437–5449. [CrossRef]
13. Nader, F.H. *Multi-Scale Quantitative Diagenesis and Impacts on Heterogeneity of Carbonate Reservoir Rocks*; Springer: Berlin/Heidelberg, Germany, 2017.
14. Frazer, G.W.; Wulder, M.A.; Niemann, K.O. Simulation and quantification of the fine-scale spatial pattern and heterogeneity of forest canopy structure: A lacunarity-based method designed for analysis of continuous canopy heights. *For. Ecol. Manag.* **2005**, *214*, 65–90. [CrossRef]
15. Fitch, P.J.R.; Lovell, M.A.; Davies, S.J.; Pritchard, T.; Harvey, P.K. An integrated and quantitative approach to petrophysical heterogeneity. *Mar. Pet. Geol.* **2015**, *63*, 82–96. [CrossRef]
16. Cooper, S.D.; Barmuta, L.; Sarnelle, O.; Kratz, K.; Diehl, S. Quantifying Spatial Heterogeneity in Streams. *Freshw. Sci.* **1997**, *16*, 174–188. [CrossRef]
17. Tavakoli, V. *Carbonate Reservoir Heterogeneity: Overcoming the Challenges*; Springer International Publishing AG: Cham, Switzerland, 2019.
18. Bear, J. *Dynamics of Fluids in Porous Media*; American Elsevier Publishing Company: Princeton, NJ, USA, 1972.
19. Brown, G.O.; Hsieh, H.T.; Lucero, D.A. Evaluation of laboratory dolomite core sample size using representative elementary volume concepts. *Water Resour. Res.* **2000**, *36*, 1199–1207. [CrossRef]
20. Rozenbaum, O.; du Roscoat, S.R. Representative elementary volume assessment of three-dimensional x-ray microtomography images of heterogeneous materials: Application to limestones. *Phys. Rev. E* **2014**, *89*, 053304. [CrossRef] [PubMed]
21. Yio, M.H.N.; Wong, H.S.; Buenfeld, N.R. Representative elementary volume (REV) of cementitious materials from three-dimensional pore structure analysis. *Cem. Concr. Res.* **2017**, *102*, 187–202. [CrossRef]
22. Shah, S.M.; Crawshaw, J.P.; Gray, F.; Yang, J.; Boek, E.S. Convex hull approach for determining rock representative elementary volume for multiple petrophysical parameters using pore-scale imaging and Lattice–Boltzmann modelling. *Adv. Water Resour.* **2017**, *104*, 65–75. [CrossRef]
23. Wu, M.; Wu, J.; Wu, J. A three-dimensional model for quantification of the representative elementary volume of tortuosity in granular porous media. *J. Hydrol.* **2018**, *557*, 9. [CrossRef]
24. Wang, Y.; Wang, L.; Wang, J.; Jiang, Z.; Wang, C.-C.; Fu, Y.; Song, Y.-F.; Wang, Y.; Liu, D.; Jin, C. Multiscale characterization of three-dimensional pore structures in a shale gas reservoir: A case study of the Longmaxi shale in Sichuan basin, China. *J. Nat. Gas Sci. Eng.* **2019**, *66*, 207–216. [CrossRef]
25. Wu, H.; Yao, Y.; Zhou, Y.; Qiu, F. Analyses of representative elementary volume for coal using X-ray μ-CT and FIB-SEM and its application in permeability predication model. *Fuel* **2019**, *254*, 115563. [CrossRef]
26. Vik, B.; Bastesen, E.; Skauge, A. Evaluation of representative elementary volume for a vuggy carbonate rock-Part: Porosity, permeability, and dispersivity. *J. Pet. Sci. Eng.* **2013**, *112*, 36–47. [CrossRef]
27. Gonzalez, J.L.; de Faria, E.L.; Albuquerque, M.P.; Albuquerque, M.P.; Bom, C.R.; Freitas, J.C.C.; Cremasco, C.W.; Correia, M.D. Representative elementary volume for NMR simulations based on X-ray microtomography of sedimentary rock. *J. Pet. Sci. Eng.* **2018**, *166*, 906–912. [CrossRef]
28. Singh, A.; Regenauer-Lieb, K.; Walsh, S.D.C.; Armstrong, R.T.; van Griethuysen, J.J.; Mostaghimi, P. On Representative Elementary Volumes of Grayscale Micro-CT Images of Porous Media. *Geophys. Res. Lett.* **2020**, *47*, e2020GL088594. [CrossRef]
29. Shahin, G.; Desrues, J.; Pont, S.D.; Combe, G.; Argilaga, A. A study of the influence of REV variability in double-scale FEM × DEM analysis. *Int. J. Numer. Methods Eng.* **2016**, *107*, 882–900. [CrossRef]
30. Norris, R.J.; Lewis, J.J.M. The Geological Modeling of Effective Permeability in Complex Heterolithic Facies. In Proceedings of the SPE Annual Technical Conference and Exhibition, SPE-22692-MS, Dallas, TX, USA, 6–9 October 1991. [CrossRef]
31. Nordahl, K.; Ringrose, P.S. Identifying the Representative Elementary Volume for Permeability in Heterolithic Deposits Using Numerical Rock Models. *Math. Geosci.* **2008**, *40*, 753–771. [CrossRef]
32. Katagiri, J.; Kimura, S.; Noda, S. Significance of shape factor on permeability anisotropy of sand: Representative elementary volume study for pore-scale analysis. *Acta Geotech.* **2020**, *15*, 2195–2203. [CrossRef]
33. Borges, J.A.R.; Pires, L.F.; Cássaro, F.A.M.; Roque, W.L.; Heck, R.J.; Rosa, J.A.; Wolf, F.G. X-ray microtomography analysis of representative elementary volume (REV) of soil morphological and geometrical properties. *Soil Tillage Res.* **2018**, *182*, 112–122. [CrossRef]
34. Razavi, M.R.; Muhunthan, B.; Al Hattamleh, O. Representative Elementary Volume Analysis of Sands Using X-Ray Computed Tomography. *Geotech. Test. J.* **2007**, *30*, 212–219. [CrossRef]
35. Costanza-Robinson, M.S.; Estabrook, B.D.; Fouhey, D.F. Representative elementary volume estimation for porosity, moisture saturation, and air-water interfacial areas in unsaturated porous media: Data quality implications. *Water Resour. Res.* **2011**, *47*, W07513. [CrossRef]
36. Al-Raoush, R.; Papadopoulos, A. Representative elementary volume analysis of porous media using X-ray computed tomography. *Powder Technol.* **2010**, *200*, 69–77. [CrossRef]
37. Zhang, J.; Yu, L.; Jing, H.; Liu, R. Estimating the Effect of Fractal Dimension on Representative Elementary Volume of Randomly Distributed Rock Fracture Networks. *Geofluids* **2018**, *2018*, 1–13. [CrossRef]

38. Wei, C.; Tian, C.; Zheng, J.; Cai, K.; Du, D.; Song, B.; Hu, Y. Heterogeneity Characteristics of Carbonate Reservoirs: A Case Study using Whole Core Data. In Proceedings of the SPE Reservoir Characterisation and Simulation Conference and Exhibition, Abu Dhabi, United Arab Emirates, 14–16 September 2015; p. D021S004R002. [CrossRef]
39. Clausnitzer, V.; Hopmans, J.W. Determination of phase-volume fractions from tomographic measurements in two-phase systems. *Adv. Water Resour.* **1999**, *22*, 577–584. [CrossRef]
40. Bažant, Z.P. Stochastic models for deformation and failure of quasibrittle structures: Recent advances and new directions. In *Computational Modelling of Concrete Structures*; A.A. Balkema Publisher: Lisse, The Netherlands, 2003; pp. 583–598.
41. Bažant, Z.P.; Pijaudier-Cabot, G. Measurement of Characteristic Length of Nonlocal Continuum. *J. Eng. Mech.* **1989**, *115*, 755–767. [CrossRef]
42. Lai, J.; Wang, G.; Wang, Z.; Chen, J.; Pang, X.; Wang, S.; Zhou, Z.; He, Z.; Qin, Z.; Fan, X. A review on pore structure characterization in tight sandstones. *Earth-Sci. Rev.* **2018**, *177*, 436–457. [CrossRef]
43. Moslemipour, A.; Sadeghnejad, S. Dual-scale pore network reconstruction of vugular carbonates using multi-scale imaging techniques. *Adv. Water Resour.* **2021**, *147*, 103795. [CrossRef]
44. Razavifar, M.; Mukhametdinova, A.; Nikooee, E.; Burukhin, A.; Rezaei, A.; Cheremisin, A.; Riazi, M. Rock Porous Structure Characterization: A Critical Assessment of Various State-of-the-Art Techniques. *Transp. Porous Media* **2021**, *136*, 431–456. [CrossRef]
45. Vogel, H.J.; Weller, U.; Schlüter, S. Quantification of soil structure based on Minkowski functions. *Comput. Geosci.* **2010**, *36*, 1236–1245. [CrossRef]
46. Cnudde, V.; Boone, M.N. High-resolution X-ray computed tomography in geosciences: A review of the current technology and applications. *Earth-Sci. Rev.* **2013**, *123*, 1–17. [CrossRef]
47. Qajar, J.; Arns, C.H. Characterization of reactive flow-induced evolution of carbonate rocks using digital core analysis- part 1: Assessment of pore-scale mineral dissolution and deposition. *J. Contam. Hydrol.* **2016**, *192*, 60–86. [CrossRef] [PubMed]
48. Chaves, J.M.; Moreno, R.B. Low- and High-Resolution X-Ray Tomography Helping on Petrophysics and Flow-Behavior Modeling. *SPE J.* **2021**, *26*, 206–219. [CrossRef]
49. Pini, R.; Madonna, C. Moving across scales: A quantitative assessment of X-ray CT to measure the porosity of rocks. *J. Porous Mater.* **2015**, *23*, 325–338. [CrossRef]
50. Ye, N.; Zhang, S.; Qing, H.; Li, Y.; Huang, Q.; Liu, D. Dolomitization and its impact on porosity development and preservation in the deeply burial Lower Ordovician carbonate rocks of Tarim Basin, NW China. *J. Pet. Sci. Eng.* **2019**, *182*, 106303. [CrossRef]
51. Chen, L.; Zhang, H.; Cai, Z.; Hao, F.; Xue, Y.; Zhao, W. Petrographic, mineralogical and geochemical constraints on the fluid origin and multistage karstification of the Middle-Lower Ordovician carbonate reservoir, NW Tarim Basin, China. *J. Petrol. Sci. Eng.* **2022**, *208*, 109561. [CrossRef]
52. Liu, P.X.; Deng, S.B.; Guan, P.; Jin, Y.Q.; Wang, K.; Chen, Y.Q. The nature, type, and origin of diagenetic fluids and their control on the evolving porosity of the Lower Cambrian Xiaoerbulak Formation dolostone, northwestern Tarim Basin, China. *Pet. Sci.* **2020**, *17*, 873–895. [CrossRef]
53. Zhang, D.; Bao, Z.; Pan, W.; Hao, Y.; Cheng, Y.; Wang, J.; Zhang, Y.; Lai, H. Characteristics and forming mechanisms of evaporite platform dolomite reservoir in Middle Cambrian of Xiaoerbulake section, Tarim Basin. *Nat. Gas Geosci.* **2014**, *25*, 498–507.
54. Deng, S.; Li, H.; Zhang, Z.; Zhang, J.; Yang, X. Structural characterization of intracratonic strike-slip faults in the central Tarim Basin. *AAPG Bull.* **2019**, *103*, 109–137. [CrossRef]
55. Shen, A.; Zheng, J.; Chen, Y.; Ni, X.; Huang, L. Characteristics, origin and distribution of dolomite reservoirs in Lower-Middle Cambrian, Tarim Basin, NW China. *Pet. Explor. Dev.* **2016**, *43*, 375–385. [CrossRef]
56. Zheng, J.P.; Wenqing, P.; Anjiang, S. Reservoir geological modeling and significance of Cambrian Xiaoerblak Formation in Keping outcrop area, Tarim Basin, NW China. *Pet. Explor. Dev.* **2019**, *47*, 392–402. [CrossRef]
57. Zhang, J.; Song, A. Application of Wavelet Analysis in Sequence Stratigraphic Division of Glutenite Sediments. In Proceedings of the 2010 International Conference on Challenges in Environmental Science and Computer Engineering, Wuhan, China, 6–7 March 2010.
58. Kadkhodaie, A.; Rezaee, R. Intelligent sequence stratigraphy through a wavelet-based decomposition of well log data. *J. Nat. Gas Sci. Eng.* **2017**, *40*, 38–50. [CrossRef]
59. Allshorn, S.L.; Dawe, R.A.; Grattoni, C.A. Implication of heterogeneities on core porosity measurements. *J. Pet. Sci. Eng.* **2019**, *174*, 486–494. [CrossRef]
60. Ostoja-Starzewski, M. Material spatial randomness: From statistical to representative volume element. *Probabilistic Eng. Mech.* **2006**, *21*, 112–132. [CrossRef]
61. Gitman, I.M.; Askes, H.; Sluys, L.J. Representative volume: Existence and size determination. *Eng. Fract. Mech.* **2007**, *74*, 2518–2534. [CrossRef]
62. Archie, G.E. Classification of Carbonate Reservoir Rocks and Petrophysical Considerations. *AAPG Bull.* **1952**, *36*, 278–298.
63. Choquette, P.W.; Pray, L.C. Geologic Nomenclature and Classification of Porosity in Sedimentary Carbonates. *AAPG Bull.* **1970**, *54*, 207–250.
64. Lønøy, A. Making sense of carbonate pore systems. *AAPG Bull.* **2006**, *90*, 1381–1405. [CrossRef]
65. Lucia, F.J. Petrophysical Parameters Estimated From Visual Descriptions of Carbonate Rocks: A Field Classification of Carbonate Pore Space. *J. Pet. Technol.* **1983**, *35*, 629–637. [CrossRef]

66. Lucia, F.J. Rock-Fabric/Petrophysical Classification of Carbonate Pore Space for Reservoir Characterization1. *AAPG Bull.* **1995**, *79*, 1275–1300.
67. Lucia, F.J. *Carbonate Reservoir Characterization*; Springer: Berlin/Heidelberg, Germany, 2007. [CrossRef]
68. Luo, P.; Machel, H.G. Pore size and pore throat types in a heterogeneous dolostone reservoir, Devonian Grosmont Formation, Western Canada sedimentary basin. *AAPG Bull.* **1995**, *79*, 1698–1720.
69. Li, B.; Tan, X.; Wang, F.; Lian, P.; Gao, W.; Li, Y. Fracture and vug characterization and carbonate rock type automatic classification using X-ray CT images. *J. Pet. Sci. Eng.* **2017**, *153*, 88–96. [CrossRef]
70. Archilla, N.L.; Missagia, R.M.; Hollis, C.; de Ceia, M.A.R.; McDonald, S.A.; Lima Neto, I.A.; Eastwood, D.S.; Lee, P. Permeability and acoustic velocity controlling factors determined from x-ray tomography images of carbonate rocks. *AAPG Bull.* **2016**, *100*, 1289–1309. [CrossRef]
71. Oliveira, G.L.P.; Ceia, M.A.R.; Missagia, R.M.; Lima Neto, I.; Santos, V.H.; Paranhos, R. Core plug and 2D/3D-image integrated analysis for improving permeability estimation based on the differences between micro- and macroporosity in Middle East carbonate rocks. *J. Pet. Sci. Eng.* **2020**, *193*, 107335. [CrossRef]

Article

Quantitative Evaluation of Water-Flooded Zone in a Sandstone Reservoir with Complex Porosity–Permeability Relationship Based on J-Function Classification: A Case Study of Kalamkas Oilfield

Xuanran Li [1], Jingcai Wang [1], Dingding Zhao [2,3,*], Jun Ni [1], Yaping Lin [1], Angang Zhang [1], Lun Zhao [1] and Yuming Liu [2,3,*]

[1] PetroChina Research Institute of Petroleum Exploration & Development, Beijing 100083, China
[2] State Key Laboratory of Petroleum Resources and Prospecting, China University of Petroleum-Beijing, Beijing 102249, China
[3] College of Geosciences, China University of Petroleum-Beijing, Beijing 102249, China
* Correspondence: 2019310048@student.cup.edu.cn (D.Z.); liuym@cup.edu.cn (Y.L.)

Citation: Li, X.; Wang, J.; Zhao, D.; Ni, J.; Lin, Y.; Zhang, A.; Zhao, L.; Liu, Y. Quantitative Evaluation of Water-Flooded Zone in a Sandstone Reservoir with Complex Porosity–Permeability Relationship Based on J-Function Classification: A Case Study of Kalamkas Oilfield. *Energies* **2022**, *15*, 7037. https://doi.org/10.3390/en15197037

Academic Editor: Reza Rezaee

Received: 12 August 2022
Accepted: 21 September 2022
Published: 25 September 2022

Abstract: The water-flooded zone in a sandstone reservoir with a complex porosity–permeability relationship is difficult to interpret quantitatively. Taking the P Formation of Kalamkas Oilfield in Kazakhstan as an example, this paper proposed a reservoir classification method that introduces the J-function into the crossplot of resistivity and oil column height to realize the classification of sandstone reservoirs with a complex porosity–permeability relationship. Based on the classification results, the initial resistivity calculation models of classified reservoirs were established. The oil–water seepage experiment was performed for classified reservoirs to measure the lithoelectric parameters and establish the relationship between water production rate and resistivity for these reservoirs, and then water production was quantitatively calculated according to the difference between the inverted initial resistivity and the measured resistivity. The results show that the reservoirs with an unclear porosity–permeability relationship can be classified by applying the J-function corresponding to grouped capillary pressure curves to the crossplot of oil column height and resistivity, according to the group average principle of capillary pressure curves. This method can solve the problem that difficult reservoir classification caused by a weak porosity–permeability correlation. Moreover, based on the results of reservoir classification, the water production rate and resistivity model of classified reservoirs is established. In this way, the accuracy of quantitative interpretation of the water-flooded zone in the reservoir can be greatly improved.

Keywords: complex porosity–permeability relationship; water-flooded zone; oil column height; reservoir classification; capillary pressure; resistivity; water production rate

1. Introduction

Kalamkas Oilfield is a typical high water-cut layered sandstone reservoir, which has been developed for 42 years, in North Ustyurt Basin, Kazakhstan. The water cut has reached 94%, while only 25% of the original oil in place (OOIP) has been recovered. Most wells suffered from water-out, which severely affected the enhanced oil recovery (EOR) [1–3]. The target strata in the Kalamkas Oilfield are characterized by complex lithology (including coarse sandstone, silty-fine sandstone, and argillaceous sandstone) and diverse reservoir fluids, consisting of gas zone, low-resistivity oil zone, normal oil zone, and oil–water zone, as well as water-flooded zone and water zone. The sandstone reservoir is highly heterogeneous, with the porosity and permeability not clearly correlated, making the permeability calculation and reservoir classification very challenging [4]. Currently, most calculations of water cuts in water-flooded zones are less precise, or quantitative classifications of water-flooded zones involve too many procedures in a long period, so they are far behind the field applications [5,6]. Accordingly, a quantitative evaluation of the water-flooding intensity of such high water-cut reservoirs with complex porosity–permeability relationship is fundamental for predicting and recovering the remaining oil in many similar high water-cut sandstone oilfields.

A reasonable reservoir classification is crucial to the evaluation of water-flooded zones in reservoirs with weak porosity–permeability relationships and strong heterogeneity [7–10]. There are mainly three reservoir classification methods [11–15]: (1) porosity and permeability, the key physical properties of reservoirs, are taken as the main parameters to classify the reservoirs as, for instance, high-porosity and high-permeability reservoirs, medium-porosity and medium-permeability reservoirs, and low-porosity and low-permeability reservoirs; (2) porosity and permeability are combined with microscopic parameters (e.g., pore structure) to classify the reservoirs through probability statistics of numerous physical property parameters, which is a multivariate evaluation method [16]; and (3) the concept of flow unit is followed, that is, the identical flow units have similar physical features and flow capacity, generally leading to similar water-flooding and remaining oil distribution characteristics. The first and third methods are essentially rooted in the function of porosity and permeability, but they cannot work well when there is no clear correlation between the two parameters. The second method yields relatively low-accurate results in reservoirs with strong heterogeneity and complex pore structure and requires a vast amount of data that can reflect the pore structure, such as grain size, pore throat radius, and sorting.

In recent years, with the development of computer technology, the combination of well logging curves and artificial intelligence methods has also been widely used to identify water-flooded zones. These methods include: the fuzzy neural network method, general neural network method, hybrid computing neural network method, and integrated classifier method [17–20]. In these methods, the fuzzy neural network needs multiple factors to make a comprehensive judgment, and too many input factors are likely to limit application; the general neural network has low convergence speed, and it is easy to fall into local optimal solution; the hybrid computing neural network method has higher requirements for the original data as a whole. The process neural network introduces the original form of the well logging curve as the sample input, which has improved the recognition efficiency to a certain extent. However, the interpolation fitting will produce fitting errors, resulting in large cumulative errors. Due to some limitations of the above methods, there are still problems of lower recognition accuracy and efficiency [21–23].

With P Formation in Kalamkas Oilfield as an example, based on the knowledge that reservoir resistivity is the comprehensive reflection of pore structure and oil column height, the resistivity vs. oil column height crossplot was established, and the J-function of capillary pressure curves was established and incorporated into the resistivity vs. oil column height crossplot. On this basis, the reservoirs with the complex porosity–permeability relationship were classified. For each class of reservoirs, the initial resistivities under different oil column heights were inverted by using the fitted relationships of multiple parameters; the oil–water flow experiment was performed to determine the oil–water relative permeability, which was then combined with the Archie formula to build the water cut and resistivity model for dividing the water-flooding levels of reservoirs. The initial and current resistivities were compared to fix the decline of resistivity, by which the water-flooding intensity of reservoirs was quantitatively evaluated. This paper proposes a simple method for reservoir classification by using only conventional well logging data. In particular, this method has a good application effect in the reservoirs with the worse porosity–permeability relationship. It requires a large number of capillary pressure test data and is not applicable to the oilfields with few coring data and incomplete capillary pressure tests.

2. Regional Geology

The Middle Jurassic P Formation, the major producing system in Kalamkas Oilfield, is a layered, unsaturated, stratigraphically-unconformable reservoir with a gas cap and edge water. The P Formation pay zones are mainly composed of fine sandstone and siltstone, with a high shale content (20–35%). The core porosity experiments show that the reservoir has a medium-high porosity (avg. 28.6%) and a medium-high permeability (avg. 357.4 mD), and it is a kind of clastic rock reservoir common in Central Asia. The P Formation was initially developed by water injection in 1980. Currently, it is in the stage of development with a high water cut (93%), with 20.4% of geological reserves recovered, and daily oil production of 0.5–55 t, and a water cut of 0–98% for new wells, indicating greatly different water flooding degrees in the reservoir. A large number of core analysis and production data show that the P Formation sandstone reservoir in Kalamkas Oilfield is complex and diverse in pore structure, obviously different in reservoir quality [24,25], and very strong in heterogeneity. The core grain size analysis reveals (Figure 1) that the reservoir rocks contain a generally high content of fine particles, of which, 37.4% exhibit the components with grain size less than 0.01 mm, and which are mainly clay mineral particles, except for a small part of fine silts. According to the porosity–permeability relationship, the lithology and pore structure are complex, and the correlation between porosity and permeability is very weak (Figure 2). After long-term water flooding, fine particles, such as clay minerals, block the port throats, thereby aggravating the reservoir heterogeneity, so the water flooding law is very complex [26,27]. With the further development, the quantitative research on water-flooded zone is particularly important.

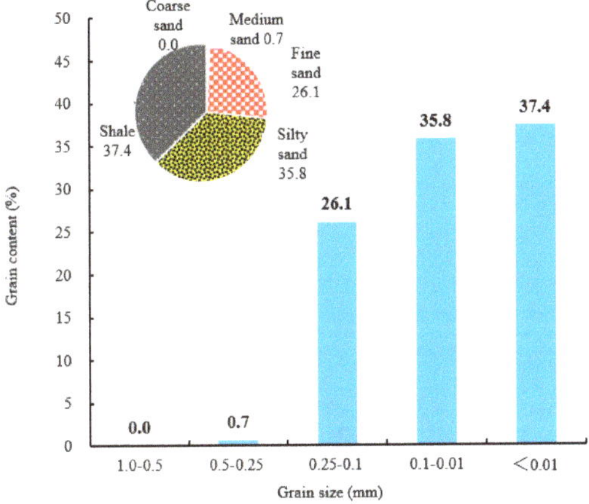

Figure 1. Grain size analysis of P Formation in Kalamkas Oilfield.

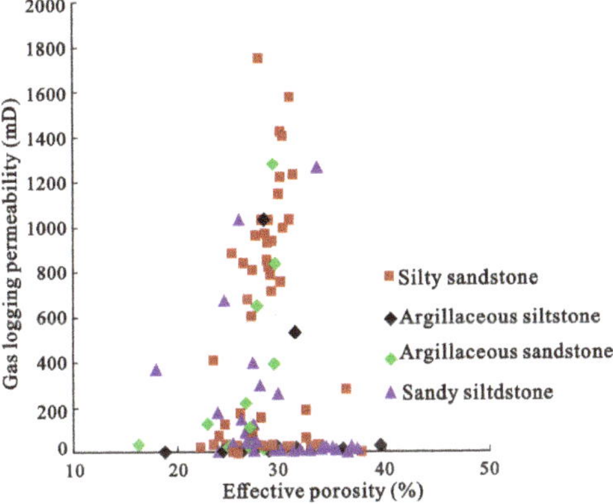

Figure 2. Porosity–permeability relationship of P Formation in Kalamkas Oilfield.

3. Research Method and Data Source

3.1. Research Method

The research is completed in five steps (Figure 3).

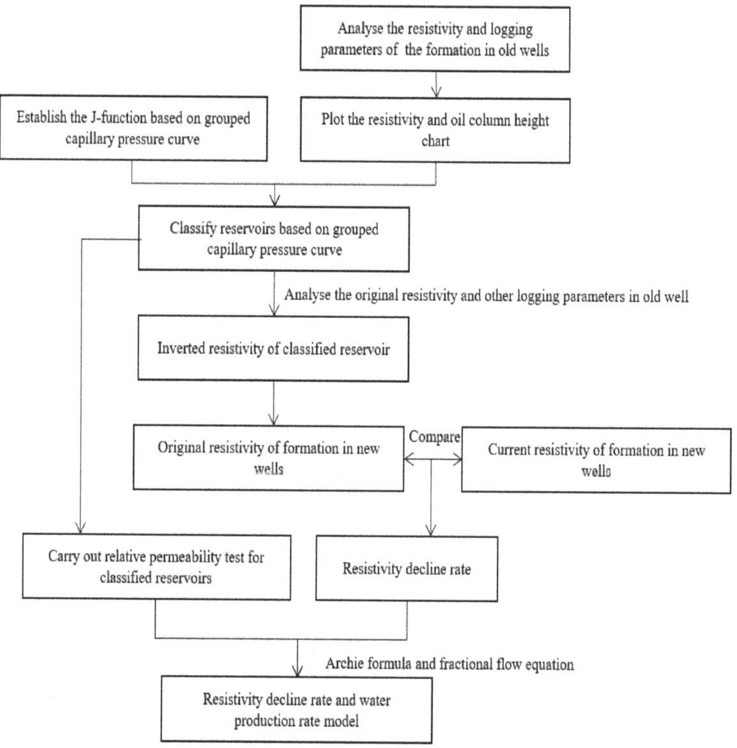

Figure 3. The quantitative evaluation process of the water-flooded zone in sandstone reservoir with complex porosity–permeability relationship.

Step 1: Group the capillary pressure curves and calculate the average value of each group (J-function). Capillary pressure curves were obtained on a CPPP-300 group capillarimeter using the semi-permeable membrane method.

Step 2: Collect the logging data and oil test data of existing old wells, and establish the crossplot of oil column height and resistivity; incorporate the grouped capillary pressure curves to the crossplot, and classify the reservoirs with the complex porosity–permeability relationship.

Step 3: Collect the conventional logging data of existing old wells, analyze the laws of initial reservoir resistivity, oil column height, and natural gamma, and establish the expression of the relation between initial reservoir resistivity, oil column height, and natural gamma by using multiple regression methods.

Step 4: Obtain the initial resistivity of each reservoir by using the relational expression in the above step for newly drilled production wells, and determine the resistivity decline rate from the current resistivity obtained from new well logging and the calculated initial resistivity.

Step 5: Perform a relative permeability test for each class of reservoirs, and establish the water cut and resistivity decline rate model of classified reservoirs by using the Archie formula and fractional flow equation, to accurately predict the water-flooding degree of each reservoir of the new well.

3.2. Data Source

This study used the capillary pressure test data of 33 samples of P Formation in Kalamkas Oilfield (Table 1). It is found that the irreducible water saturation is 11–41%, the saturated median pressure is 0.02–0.35 MPa, the gas logging porosity is 21–37%, and the gas logging permeability is $(1.88–1140) \times 10^{-3} \ \mu m^2$. The grain size analysis was made on 161 samples from 10 coring wells, basically covering all target horizons, with a grain density of 2.47–2.96 g/cm^3 and a rock density of 1.67–2.19 g/cm^3 (Table 2). The reservoir classification was completed with a logging interpretation data of 502 wells and a resistivity of 2.3–40 ohm (avg. 6.52 ohm). It was determined that the porosity is 23–38% (avg. 27.5%), the permeability is 4.8–3520 mD (avg. 425 mD), and the shale content is 3–45% (avg. 18%) (Table 3).

Table 1. Capillary pressure test data of P Formation core samples.

Core No.	Core Length/cm	Core Diameter/cm	Porosity (Water)/%	Porosity (Gas)/%	Gas Logging Permeability/ $10^{-3} \ \mu m^2$	RQI	Saturated Median Pressure/Mpa	Minimum Wet Phase Saturation (Irreducible Water Saturation)/%
486	5.3	3.8	25.6	25.7	18.4	8.5	0.1	25.0
488	5.4	3.4	25.1	25.1	1.9	2.7	0.1	33.0
489	5.2	3.7	21.0	21.1	2.7	3.6	0.4	41.0
491	5.5	3.8	22.8	22.9	5.3	4.8	0.3	11.0
495	5.2	3.5	26.6	26.8	15.0	7.5	0.2	30.0
41	5.2	3.8	25.9	25.9	12.3	6.9	0.2	37.0
496	5.0	3.5	29.7	29.7	51.6	13.2	0.2	25.0
497	5.1	3.8	30.7	30.8	53.6	13.2	0.2	24.0
492	5.2	3.6	29.7	29.8	81.4	16.5	0.2	26.0
493	5.5	3.1	30.1	30.2	60.4	14.2	0.2	26.0
494	5.5	3.3	29.8	30.1	58.5	13.9	0.1	22.0
487	5.2	3.5	29.5	29.6	36.8	11.2	0.1	26.0
503	5.2	3.8	33.6	33.8	533.8	39.7	0.1	21.0
513	4.9	3.5	32.9	33.2	475.4	37.9	0.2	22.0
514	5.2	3.2	32.1	32.1	404.7	35.5	0.1	21.0
515	5.3	3.4	32.0	32.1	309.7	31.1	0.1	20.0
516	5.5	3.5	32.3	32.4	482.2	38.6	0.1	21.0
517	5.6	3.5	30.8	30.8	345.2	33.5	0.2	24.0
527	5.3	3.8	33.0	33.1	448.4	36.8	0.1	21.0
528	5.3	3.8	32.6	32.7	414.6	35.6	0.1	20.0
518	5.2	3.8	32.2	32.2	1020.0	56.3	0.2	21.0
519	5.2	3.8	34.1	34.2	845.4	49.7	0.1	17.0
520	5.4	3.7	30.6	30.7	1140.0	60.9	0.1	18.0
521	5.0	3.8	30.8	30.8	721.5	48.4	0.2	23.0
522	5.3	3.8	32.2	32.3	1120.0	58.9	0.1	18.0
523	5.2	3.7	30.9	31.0	758.5	49.5	0.1	21.0
524	5.1	3.5	32.6	32.7	1070.0	57.2	0.1	16.0
525	4.8	3.2	33.5	33.5	880.9	51.3	0.1	18.0
526	5.3	3.1	31.0	31.1	951.4	55.3	0.1	19.0
3	5.2	3.6	36.9	36.9	821.9	47.2	0.0	18.0
6	5.7	3.2	30.8	30.8	394.7	35.8	0.0	22.0
23	4.9	3.1	35.1	35.1	654.9	43.2	0.2	19.0
28	5.7	3.3	35.8	35.8	522.4	38.2	0.2	19.0

Table 2. Grain size analysis results of P Formation.

Formation	>1	1.0–0.5	0.5–0.25	0.25–0.1	0.1–0.01	<0.01
		Coarse sand	Medium sand	Fine sand	Silty sand	Mud
P	0.00	0.01	0.66	26.13	35.81	37.38

Table 3. Partial logging interpretation data of 502 wells in the P Formation.

Well Name	Well Type	Start Depth (m)	End Depth (m)	GZ3 Average	Oil Column Height	GR Relative Value	Inverted Resistivity	Class of Reservoir	GR	VSH (V/V)	SW (V/V)	PERM
XX84	Production well	732.8	738.8	9.0	111.5	83.8	5.4	III	155.0	0.1	0.3	328.0
XX57	Production well	742.1	744.6	7.6	104.4	81.3	5.2	III	130.0	0.1	0.4	444.5
XX24	Production well	754.5	757.3	6.9	92.9	82.4	4.8	III	136.0	0.2	0.1	310.5
XX67	Production well	740.9	745.8	6.8	103.3	80.0	5.2	III	124.0	0.2	0.3	160.3
XX72	Production well	745.3	747.1	6.2	101.8	85.7	5.0	III	150.0	0.2	0.3	11.6
XX63	Production well	744.3	749.1	19.5	101.2	73.8	11.4	II	104.0	0.1	0.2	283.9
XX73	Production well	740.4	748.4	18.3	101.6	81.5	10.6	II	106.0	0.1	0.2	138.9
XX46	Water injection well	745.7	751.4	15.6	97.5	80.0	10.4	II	124.0	0.2	0.2	301.9
XX39	Production well	760.3	764.8	15.5	83.9	77.9	9.6	II	113.0	0.2	0.2	290.9
XX15	Production well	763.9	767.7	14.6	82.4	80.6	9.1	II	141.0	0.2	0.2	167.5
XX54	Production well	807.4	811.0	28.3	37.6	81.5	12.2	I	137.0	0.1	0.1	1537.5
XX72	Water injection well	773.3	782.2	27.9	68.2	76.5	17.9	I	146.8	0.2	0.2	229.3
XX21	Production well	812.6	814.5	27.1	36.4	70.3	14.8	I	116.0	0.1	0.2	206.8
XX54	Production well	774.5	784.6	26.3	65.4	81.2	16.3	I	134.0	0.1	0.1	317.2
XX76	Production well	747.7	758.2	24.6	88.9	90.0	17.6	I	162.0	0.1	0.1	482.5

4. Result

4.1. Logging Responses of Water-Flooded Zone

After long-term water injection development, the water cut of the sandstone reservoir increases continuously. After entering the reservoir, the injected water interacts with the reservoir, changing the fluid properties, pore structure, rock physicochemical properties, and oil–water distribution of the reservoir to a certain extent. This change will cause the variation of logging curves. Determining the logging responses is fundamental for locating the water-flooded point and confirming the water-flooding degree [28–30].

4.1.1. Resistivity Logging

Resistivity is an important parameter reflecting the fluid properties of the reservoir. The P Formation has been developed for 41 years by reinjecting the waste water. The salinity of injected water is close to the initial salinity of the formation, being about 105,000–150,000 mg/L. The water saturation during water flooding is a gradual process, which can be roughly divided into the early stage with low water cut, the middle stage with medium water cut, and the late stage with high water cut. In the early stage with low water cut, injected water displaces the movable oil in the reservoir and exchanges ions with the initial formation water in the swept zone. Since the P Formation reservoir is developed by waste water reinjection, and the initial formation water is close to the injected water in salinity, the reservoir fluid can reach dynamic balance very quickly, and the reservoir resistivity decreases with the increase in water saturation Sw. At this time, there is no water at the outlet end. With the progress of development, the oilfield enters the middle stage with a medium water cut, resulting in a water breakthrough at the outlet end. The injected water continues to drive out the movable oil and further mixes with the liquid mixture in the swept zone. In this process, the resistivity of the liquid mixture changes greatly. In the late stage with high water cut, the reservoir is completely flooded, the injected water can only drive out a small amount of oil, and the resistivity of the liquid mixture almost reaches the resistivity of the injected water [28,29]. The resistivity generally decreases with the increase in water saturation. The resistivity of the water-flooded zone in the P Formation shows an obvious downward trend. When water flooding is serious, the resistivity of the water-flooded zone is very close to that of the water zone (Figure 4).

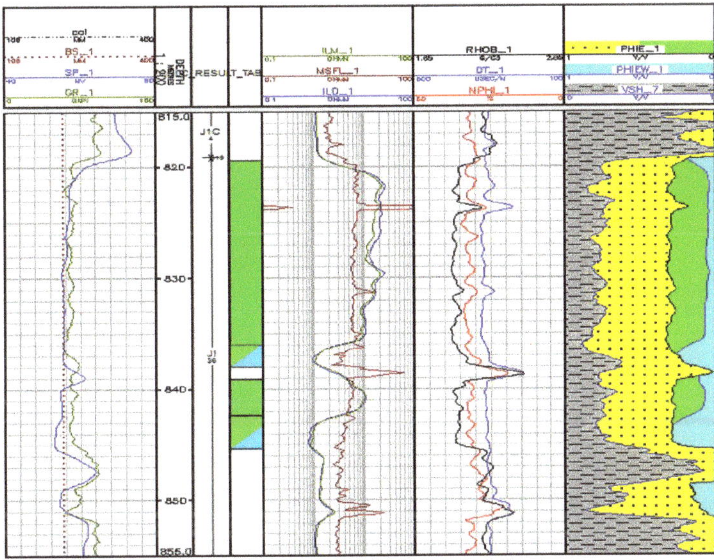

Figure 4. Interpretation of water-flooded zone of Well xx36 in P Formation.

4.1.2. Spontaneous Potential Logging

The spontaneous potential (*SP*) logging response was analyzed to locate the water-flooded point in the reservoir. According to the previous study [24], in the interpretation of water-flooded zones in the sandstone reservoir, the comprehensive influence of injected water and oil saturation on *SP* can be expressed by Formula (1) below. Since the salinity of the injected water in the P Formation is close to that of the initial formation water, the baseline shift of *SP* and the amplitude change of *SP* are not obvious after water flooding of some reservoirs (Figure 4).

$$SP = -K_C \times \log \frac{R_{mf}}{R_w} + K \times \log \frac{R'_w}{R_w} = -K_C \times \log \frac{R_{mf}}{R_w} - K \times \log S_w \tag{1}$$

where the first term represents the influence of injected water on spontaneous potential (*SP*), and the second term represents the influence of oil saturation on *SP*. The change of *SP* log in water flooding results from the stacking of these two processes.

4.2. Classification of Sandstone Reservoirs with a Complex Porosity–Permeability Relationship Based on Average Capillary Pressure (J-Function)

When logging data are used to identify water-flooded zones, reservoir classification is an effective way to improve the identification accuracy of water-flooded zones in heterogeneous reservoirs. Reasonable reservoir classification is particularly critical in the interpretation of water-flooded zones in reservoirs with no obvious porosity–permeability relationship. Many scholars classify reservoirs according to the principle that the same flow units have similar physical properties and flow capacity. Wang et al. [14] classified the reservoirs with permeability as the primary parameter and porosity as the second parameter. Yang et al. [10] proposed a multivariate evaluation method combining porosity and permeability with microscopic parameters. Gunter et al. [13] classified the reservoirs according to the concept of flow unit. Essentially, these methods are based on the function of porosity and permeability, but are limited for reservoirs with a very unclear porosity–permeability relationship. According to the conventional core analysis, capillary pressure curve shape, logging responses, and analysis of logging facies and lithofacies, the capillary pressure curves of Kalamkas Oilfield are divided into three groups (Figure 5): Type 1, Type 2, and Type 3. Type 1 capillary pressure curves are the longest in the middle gentle segment and the lowest in position, and represent the reservoirs with the best sorting and the largest throat radius, being the reservoirs with the best physical properties. Type 2 capillary pressure curves have a slightly shorter middle gentle segment and higher position than Type 1, and represent the reservoirs with moderate physical properties. Type 3 capillary pressure curves have the shortest and highest gentle segment, and represent the reservoirs with the worst physical properties. Accordingly, the average value of each group of capillary pressure curves (J-function) was obtained. Figure 6 shows the J-function of the P Formation in Kalamkas Oilfield. It can be seen that the data points are concentrated, indicating that the grouping of capillary pressure curves is reasonable.

Oil column height, porosity, pore connectivity, and oil–water density difference are the main factors affecting the initial resistivity of the reservoir. The higher the oil column height, the better the pore structure, the stronger the hydrocarbon charging capacity, and the higher the resistivity [30–33]. Accordingly, the crossplot of resistivity and oil column height can be established to characterize the pore structure and reflects the physical properties gradually deteriorating from the data point to the right. According to the Archie formula [33], there is a direct relationship between reservoir resistivity and water saturation. According to the concept of capillary force, the capillary pressure is directly proportional to the rising height of the wetting phase in the capillary. Therefore, the height of the oil column in the reservoir is a direct reflection of the value of capillary pressure. According to the above principle, the crossplot of resistivity and oil column height can be combined with the capillary pressure curves. According to the capillary pressure measured in the laboratory test of cores, the J-function (average capillary pressure) of three classes of reservoirs (good, moderate, and poor) was established, as shown in Figures 5 and 6. Based on the J-function of existing coring wells, the continuous capillary pressure curves can be reconstructed for the non-coring intervals [34,35]. The reconstructed multiple capillary pressure curves are applied to the crossplot. The reservoirs of the P Formation can be classified as I, II, and III (Figure 7), corresponding, respectively, to good, moderate, and poor porosity and permeability, which are arranged in turn from left to right on the crossplot. This classification can eliminate the influences of permeability and porosity and realize the classification of highly heterogeneous reservoirs with no obvious porosity–permeability relationship.

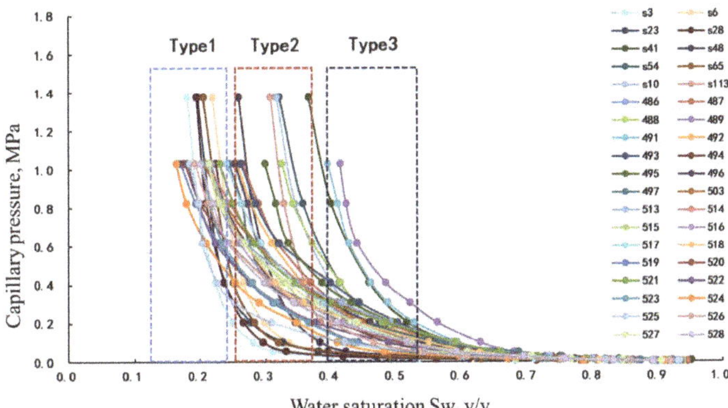

Figure 5. Grouping of capillary pressure curves.

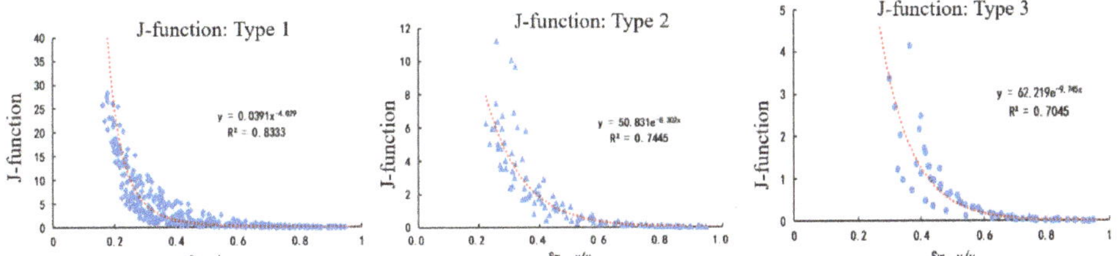

Figure 6. J-functions of grouped capillary pressure curves.

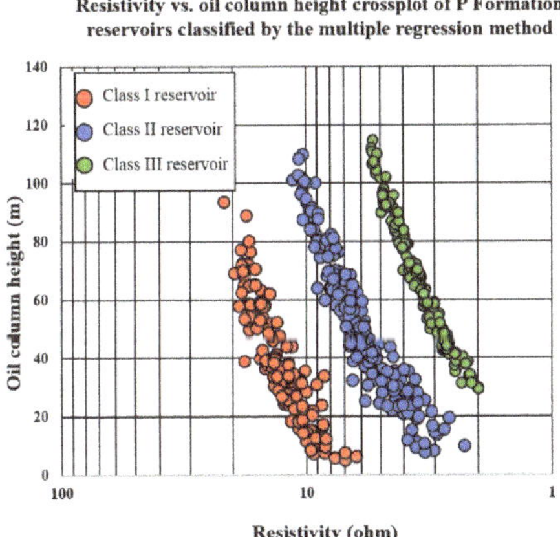

Figure 7. Crossplots of resistivity and oil column height of P Formation reservoirs.

According to the analysis of logging interpretation results of Classes I, II, and III reservoirs (Table 4), Class I reservoirs, with the porosity of more than 28% and the relative GR value of less than 76%, are mainly composed of fine sandstone, with a small amount of medium and coarse sandstone, and contain the sedimentary facies dominated by underwater distributary channel of delta front; Class II reservoirs, with the porosity of 22–28% and the relative GR value of 76–85%, are mainly composed of siltstone, and contain both distributary channel and estuarine bar; and Class III reservoirs, with the porosity of less than 22% and the relative GR value of greater than 85%, are mainly composed of argillaceous sandstone, and contain a distal bar and bar margin deposits.

Table 4. Logging interpretation results of reservoirs.

Class of Reservoirs	Porosity (%)	Permeability (mD)	Relative GR Value (%)	Shale Content (%)	Lithology
I	>28	>600	<76	<21	Mainly sandstone and fine sandstone
II	22~28	30~600	76~85	21~32	Mainly siltstone
III	<22	<30	>85	>32	Mainly argillaceous sandstone

4.3. Initial Resistivity Inversion and Water Production Rate Calculation

The resistivity is sensitive to the water flooding degree. The change of resistivity is the main parameter to identify the water-flooded zones. The water-flooded zones exhibit the decline of resistivity to different degrees, while the non-flooded zones basically do not have a change in resistivity. Therefore, the water-flooded zone can be identified according to the difference between the initial resistivity and the current resistivity.

4.3.1. Initial Resistivity Inversion of Classified Reservoirs

Archie [33] discussed the relationship between resistivity, water saturation, and porosity under the condition that the rock skeleton is not conducive. Generally, the larger the porosity and the better the pore connectivity, the stronger the oil charging capacity. When the oil saturation and formation water resistivity are higher, the resistivity of the reservoir is higher and the oil column height is larger. When the oil saturation and formation water resistivity are lower, the resistivity of the reservoir is lower and the oil column height is smaller. Through the analysis of the correlation between resistivity, oil column height, relative GR value, and shale content of P Formation reservoirs in old wells, it is found that the correlation between relative GR value, oil column height, and resistivity of the reservoir is good (Figure 8), which shows that oil column height and pore structure are the key parameters to control reservoir resistivity. For the classified reservoirs, the initial resistivity was calculated by multiple regression of oil column height and shale content. Based on the data of 141 old wells (these representative data basically cover all target horizons both vertically and horizontally), multiple regression was conducted for the classified reservoirs with the formulas as follows:

$$\text{Class I: reservoirs: } R_i = 26.6858 + H \times 0.145736 - DGR \times 0.2447 \tag{2}$$

$$\text{Class II: reservoirs: } R_i = 11.75654 + H \times 0.078969 - DGR \times 0.11312 \tag{3}$$

$$\text{Class III: reservoirs: } R_i = 3.53265 + H \times 0.039465 - DGR \times 0.02916 \tag{4}$$

$$DGR = 100 \times (GR/GR_{max}) \tag{5}$$

where R_i is the inverted initial resistivity of the reservoir, ohm; H is the oil column height, m; DGR is the relative GR value, %; GR is the measured natural gamma value, API; and GR_{max} is the value of mudstone marker layer.

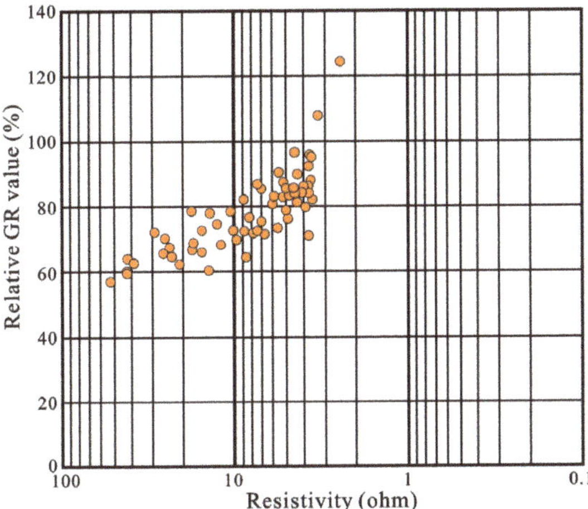

Figure 8. Crossplots of resistivity and relative GR value of P Formation reservoirs.

4.3.2. Calculation of Water Production Rate of Water-Flooded Zone

Relative permeability is one of the basic parameters for analyzing the multiphase flow in a reservoir. The water production rate can be calculated directly by using the relative permeability [32]. The oil–water relative permeability ratio in the relative permeability curve is relative to the water saturation. The relationship between resistivity and water saturation can be obtained through the Archie formula. Thus, the relationship between resistivity and water production rate can be established. In particular, m, n, a, and b in the Archie formula are determined from core litho-electric experiment and R_w from water analysis data statistics. In this study, $a = 1.1$, $b = 1$, $m = 1.77$, $n = 1.9$, formation water resistivity $R_w = 0.05$ ohm.

$$F_w = Q_w / (Q_0 + Q_w) = 1 / \left(1 + B \times \frac{k_{ro}}{k_{rw}} \times \frac{\mu_w}{\mu_o} \right) \tag{6}$$

where K_{ro} is the oil relative permeability, 10^{-3} µm²; K_{rw} is the water relative permeability, 10^{-3} µm²; and μ_w / μ_o is the viscosity ratio of water to oil.

$$S_w = \left[\frac{abR_w}{R_t \Phi^m} \right]^{\frac{1}{n}} \tag{7}$$

where R_t is the resistivity of undisturbed formation, Ω·m; S_w is water saturation, decimal; a, b, m, and n are lithology coefficient, cementation index, and saturation index in litho-electric parameters, respectively; Φ is the effective porosity, decimal; and R_w is the formation water resistivity, Ω·m.

As classified, the reservoirs with the porosity $\varnothing < 23$ are Class III reservoirs, the reservoir with the porosity of $23 < \varnothing < 27$ is a Class II reservoir, and the reservoir with the porosity $\varnothing > 27$ is a Class I reservoir. The relative permeability test was carried out for each class of reservoirs (Figure 9). In Figure 9a, the oil–water two-phase flow area is wide, the endpoint permeability is high, and the irreducible water saturation is low, indicative of reservoirs with large and well-connected pores. In Figure 9b, the oil–water two-phase flow area is narrower, the endpoint permeability is lower, and the irreducible water saturation is higher than that in Figure 9a, indicative of reservoirs with relatively poor physical properties, and small but moderately-connected pores. In Figure 9c, the oil–water two-phase flow area is narrow and the endpoint permeability is low, indicative of sandstone reservoirs with high shale content and poor connectivity. According to the Specification for Logging Data Processing and Interpreting of Water-flooded Zone (SY/T 6178-2017) [31], the water cut of the water-flooded zone can be divided into: $f_\text{w} \leq 10\%$ (clastic rocks), $10\% < f_\text{w} \leq 40\%$ (low level), $40\% < f_\text{w} \leq 80\%$ (moderate level), $80\% < f_\text{w} < 90\%$ (high level), and $f_\text{w} \geq 90\%$ (ultra-high level); accordingly, by the water-flooding intensity, each class of reservoir can be divided into non-water-flooded (oil zone), weakly water-flooded, moderately water-flooded, and highly water-flooded. Based on the relative permeability test, the relationship between water cut and water saturation is established for each class of reservoirs, and the relationship between water saturation and resistivity (or resistivity decline rate or RDR) under the experimental conditions is calculated from the Archie formula. Therefore, the relational expression between water cut and RDR for each class of reservoirs can be obtained (Figure 10a–c), and then the quantitative evaluation model of water-flooded zones can be built. According to Formulas (2)–(5), the initial resistivity can be determined from the basic logging parameters. Then, the difference between the initial resistivity and the measured resistivity after water flooding is obtained, and the difference (i.e., RDR) is brought into the relational expression between RDR and water cut (Figure 10), so as to realize the quantitative evaluation of water-flooded zones. In Class I reservoirs, those with RDR > 80% are extremely highly water-flooded zones, those with 69% < RDR < 80% are highly water-flooded zones, those with 41% < RDR < 60% are moderately water-flooded zones, and those with RDR < 41% are weakly water-flooded or oil zones. In Class II reservoirs, those with RDR > 68% are extremely highly water-flooded zones, those with 62% < RDR < 68% are highly water-flooded zones, those with 43% < RDR < 62% are moderately water-flooded zones, and those with RDR < 43% are weakly water-flooded or oil zones. In Class III reservoirs, those with RDR > 57% are extremely highly water-flooded zones, those with 49% < RDR < 57% are highly water-flooded zones, those with 37% < RDR < 49% are moderately water-flooded zones, and those with RDR < 37% are weakly water-flooded or oil zones.

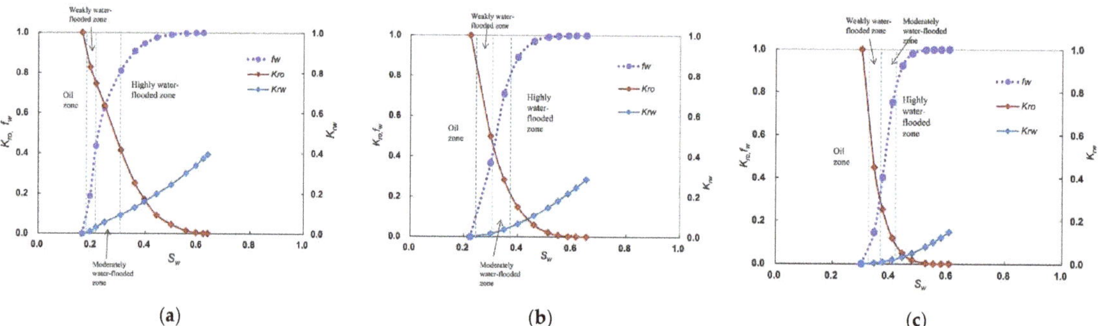

Figure 9. Oil–water relative permeability of classified reservoirs. (**a**) Class I reservoirs, (**b**) Class II reservoirs, and (**c**) Class III reservoirs.

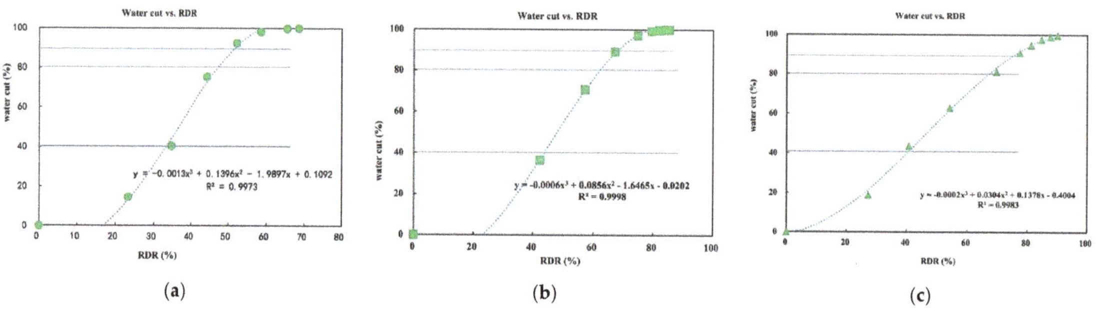

Figure 10. Relationship between resistivity decline rate (RDR) and water production rate of classified reservoirs. (**a**) Class I reservoirs, (**b**) Class II reservoirs, and (**c**) Class III reservoirs.

According to the above classification criteria and the latest model established, water-flooded zones in 177 reservoirs of 37 new wells in the P Formation of Kalamkas Oilfield were identified. It is found that Class I reservoirs account for about 54%, Class II reservoirs account for about 39%, and Class III reservoirs account for about 7%. The remaining oil in the P Formation is mainly distributed in relatively poor Class III and Class II reservoirs (Table 5).

Table 5. Water-flooded zones in classified reservoirs in P Formation, Kalamkas Oilfield.

Class of Reservoirs	Proportion	Highly Water-Flooded	Moderately Water-Flooded	Weakly Water-Flooded	Non-Water-Flooded
I	54%	60%	23%	7%	9%
II	39%	25%	31%	9%	33%
III	7%	2%	19%	15%	63%

5. Application

According to the research results, water-flooded zones in 315 reservoirs of 75 new wells in the study area from 2019 to 2021 were evaluated, and the coincidence rate (Table 6) of quantitative calculation of water production rate of water-flooded zones is as high as 91%. The interpretation results of some wells are inconsistent with the actual production data, which is believed to attribute to the fact that the error of oil column height is amplified in the zone close to the oil–water contact, so the calculation error of initial resistivity of oil zone is large to affect the calculation accuracy of water cut.

Table 6. Interpretation results of some new wells in the P Formation.

Well	Depth_T	Depth_B	Class of Reservoir	Regression Resistivity	Calculated Water Production Rate, %	Water-Flooding Level	Well Production Data
XX28	790.9	793.2	II	7.74	32.2	Moderately water-flooded	$f_w = 62.9\%$
XX28	794.7	801.2	II	7.02	85.0	Highly water-flooded	
XX36	800.4	804.5	I	17.94	96.9	Highly water-flooded	
XX36	804.5	806.6	I	18.07	99.5	Highly water-flooded	$f_w = 94.1\%$
XX36	811.0	812.7	I	17.21	92.5	Highly water-flooded	
XX36	810.7	814.0	II	12.31	99.5	Highly water-flooded	
XX46	816.3	817.6	II	6.08	12.2	Weakly water-flooded	
XX46	818.9	824.4	I	9.15	64.0	Moderately water-flooded	
XX46	826.0	826.6	II	9.29	75.5	Moderately water-flooded	$f_w = 57.6\%$
XX46	829.6	838.9	I	12.83	99.2	Highly water-flooded	
XX46	840.5	841.0	II	4.42	67.8	Moderately water-flooded	
XX46	842.0	842.8	III	2.50	−55.3	Non-water-flooded	
XX37	787.2	790.7	II	8.28	89.05	Highly water-flooded	$f_w = 79.6\%$
XX37	791.9	794.4	I	8.05	73.59	Moderately water-flooded	

6. Conclusions

The resistivity is directly related to the water saturation, and the oil column height is the reflection of the capillary pressure. The J-functions of grouped capillary pressure curves are applied to the resistivity vs. oil column height crossplot to realize the classification of reservoirs with an unclear porosity–permeability relationship. Oil column height and GR are the key controls on the initial resistivity of the reservoirs in the study area. The initial resistivity can be reconstructed by using the multiple regression method. The relative permeability test is performed for the reservoirs. For each class of reservoirs, the relationship between water cut and resistivity decline rate is established, and the water-flooding intensity is finely divided. This method can help improve the accuracy of the quantitative evaluation of water-flooded zones.

Field application reveals that the classification of sandstone reservoirs with a complex porosity–permeability relationship and quantitative evaluation of water-flooded zones contribute a coincidence rate of more than 90%, which meets the required interpretation accuracy of water-flooded zones in oilfields. However, this method yields a relatively low accuracy in evaluating the water-flooded zones close to the oil–water contact.

Author Contributions: Conceptualization and methodology, D.Z. and Y.L. (Yuming Liu); investigation, J.N. and A.Z.; data curation, Y.L. (Yaping Lin) and L.Z.; writing—original draft preparation, J.W.; writing—review and editing, X.L. All authors have read and agreed to the published version of the manuscript.

Funding: This work is supported by the Major scientific and technological projects of PetroChina (2021DJ3201).

Institutional Review Board Statement: Not applicable.

Informed Consent Statement: Not applicable.

Data Availability Statement: Not applicable.

Conflicts of Interest: The authors declare no conflict of interest.

References

1. Sagyndikov, M.; Mukhambetov, B.; Orynbasar, Y.; Nurbulatov, A.; Aidarbayev, S. Evaluation of polymer flooding efficiency at brownfield development stage of giant Kalamkas oilfield, western Kazakhstan. In Proceedings of the SPE Annual Caspian Technical Conference and Exhibition, Astana, Kazakhstan, 31 October–2 November 2018.
2. Muratov, I.M. Structure of the oil and gas Pools of the Jurassic Horizons of the Kalamkas field. *Petrol. Geol. A Dig. Russ. Lit. Pet. Geol.* **1984**, *21*, e511–e512.
3. Li, X.; Jin, R.; Fu, L.; Xu, B.; Zhang, Z. Water-Out Characteristics and Remaining Oil Distribution of Delta Front Reservoir—Take J-2C Reservoir of Kalamkas Oilfield in Kazakhstan as an Example. In Proceedings of the International Field Exploration and Development Conference 2018, Xi'an, China, 18–20 September 2018; Springer: Singapore, 2018; pp. 1527–1535.
4. Ni, J.; Zhao, D.; Liao, X.; Li, X.; Fu, L.; Chen, R.; Xia, Z.; Liu, Y. Sedimentary Architecture Analysis of Deltaic Sand Bodies Using Sequence Stratigraphy and Seismic Sedimentology: A Case Study of Jurassic Deposits in Zhetybay Oilfield, Mangeshrak Basin, Kazakhstan. *Energies* **2022**, *15*, 5306. [CrossRef]
5. Lin, Q. Technology and Practice of stabilizing oil production and controlling water cut in Kalamkas Oilfield in central Asia. *China Oil Gas J.* **2015**, *22*, e48–e54.
6. Tangparitkul, S.; Saul, A.; Leelasukseree, C.; Yusuf, M.; Kalantariasl, A. Fines migration and permeability decline during reservoir depletion coupled with clay swelling due to low-salinity water injection: An analytical study. *J. Pet. Sci. Eng.* **2020**, *194*, 107448. [CrossRef]
7. Wang, X.; Zhang, F.; Li, S.; Dou, L.; Liu, Y.; Ren, X.; Chen, D.; Zhao, W. The Architectural Surfaces Characteristics of Sandy Braided River Reservoirs, Case Study in Gudong Oil Field, China. *Geofluids* **2021**, *2021*, 8821711. [CrossRef]
8. Xu, L.; Zhang, J.; Ding, J.; Liu, T.; Shi, G.; Li, X.; Dang, W.; Cheng, Y.; Guo, R. Evaluation and optimization of flow unit division methods. *Lithol. Reserv.* **2015**, *27*, 74–80.
9. Hearn, C.L.; Ebanks, W., Jr.; Tye, R.S. Geological factors in influencing reservoir performance of the Hartzog Draw Field, Wyoming. *Petrol. Tech.* **1984**, *36*, 1335–1344. [CrossRef]
10. Yang, J.; Fan, T.; Ma, H. On Log Interpretation of water flooded layer with reservoir classification method. *Well Logging Technol.* **2010**, *34*, 238–241.
11. Hui, G.; Wang, Y.; Yan, L.; Li, J.; Liu, L. Method for water-flooded layer evaluation in extra-low permeability reservoir in WY block of AS oilfield flooded by fresh-water. *J. Southwest Pet. Univ. (Sci. Technol. Ed.)* **2016**, *38*, 138.
12. Wang, Z.; He, G. The classification method and application of reservoir flow units. *Nat. Gas Geosci.* **2010**, *21*, 362–366.
13. Gunter, G.W.; Finneran, J.M.; Hartmann, D.J.; Miller, J.D. Early determination of reservoir flow units using an integrated petrophysical method. In Proceedings of the SPE Annual Technical Conference and Exhibition, San Antonio, TX, USA, 5–8 October 1997.
14. Wang, S.; Wu, S.; Cai, F.; Li, Y.N.; Chen, Y.X. Sedimentary Facies of Kumkol Formation of Upper Jurassic in Kumkol Oilfield, South Turgay Basin. *Xi'an Shiyou Univ. (Nat. Sci. Ed.)* **2012**, *27*, 15–20.
15. Chen, C.; Song, X.; Li, J. Dominant flow channels of point-bar reservoirs and their control on the distribution of remaining oils. *Acta Pet. Sin.* **2012**, *33*, 257–263.
16. Huang, H.; Li, R.; Xiong, F.; Hu, H.; Sun, W.; Jiang, Z.; Chen, L.; Wu, L. A method to probe the pore-throat structure of tight reservoirs based on low-field NMR: Insights from a cylindrical pore model. *Mar. Pet. Geol.* **2020**, *117*, 104344. [CrossRef]
17. Pirrone, M.; Battigelli, A.; Ruvo, L. Lithofacies Classification of Thin Layered Reservoirs Through the Integration of Core Data and Dielectric Dispersion Log Measurements. In Proceedings of the SPE Annual Technical Conference and Exhibition, Amsterdam, The Netherlands, 27–29 October 2014.
18. Al-Mudhafar, W.J. Integrating machine learning and data analytics for geostatistical characterization of clastic reservoirs. *J. Pet. Sci. Eng.* **2020**, *195*, 107837. [CrossRef]

19. Lee, S.H.; Datta-Gupta, A. Electrofacies characterization and permeability predictions in carbonate reservoirs: Role of multivariate analysis and nonparametric regression. In Proceedings of the SPE Annual Technical Conference and Exhibition, Houston, TX, USA, 3–6 October 1999.

20. Ameur-Zaimeche, O.; Zeddouri, A.; Heddam, S.; Kechiched, R. Lithofacies prediction in non-cored wells from the Sif Fatima oil field (Berkine basin, southern Algeria): A comparative study of multilayer perceptron neural network and cluster analysis-based approaches. *J. Afr. Earth Sci.* **2020**, *166*, 103826. [CrossRef]

21. Bressan, T.S.; de Souza, M.K.; Girelli, T.J.; Junior, F.C. Evaluation of machine learning methods for lithology classification using geophysical data. *Comput. Geosci.* **2020**, *139*, 104475. [CrossRef]

22. Tang, H.; White, C.; Zeng, X.; Gani, M.; Bhattacharya, J. Comparison of multivariate statistical algorithms for wireline log facies classification. *AAPG Annu. Meet. Abstr.* **2004**, *88*, 13.

23. Ren, X.; Hou, J.; Song, S.; Liu, Y.; Chen, D.; Wang, X.; Dou, L. Lithology identification using well logs: A method by integrating artificial neural networks and sedimentary patterns. *J. Pet. Sci. Eng.* **2019**, *182*, 106336. [CrossRef]

24. Utepov, M.S.; Sarbopeyev, O.K. Improvement of the efficiency of repair and insulation works to limit water product from reservoir, using new technologies in Kalamkas oilfield. *Kazakhstan J. Oil Gas Ind.* **2020**, *2*, 70–78. [CrossRef]

25. Sagyndikov, M.; Satbayev University; Salimgarayev, I.; Ogay, E.; Seright, R.; Kudaibergenov, S. Assessing polyacrylamide solution chemical stability during a polymer flood in the Kalamkas field, Western Kazakhstan. *Bull. Karaganda Univ. Chem. Ser.* **2022**, *105*, 99–112. [CrossRef]

26. Nguyen, C.; Loi, G.; Russell, T.; Shafian, S.M.; Zulkifli, N.; Chee, S.; Razali, N.; Zeinijahromi, A.; Bedrikovetsky, P. Well inflow performance under fines migration during water-cut increase. *Fuel* **2022**, *327*, 124887. [CrossRef]

27. Bedrikovetsky, P.; de Siqueira, F.D.; Furtado, C.A.; Souza, A.L.S. Modified particle detachment model for colloidal transport in porous media. *Transp. Porous Media* **2011**, *86*, 353–383. [CrossRef]

28. Tian, Z.; Mu, L.; Sun, D.; Lv, L.H. Logging attributes and mechanism study of grit water-flooding reservoir. *Acta Pet. Sin.* **2002**, *23*, 50–55.

29. Liu, Y.; Hou, J.; Wang, L.; Xue, J.; Liu, X.; Fu, X. Architecture analysis of braided river reservoir. *J. China Univ. Pet. Ed. Nat. Sci.* **2009**, *30*, 7–11.

30. Zhao, W. Experiment research on the rock resistivity properties of watered-out formation. *Oil Gas Recovery Technol.* **1995**, *2*, 32–39.

31. National Energy Administration. *Specification for Logging Data Processing and Interpreting of Water-Flooded Zone: SY/T 6178-2017[S]*; Petroleum Industry Press: Beijing, China, 2017.

32. Li, C.; Zhu, S. Some topics about water cut rising rule in reservoirs. *Lithol. Reserv.* **2016**, *28*, 1–5.

33. Archie, G.E. The electrical resistivity log as an aid in determining some reservoir characteristics. *Pet. Trans. Aime* **1942**, *146*, 54–61. [CrossRef]

34. Tan, F.; Li, H.; Xu, C.; Li, Q.; Peng, S. Quantitative evaluation methods for water-flooded layers of conglomerate reservoir based on well logging data. *Pet. Sci.* **2010**, *7*, 485–493. [CrossRef]

35. Lin, Y.; Guo, Z.; Sheng, S.; Zheng, J.; Luo, M.; Liang, H. A New Method for Flooded Layer Evaluation by Logging with Strong Heterogeneous in Heavy Oil Reservoirs, K Field, Central Asia. In Proceedings of the International Field Exploration and Development Conference 2018, Xi'an, China, 18–20 September 2018; Springer: Singapore, 2018; pp. 328–339.

Article

High-Resolution Seismic Characterization of Gas Hydrate Reservoir Using Wave-Equation-Based Inversion

Jie Shao [1,2], Yibo Wang [1,2,*], Yanfei Wang [1,2] and Hongyong Yan [1,2]

[1] Key Laboratory of Petroleum Resource Research, Institute of Geology and Geophysics, Chinese Academy of Sciences, Beijing 100029, China
[2] Innovation Academy for Earth Science, Chinese Academy of Sciences, Beijing 100029, China
* Correspondence: wangyibo@mail.iggcas.ac.cn

Abstract: The high-resolution seismic characterization of gas hydrate reservoirs plays an important role in the detection and exploration of gas hydrate. The conventional AVO (amplitude variation with offset) method is based on a linearized Zoeppritz equation and utilizes only the reflected wave for inversion. This reduces the accuracy and resolution of the inversion properties and results in incorrect reservoir interpretation. We have studied a high-resolution wave-equation-based inversion method for gas hydrate reservoirs. The inversion depends on the scattering integral wave equation that describes a nonlinear relationship between the seismic wavefield and the elastic properties of the subsurface medium. In addition to the reflected wave, it considers more wavefields including the multiple scattering and transmission during inversion to improve the subsurface illumination, so as to enhance the accuracy and resolution of the inversion properties. The results of synthetic data from Pearl River Mouth Basin, South China Sea, demonstrate the validity and advantages of the wave-equation-based inversion method. It can effectively improve the resolution of inversion results compared to the conventional AVO method. In addition, it has good performance in the presence of noise, which makes it a promising method for field data.

Keywords: gas hydrate reservoir; wave-equation-based inversion; scattering integral theory; high-resolution

Citation: Shao, J.; Wang, Y.; Wang, Y.; Yan, H. High-Resolution Seismic Characterization of Gas Hydrate Reservoir Using Wave-Equation-Based Inversion. *Energies* **2022**, *15*, 7652. https://doi.org/10.3390/en15207652

Academic Editor: Hossein Hamidi

Received: 6 September 2022
Accepted: 12 October 2022
Published: 17 October 2022

Publisher's Note: MDPI stays neutral with regard to jurisdictional claims in published maps and institutional affiliations.

1. Introduction

Gas hydrate is an ice-like solid formed of water and gas. It is composed of a methane molecule enclosed within a crystalline structure of water molecules [1–4]. There is abundant methane in gas hydrate. The carbon stored in gas hydrate is about twice the total carbon content of all fossil fuels (including coal, oil, and natural gas) [5]. The advantages of gas hydrate make it a potential energy resource in the future. However, from the environmental aspects, the methane in gas hydrate is a powerful greenhouse gas. It is 20–30 times more potent at trapping heat in the atmosphere than carbon dioxide [6,7]. Climate and ocean warming may reduce the stability of gas hydrates, leading to hydrate dissociation and thus the release of methane into the ocean and overlying sediments. The released methane may eventually reach the atmosphere and aggravate the greenhouse effect [6]. Thus, gas hydrate reservoirs have a possible impact on climate change. The detection and exploration of gas hydrate reservoirs are important from both the energy and environmental perspective.

The occurrence of gas hydrate is controlled by the geological environment. Sufficient concentrations of methane are necessary to form the gas hydrate reservoir. The methane may be generated by biological activities in sediments, and it may also migrate from the organic matter at depth. Therefore, natural gas hydrate is most likely to be formed at locations wherein active upward fluid migration occurs, such as oceanic and lacustrine sediments [8]. In addition, gas hydrate in nature is usually found in areas with high pressure and low temperature, such as the seafloor and permafrost sediments. The pressure and temperature conditions in these areas can keep gas hydrate stable [9]. According to the geological environment and physical properties, the geological gas hydrate deposits can be categorized into five major types [10]. The regionally disseminated low-concentration hydrate is primarily found in mostly impermeable clays. In this case, methane hydrate fills pores and/or displaces sediment grains to form crystals or nodules from a few tens of meters below the seafloor to the base of the gas hydrate stability zone (GHSZ). The saturation of methane hydrate is less than ~10%. The fracture-filling hydrate is usually found in clay-dominated fracture sediments at non-vent sites. This type of hydrate is distributed at a shallow depth (e.g., 50–300 mbsf) below seafloor with a low-to-moderate saturation. In addition, the hydrate may be enriched at the base of the GHSZ in muddy sediments. The saturation of this type of hydrate commonly increases abruptly, with depth from the background value in the muddy sediments, to more than 10% near the base of the GHSZ. The fourth type is concentrated hydrate at vent sites. This type of hydrate may exist from near the seafloor to approximately 50 mbsf, or it may be as deep as ~160 mbsf. The saturation ranges from 40% to >90%. The final type is concentrated hydrate in sandy sediments. It may be found above the base of the GHSZ and is composed of gas hydrate in thin sandy or silty beds bounded by sandy sediments. It may also be found in thick sandy sediments that cross or near the base of the GHSZ.

Seismic technology plays an important role in the detection and exploration of gas hydrate reservoirs. Seismic data have been successfully applied in the identification of gas hydrate reservoirs together with other geophysical methods, including seismic facies analysis. Yoo et al. have studied multichannel seismic reflection and well-log data from the Ulleung Basin, East Sea [11]. These data have revealed several seismic features indicative of gas hydrate occurrence, including the bottom-simulating reflector (BSR), seismic chimneys, acoustic blanking, enhanced reflection below the BSR, and seafloor gas-escape features. The BSR is formed by a strong impedance contrast between the overlying sediments containing gas hydrate and the underlying sediments containing free gas [12,13]. It is usually parallel to the seafloor and has high amplitude and reversed polarity with respect to the seafloor reflection. However, the BSR does not necessarily translate into the existence of gas hydrate because it might be present in the sediments without gas hydrate [14]. Seismic chimneys are characterized by low-to-high, upward-bending internal reflections. The velocity inside the chimney is higher than that in the surrounding sediments, which is caused by the active migration of fluid gas into the GHSZ. The acoustic blanking in the seismic profile may be attributed to the energy attenuation due to the presence of free gas or the poor seismic energy penetration due to the strong reflection from a layer of gas hydrate. The enhanced reflection below the BSR is correlated with the strong impedance contrasts due to free gas accumulation below the BSR. When the upward-migrating gas escapes into the water column through the seafloor pockmarks and mud mounds, gas seepage at the seafloor can be found [15]. Riedel et al. have added an additional element into a regional assessment strategy of gas hydrate occurrence by including the depositional environment defined through seismic facies classes [16]. The seismic facies classification is attempted using regional 2D seismic data and a 3D seismic volume, as well as core and log data from two gas hydrate drilling expeditions carried out in the Ulleung Basin, East Sea, to conduct a fully integrated gas hydrate assessment. Wu et al. have analyzed the drilling results in the Shenhu Area, South China Sea [17]. In the case that free gas exists beneath hydrate deposits, the frequency of the hydrate deposits will be noticeably attenuated, with the attenuation degree mainly affected by pore development and free gas content. Thus, frequency can be used as an important seismic attribute to identify hydrate reservoirs. These parameters could indicate the occurrence of gas hydrate; however, they fail to quantify it.

Amplitude variation with offset (AVO) is a conventional technique for quantitative reservoir characterization [18,19]. It can predict the elastic parameters of formation (P-wave velocity, S-wave velocity, and density) from the seismic data. Then, the rock and fluid properties of the reservoir (lithology, porosity, permeability, and saturation) can be estimated from the predicted elastic parameters based on rock physics models. Chen et al. applied AVO inversion to calculate the P- and S-wave velocities and density and then to estimate the gas hydrate and free-gas concentrations above and below the BSR interface. The estimated gas hydrate and free-gas concentrations are at a 90% credibility level. The results indicate that this method cannot provide enough accuracy for resolving the gas hydrate and free-gas concentrations independently [20]. Ojha and Sain performed an amplitude versus angle (AVA) modeling of seismic data from a BSR to calculate the P- and S-wave velocities and then derived the saturation of gas hydrate with rock physics modeling. The results can help to understand the origin of BSR [21]. Zhang et al. performed AVO forward modeling and AVO attribute inversion to the seismic data from the Shenhu area [22]. Their results confirmed that the AVO attributes depend on the content of gas hydrate and free gas. However, most of the current AVO method only considers the reflected wave for inversion. It is built based on the approximation of the Zoeppritz equation [23,24]. The simplified equation indicates a linear relationship between the reflected wavefield and the contrast variables of P- and S-wave velocities and density (i.e., $\Delta V_P / V_P$, $\Delta V_S / V_S$, and $\Delta\rho/\rho$). In fact, the seismic wavefield is usually non-linear with respect to the elastic properties of the subsurface medium. Therefore, the conventional AVO method based on the linearized Zoeppritz equation reduces the accuracy and resolution of the inversion properties, resulting in incorrect reservoir interpretation.

Different methods have been proposed to improve the accuracy and resolution of the conventional seismic inversion method. Alemie and Sacchi proposed a high-resolution three-term AVO inversion by introducing a Trivariate Cauchy probability distribution. This distribution can model the prior distribution of the AVO parameters with sparsity, thus leading to a high-resolution estimate of subsurface models [25]. Zhang et al. introduced low-frequency information to improve the resolution [26]. Niu et al. proposed a data-driven method to improve the linear approximation of the conventional AVO inversion method. Well-logging data were used to correct the inaccurate linearized AVO operators. The results of synthetic and field data demonstrated that the accuracy and resolution of the inverted results were improved using the proposed method [24]. Yi et al. proposed a new method using stepwise seismic inversion and 3D seismic datasets with two different resolutions [27]. The proposed method can track a thin gas hydrate-bearing sand layer compared with the conventional seismic inversion method with a maximum resolution of ~10 m. The gas hydrate distribution around the UBGH2-6 well in Ulleung Basin was estimated successfully using their method.

In this study, we have studied a high-resolution seismic characterization of a gas hydrate reservoir using a wave-equation-based method. Besides the reflected wave, more wavefields including multiple scattering and transmission are considered in the inversion process. We first present theories of the wave-equation-based inversion method. Then, a synthetic model from Pearl River Mouth Basin, South China Sea, is used to demonstrate the performance of this method in the high-resolution seismic characterization of the gas hydrate reservoir. Finally, we discuss the obtained results and the advantages of this method and provide the conclusions.

2. Methods

2.1. The AVO Inversion Method

The conventional AVO (amplitude variation with offset) inversion method is based on the approximation of the Zoeppritz equation, which describes the change of reflected amplitude with incident angle. The Zoeppritz equation is derived for the case of two half-space media separated by a horizontal interface [28]. As we know, the half-space media can be characterized by three elastic properties, namely P-wave velocity (V_P), S-wave velocity (V_S), and density (ρ). When an incident P-wave hits the interface, it is split into the reflected P- and S- waves and transmitted to the P- and S- waves. According to the Zoeppritz equation, the reflection and transmission coefficients are functions of the incident angle and the three elastic properties. Therefore, the elastic properties can be inverted from the observed reflection coefficient, which is the basis of the AVO inversion method. Considering the complexity of the Zoeppritz equation, some approximations are usually used to simplify the Zoeppritz equation and provide an intuitive understanding of the relationship between reflection amplitude and elastic properties. One of the commonly used approximations, the Aki and Richards approximation [28], is shown in Equation (1). It assumes that the perturbation in the elastic properties is small.

$$R_{PP}(\theta) \approx \frac{1}{2}(1 - 4\overline{\gamma}^2 \sin^2 \overline{\theta})\frac{\Delta\rho}{\overline{\rho}} + \frac{1}{2}\sec^2 \overline{\theta}\frac{\Delta V_P}{\overline{V}_P} - 4\overline{\gamma}^2 \sin^2 \overline{\theta}\frac{\Delta V_S}{\overline{V}_S} \tag{1}$$

where $\overline{V}_P$, $\overline{V}_S$, $\overline{\rho}$, and $\overline{\theta}$ represent the average V_P, V_S, ρ, and incident angle, respectively, across the interface. ΔV_P, ΔV_S, and $\Delta\rho$ represent the change of V_P, V_S, and ρ, respectively, across the interface. $\overline{\gamma}$ represents the ratio of $\overline{V}_S$ to $\overline{V}_P$.

Equation (1) can be expressed in the following matrix form.

$$\begin{bmatrix} R_{PP}(\theta_1) \\ \vdots \\ R_{PP}(\theta_M) \end{bmatrix} = \begin{bmatrix} \sec^2 \overline{\theta}_1 & -8\overline{\gamma}^2 \sin^2 \overline{\theta}_1 & (1 - 4\overline{\gamma}^2 \sin^2 \overline{\theta}_1) \\ \vdots & \vdots & \vdots \\ \sec^2 \overline{\theta}_M & -8\overline{\gamma}^2 \sin^2 \overline{\theta}_M & (1 - 4\overline{\gamma}^2 \sin^2 \overline{\theta}_M) \end{bmatrix} \begin{bmatrix} \frac{1}{2}\frac{\Delta V_P}{\overline{V}_P} \\ \frac{1}{2}\frac{\Delta V_S}{\overline{V}_S} \\ \frac{1}{2}\frac{\Delta\rho}{\overline{\rho}} \end{bmatrix} \tag{2}$$

where M represents the number of incident angles. As seen from Equation (2), it describes a linear relationship between the reflection coefficient and elastic properties. It can be rewritten as

$$\mathbf{Gx} = \mathbf{d} \tag{3}$$

where $\mathbf{G}$ is the linear operator defined by Equation (2), $\mathbf{x}$ the unknown elastic properties, and $\mathbf{d}$ the input seismic data.

The objective function of the conventional AVO inversion method is built based on Equation (3) and is shown as follows:

$$\mathbf{x} = \mathrm{argmin}\|\mathbf{d} - \mathbf{Gx}\|_2^2 \tag{4}$$

The elastic properties are easily obtained by solving Equation (4) using the least squares algorithm.

As seen from Equation (2), the conventional AVO method is built based on the approximation of the Zoeppritz equation [23,24]. The simplified equation considers only the reflected wave for inversion. It indicates a linear relationship between the reflected wavefield and the contrast variables of P- and S-wave velocities and density (i.e., $\Delta V_P / V_P$, $\Delta V_S / V_S$, and $\Delta\rho/\rho$, respectively). In fact, the seismic wavefield is usually non-linear with respect to the elastic properties of the subsurface medium. Therefore, the conventional AVO method based on the linearized Zoeppritz equation reduces the accuracy and resolution of the inversion properties, resulting in incorrect reservoir interpretation.

2.2. The Wave-Equation-Based Inversion Method

The wave-equation-based inversion is employed to realize the high-resolution seismic characterization of the gas hydrate reservoir [29]. Given an initial subsurface model, the wave-equation-based inversion method first simulates the seismic wavefields using the given model and updates the model from the differences between the simulated wavefields and real data. The seismic wavefields are simulated based on the scattering integral equation for elastic waves [30,31]. As we know, the three elastic properties, namely V_P, V_S, and ρ, can be expressed in terms of the elastic moduli as follows:

$$V_P = \sqrt{\frac{1}{\rho}(\frac{1}{\kappa} + \frac{4}{3M})} \tag{5}$$

$$V_S = \sqrt{\frac{1}{M\rho}} \tag{6}$$

where the compressibility κ is related to the bulk modulus K as $\kappa = \frac{1}{K}$, and the shear compliance M is related to the shear modulus μ as $M = \frac{1}{\mu}$. Assuming that the smooth background medium (κ_0, M_0, ρ_0) is known, the contrasts against the background $(\chi_\kappa, \chi_M,$ and $\chi_\rho)$ are formulated as

$$\chi_\kappa = \frac{\kappa - \kappa_0}{\kappa_0} \tag{7}$$

$$\chi_M = \frac{M - M_0}{M_0} \tag{8}$$

$$\chi_\rho = \frac{\rho - \rho_0}{\rho_0} \tag{9}$$

Therefore, the scattering integral equation for elastic waves in the frequency domain is expressed by the contrast functions as

$$\mathbf{p}(z, z_s, \omega) = \mathbf{p}_0(z, z_s, \omega) + \int_D \mathbf{G}(z, z', \omega)\chi(z')\mathbf{p}(z', z_s, \omega)dz' \tag{10}$$

where $\mathbf{p}(z, z_s, \omega)$ represents the total seismic wavefield propagating in the true medium, and $\mathbf{p}_0(z, z_s, \omega)$ the incident wavefield propagating in the background medium. They are excited by a source at z_s and recorded at each depth z in the subsurface medium. The second term on the right of the equal sign in Equation (10) represents the scattered wavefield field. $\mathbf{G}(z, z', \omega)$ is the Green's function. D defines the objective domain of interest. $\mathbf{p}_0(z, z_s, \omega)$ and $\mathbf{G}_0(z, z', \omega)$ are pre-calculated using the known background medium. As seen from Equation (10), the relationship between the total seismic wavefield and the elastic properties of the subsurface medium is nonlinear. This nonlinearity indicates that multiple scattering and transmission are considered for the inversion of medium properties, not only the reflected wavefield. Therefore, the inversion method based on Equation (10) can improve the accuracy and resolution of the inversion properties compared with the conventional AVO method based on Equation (2). However, it has a higher computational cost.

According to Equation (10), the total wavefield $\mathbf{p}(z, z_s, \omega)$ is obtained once the contrast function $\chi(z')$ is known. After the total wavefield $\mathbf{p}(z, z_s, \omega)$ is calculated, the seismic data $\mathbf{p}_d(z_r, z_s, \omega)$ recorded at the receiver z_r are given as follows:

$$\mathbf{p}_d(z_r, z_s, \omega) = \int_D \mathbf{G}(z_r, z, \omega)\chi(z)\mathbf{p}(z, z_s, \omega)dz \tag{11}$$

The wave-equation-based inversion scheme is built based on the above Equations (10) and (11), and is implemented in an iterative manner by alternately updating the contrast models with a current best knowledge of the total wavefield and then updating the total wavefield with a current contrast model. The objective function of the contrast model update is based on the misfit between actual and synthetic data and is shown as follows:

$$\chi = \mathrm{argmin}\|\mathbf{d}(z_r, z_s, \omega) - \mathbf{p}_d(z_r, z_s, \omega)\|_2^2 \tag{12}$$

where $\mathbf{d}(z_r, z_s, \omega)$ is actual seismic data and $\mathbf{p}_d(z_r, z_s, \omega)$ is synthetic data calculated by Equation (11). The total wavefield in Equation (11) is the incident wavefield at the first iteration and fixed at the current best estimate in the subsequent iterations.

The inversion process defined by the objective function in Equation (12) is unstable because there are always some forms of noise in the actual seismic data. Therefore, a regularization term is introduced to stabilize the inversion process. There are many different regularization methods, such as Tikhonov regularization [32,33] and total variation (TV) regularization [34] et al. We use a Sobolev norm-based regularization [35,36] which is shown as follows:

$$W_p^1 = \sum_z (\nabla\chi_z \cdot \nabla\chi_z + \varepsilon)^{p/2}, \varepsilon > 0 \tag{13}$$

Equation (13) becomes TV regularization when $p = 1$. It becomes Tikhonov regularization when $p = 2$. Therefore, the Sobolev norm-based regularization is a blend of the Tikhonov regularization and the TV regularization. In this study, we set p to decrease gradually from 2 to 1 in a logarithmic decreasing manner during the inversion process. This can ensure the smoothness of the inverted models in the early iteration of inversion and preserve the boundary of the models in the later iteration.

The regularization term can be added in an additive and multiplicative manner. Compared with additive regularization, multiplicative regularization can avoid the selection of a regularization parameter that balances the data misfit and regularization term. Therefore, we select multiplicative regularization in this study. The final objective function is defined as:

$$\chi = \mathrm{argmin}(\|\mathbf{d}(z_r, z_s, \omega) - \mathbf{p}_d(z_r, z_s, \omega)\|_2^2)(\sum_z (\nabla \chi_z \cdot \nabla \chi_z + \varepsilon)^{p/2}) \tag{14}$$

Once the contrast model is updated, the total wavefield is updated based on Equation (10) by fixing the updated contrast model. However, the update of the total wavefield is not realized at one time. Instead, it is updated iteratively by a Krylov subspace method in Equation (15):

$$\mathbf{p}^n(z, z_s, \omega) = \mathbf{p}_0(z, z_s, \omega) + \sum_{i=1}^n \alpha^i \Phi^i(z, z_s, \omega) \tag{15}$$

where α^i is the weighting coefficient, and n is the number of iterations. $\Phi^i(z, z_s, \omega)$ is the difference between two successive wavefields and is defined as follows [29]:

$$\Phi^i(z, z_s, \omega) = \int_D \mathbf{G}(z, z', \omega)[\chi^i(z')\mathbf{p}^{i-1}(z', z_s, \omega) - \chi^{i-1}(z')\mathbf{p}^{i-2}(z', z_s, \omega)]dz' \tag{16}$$

The weighing coefficients α^i are solved when the calculated wavefields by Equations (10) and (16) fit well. One more order of scattering is added in each iteration of the wavefield update. All orders of scattering are considered until the last iteration in the inversion process.

After the total wavefields are updated, they are substituted into Equations (11) and (12) again to obtain an improved estimate of the contrast models. This process is iterated until the synthetic data fits well with the real data. Finally, characteristics (V_P, V_S, and ρ) of the subsurface media are obtained by substituting the inversion models into Equations (5) and (6). Figure 1 summarizes the workflow of the wave-equation-based inversion method. As shown in Figure 1, the models and total wavefield are updated iteratively in an alternating manner during the inversion process. The subsurface models are first updated given the initial background models. Then, the wavefield is updated based on the updated models. The two processes are repeated until the misfit between the real and synthetic seismic wavefield is small. The total wavefield is regarded as the summation of the background wavefield and scattering wavefield. One more order of scattering is considered at each iteration. As more order scattering fields are included, the inversion models become closer to the real models. Finally, the total wavefields used for inversion include all multiple scattering and transmission.

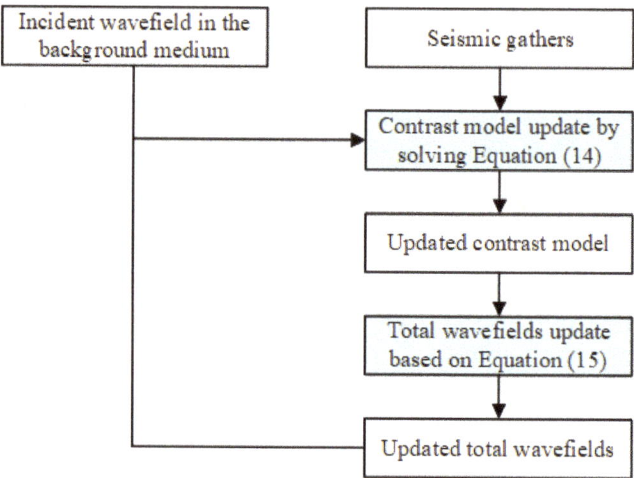

Figure 1. Workflow of the wave-equation-based inversion method.

3. Synthetic Model and Data

A synthetic model from Pearl River Mouth Basin, South China Sea, is used to verify the wave-equation-based inversion method. In this section, we first present the geological setting in the studied area and introduce how the synthetic model and data were derived.

3.1. Geological Setting

The Pearl River Mouth Basin is located in the northern part of the South China Sea. The Baiyun Sag, located in the southern Pearl River Mouth basin, is the largest and deepest subbasin in the Pearl River Mouth Basin (Figure 2). It has a water depth of 200~2000 m and an area of more than 20,000 km^2 [37–39]. The Baiyun Sag has experienced complex tectonic evolution, including rifting, depression, and fault block rise and fall. The depositional environment in the Baiyun Sag evolved gradually from continental into shallow marine and continental slope deep water facies and finally, substantial deep-water sediments were developed [40]. The formations from the base to top are the Wenchang Formation of Eocene, the Enping Formation and Zhuhai Formation of Oligocene, the Zhujiang Formation, the Hanjiang Formation, the Yuehai Formation of Miocene, the Wanshan Formation of Pliocene, and the Quaternary sediment [41–43]. The Wenchang Formation and the Enping Formation are the high-quality source rock that provide gas for the shallow gas hydrate deposits. The former is lacustrine sediment during the rifting period. The latter is lacustrine sediment during the fault-depression period. The Zhuhai Formation is a transitional deltaic deposit. It is composed of littoral sandy mudstone. The Zhujiang, Hanjiang, Yuehai, and Wanshan formations are shelf edge delta to slope-deep water deposits. They are composed of littoral mudstone, marine mudstone, and neritic sandstone and mudstone. These strata develop three sets of reservoir-cap assemblage. A large number of faults are developed in the study area. They migrate the deep gas to the shallow gas hydrate stable zones. Some thin gas hydrate reservoirs have been found in the shallow layer. They are distributed tens to hundreds of meters below the seabed.

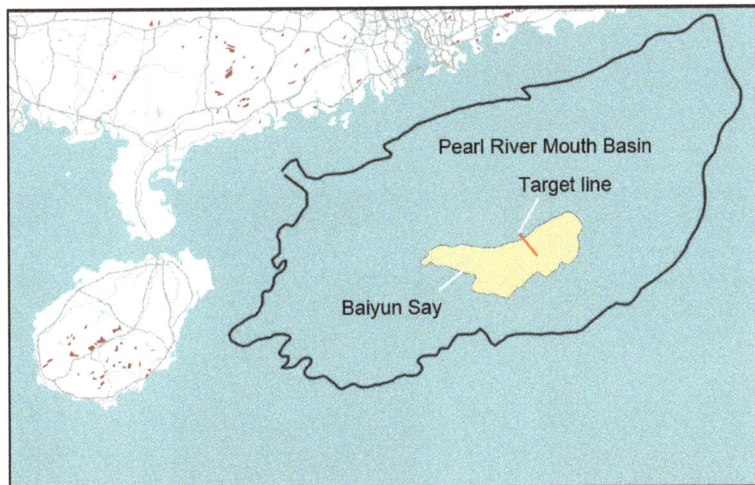

Figure 2. Location of the studied area. The red line indicates the location of the 2D seismic line that was used to build the geological model in Figure 3.

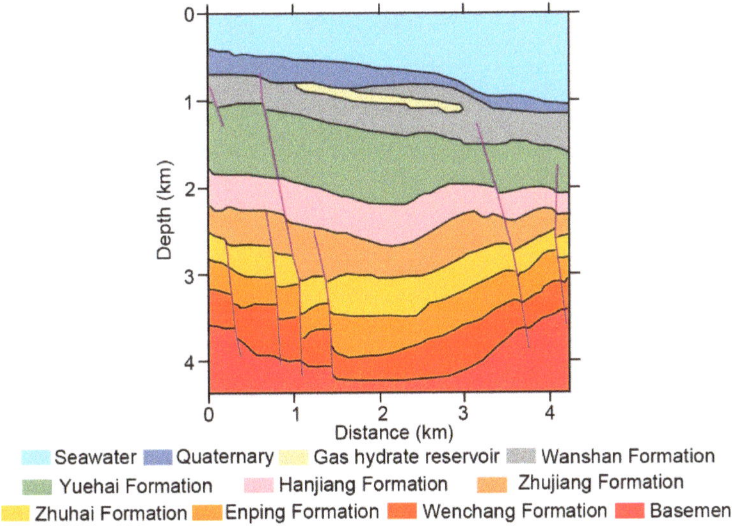

Figure 3. The geological model for synthetic data analysis. The black lines indicate the stratigraphic horizons and the red lines indicate the faults.

3.2. Synthetic Model and Data

The geological structure of the Pearl River Mouth Basin has been investigated in detail [37–43] in previous studies. Several geological models have been published. We used a geological model (in Figure 3) [44] built based on the study of a 2D seismic line indicated by the red line in Figure 1. The horizontal and vertical intervals of the model are both 5 m. The seawater depth at this location is about 400~1000 m. As mentioned above, there are nine different strata below the sea floor defined by their sedimentary environment. A large number of faults are developed in the study area, indicated by the red lines in Figure 3. The gas hydrate reservoir was formed in the Wanshan Formation and lies at a depth of 800 m according to a bottom-simulating reflector in the seismic profile. The thickness of the gas hydrate reservoir is about 15 m.

Based on the well log information and fine velocity modeling in the study area [45,46], the P-wave velocity (V_P), S-wave velocity (V_S), and density (ρ) are assigned to the geological model to simulate seismic wavefields, as shown in Figure 4. Then, seismic wavefields are modeled using the reflectivity method [47,48]. A total of 850 angle gathers are simulated. Figure 5 shows examples of obtained seismic angle gathers at 1.25 km, 2.25 km, 3.25 km, and 4.25 km.

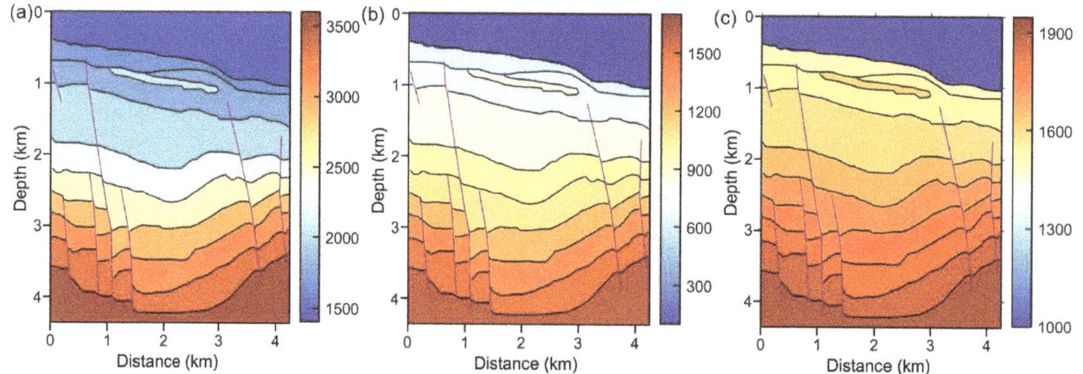

Figure 4. The true models for the synthetic data: (**a**) P-wave velocity, (**b**) S-wave velocity, and (**c**) density. The black lines indicate the stratigraphic horizons and the red lines indicate the faults.

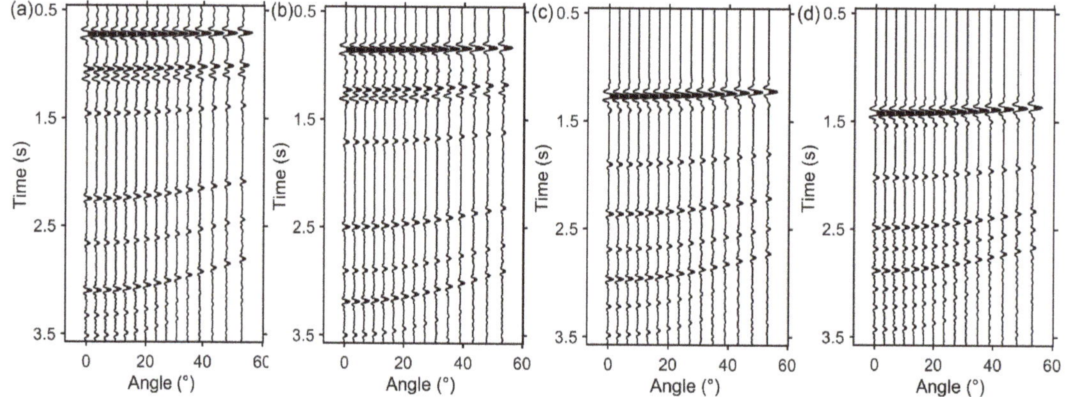

Figure 5. The noise-free seismic angle gathers at (**a**) 1.25 km, (**b**) 2.25 km, (**c**) 3.25 km, and (**d**) 4.25 km of the model.

4. Results

4.1. Inversion of the Synthetic Gas Hydrate Reservoir Model

To compare with the wave-equation-based inversion method, the conventional AVO inversion method is first applied to all the input gathers to invert the P-wave velocity, S-wave velocity, and density models. Then, the wave-equation-based inversion method is performed. The background models used in the conventional AVO inversion and the wave-equation-based inversion method are obtained by applying Gaussian smoothing to the real models in Figure 4. The stratigraphic horizons in the background models are almost unrecognizable, as shown in Figure 6.

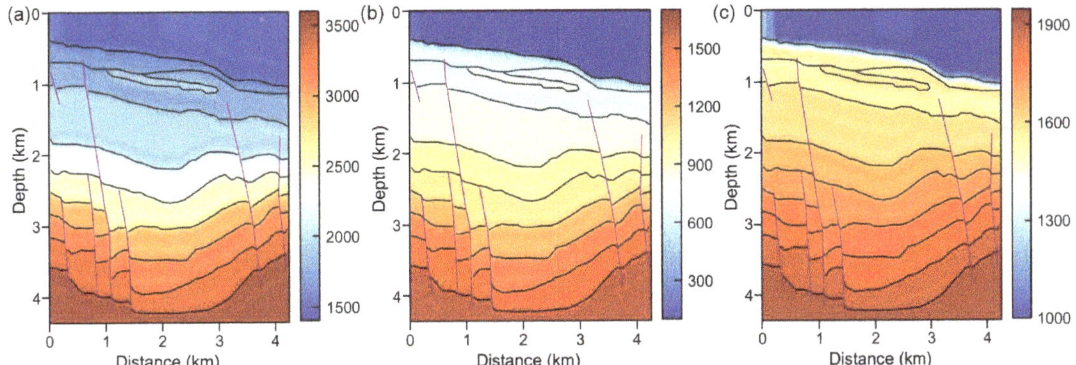

Figure 6. The background models used in the inversion method: (**a**) P-wave velocity, (**b**) S-wave velocity, and (**c**) density. The black lines indicate the stratigraphic horizons and the red lines indicate the faults.

Figure 7 shows the inverted P-wave velocity, S-wave velocity, and density models using the conventional AVO method. The stratigraphic horizons and faults indicated by the black and red lines in Figure 2 are added to the inverted models to compare with the true models in Figure 2. The black lines indicate the horizons and the red lines indicate the faults. As shown in Figure 7, the obtained models using the conventional AVO method are smooth and have a low resolution. The gas hydrate reservoir cannot be well identified from the inversion results. The inverted stratigraphic horizons indicated by the black arrows do not agree well with the real ones. Figure 8 compares the true models (red lines) with the inverted models (blue lines) at 2.25 km. As shown by the black arrows, it is difficult to identify the boundary of the gas hydrate reservoir, especially for the S-wave velocity and density models. The synthetic seismic angle gather after the final inversion at 2.25 km is shown in Figure 9a. It has a large error with the real seismic gather in Figure 5b, as shown in Figure 9b. This is because the conventional AVO inversion method only considers the reflected wavefield and a linear relationship between the reflected wavefield and the elastic properties of the subsurface medium, thus leading to lower accuracy and resolution.

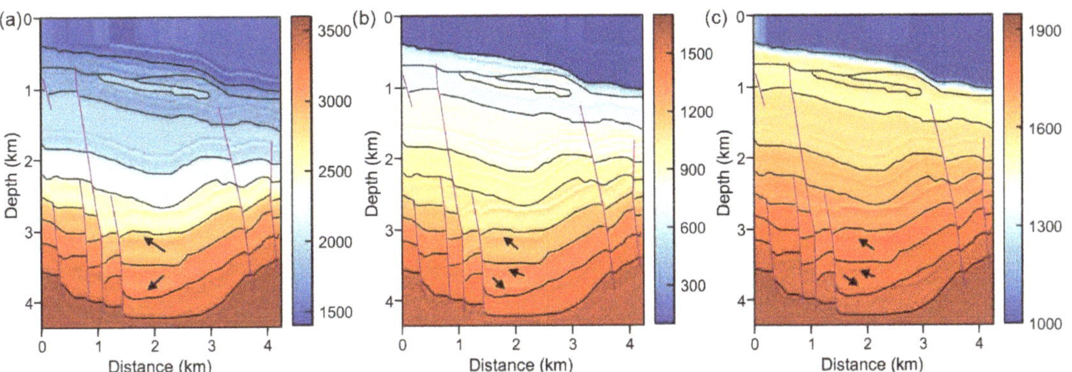

Figure 7. The inverted models from the noise-free data using the conventional AVO method: (**a**) P-wave velocity, (**b**) S-wave velocity, and (**c**) density. The black lines indicate the stratigraphic horizons and the red lines indicate the faults. The inverted stratigraphic horizons indicated by the black arrows do not agree well with the real ones.

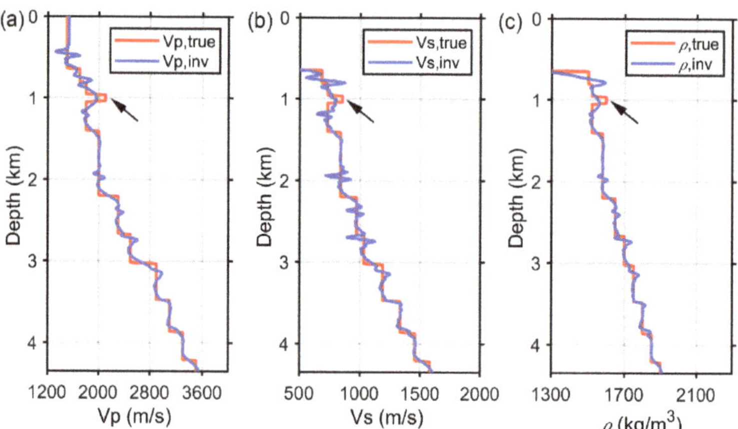

Figure 8. Comparison between the true model at 2.25 km and the inverted model from the noise-free data using the conventional AVO method: (**a**) P-wave velocity, (**b**) S-wave velocity, and (**c**) density. The black arrow indicates the location of the gas hydrate reservoir.

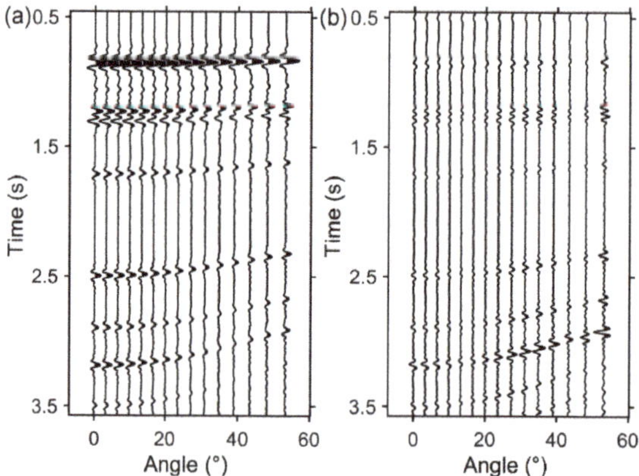

Figure 9. (**a**) Synthetic seismic angle gather at 2.25 km after the conventional AVO method, (**b**) difference between the real seismic gather in Figure 5b and synthetic seismic gather in (**a**).

The inverted models using the wave-equation-based inversion method are shown in Figure 10. The gas hydrate reservoir is clearly reconstructed with high accuracy and resolution. More details are shown in the inverted models than in the conventional AVO inversion results in Figure 7. The inverted stratigraphic horizons agree well with the real ones. Figure 11 compares the true models (red lines) with the inverted models (blue lines) at 2.25 km. The inversion results are almost the same as the true models (red lines). The synthetic seismic angle gather for the final inverted models in Figure 12a exhibits a good agreement with the real seismic gather in Figure 5b. The error between the two gathers is small, as shown in Figure 12b. As described in the theory of the wave-equation-based method, the total wavefields including all multiple scattering and transmission are used for inversion. This contributes to improving the subsurface illumination by considering more wavefields. Therefore, the accuracy and resolution of the inversion results are improved.

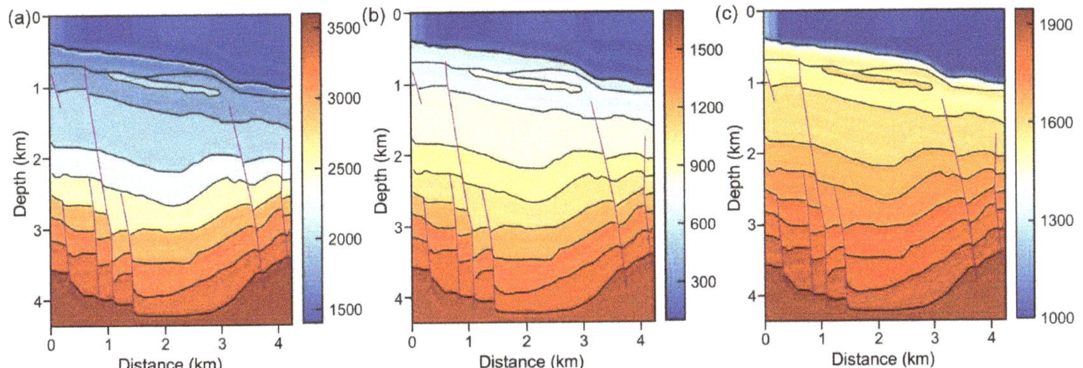

Figure 10. The inverted models from the noise-free data using the wave-equation-based inversion method: (**a**) P-wave velocity, (**b**) S-wave velocity, and (**c**) density. The black lines indicate the stratigraphic horizons and the red lines indicate the faults.

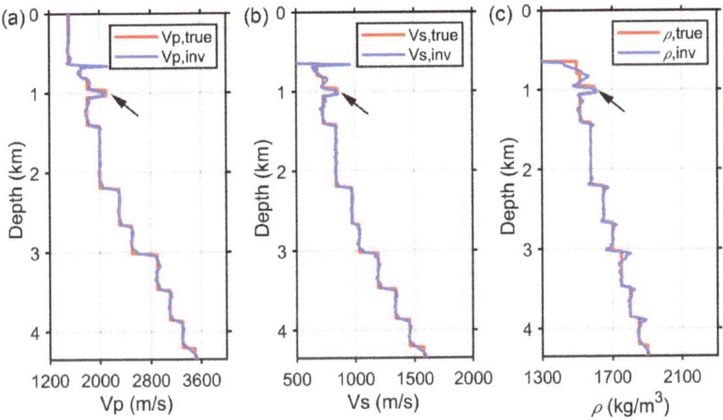

Figure 11. Comparison between the true model at 2.25 km and the inverted model from the noise-free data using the wave-equation-based inversion method: (**a**) P-wave velocity, (**b**) S-wave velocity, and (**c**) density. The black arrow indicates the location of the gas hydrate reservoir.

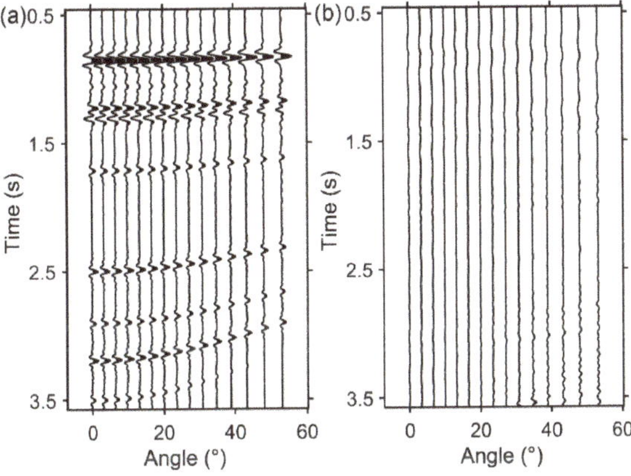

Figure 12. (**a**) Synthetic seismic angle gathers at 2.25 km after the wave-equation-based inversion method, (**b**) difference between the real seismic gather in Figure 3b and synthetic seismic gather in (**a**).

4.2. Reliability Analysis for the Noisy Data

Due to the sensitivity to noise of inversion methods, the reliability analysis for the noisy data is studied to test the influence of noise on the inversion. The noisy seismic gathers are generated by adding Gaussian random noise. The signal-to-noise ratio of noisy gathers in Figure 13 is of 10 dB. Then, the wave-equation-based inversion method is applied to the noisy data. Figure 14 shows the final inverted 2D models. As seen from the results, the models are still well inverted in the presence of noise. The gas hydrate reservoir is easy to identify from the inverted models. The inverted stratigraphic horizons still agree well with the real ones. Figure 15 shows the true (red lines) and inverted (blue lines) models for the seismic angle gather at 2.25 km. The boundary of the gas hydrate reservoir is clearly characterized. The synthetic seismic angle gathers after the final inversion in Figure 16a are similar to the real seismic angle gather in Figure 5b. The noise in the input angle gather is not fitted, as shown in Figure 16b.

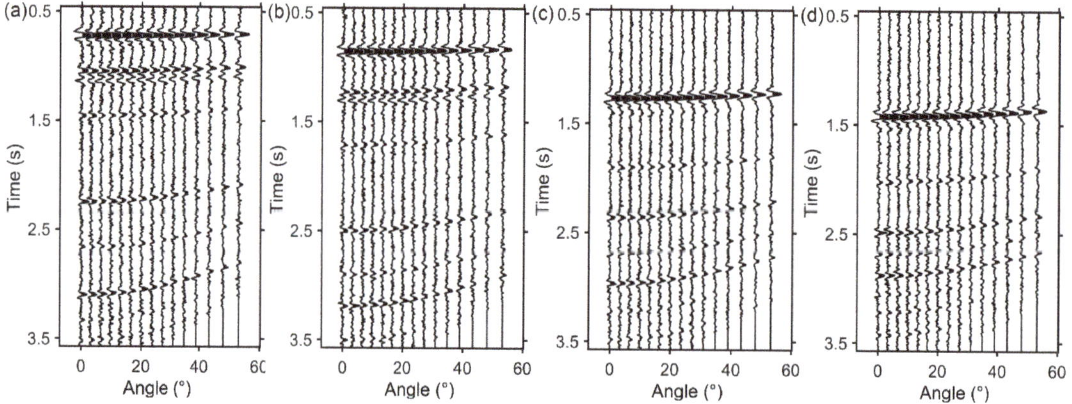

Figure 13. The noisy seismic angle gathers at (**a**) 1.25 km, (**b**) 2.25 km, (**c**) 3.25 km, and (**d**) 4.25 km of the model.

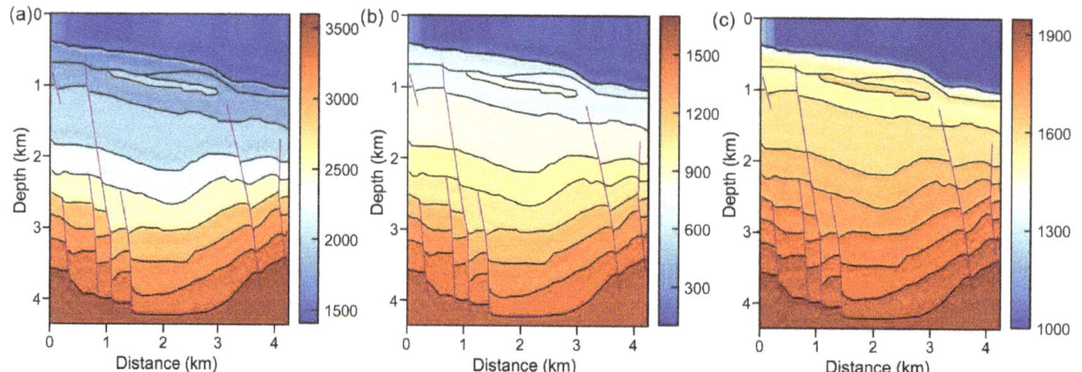

Figure 14. The inverted models from the noisy data using the wave-equation-based inversion method: (**a**) P-wave velocity, (**b**) S-wave velocity, and (**c**) density. The black lines indicate the stratigraphic horizons and the red lines indicate the faults.

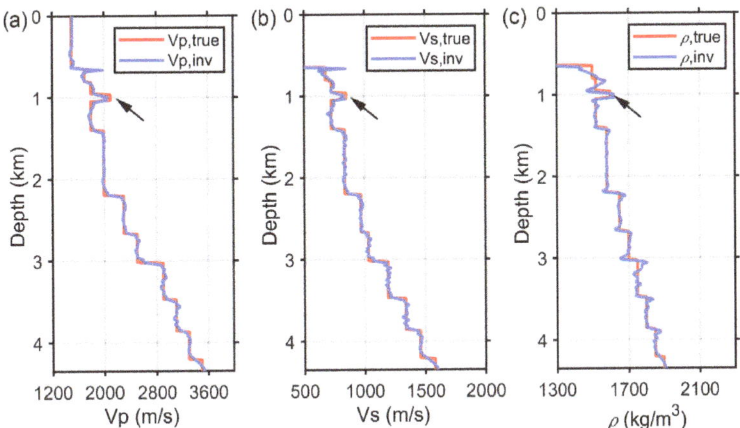

Figure 15. Comparison between the true model at 2.25 km and the inverted model from the noisy data using the wave-equation-based inversion method: (**a**) P-wave velocity, (**b**) S-wave velocity, and (**c**) density. The black arrow indicates the location of the gas hydrate reservoir.

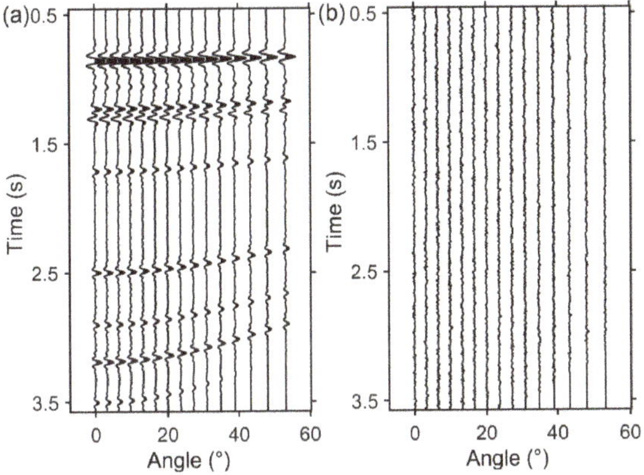

Figure 16. (**a**) Synthetic seismic angle gathers at 2.25 km after the wave-equation-based inversion method, (**b**) difference between the real seismic gather in Figure 10b and synthetic seismic gather in (**a**).

5. Conclusions

The high-resolution seismic characterization of gas hydrate reservoirs is important for the accurate evaluation of gas hydrate resources, especially for thin gas hydrate reservoirs. We have studied a wave-equation-based inversion method for gas hydrate reservoirs in this work. It is based on the scattering integral wave equation, and a Sobolev norm-based regularization is considered during the inversion. Compared to the conventional AVO method, the wave-equation-based inversion method can characterize the gas hydrate reservoir with high resolution by taking into account all the multiple scattering and transmission. The advantages of this method are validated by a synthetic model from Pearl River Mouth Basin, South China Sea. Thin gas hydrate reservoirs are distributed in the studied area. They are buried shallowly and distributed tens to hundreds of meters below the seabed. The results demonstrate that the wave-equation-based inversion method can provide a higher resolution and accuracy than the conventional AVO inversion method. It shows good performance in the presence of noise. This makes it a promising method for the accurate evaluation of gas hydrate resources, especially for thin gas hydrate reservoirs.

Author Contributions: Conceptualization, J.S. and Y.W. (Yibo Wang); methodology, J.S. and Y.W. (Yibo Wang); software, J.S. and Y.W. (Yibo Wang); validation, J.S. and Y.W. (Yibo Wang); data curation, H.Y.; writing—original draft preparation, J.S. and Y.W. (Yibo Wang); writing—review and editing, J.S., Y.W. (Yibo Wang), Y.W. (Yanfei Wang) and H.Y.; project administration, Y.W. (Yibo Wang); funding acquisition, Y.W. (Yibo Wang). All authors have read and agreed to the published version of the manuscript.

Funding: This research was funded by The Key Research Program of the Institute of Geology and Geophysics, CAS: IGGCAS-201903109.

Data Availability Statement: Not applicable.

Conflicts of Interest: The authors declare no conflict of interest.

References

1.   Hassanpouryouzband, A.; Joonaki, E.; Farahani, M.V.; Takeya, S.; Ruppel, C.; Yang, J.; English, N.J.; Schicks, J.M.; Edlmann, K.; Mehrabian, H.; et al. Gas hydrates in sustainable chemistry. *Chem. Soc. Rev.* **2020**, *49*, 5225–5309. [CrossRef]
2.   Demirbas, A. Methane from gas hydrates in the black sea. *Energy Sources Part A Recovery Util. Environ. Eff.* **2009**, *32*, 165–171. [CrossRef]
3.   Aregbe, A.G. Gas hydrate—Properties, formation and benefits. *Open J. Yangtze Oil Gas* **2017**, *2*, 27–44. [CrossRef]
4.   Chazallon, B.; Rodriguez, C.T.; Ruffine, L.; Carpentier, Y.; Donval, J.P.; Ker, S.; Riboulot, V. Characterizing the variability of natural gas hydrate composition from a selected site of the Western Black Sea, off Romania. *Mar. Pet. Geol.* **2021**, *124*, 104785. [CrossRef]
5.   Park, Y.; Kim, D.Y.; Lee, J.W.; Huh, D.G.; Park, K.P.; Lee, J.; Lee, H. Sequestering carbon dioxide into complex structures of naturally occurring gas hydrates. *Proc. Natl. Acad. Sci. USA* **2006**, *103*, 12690–12694. [CrossRef] [PubMed]
6.   Farahani, M.V.; Hassanpouryouzband, A.; Yang, J.; Tohidi, B. Insights into the climate-driven evolution of gas hydrate-bearing permafrost sediments: Implications for prediction of environmental impacts and security of energy in cold regions. *RSC Adv.* **2021**, *11*, 14334–14346. [CrossRef] [PubMed]
7.   Farahani, M.V.; Hassanpouryouzband, A.; Yang, J.; Tohidi, B. Development of a coupled geophysical–geothermal scheme for quantification of hydrates in gas hydrate-bearing permafrost sediments. *Phys. Chem. Chem. Phys.* **2021**, *23*, 24249–24264. [CrossRef]
8.   Hovland, M.; Gallagher, J.W.; Clennell, M.B.; Lekvam, K. Gas hydrate and free gas volumes in marine sediments: Example from the Niger Delta front. *Mar. Pet. Geol.* **1997**, *14*, 245–255. [CrossRef]
9.   Schicks, J.M. Gas hydrates in nature and in the laboratory: Necessary requirements for formation and properties of the resulting hydrate phase. *ChemTexts* **2022**, *8*, 13. [CrossRef]
10. You, K.; Flemings, P.B.; Malinverno, A.; Collett, T.S.; Darnell, K. Mechanisms of methane hydrate formation in geological systems. *Rev. Geophys.* **2019**, *57*, 1146–1196. [CrossRef]
11. Yoo, D.G.; Kang, N.K.; Yi, B.Y.; Kim, G.Y.; Ryu, B.J.; Lee, K.; Lee, G.H.; Riedel, M. Occurrence and seismic characteristics of gas hydrate in the Ulleung Basin, East Sea. *Mar. Pet. Geol.* **2013**, *47*, 236–247. [CrossRef]
12. Mosher, D.C. A margin-wide BSR gas hydrate assessment: Canada's Atlantic margin. *Mar. Pet. Geol.* **2011**, *28*, 1540–1553. [CrossRef]
13. Madrussani, G.; Rossi, G.; Camerlenghi, A. Gas hydrates, free gas distribution and fault pattern on the west Svalbard continental margin. *Geophys. J. Int.* **2010**, *180*, 666–684. [CrossRef]
14. Ker, S.; Thomas, Y.; Riboulot, V.; Sultan, N.; Bernard, C.; Scalabrin, C.; Ion, G.; Marsset, B. Anomalously deep BSR related to a transient state of the gas hydrate system in the western Black Sea. *Geochem. Geophys. Geosyst.* **2019**, *20*, 442–459. [CrossRef]
15. Hovland, M.; Svensen, H. Submarine pingoes: Indicators of shallow gas hydrates in a pockmark at Nyegga, Norwegian Sea. *Mar. Geol.* **2006**, *228*, 15–23. [CrossRef]
16. Riedel, M.; Bahk, J.J.; Kim, H.S.; Yoo, D.G.; Kim, W.S.; Ryu, B.J. Seismic facies analyses as aid in regional gas hydrate assessments. Part-I: Classification analyses. *Mar. Pet. Geol.* **2013**, *47*, 248–268. [CrossRef]
17. Wu, S.Y.; Liu, J.; Xu, H.N.; Liu, C.L.; Ning, F.L.; Chu, H.X.; Wu, H.R.; Wang, K. Application of frequency division inversion in the prediction of heterogeneous natural gas hydrates reservoirs in the Shenhu Area, South China Sea. *China Geol.* **2022**, *5*, 251–266. [CrossRef]
18. Hampson, D. AVO inversion, theory and practice. *Lead. Edge* **1991**, *10*, 39–42. [CrossRef]
19. Buland, A.; Omre, H. Bayesian linearized AVO inversion. *Geophysics* **2003**, *68*, 185–198. [CrossRef]
20. Chen MA, P.; Riedel, M.; Hyndman, R.D.; Dosso, S.E. AVO inversion of BSRs in marine gas hydrate studies. *Geophysics* **2007**, *72*, C31–C43. [CrossRef]
21. Ojha, M.; Sain, K. Seismic velocities and quantification of gas-hydrates from AVA modeling in the western continental margin of India. *Mar. Geophys. Res.* **2007**, *28*, 101–107. [CrossRef]
22. Zhang, X.; Yin, C.; Zhang, G. Confirmation of AVO Attribute Inversion Methods for Gas Hydrate Characteristics Using Drilling Results from the Shenhu Area, South China Sea. *Pure Appl. Geophys.* **2021**, *178*, 477–490. [CrossRef]
23. Wang, Y. Approximations to the Zoeppritz equations and their use in AVO analysis. *Geophysics* **1999**, *64*, 1920–1927. [CrossRef]

24. Niu, L.; Geng, J.; Wu, X.; Zhao, L.X.; Zhang, H. Data-driven method for an improved linearised AVO inversion. *J. Geophys. Eng.* **2021**, *18*, 1–22. [CrossRef]
25. Alemie, W.; Sacchi, M.D. High-resolution three-term AVO inversion by means of a Trivariate Cauchy probability distribution. *Geophysics* **2011**, *76*, R43–R55. [CrossRef]
26. Zhang, Y.P.; Zhou, H.; Zhang, M.Z.; Yu, B.; Wang, L.Q. High-resolution AVO inversion based on low-frequency information constraint. In *SEG Technical Program Expanded Abstracts 2019*; Society of Exploration Geophysicists: Houston, TX, USA, 2019; pp. 714–718.
27. Yi, B.Y.; Yoon, Y.H.; Kim, Y.J.; Kim, G.Y.; Joo, Y.H.; Kang, N.K.; Kim, J.K.; Chun, J.H.; Yoo, D.G. Characterization of thin gas hydrate reservoir in ulleung basin with stepwise seismic inversion. *Energies* **2021**, *14*, 4077. [CrossRef]
28. Aki, K.; Richards, P.G. *Quantitative Seismology: Theory and Methods*; WH Freeman and Co.: New York, NY, USA, 1980.
29. Gisolf, D.; Haffinger, P.R.; Doulgeris, P. Reservoir-oriented wave-equation-based seismic amplitude variation with offset inversion. *Interpretation* **2017**, *5*, SL43–SL56. [CrossRef]
30. Yang, J.Q.; Abubakar, A.; van den Berg, P.M.; Habashy, T.M.; Reitich, F. A CG-FFT approach to the solution of a stress-velocity formulation of three-dimensional elastic scattering problems. *J. Comput. Phys.* **2008**, *227*, 10018–10039. [CrossRef]
31. Noor, M.A.; Noor, K.I.; Al-Said, E. New iterative methods for solving integral equations. *Int. J. Mod. Phys. B* **2011**, *25*, 4655–4660. [CrossRef]
32. Golub, G.H.; Hansen, P.C.; O'Leary, D.P. Tikhonov regularization and total least squares. *SIAM J. Matrix Anal. Appl.* **1999**, *21*, 185–194. [CrossRef]
33. Borsdorff, T.; Hasekamp, O.P.; Wassmann, A.; Landgraf, J. Insights into Tikhonov regularization: Application to trace gas column retrieval and the efficient calculation of total column averaging kernels. *Atmos. Meas. Tech.* **2014**, *7*, 523–535. [CrossRef]
34. Strong, D.; Chan, T. Edge-preserving and scale-dependent properties of total variation regularization. *Inverse Probl.* **2003**, *19*, S165. [CrossRef]
35. Hu, W.; Abubakar, A.; Habashy, T.M. Simultaneous multifrequency inversion of full-waveform seismic data. *Geophysics* **2009**, *74*, R1–R14. [CrossRef]
36. Brezis, H.; Van Schaftingen, J.; Yung, P.L. A surprising formula for Sobolev norms. *Proc. Natl. Acad. Sci. USA* **2021**, *118*, e2025254118. [CrossRef]
37. Wu, N.Y.; Zhang, H.Q.; Yang, S.X.; Zhang, G.X.; Liang, J.Q.; Lu, J.A.; Su, X.; Schultheiss, P.; Holland, M.; Zhu, Y.H. Gas Hydrate System of Shenhu Area, Northern South China Sea: Geochemical Results. *J. Geol. Res.* **2011**, *2011*, 370298. [CrossRef]
38. Liu, C.L.; Ye, Y.G.; Meng, Q.G.; He, X.G.; Lu, H.L.; Zhang, J.; Liu, J.; Yang, S.X. The characteristics of gas hydrates recovered from Shenhu Area in the South China Sea. *Mar. Geol.* **2012**, *307*, 22–27. [CrossRef]
39. Liang, J.; Meng, M.M.; Liang, J.Q.; Ren, J.F.; He, Y.L.; Li, T.W.; Xu, M.J.; Wang, X.X. Drilling Cores and Geophysical Characteristics of Gas Hydrate-Bearing Sediments in the Production Test Region in the Shenhu sea, South China sea. *Front. Earth Sci.* **2022**, *10*, 911123. [CrossRef]
40. Pang, X.; Ren, J.Y.; Zheng, J.Y.; Liu, J.; Peng, Y.; Liu, B.J. Petroleum geology controlled by extensive detachment thinning of continental margin crust: A case study of Baiyun sag in the deep-water area of northern South China Sea. *Pet. Explor. Dev.* **2018**, *45*, 29–42. [CrossRef]
41. Chen, D.X.; Wu, S.G.; Dong, D.D.; Mi, L.J.; Fu, S.Y.; Shi, H.S. Focused fluid flow in the Baiyun Sag, northern South China Sea: Implications for the source of gas in hydrate reservoirs. *Chin. J. Oceanol. Limnol.* **2013**, *31*, 178–189. [CrossRef]
42. Lin, H.; Shi, H. Hydrocarbon accumulation conditions and exploration direction of Baiyun–Liwan deep water areas in the Pearl River Mouth Basin. *Nat. Gas Ind. B* **2014**, *1*, 150–158. [CrossRef]
43. Gao, G.; Gang, W.; Zhang, G.; He, W.; Cui, X.; Shen, H.; Miao, S. Physical simulation of gas reservoir formation in the Liwan 3-1 deep-water gas field in the Baiyun sag, Pearl River Mouth Basin. *Nat. Gas Ind. B* **2015**, *2*, 77–87. [CrossRef]
44. Su, P.B.; Liang, J.Q.; Sha, Z.B.; Fu, S.Y.; Lei, H.Y.; Gong, Y.H. Dynamic simulation of gas hydrate reservoirs in the Shenhu area, the northern South China Sea. *Acta Pet. Sin.* **2011**, *32*, 226.
45. Liang, J.; Wang, M.J.; Lu, J.A.; Liang, J.Q.; Wang, H.B.; Kuang, Z.G. Characteristics of sonic and seismic velocities of gas hydrate bearing sediments in the Shenhu area, northern South China Sea. *Nat. Gas Ind.* **2013**, *33*, 29–35.
46. Xue, H.; Du, M.; Wen, P.F.; Zhang, R.W.; Xu, Y.X.; Chen, X. Research and application of fine velocity modeling to gas hydrate testing development in the Shenhu area of South China Sea. *Mar. Geol. Front.* **2019**, *35*, 8–17.
47. Kennett, B.L.N. *Seismic Wave Propagation in Stratified Media*; Cambridge University Press: Cambridge, UK, 1983.
48. Mavko, G.; Mukerji, T.; Dvorkin, J. *The Rock Physics Handbook*; Cambridge University Press: Cambridge, UK, 1998.

Article

Along-Strike Reservoir Development of Steep-Slope Depositional Systems: Case Study from Liushagang Formation in the Weixinan Sag, Beibuwan Basin, South China Sea

Sheng Liu [1,2], Hongtao Zhu [1,2,*], Qianghu Liu [1,2], Ziqiang Zhou [1,2] and Jiahao Chen [1,2]

[1] Key Laboratory of Tectonics and Petroleum Resources, China University of Geosciences (Wuhan), Ministry of Education, Wuhan 430074, China
[2] School of Earth Resources, China University of Geosciences (Wuhan), Wuhan 430074, China
*   Correspondence: zhuoscar@sohu.com

Abstract: Seismic, core, drilling, logging, and thin-section data are considered to analyze the reservoir diversity in the east, middle, and west fan of the Liushagang Formation in the steep-slope zone of the Weixinan Sag, Beibuwan Basin. Three factors primarily affect the reservoir differences for steep-slope systems: (1) Sedimentary factors mostly control reservoir scales and characteristics and the drainage system and microfacies. Massive high-quality reservoirs have shallow burial depths. Channel development and sediment supply favor the formation of these reservoirs. The sedimentary microfacies suggest fan delta plain distributary channels. (2) Lithofacies factors primarily control reservoir types and evolution. The diagenesis of high-quality reservoirs is weak, and a weak compaction–cementation diagenetic facies and medium compaction–dissolution diagenetic facies were developed. (3) Sandstone thickness factors primarily control the oil-bearing properties of reservoirs. The average porosity and permeability of high-quality reservoirs are large, the critical sandstone thickness is small, the average sandstone thickness is large, and the oil-bearing capacity is high. Furthermore, the reservoir prediction models are summarized as fan delta and nearshore subaqueous fan models. The high-quality reservoir of the fan delta model is in the fan delta plain, and the lithology is medium–coarse sandstone. The organic acid + meteoric freshwater two-stage dissolution is developed, various dissolved pores are formed, and a Type I reservoir is developed. The high-quality reservoir of the nearshore subaqueous fan model is in the middle fan, and the lithology is primarily medium–fine sandstone. Only organic acid dissolution, dissolution pores, and Type I–II reservoirs are developed. Regarding reservoir differences and models, the high-quality reservoir of the steep-slope system is shallow and large-scale, and the reservoir is a fan delta plain distributary channel microfacies. Weak diagenetic evolution, good physical properties, thick sandstone, and good oil-bearing properties developed a Type I reservoir. The study of reservoir control factors of the northern steep-slope zone was undertaken in order to guide high-quality reservoir predictions. Further, it provides a reference for high-quality reservoir distribution and a prediction model for the steep-slope system.

Keywords: Weixinan Sag; reservoir diversity; Liushagang Formation; northern steep-slope

Citation: Liu, S.; Zhu, H.; Liu, Q.; Zhou, Z.; Chen, J. Along-Strike Reservoir Development of Steep-Slope Depositional Systems: Case Study from Liushagang Formation in the Weixinan Sag, Beibuwan Basin, South China Sea. *Energies* **2023**, *16*, 804. https://doi.org/10.3390/en16020804

Academic Editors: Yuming Liu and Bo Zhang

Received: 23 November 2022
Revised: 29 December 2022
Accepted: 6 January 2023
Published: 10 January 2023

1. Introduction

Since the concept was first proposed in the 1960s, the fan delta depositional system has received increasing attention. With the development of research, studying the fan delta has gradually deepened from the initial study of sediment characteristics and outcrops to the sedimentary model and fan delta reservoirs [1–6]. With the gradual deepening of studies on the fan delta depositional system, we found that the fan delta depositional system is widely developed in the continental lacustrine basins in the early stage of structural development and belongs to a type of accumulation of coarse debris [7–11]. Furthermore, the nearshore subaqueous fan comes from the deep-water fan, primarily manifesting as a submarine fan, with coarse-grained sediment and developed in the lowstand system tract [12–15]. Research on the nearshore subaqueous fan depositional system is becoming increasingly hot and can be divided into inner, middle, and outer fans and shows different characteristics according to the lithology, grain size, and structure [16–19]. The fan delta and the nearshore subaqueous fan have similar sedimentation; however, as an unconventional reservoir, some reservoir diversity problems of the nearshore subaqueous fan and fan delta exist [20–25]. Based on the different sedimentary reservoir factors [26–29], this is a comprehensive study on the reservoir diversity in the northern steep-slope zone of the Weixinan Sag, Beibuwan Basin.

The Beibuwan Basin is a Cenozoic-faulted sedimentary basin under the background of a Mesozoic regional uplift. After more than 40 years of exploration and development, the Weixinan Sag of the Beibuwan Basin is currently proven to be a hydrocarbon-rich sag [30–32]. Located on the northwest edge of the Beibuwan Basin, Weixinan Sag is a primary battlefield for oil exploration and development in the western South China Sea. The production situation of existing oil fields is challenging, and an urgent need exists to find large-scale, high-quality reserves. Studies have defined the sedimentary system and characteristics of the steep northern slope of the Liushagang Formation in the Weixinan Sag. The Weixinan Sag was in the early stage of tectonic evolution during the Liushagang Formation and developed a fan delta and nearshore subaqueous fan sedimentary system [33]. However, different areas in the northern steep-slope zone of the No. 1 fault in the Weixinan Sag are affected by large differences in burial depth, differences in the overall sedimentary environment, and varying oil and gas reservoir types [34–36]. There are still many problems in the prediction of high-quality reservoirs in the northern steep-slope zone. The specific influencing factors and the distribution law of high-quality reservoirs are still unclear.

Fan delta and nearshore subaqueous fan deposits are developed in the Liushagang Formation of the steep-slope belt to the north of the study area. Under the constraints of general sedimentary facies, a series of high-quality reservoir distribution problems exist in the northern steep-slope zone of the Weixinan Sag in the Beibuwan Basin. In this contribution, we choose the slope belt of a faulted lacustrine basin to dissect the high-quality reservoir and origin. (1) The abundant data show differences in reservoir development in the northern steep-slope zone's western, central, and eastern areas. (2) Through the reservoir differences of different areas in the northern steep-slope zone, the reasons for controlling the reservoir differences are clarified, and it is considered that the sedimentary, lithofacies and sandstone thickness factors jointly control the reservoir differences. (3) By studying the controlling factors of high-quality reservoirs in different areas, two high-quality reservoir development models, a fan delta model and a nearshore subaqueous fan model, are summarized to provide a corresponding basis for studying high-quality reservoir distribution.

2. Geological Setting

The Beibuwan Basin to the north of Hainan Island, south of Guangxi, and connected with the Yinggehai Basin in the west, is the primary petroleum-bearing basin north of the south China Sea, with an area of approximately 39,000 km^2 (Figure 1A). The entire basin contains eight sags and three uplifts. The Weixinan Sag northwest of the Beibuwan Basin has an area of 3000 km^2 (Figure 1B). It is bounded by the Yuegui Uplift to the northwest, the Weixinan Uplift to the southwest, and the Qixi Uplift to the east and southeast. The Weixinan Sag is surrounded by mountains to the north and south and connects the east to the west. The Weixinan Sag can be divided into three subsags: the A subsag to the north, the B subsag in the middle, and the C subsag to the west [37–39]. In the Cenozoic era, the Weixinan Sag experienced complex tectonic evolution activities and can be divided into two stages of tectonic evolution: the rifting stage (the Changliu Formation to the Weizhou Formation) and the depression stage (the Weizhou Formation to the Wanglougang Formation) [40–43]. The fault activity is noticeable during the rifting period, and faults control the basin's development. However, during the depression period, the fault activity weakened or disappeared, and sedimentation controlled the basin development. Faults are widely developed in the Weixinan Sag. The No. 1 fault system is developed in the northern boundary area, and the No. 2 fault system is in the basin's center (comprising many en-echelon faults) (Figure 1C).

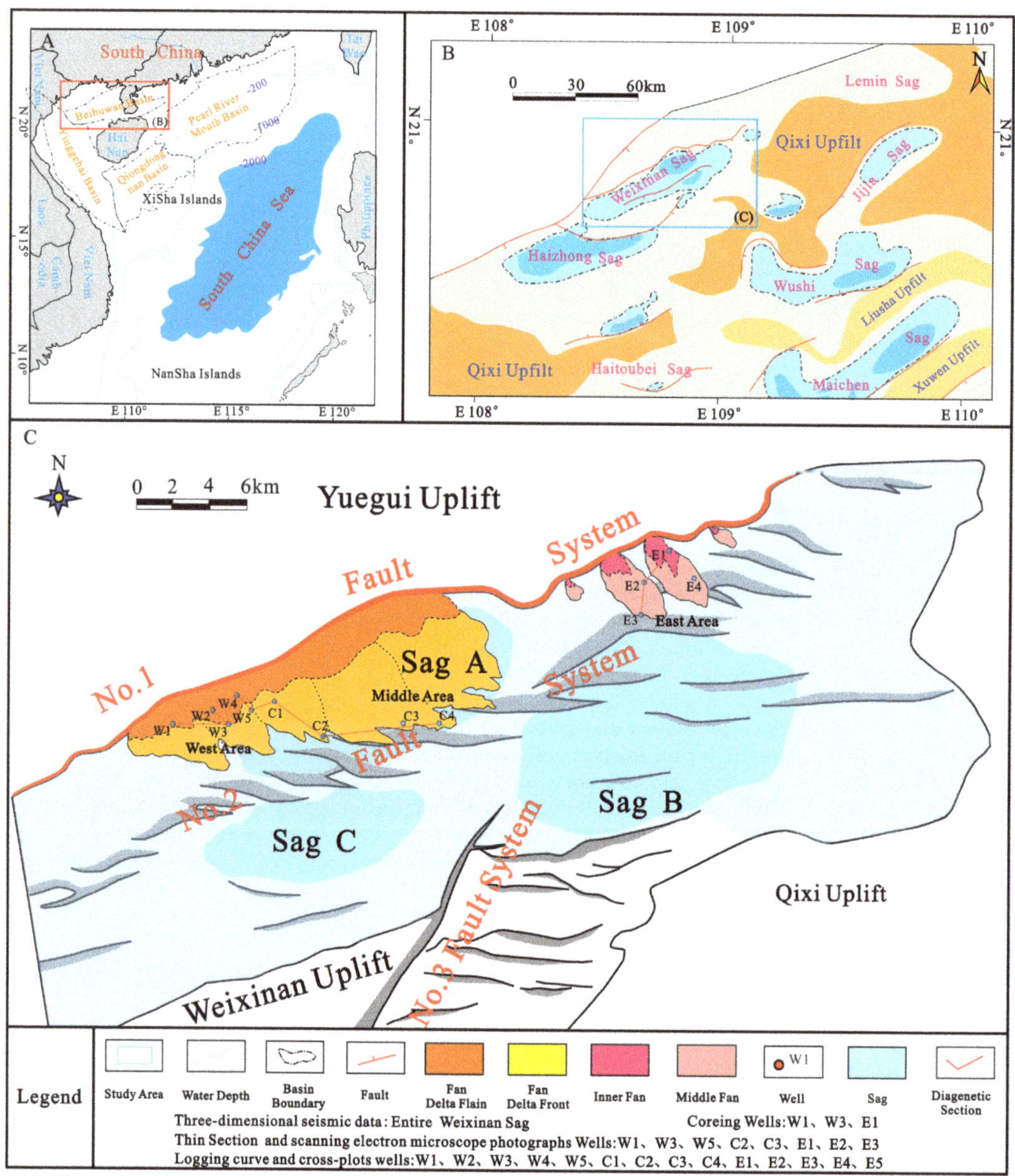

Figure 1. (**A**). Location map of sedimentary basins in the south China Sea. (**B**). Location map of the Weixinan Sag and other sags in the Beibuwan Basin. The Beibuwan Basin is shown in (**A**). (**C**). The division of the specific depositional system of the Liushagang Formation in the northern steep-slope zone in the Weixinan Sag. The Weixinan Sag is shown in (**B**).

The entire Cenozoic strata are presented in the Weixinan Sag, with a thickness of 6700 m, including continental sedimentary strata in the Paleogene and marine sedimentary strata in the Neogene and Quaternary.

From the bottom to the top, the Changliu, Liushagang, and Weizhou Formations are developed in the Paleogene, and continental sediments, such as lake and delta facies, are primarily developed (Figure 2). The Changliu Formation (65.5–55.8 Ma) is typically less than 300 m. The lithology is brownish-red and purplish-red sandy mudstone, mudstone, sandy conglomerate, and pebbly sandstone. Alluvial fluvial facies deposits are developed. In the early rifting stage, the Liushagang Formation's strata (55.8–35 Ma) have a thickness of approximately 2000 m [38,41]. Sequence stratigraphy analysis shows that the Liushagang Formation can be divided into three members. The lower sequence is the Liushagang Formation's third member, the middle sequence is the second member, and the upper sequence is the first member. In the Liushagang Formation's third member (T90–T86), the lake level was low, and the lithology was pebbly sandstone mixed with thin mudstone. A set of fan delta and shore shallow lake deposits developed. In the Liushagang Formation's second member (T86–T83), the lake level rose, and the lithology was dark mudstone, oil shale, and thin sandstone, and a set of lacustrine deposits developed. In the Liushagang Formation's first member (T83–T80), the lake level dropped again. The lithology is medium–fine sandstone mixed with mudstone, and a set of braided river delta deposits developed [38,44–46]. The thickness of the Weizhou Formation (35–23 Ma) is large and the maximum is over 3000 m. The lithology is the interbedding of sandstone, conglomerate, and mudstone, and the meandering river delta depositional system developed (Figure 2). The study area is in the eastern, central, and western areas of the A subsag northwest of the No.1 fault system. A set of fan delta sedimentary systems from the Yuegui Uplift developed.

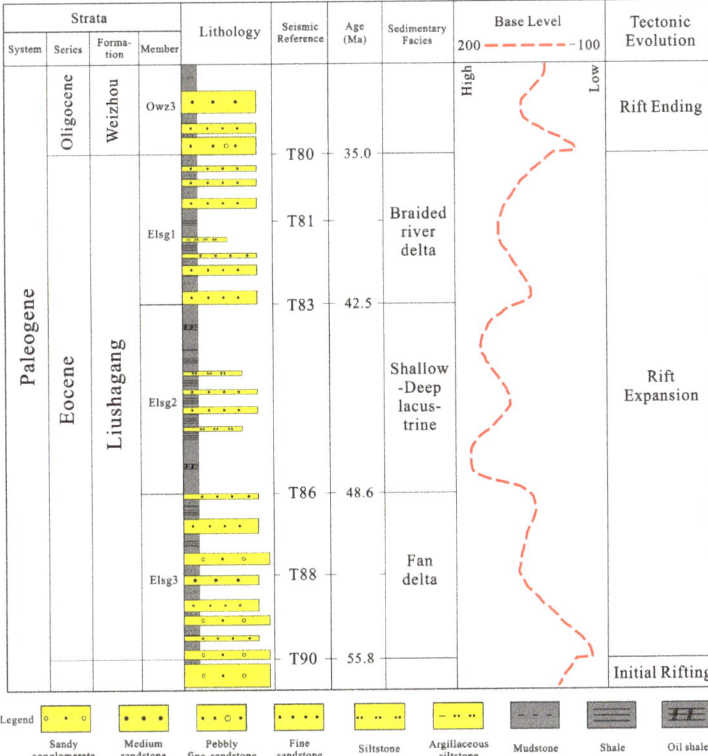

Figure 2. Stratigraphic evolution histogram of the Weixinan Sag, Beibuwan Basin.

3. Materials and Methods

Much three-dimensional seismic, core, logging, and thin-section data from identification, analysis, and testing are used to analyze the reservoir diversity in the northern steep-slope zone of the Weixinan Sag. All data are from the CNOOC Hainan Branch.

The three-dimensional seismic data cover 500 km^2 of the entire northern steep-slope zone, and the dominant seismic frequency is 30–35 Hz. The data are used for sequence stratigraphic analysis, restoring paleogeomorphology and the drainage system, and studying the scale of the sedimentary channel and system in the northern steep-slope zone of the Weixinan Sag to distinguish the differences between the west, middle, and east areas [47]. Therefore, based on the identification of denudation areas, denudation/downlap areas, and downlap areas, the source and sink areas should be considered in the restoration process. The specific steps include: (1) eliminating the subsidence difference caused by post-rifting tectonic movement; (2) dividing the source-to-sink systems; (3) recovering the denudation volume in the denudation/downlap area of each source-to-sink system; (4) calculating the denudation volume in the denudation area of each source-to-sink system; and (5) restoring the geomorphology.

This area has 3 coring wells (Wells W1, W3, E1), and the overall coring length of the Liushagang Formation section is more than 50 m. The analysis of typical core photos is used for the lithologic discrimination and fine description of different sedimentary microfacies to analyze the oil–gas properties of high-quality reservoirs. The Liushagang Formation has more than 10 wells drilled. The statistics of lithology combination and sand content for several wells are used for analyzing the sedimentary microfacies and the statistics of reservoir sandstone thickness, clarifying the critical sandstone thickness for reservoirs.

Fourteen wells in the Liushagang Formation were identified by thin sections and scanned by an electron microscope (Wells W1, W3, W5, C2, C3, E1, E2, E3) to analyze the reservoir mineral type, reservoir type, reservoir diagenesis, and reservoir diagenetic facies distribution (Wells W1, W2, W3, W4, W5, C1, C2, C3, C4, E1, E2, E3, E4, E5).

A few wells were analyzed for reservoir oil–gas properties, porosity, and permeability. By combining various data using the theories of sedimentology, sequence stratigraphy, and sedimentary reservoirs, this study summarizes the control factors of the differences between high-quality reservoirs in the study area and predicts the development model of high-quality reservoirs.

4. Results and Interpretations

Analysis of the exploration data on Weixinan Sag confirms noticeable differences between the reservoirs in the western, central, and eastern areas of the northern steep-slope zone of Weixinan Sag, including the following three sections:

4.1. Catchment-Fan Systems along the Steep-Slope Zone

(1) Drainage system

By combining 3D seismic data and denudation restoration, we restored the Wanshan Uplift's landform and drainage system [47]. The western and central areas have a large source area and adequate material supply. The short-axis steep-slope source-to-sink and drainage systems are developed. It is a composite drainage system, and the entire source area is connected, showing a composite linear source.

The eastern source area is small, and the material supply is inadequate. The short-axis slope source-to-sink system is developed, and the drainage system is undeveloped. It is a single drainage system, and the entire source area is isolated, showing a single-point source (Figure 3).

(2) Supply flux

Four aspects of the supply flux are counted: channel (slope), valley (width, depth, and area), fault (activity), and sedimentary system (burial depth, sedimentary area, thickness, and extension length). The differences between the west, middle, and east areas are compared (Figure 4 and Table 1).

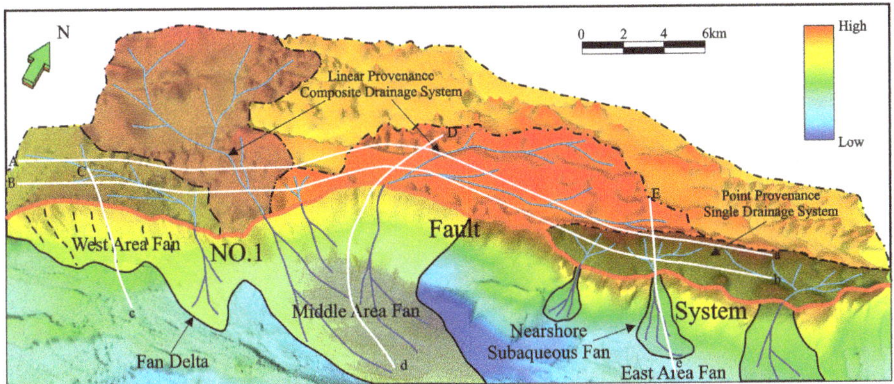

Figure 3. Distribution map of paleogeomorphology and drainage system in different areas of the northern steep-slope zone of the Weixinan Sag (The Aa, Bb, Cc, Dd and Ee are seismic section).

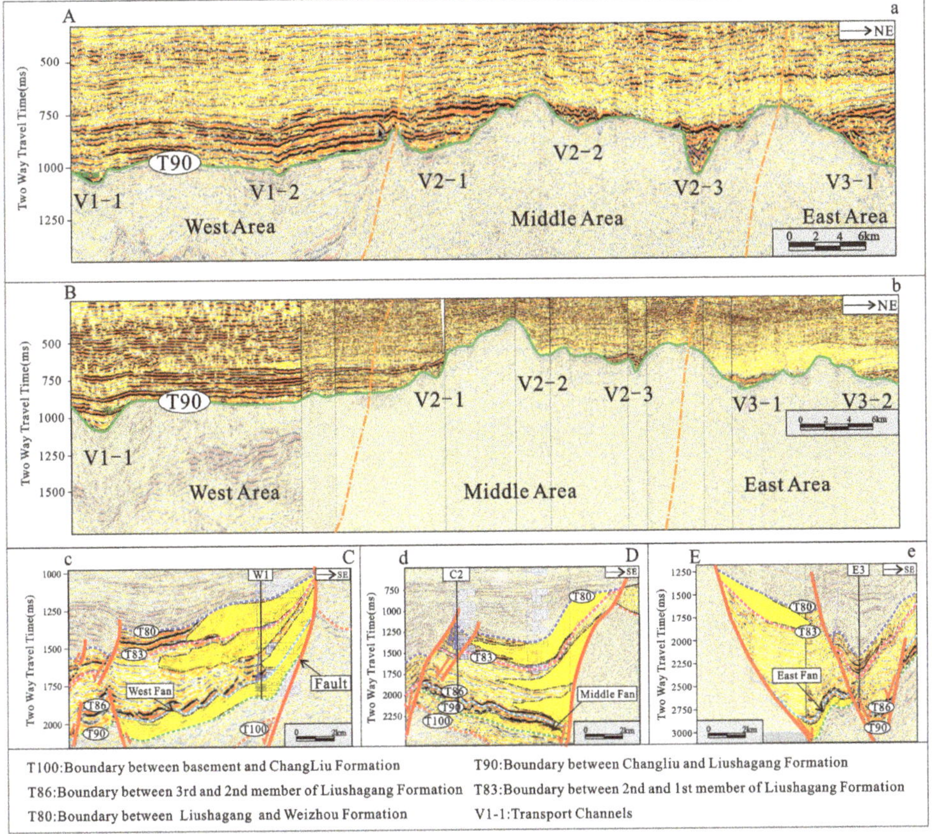

Figure 4. Interpreted seismic section Aa and Bb (SW–NE) of the gully and interpreted seismic section Cc, Dd, and Ee (NW–SE) of the sedimentary system in the northern steep-slope zone of the Weixinan Sag, Beibuwan Basin (the location of the section is shown in Figure 3).

Table 1. The channel information of the west, middle, and east areas in the northern steep-slope zone of the Weixinan Sag.

Area	Number	Channel		Valley		Fault	Deposition System			
		Slpoe/°	Depth/m	Width/m	Area/m²	Activity/m/Ma	Deposition Depth/ms	Deposition Area/m²	Thickness/m	Extend/m
West Area	V1-1	4	0.2	4	0.6	500	1850–2150	200	0.6	15.5
	V1-2	3	0.15	3	0.5	550				
Middle Area	V2-1	6	0.2	4	0.7	300	2000–2250	300	0.45	25
	V2-2	7	0.2	6	0.65	200				
	V2-3	8	0.3	4	0.8	150				
East Area	V3-1	8	0.25	3	0.5	700	2750–3000	125	0.3	10
	V3-2	7	0.23	2.5	0.4	610				

The western area is characterized by a small slope, medium gully, strong fault activity, shallow burial depth of the sedimentary system, and medium-scale development. The central area is characterized by a medium slope, large gully, medium fault activity, medium burial depth of the sedimentary system, and large-scale development. The eastern area is characterized by a large slope, small gully, strong fault activity, large burial depth of the sedimentary system, and small-scale development.

The drainage system in the western area is developed, the sediment supply is large, and the reservoir scale is medium to large. The central area is far from the fault, the sediment supply is weak, and the reservoir scale is large. The eastern area has an undeveloped drainage system, a small sediment supply, and a small reservoir scale. (Figure 4 and Table 1).

(3) Sedimentary facies

Different sedimentary facies represent different sedimentary environments and control the reservoir types [6,10,13,14,26,27]. Fan delta and nearshore subaqueous fan depositional systems are developed in the study area.

In the Liushagang Formation of well W1 in the west, fan delta plain distributary channel microfacies are developed. A large set of thick sandstone developed between 2100 and 2200 m (Figure 5). The core shows that the lithology is grayish-brown massive oil-bearing medium sandstone. The reservoir's thin section at 2118.1 m shows that the content of matrix and muddy is low, and many primary and secondary pores are developed (Figure 5). It is a suitable reservoir type.

The fan delta front underwater distributary bay microfacies are developed in the middle area. The overall lithology is fine, the content of muddy and matrix is high, the primary porosity is reduced, and the secondary porosity is increased, making it a medium reservoir type.

During the Liushagang Formation of well E1 in the east, the middle fan branch channel microfacies are developed (Figure 5). The core at 3049.37 m shows that the lithology is gray, massive siltstone. The reservoir's thin section at 3062 m shows that the matrix and muddy content is high, and the proportion of primary pores is reduced and undeveloped. The proportion of secondary pores is high, and a small amount is developed, making it a poor reservoir type (Figure 5).

Fan delta plain distributary channel microfacies are developed in the west and are the primary reservoirs. Fan delta front distributary bay microfacies are developed in the central area, making them medium reservoirs. In the eastern area, nearshore subaqueous fan and middle fan branch channel microfacies are developed, making it the worst reservoir.

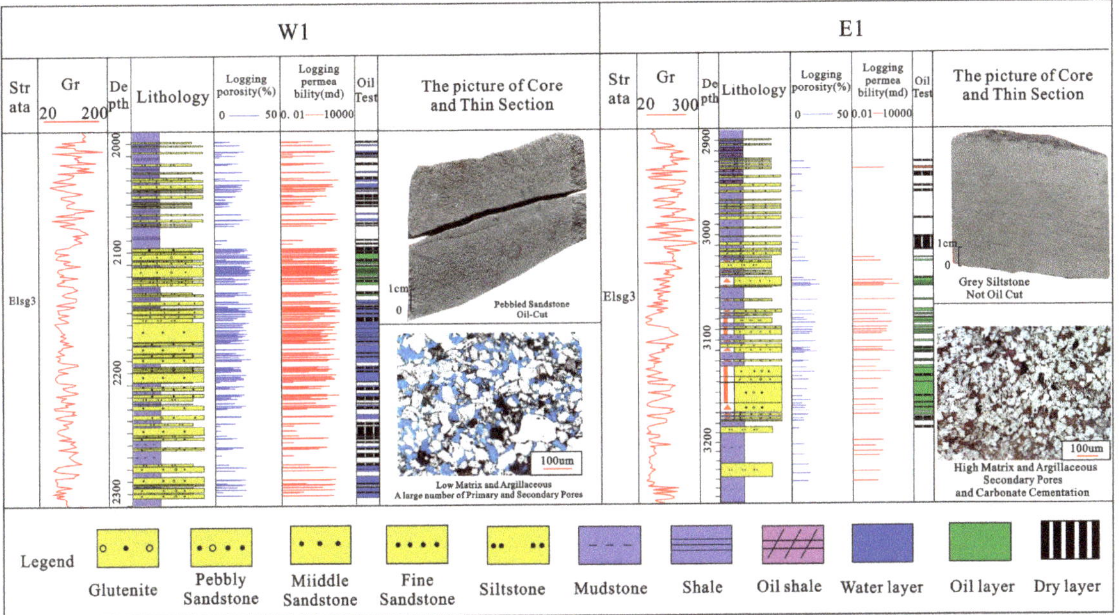

Figure 5. Single-well cores, logging, and reservoir thin sections in the west and east areas (well W1 (the well location is shown in Figure 1C) in the west, with good physical properties, and well E1 (the well location is shown in Figure 1C) in the east, with poor physical properties).

4.2. Diagenetic Processes and Facies

(1)   Diagenesis type

Various reservoir diagenesis types exist [48,49]. This study clarifies the differences between diagenesis in the east, middle, and west areas regarding compaction, cementation, and dissolution and guides the study of diagenetic facies. Through the analysis of typical well thin sections in the northern steep-slope zone of the Liushagang Formation, the west area is shallow buried and has weak compaction, and the detrital grains contact is primarily the point contact (Figure 6A); interstitial materials are argillaceous cementation (Figures 6F and 7A), and carbonate and siderite cementation also developed. The primary pores are well developed, two-stage acid corrosion of meteoric freshwater + organic acid is developed, and the corrosion pores are developed (Figure 6G).

The middle area is moderately buried and moderately compacted, and the detrital grains are the line contact (Figure 6B); the interstitial materials are cemented by argillaceous, carbonate, and clay minerals (Figures 6F and 7B), and the primary and secondary pores are developed. The late organic acid dissolution is primary, and the dissolution pores are medium (Figure 6H).

The east area is deeply buried, strongly compacted, and the detrital grain contact is concave–convex (Figure 6C). The interstitial materials are cemented by argillaceous and carbonate minerals (Figures 6F and 7C), and the secondary pores are primary. The dissolution of meteoric freshwater is limited, and the dissolution of late organic acids is primary, and the dissolution pores are small (Figure 6I).

(2)   Diagenetic facies

Different diagenetic facies control different reservoir types and predict high-quality reservoir development [27,28,50]. Diagenetic facies types were divided by thin section observation, logging curve classification, and cross-plots of different logging curves. Five types of diagenetic facies occur in the study area: Type I is weak compacted and cemented diagenetic facies, Type II is medium compaction and dissolution diagenetic facies, Type III is strong compaction and medium dissolution diagenetic facies, Type IV is compaction and argillaceous filling diagenetic facies, and Type V is dense compaction diagenetic facies.

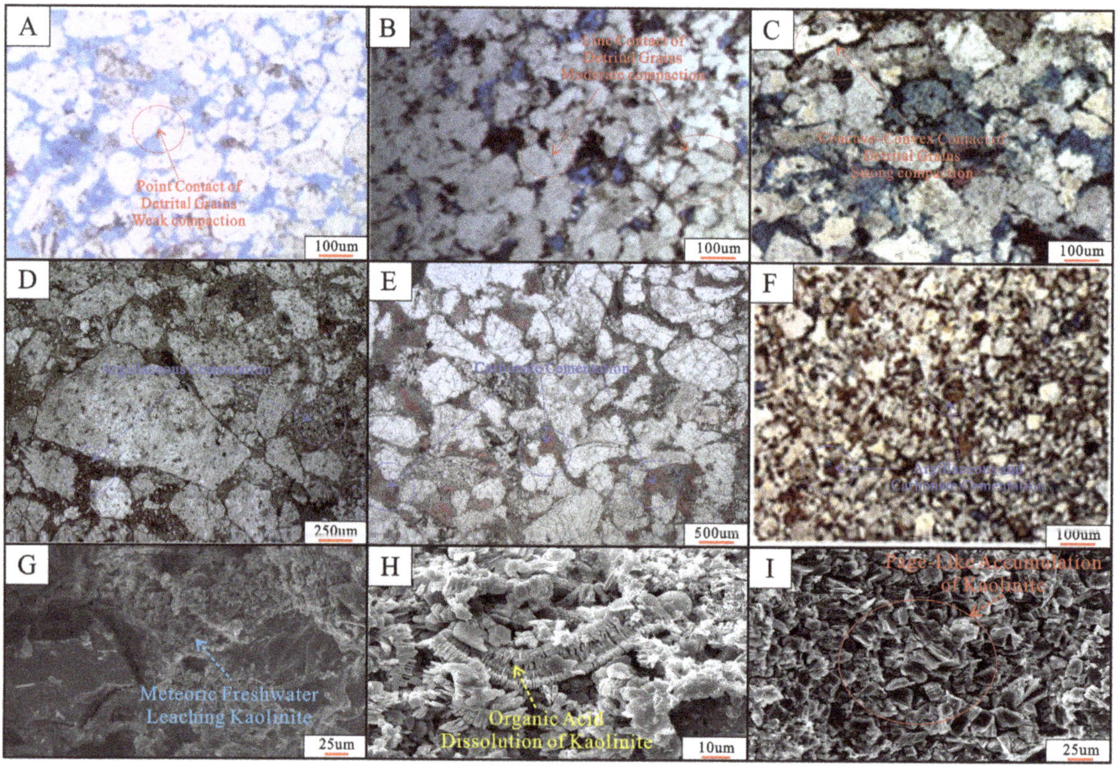

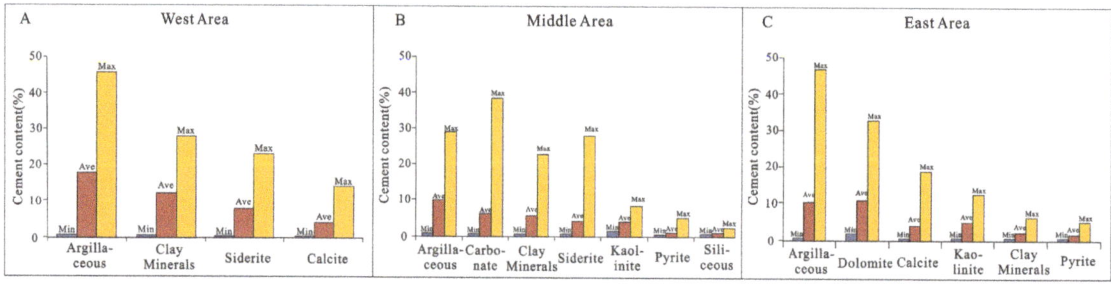

Figure 6. The thin section and scanning electron microscope photographs of the diagenesis of a typical well in the northern steep-slope zone of the Weixinan Sag. (**A**). Well W5, 2053.2 m, the point contact of detrital grains. (**B**). Well C3, 3158.44 m, the line contact of detrital grains. (**C**). Well E3, 3566.74 m, the concave–convex contact of detrital grains. (**D**). Well W3, 1935.9 m, the argillaceous cementation. (**E**). Well C3, 3229.31 m, the carbonate cementation. (**F**). Well E3, 3571.82 m, the argillaceous and carbonate cementation. (**G**). Well W1, 2152.65 m, the early meteoric freshwater leaching kaolinite. (**H**). Well C2, 2677.16 m, the late organic acid dissolution of kaolinite. (**I**). Well E2, 3047.85 m, the page-like accumulation of kaolinite and typical late dissolution.

Figure 7. The cement types and contents of west area (**A**), Middle area (**B**) and East area (**C**) of the Liushagang Formation in the northern steep-slope zone of the Weixinan Sag.

We mainly used two methods to divide diagenetic facies:

1. Typical thin sections and quantitative statistics

Through thin-section observation of typical wells, we have summarized the characteristics of five types of diagenetic facies: Type I, weak compacted and cemented diagenetic facies (weak compacted, many primary pores are developed, porosity > 20%); Type II, medium compaction and dissolution diagenetic facies (the primary and secondary pores are developed, the porosity ranges from 15% to 20%); Type III, strong compaction and medium dissolution diagenetic facies (strong compacted, the secondary pores are developed, the porosity ranges from 10% to 15%); Type IV, compaction and argillaceous filling diagenetic facies (strong compacted, the content of matrix and muddy is high, the porosity ranges from 6% to 10%); and Type V, dense compaction diagenetic facies (strong cementation, a small amount of secondary pores are developed, porosity < 6%).

2. Logging curve and cross-plot identification (GR, RD, DEN, CNC, AC)

The range of logging curve values of different diagenetic facies types can be determined through the cross-plots of logging curves. In combination with AC, CNC, DEN, RD, and GR logging curves, we divided the diagenetic facies in the study area (Table 2).

Table 2. Typical logging curves of reservoir diagenetic facies.

Diagenetic Facies Type	AC/(um/S)	CNC/(v/v)	DEN/(g/m^3)	RD/(Ω/m)	GR/(API)
Type I weak compacted and cemented diagenetic facies	100–130	0.3–0.5	2.0–2.3	0.7–6	40–60
Type II medium compaction and dissolution diagenetic facies	70–90	0.1–0.2	2.3–2.5	5–20	50–75
Type III strong compaction and medium dissolution diagenetic facies	60–70	0.05–0.15	2.4–2.6	20–45	70–85
Type IV is compaction and argillaceous filling diagenetic facies	70–90	0.1–0.2	2.3–2.5	5–20	80–95
Type V is dense compaction diagenetic facies	60–70	0.05–0.15	2.4–2.6	20–45	90–110

1. Type I weak compaction and weak cementation facies: high GR, AC, and CNC and low RD and DEN.
2. Type II medium compaction medium dissolution facies: high GR, AC, and CNC and low RD and DEN.
3. Type III strong compaction medium strong dissolution facies: high GR, AC, and CNC and low RD and DEN.
4. Type IV compaction argillaceous filling facies: low GR, CNC, and AC and high RD and DEN.
5. Type V tight cementation facies: low GR, CNC, and AC and high RD and DEN (Table 2).

Combined with various diagenetic facies, we studied the diagenetic facies in different areas by connecting wells. The west area primarily develops thick massive pebbly sandstone with Type I weak compaction and cementation diagenetic facies, Type IV compaction argillaceous filling diagenetic facies at the root, and Type V dense cementation diagenetic facies at the thin-front sandstone (Figure 8).

The middle area is dominated by Type II medium compaction and dissolution diagenetic facies and Type III strong compaction and medium dissolution diagenetic facies. Unlike the west area, Type V tight cemented diagenetic facies are more developed (Figure 8).

The east area is dominated by Type III strong compaction and medium dissolution diagenetic facies of an underwater distributary channel sandstone reservoir (Figure 9). The thin layer primarily comprises Type V dense cemented diagenetic facies.

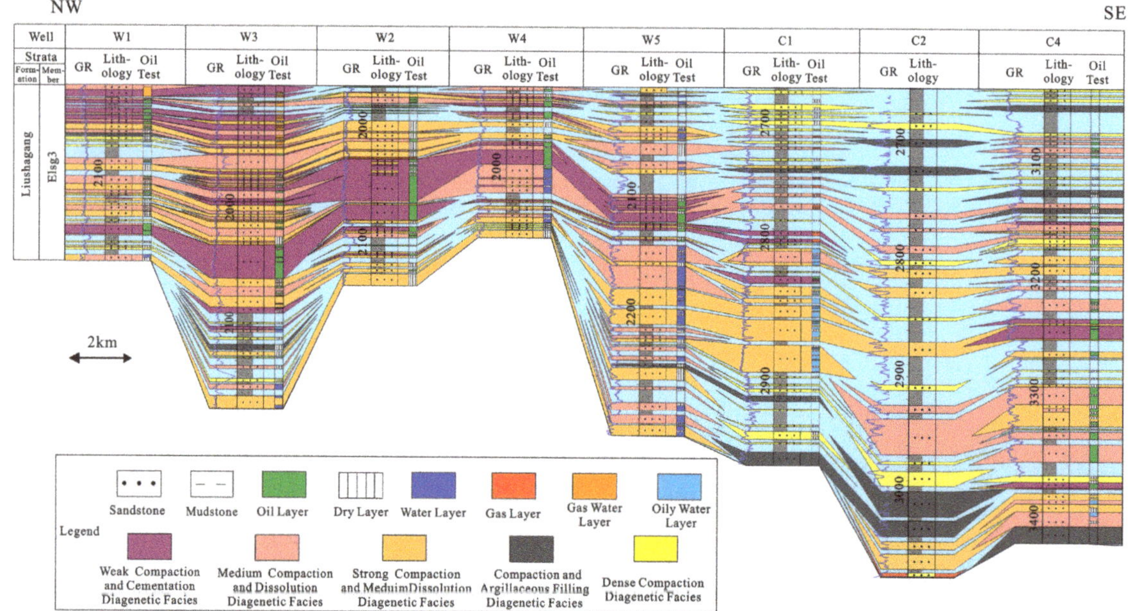

Figure 8. The diagenetic facies correlation section from wells W1 to C4 showing the diagenetic evolution of the west and middle areas in the Weixinan Sag (the section location is shown in Figure 1).

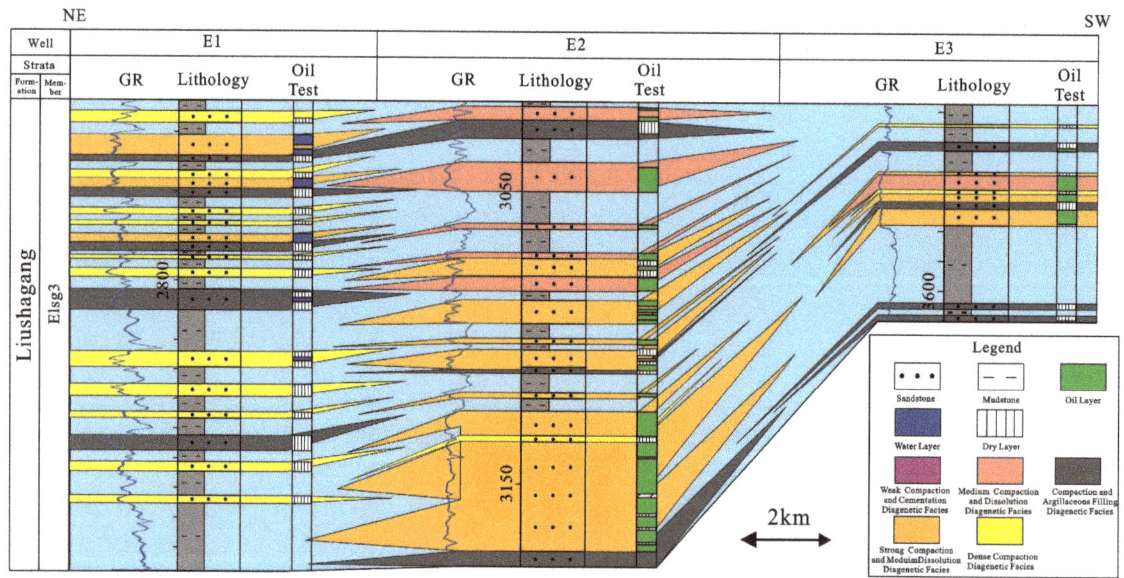

Figure 9. The diagenetic facies correlation section from wells E1 to E3 showing the diagenetic evolution of the east area in the Weixinan Sag (the section location is shown in Figure 1).

4.3. Porosity, Permeability, and Oil Saturation

(1) Porosity and permeability

Through physical property statistics, the physical property characteristics of the three areas in the steep-slope zone of the Weixinan Sag are as follows.

The porosity of the western area ranges from 0.6% to 28.5%, averaging 15.1% (Figure 10A). The permeability is between 0.01 and 4481 mD, averaging 184.4 mD (Figure 10B). The porosity of the central area ranges from 0.3% to 25.6%, averaging 12.7% (Figure 10A), and the permeability is between 0.01 and 2335 mD, averaging 62.1 mD (Figure 10B). The porosity of the eastern area ranges from 1.7% to 13.6%, averaging 8.8% (Figure 10A), and the permeability is between 0.06 and10.2 mD, averaging 0.97 mD (Figure 10B).

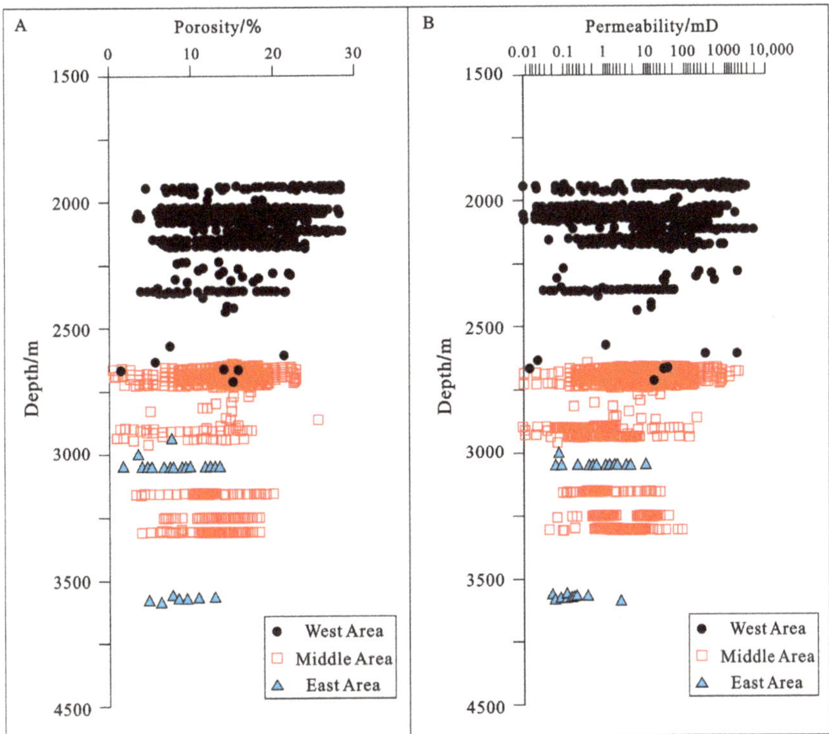

Figure 10. The relationship between (**A**) porosity and (**B**) permeability and depth in the western, central, and eastern areas.

Further, we divided these reservoirs into different types by adopting the following criteria:

1. Type I conventional reservoir: porosity > 12% and permeability > 10 mD;
2. Type II low permeability reservoir: porosity range from 6% to 12% and the permeability is between 1 and 10 mD;
3. Type III tight reservoir: porosity < 6%, permeability < 1 mD (Table 3).

Table 3. The reservoir classification standards table.

Reservoir Type	Porosity/(%)	Permeability/(mD)
Type I Conventional Reservoirs	>12%	>10 mD
Type II Low Permeability Reservoirs	6–12%	1 mD–10 mD
Type III Tight Reservoirs	<6%	<1 mD

(2)   Oil saturation

First, the reservoir grade controls the reservoir's oil–gas properties, and we clarified the relationship between different reservoirs and oil saturation. The western region primarily developed Type I conventional reservoirs, with oil saturation from 35 to 90%. In the central area, Type II low permeability reservoirs are developed, with Type III tight reservoirs developed around them. The oil saturation is 20–68%. In the eastern region, the range of Type I conventional reservoirs is small, and most areas develop Type II low permeability reservoirs and Type III tight reservoirs with oil saturation of 10–45%. (Figure 11).

Sandstone thickness is another crucial factor controlling oil saturation. In the northern steep-slope zone of the Weixinan Sag, the critical sandstone thickness (the oil saturation in most areas exceeding the critical thickness is more than 50%) in different areas is determined from the statistical analysis of sandstone thickness and oil saturation in the western, central, and eastern areas. The critical sandstone thickness in the west area is the smallest at approximately 3 m, while that in the middle area is medium at approximately 5 m and that in the east area is the largest at approximately 8 m (Figure 12A,B).

Simultaneously, we studied the distribution of the average sandstone thickness in different areas. The sandstone thickness in the west is the largest (average sandstone thickness 40 m), and the sandstone thickness in the middle area is medium (average sandstone thickness 25 m). The sandstone thickness in the east is the thinnest (average sandstone thickness < 20 m) (Figure 13).

The critical sandstone thickness in the western region is small, the average sandstone thickness is large, and the oil-bearing property is the best. The critical sandstone thickness in the central region is medium, the average sandstone thickness is medium, and the oil-bearing property is medium. The critical sandstone thickness in the eastern region is large, the average sandstone thickness is small, and the physical property is the worst.

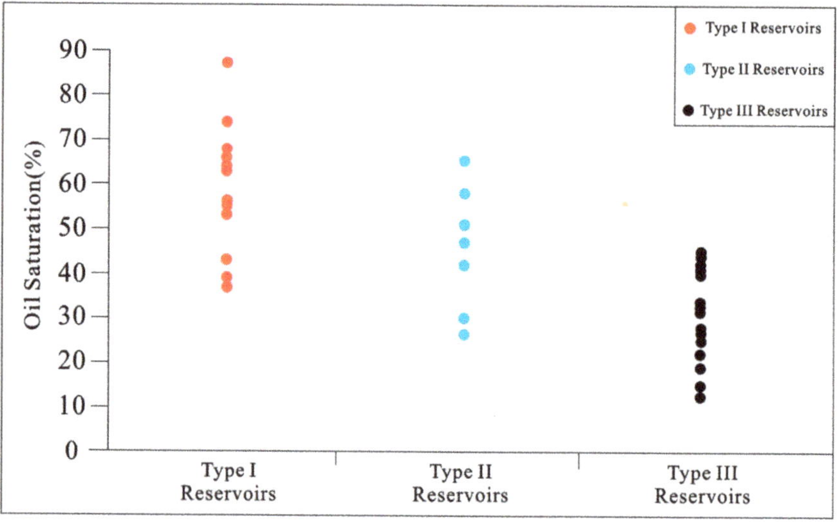

Figure 11. The oil saturation characteristics of different reservoirs.

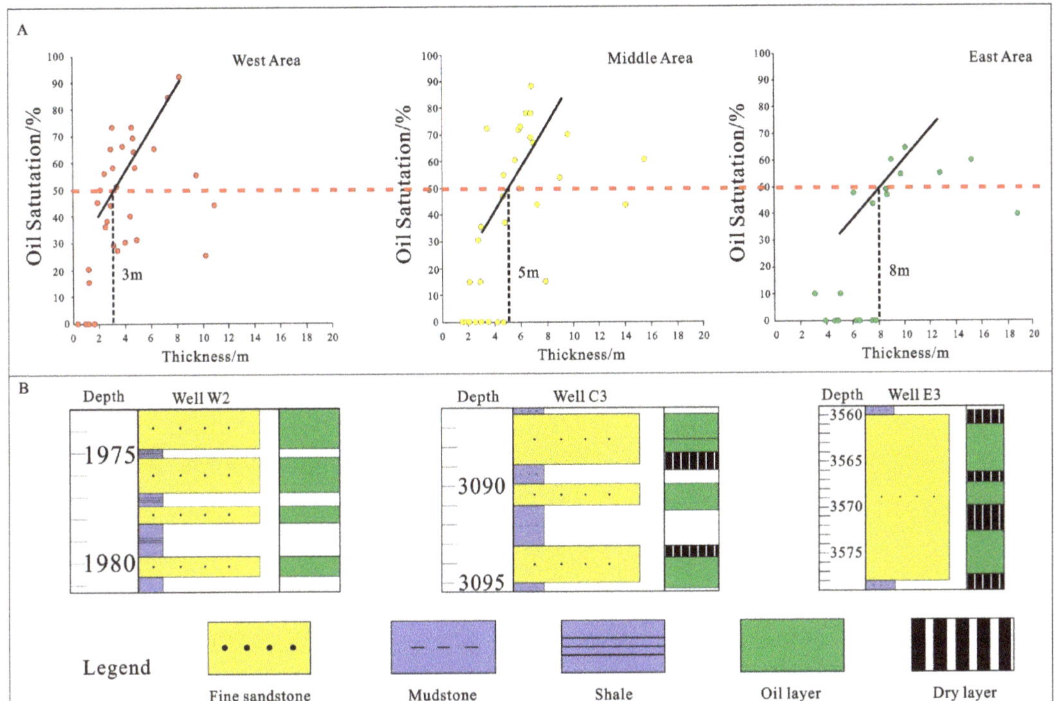

Figure 12. (**A**) Statistical relationship between critical sandstone thickness and oil saturation in different areas of the northern steep-slope zone of the Weixinan Sag. (**B**) Sandstone thickness and oil–gas bearing analysis of typical well logging in different areas.

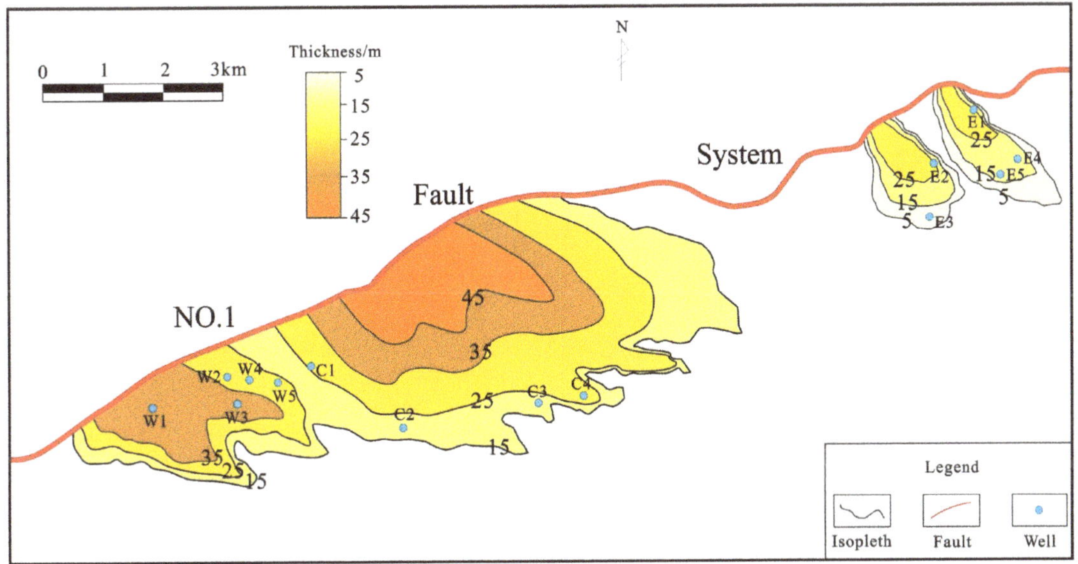

Figure 13. Average sandstone thickness in different areas of the Liushagang Formation in the northern steep-slope zone of the Weixinan Sag (with an average of 40 m in the west, 25 m in the middle, and 20 m in the east).

5. Discussion

Combined with the reservoir differences in different areas, the control factors and models of high-quality reservoirs were studied, and finally, the distribution of high-quality reservoirs was predicted.

5.1. Factors Controlling Reservoir Quality

For the steep-slope sedimentary system, comparing the reservoir differences in the western, central, and eastern areas of the Weixinan Sag, macro-to-micro-reservoir control factors are summarized. It is considered that sedimentary, lithofacies and sandstone thickness factors jointly control the differences between reservoirs.

Sedimentary factors are macroscopic aspects of reservoir development. Different sedimentary environments control different reservoir scales and types, primarily including the drainage system and sedimentary microfacies. The dominant sedimentary characteristics of the steep-slope sedimentary system are, to a considerable extent, shallow burial depth, adequate material supply, large reservoir scale, and the development of microfacies in the distributary channel of fan delta plain.

Lithofacies factors play a decisive role in high-quality reservoirs, primarily controlling high-quality reservoirs from microscopic distribution, including diagenesis and diagenetic evolution. The dominant lithofacies characteristics of high-quality reservoirs in the steep-slope sedimentary system are weak diagenesis and diagenetic evolution, developing Type I reservoirs.

The sandstone thickness factor controls the reservoir's oil-bearing property, including porosity, permeability, and oil saturation. The dominant sandstone thickness characteristics of the steep-slope sedimentary system are larger average sandstone thickness, smaller critical sandstone thickness, and higher oil-bearing properties.

5.2. Models of High-Quality Reservoirs

In combination with the reservoir differences and controlling factors in different areas, high-quality reservoir prediction models for different sedimentary systems were established. Fan delta sedimentary systems were developed in the western and central areas, and fan delta reservoir prediction modes were developed.

The characteristics of the fan delta reservoir model are as follows. It is a Type I reservoir in the fan delta plain. The lithology is medium–coarse sandstone containing a small amount of gravel. The two-stage acid fluid of organic acid + meteoric freshwater is active, the dissolution is strong, and the dissolved material migrates out of the system. Intergranular, intragranular, and matrix dissolved pores are developed, and the physical properties are the best. The Type II reservoir is in the fan delta front, the lithology is medium–fine sandstone, and argillaceous fine sandstone can be seen in the outer front area. The clay matrix and debris contents are high, and only organic acid dissolution is developed. The Type III reservoir is in the pre-delta, comprising mudstone, fine grain size, poor sorting, high shale content, compaction, and mostly tight layers (Figure 14A).

The nearshore subaqueous fan sedimentary system in the eastern areas belongs to the nearshore subaqueous fan prediction model. The reservoir characteristics of this model are as follows. The Type I reservoir is located in the middle fan and is dominated by medium–fine sandstone, with only organic acid fluid, intergranular, intragranular, and matrix dissolution pores developing. The physical properties are the best. The Type II reservoir is located in the inner and middle fan edge and is dominated by medium–fine sandstone and argillaceous fine sandstone. Organic acid fluid dissolution is weakened, argillaceous and mica contents are high, authigenic clay mineral content is high, and clay mineral intercrystalline pore development and the physical properties are moderate. The Type III reservoir is in the outer fan area, primarily mudstone, with a fine grain size, poor sorting, high shale content, compaction, and mostly dense layers (Figure 14B).

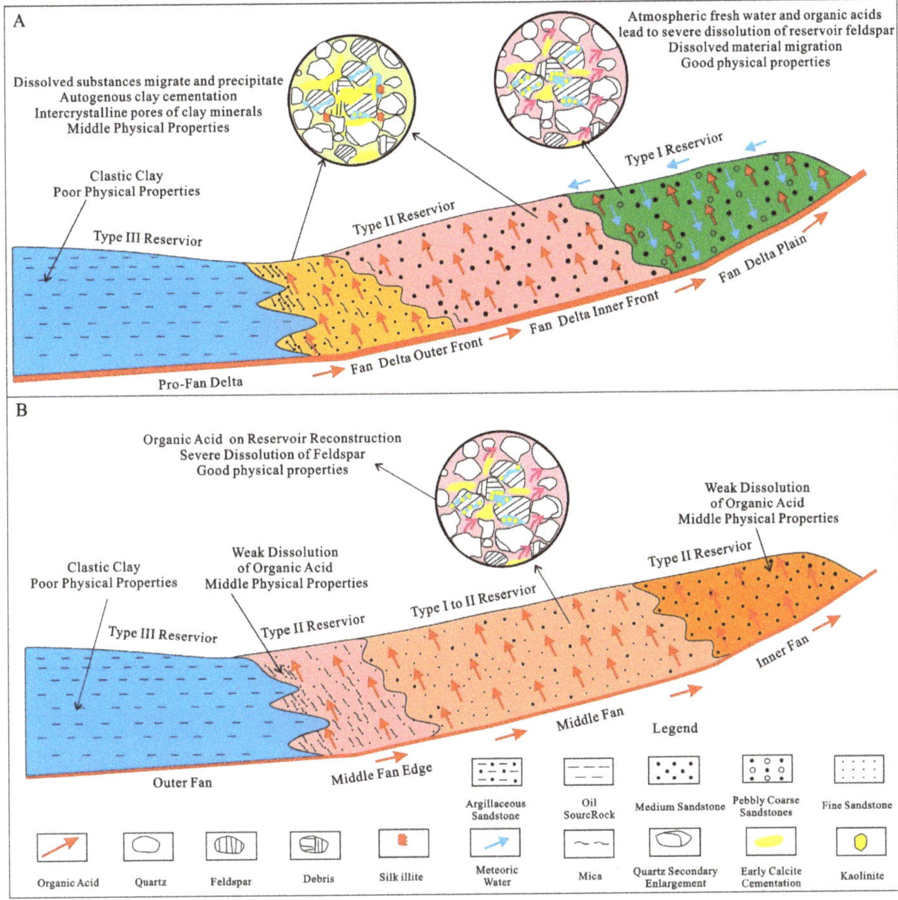

Figure 14. Reservoir prediction model of the (**A**) fan delta and (**B**) nearshore subaqueous fan in the Weixinan Sag.

5.3. Implications for Reservoir Development along the Steep-Slope Zone

Through the main controlling factors and development models of reservoirs in different areas, we predicted the distribution characteristics of high-quality reservoirs in the study area. In the western area, sweet spot reservoirs are distributed surrounded by wells W1 and W3, and Type I conventional reservoirs and Type II low permeability reservoirs are developed around them. A small number of Type III tight reservoirs is distributed in the edge area, mainly developing oil layers and dry layers.

In the central area, well C3 in the sweet spot reservoir's distribution area is small. The Type I conventional reservoir and Type II low permeability reservoir are widely distributed, and the Type III tight reservoir is slight. It develops oil, oil–water, and dry layers.

In the eastern area, with well E5 and E3 as the center, sweet spot reservoirs and Type I conventional reservoirs are developed around them, and Type II low permeability reservoirs are developed around them. Type II low permeability reservoirs account for the largest proportion, and Type III tight reservoirs are less distributed in the marginal area, developing oil, water, and dry layers (Figure 15).

The shallowly buried western area is dominated by Type I reservoirs with high oil-bearing properties, the moderately buried central area is dominated by Type II reservoirs with medium oil-bearing properties, and the deep-buried eastern area is dominated by Type II and III reservoirs with low oil-bearing properties.

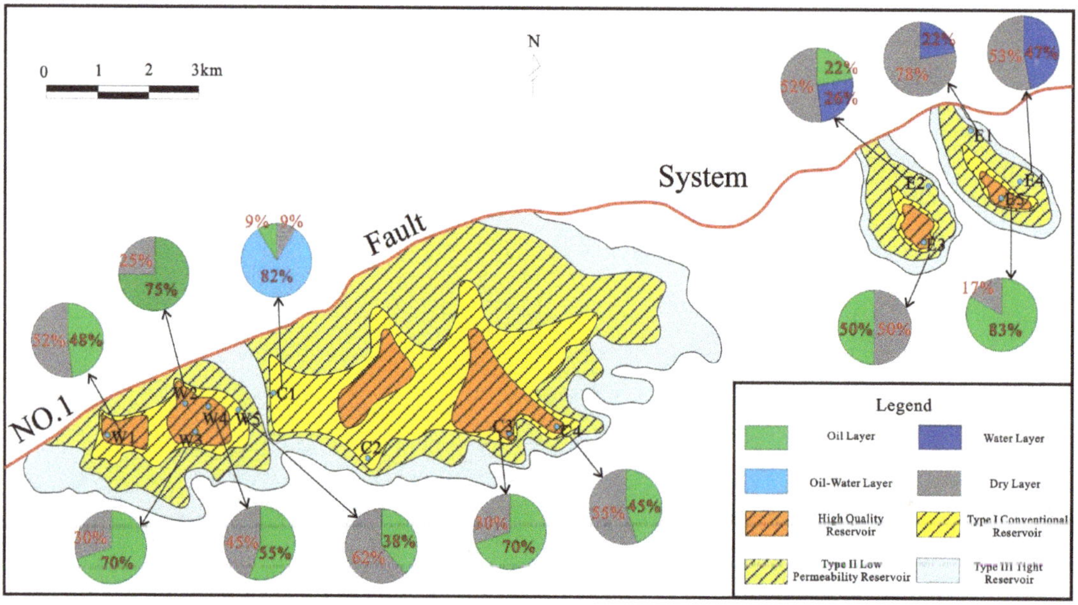

Figure 15. The high-quality reservoirs in the northern steep-slope zone of the Weixinan Sag.

6. Conclusions

1. The Liushagang Formation in the Weixinan Sag of Beibuwan Basin develops a steep-slope sedimentary system with different reservoir characteristics in the west, middle, and east areas. Integrating the reservoir variety in different areas, we summarize the controlling factors and development models for high-quality reservoirs and finally predict the location of high-quality reservoirs.

2. Combined with the reservoir differences between the western, central, and eastern areas, the macro-to-micro-reservoir controlling factors of the steep-slope system are defined. Sedimentary, lithofacies, and sandstone thickness factors jointly control the reservoir differences. The sedimentary factor is a prerequisite for affecting reservoirs, controlling their scale, material source, and type. Lithofacies factors play a decisive role in high-quality reservoirs. Different lithofacies control the diagenesis and diagenetic evolution of reservoirs. Sandstone thickness plays a significant role in the exploration of high-quality reservoirs, with different sandstone thicknesses controlling the oil-bearing properties. High-quality reservoirs in the steep-slope systems are characterized by shallow burial depths, adequate material supply, weak diagenesis, shallow diagenetic evolution, large sandstone thickness, and developing Type I reservoirs.

3. The reservoir prediction models for fan delta and nearshore subaqueous fans are summarized. The high-reservoir for the fan delta model is the fan delta plain. Organic acid + meteoric freshwater dissolution is developed, all types of dissolved pores are developed, and it has the best physical properties. The high-reservoir of the nearshore subaqueous fan model in the middle fan develops only organic acid dissolution, all types of dissolved pores develop, and the physical properties are moderate.

Author Contributions: Conceptualization, investigation, and data curation, S.L., Z.Z. and J.C.; methodology, H.Z. and Q.L.; formal analysis, validation, software, resources, and visualization, S.L.; writing—original draft preparation, S.L.; writing—review and editing, S.L. and H.Z.; supervision, H.Z.; project administration and funding acquisition, H.Z. All authors have read and agreed to the published version of the manuscript.

Funding: This work was supported by Major Outsourcing Projects of China National Offshore Oil Corporation (Hainan) (Grant Nos. CCL2020ZJFN0343).

Data Availability Statement: The data that support the findings of this study are available on request from the corresponding authors. The data are not publicly available due to confidentiality restrictions.

Acknowledgments: We thank the CNOOC (Hainan) for the release of all the data. We thank the two anonymous reviewers for reviewing this manuscript and the editorial department for editorial handling and helpful comments.

Conflicts of Interest: The authors declare no conflict of interest.

References

1. Galloway, W.E.; Hobday, D.K. *Terrigenous Clastic Depositional Systems: Applications to Petroleum, Coal, and Uranium Exploration*; Springer: New York, NY, USA, 1983; pp. 25–111.
2. Patranabis, D.S.; Chaudhuri, A.K. A retreating fan-delta system in the Neoproterozoic Chattisgarh rift basin, central India: Major controls on its evolution. *AAPG Bull.* **2007**, *91*, 785–808. [CrossRef]
3. Rohais, S.; Eschard, R.; Guillocheau, F. Depositional model and stratigraphyic architecture of rift climax Gilbert-type fan deltas (Gulf of Corinth, Greece). *Sediment. Geol.* **2008**, *210*, 132–145. [CrossRef]
4. Tang, Y.; Xu, Y.; Qu, J.H.; Meng, X.C.; Zou, Z.W. Fan-delta group characteristicsand its distribution of the Triassic Baikouquan reservoirs in Mahu Sag of Junggar Basin. *Xingjiang Pet. Geol.* **2014**, *35*, 628–635.
5. Jia, H.B.; Ji, H.C.; Wang, L.S.; Gao, Y.; Li, X.W.; Zhou, H. Reservoir quality variations within a conglomeratic fan-delta system in the Mahu sag, northwestern Junggar Basin: Characteristics and controlling factors. *J. Pet. Sci. Eng.* **2017**, *152*, 165–181. [CrossRef]
6. He, L.; Amorosi, A.; Ye, S.Y.; Xue, C.T.; Yang, S.X.; Laws, E.A. River avulsions and sedimentary evolution of the Luanhe fan-delta system (North China) since the late Pleistocene. *Mar. Geol.* **2020**, *425*, 106194. [CrossRef]
7. Coleman, J.M.; Wright, L.D. *Modern River Deltas: Variability of Processes and Sand Bodies*; Houston Geological Society: Houston, TX, USA, 1975; pp. 99–149.
8. Nemec, W.; Steel, R.J.; Porebski, S.J.; Spinnangr, A. Domba conglomerate, Devonian, Norway: Process and lateral variability in a mass flow-dominated, lacustrine fan-delta. In *Sedimentology of Gravels and Conglomerates*; Koster, E.H., Steel, R.J., Eds.; Canadian Society of Petroleum Geologists Memoir 10: Calgary, AB, Canada, 1984; pp. 295–320.
9. Orton, G.J.; Reading, H.G. Variability of deltaic processes in terms of sediment supply, with particular emphasis on grain size. *Sedimentology* **1993**, *40*, 475–512. [CrossRef]
10. Lin, C.F.; Liu, S.F.; Zhuang, Q.T.; Steel, R.J. Sedimentation of Jurassic fan-delta wedges in the Xiahuayuan basin reflecting thrust-fault movements of the western Yanshan fold-and-thrust belt, China. *Sediment. Geol.* **2018**, *368*, 24–43. [CrossRef]
11. Sendra, J.; Reolid, M.; Reolid, J. Palaeoenvironmental interpretation of the Pliocene fan-delta system of the Vera Basin (SE Spain): Fossil assemblages, ichnology and taphonomy. *Palaeoworld* **2020**, *29*, 769–788. [CrossRef]
12. Liu, Z.J. Lacus nearshore subaqueous fans sedimentary characteristics and influence factor d a case study of Shuangyang formation in Moliqing fault subsidence of the Yitong Basin. *Acta Sedimentol. Sin.* **2003**, *21*, 148–153. (In Chinese with English Abstract)
13. Cao, Y.C.; Wang, Y.Z.; Gluyas, J.G.; Liu, H.M.; Liu, H.N.; Song, M.S. Depositional model for lacustrine nearshore subaqueous fans in a rift basin: The Eocene Shahejie Formation, Dongying Sag, Bohai Bay Basin, China. *Sedimentology* **2018**, *65*, 2117–2148. [CrossRef]
14. Zhang, X.; Zhu, X.M.; Lu, Z.Y.; Lin, C.S.; Wang, X.; Pan, R.; Geng, M.Y.; Xue, Y. An early Eocene subaqueous fan system in the steep slope of lacustrine rift basins, Dongying Depression, Bohai Bay Basin, China: Depositional character, evolution and geomorphology. *J. Asian Earth Sci.* **2019**, *171*, 28–45. [CrossRef]
15. Dong, D.T.; Qiu, L.W.; Ma, P.J.; Yu, G.D.; Wang, Y.Z.; Zhou, S.B.; Yang, B.L.; Huang, H.Q.; Yang, Y.Q.; Li, X. Initiation and evolution of coarse-grained deposits in the Late Quaternary Lake Chenghai source-to-sink system: From subaqueous colluvial apron (subaqueous fans) to Gilbert-type delta. *J. Palaeogeogr.* **2022**, *11*, 194–221. [CrossRef]
16. Shi, W.Z.; Zhao, Z.K.; Jiang, T.; Miao, H.B.; Wang, X.L. Identifying updip pinch-out sandstone in nearshore subaqueous fans using acoustic impedance and the instantaneous phase in the Liangjia area, Yitong Basin, China. *Mar. Pet. Geol.* **2012**, *30*, 32–42. [CrossRef]
17. Li, J.Z.; Zhang, J.L.; Sun, S.Y.; Zhang, K.; Du, D.X.; Sun, Z.Q.; Wang, Y.Y.; Liu, L.L.; Wang, G.Q. Sedimentology and mechanism of a lacustrine syn-rift fan delta system: A case study of the Paleogene Gaobei Slope Belt, Bohai Bay Basin, China. *Mar. Pet. Geol.* **2018**, *98*, 477–490. [CrossRef]
18. Chen, Y.Y.; Deng, B.; Chen, Y.F.; Wang, D.R.; Zhang, J. Holocene sedimentary evolution of a subaqueous delta off a typical tropical river, Hainan Island, South China. *Mar. Geol.* **2021**, *442*, 106664. [CrossRef]

19. Liu, E.T.; Wang, H.; Pan, S.Q.; Qin, C.Y.; Jiang, P.; Chen, S.; Yan, D.T.; Lü, X.X.; Jing, Z.H. Architecture and depositional processes of sublacustrine fan systems in structurally active settings: An example from Weixinan Depression, northern South China Sea. *Mar. Pet. Geol.* **2021**, *134*, 105380. [CrossRef]

20. Kang, X.; Hu, W.X.; Cao, J.; Wu, H.G.; Xiang, B.L.; Wang, J. Controls on reservoir quality in fan-deltaic conglomerates: Insight from the Lower Triassic Baikouquan Formation, Junggar Basin, China. *Mar. Pet. Geol.* **2019**, *103*, 55–75. [CrossRef]

21. Wu, D.; Li, H.; Jiang, L.; Hu, S.H.; Wang, Y.H.; Zhang, Y.L.; Liu, Y. Diagenesis and reservoir quality in tight gas bearing sandstones of a tidally influenced fan delta deposit: The Oligocene Zhuhai Formation, western Pearl River Mouth Basin, South China Sea. *Mar. Pet. Geol.* **2019**, *107*, 278–300. [CrossRef]

22. Yang, X.G.; Guo, S.B. Reservoirs characteristics and environments evolution of lower permian transitional shale in the Southern North China Basin: Implications for shale gas exploration. *J. Pet. Sci. Eng.* **2021**, *96*, 104282. [CrossRef]

23. Niu, D.M.; Li, Y.L.; Zhang, Y.F.; Sun, P.C.; Wu, H.G.; Fu, H.; Wang, Z.Q. Multi-scale classification and evaluation of shale reservoirs and 'sweet spot' prediction of the second and third members of the Qingshankou Formation in the Songliao Basin based on machine learning. *J. Pet. Sci. Eng.* **2022**, *216*, 110678. [CrossRef]

24. Kra, K.L.; Qiu, L.W.; Yang, Y.Q.; Yang, B.L.; Ahmed, K.S.; Camara, M.; Khan, D.; Wang, Y.L.; Kouame, M.E. Sedimentological and diagenetic impacts on sublacustrine fan sandy conglomerates reservoir quality: An example of the Paleogene Shahejie Formation (Es4s Member) in the Dongying Depression, Bohai Bay Basin (East China). *Sediment. Geol.* **2022**, *427*, 106047. [CrossRef]

25. Obafemi, S.; Oyedele, K.; Omeru, T.; Bankole, S.; Opatola, A.; Okwudili, P.N.; Akinwale, R.; Ademilola, J.; Adebayo, R.; Victor, A.L. 3D facies and reservoir property prediction of deepwater turbidite sands; case study of an offshore Niger delta field. *J. Afr. Earth Sci.* **2022**, *194*, 104633. [CrossRef]

26. Xue, Y.A.; Zhao, M.; Liu, X.J. Reservoir Characteristics and Controlling Factors of the Metamorphic Buried Hill of Bozhong Sag, Bohai Bay Basin. *J. Earth Sci.* **2021**, *32*, 919–926. [CrossRef]

27. Mao, Z.G.; Zhu, R.K.; Wang, J.H.; Luo, J.L.; Su, L. Characteristics of Diagenesis and Pore Evolution of Volcanic Reservoir: A Case Study of Junggar Basin, Northwest China. *J. Earth Sci.* **2021**, *32*, 960–971. [CrossRef]

28. Qian, W.D.; Yin, T.J.; Zhang, C.M.; Tang, H.J.; Hou, G.W. Diagenetic evolution of the Oligocene Huagang Formation in Xihu sag, the East China Sea Shelf Basin. *Sci. Rep.* **2020**, *10*, 19402. [CrossRef]

29. He, W.G.; Barzgar, E.; Feng, W.P.; Huang, L. Reservoirs Patterns and Key Controlling Factors of the Lenghu Oil & Gas Field in the Qaidam Basin, Northwestern China. *J. Earth Sci.* **2021**, *32*, 1011–1021.

30. Sun, W.Z.; Wang, C.L.; Yang, X.B. Types and favorable exploration areas of Eocene subtle traps in Weixinan Sag, BBW Basin. *Nat. Gas Geosci.* **2007**, *18*, 8488.

31. Zhu, W.L.; Jiang, W.R. Faults and oil & gas reservoirs of the Weixinan Depression in the Beibuwan Basin. *Acta Pet. Sin.* **1998**, *19*, 610.

32. Wang, J.; Cao, Y.C.; Li, J.L. Sequence structure and non-structural traps of the Paleogene in the Weixi'nan Sag, Beibuwan Basin. *Pet. Explor. Dev.* **2012**, *39*, 325–334. [CrossRef]

33. Xi, M.H.; Yu, X.B.; Huang, J.T. Paleogene stratigraphic sequence and sedimentary feature in the west of Weixinan Depression. *Offshore Oil* **2007**, *27*, 1–12.

34. Liu, P.; Xia, B.; Tang, Z.Q.; Wang, X.G.; Zhang, Y. Fluid inclusions in reservoirs of Weixinan Sag, Beibuwan Basin. *Pet. Explor. Dev.* **2008**, *35*, 164–200. [CrossRef]

35. Gao, Z.Y.; Yang, X.B.; Hu, C.H.; Wei, L.; Jiang, Z.X.; Yang, S.; Fan, Y.P.; Xue, Z.X.; Yu, H. Characterizing the pore structure of low permeability Eocene Liushagang Formation reservoir rocks from Beibuwan Basin in northern South China Sea. *Mar. Pet. Geol.* **2019**, *99*, 107–121. [CrossRef]

36. Yang, Y. Reservoir characteristics and controlling factors of the Lower paleogene sandstones in the southeast part of Jiyang Sag, BohaiBay Basin, China. *Alex. Eng. J.* **2022**, *61*, 10277–10282. [CrossRef]

37. Wu, S.G.; Han, Q.H.; Ma, Y.B.; Dong, D.D.; Lü, F.L. Petroleum system in deepwater basins of the northern South China Sea. *J. Earth Sci.* **2009**, *20*, 124–135. [CrossRef]

38. Huang, B.J.; Tian, H.; Wilkins, R.W.T.; Xiao, X.M.; Li, L. Geochemical characteristics, palaeoenvironment and formation model of Eocene organic-rich shales in the Beibuwan Basin, South China Sea. *Mar. Pet. Geol.* **2013**, *48*, 77–89. [CrossRef]

39. Liu, E.T.; Wang, H.; Li, Y.; Zhou, W.; Leonard, N.D.; Lin, Z.L.; Ma, Q.L. Sedimentary characteristics and tectonic setting of sublacustrine fans in a half-graben rift depression, Beibuwan Basin, South China Sea. *Mar. Pet. Geol.* **2014**, *52*, 9–21. [CrossRef]

40. Zhu, W.L.; Wu, G.X.; Li, M.B. Palaeolimnology and hydrocarbon potential in beibu gulf basin of south china sea. *Oceanol. Limnol. Sin.* **2004**, *35*, 8–14. (In Chinese with English Abstract)

41. Zhang, G.C.; Xie, X.J.; Wang, W.Y.; Liu, S.X.; Wang, Y.B.; Dong, W.; Shen, H.L. Tectonic types of petroliferous basins and its exploration potential in the South China Sea. *Acta Pet. Sin.* **2013**, *34*, 611–627. (In Chinese with English Abstract)

42. Xie, N.; Cao, Y.C.; Wang, J.; Jin, J.H.; Zhang, W.J.; Zhong, Z.H. Diagenesis and its control on physical property of the reservoirs in the 3rd member of the Paleogene Liushagang Formation in Weixinan Depression, Beibuwan Basin. *Nat. Gas Geosci.* **2019**, *30*, 1743–1754. (In Chinese with English Abstract)

43. Zhao, Y.P.; Wang, H.; Yan, D.T.; Jiang, P.; Chen, S.; Zhou, J.X.; Ma, J.H.; Qin, C.Y.; He, J.; Zhao, Y.Q. Sedimentary characteristics and model of gravity flows in the eocene Liushagang Formation in Weixi'nan depression, South China Sea. *J. Pet. Sci. Eng.* **2020**, *190*, 107082. [CrossRef]

44. Li, M.J.; Wang, T.G.; Liu, J.; Lu, H.; Wu, W.Q.; Gao, L.H. Occurrence and origin of carbon dioxide in the fushan depression, Beibuwan Basin, south China sea. *Mar. Pet. Geol.* **2008**, *25*, 500–513. [CrossRef]
45. Dong, G.N.; Li, J.L. Subtle hydrocarbon reservoirs in Liu-1 Member of the Weixi'nan Sag, Beibuwan Basin, China. *Pet. Explorat. Dev.* **2010**, *37*, 552560.
46. Cao, L.; Zhang, Z.H.; Li, H.Y.; Zhong, N.N.; Xiao, L.L.; Jin, X.; Li, H. Mechanism for the enrichment of organic matter in the Liushagang Formation of the Weixinan Sag, Beibuwan Basin, China. *Mar. Pet. Geol.* **2020**, *122*, 104649. [CrossRef]
47. Zhao, Q.; Zhu, H.T.; Zhang, X.T.; Liu, Q.H.; Qiu, X.W.; Li, M. Geomorphologic reconstruction of an uplift in a continental basin with a source-to-sink balance: An example from the Huizhou-Lufeng uplift, Pearl River Mouth Basin, South China sea. *Mar. Pet. Geol.* **2021**, *128*, 104984. [CrossRef]
48. Xiao, M.; Wu, S.T.; Yuan, X.J.; Xie, Z.R. Conglomerate Reservoir Pore Evolution Characteristics and Favorable Area Prediction: A Case Study of the Lower Triassic Baikouquan Formation in the Northwest Margin of the Junggar Basin, China. *J. Earth Sci.* **2021**, *32*, 998–1010. [CrossRef]
49. Yousef, I.; Morozov, V.; Sudakov, V.; Idrisov, I. Cementation Characteristics and Their Effect on Quality of the Upper Triassic, the Lower Cretaceous, and the Upper Cretaceous Sandstone Reservoirs, Euphrates Graben, Syria. *J. Earth Sci.* **2021**, *32*, 1545–1562. [CrossRef]
50. Qian, W.D.; Sun, Q.L.; Jones, S.J.; Yin, T.J.; Zhang, C.M.; Xu, G.S.; Hou, G.W.; Zhang, B. Diagenesis and controlling factors of Oligocene Huagang Formation tight sandstone reservoir in the south of Xihu sag, the East China Sea Shelf Basin. *J. Pet. Sci. Eng.* **2022**, *215*, 110579. [CrossRef]

Article

Thin Reservoir Identification Based on Logging Interpretation by Using the Support Vector Machine Method

Xinmao Zhou [1], Yawen Li [2], Xiaodong Song [2], Lingxuan Jin [2] and Xixin Wang [3,*]

1   Research Institute of Petroleum Exploration and Development, China National Petroleum Corporation, Beijing 100083, China
2   College of Geosciences, China University of Petroleum-Beijing, Beijing 102249, China
3   School of Geosciences, Yangtze University, Wuhan 430100, China
*   Correspondence: wangxixin86@hotmail.com

Abstract: A reservoir with a thickness less than 0.5 m is generally considered to be a thin reservoir, in which it is difficult to directly identify oil-water layers with conventional logging data, and the identify result coincidence rate is low. Therefore, a support vector machine method (SVM) is introduced in the field of oil-water-dry layer identification. The basic approach is to map the nonlinear problem (input space) to a new high-dimensional feature space through the introduction of a kernel function, and then construct the optimal decision surface in the high-dimensional feature space and conduct sample classification. There are plenty of thin reservoirs in Wangguantun oilfield. Therefore, 63 samples are established by integrating general logging data and oil testing data from the study area, including 42 learning samples and 21 prediction samples, which are normalized. Then, the kernel function is selected, based on previous experience, and the fluid identification model of the thin reservoir is built. The model is used to identify 21 prediction samples; 18 are correct, and the prediction accuracy reaches 85.7%. The results show that the SVM method is feasible for fluid identification in thin reservoirs.

Keywords: support vector machine; thin reservoir; fluid identification; Wangguantun oilfield

Citation: Zhou, X.; Li, Y.; Song, X.; Jin, L.; Wang, X. Thin Reservoir Identification Based on Logging Interpretation by Using the Support Vector Machine Method. *Energies* **2023**, *16*, 1638. https://doi.org/10.3390/en16041638

Academic Editor: Reza Rezaee

Received: 23 December 2022
Revised: 30 January 2023
Accepted: 31 January 2023
Published: 7 February 2023

1. Introduction

The oil fields previously developed in China are now entering the "double extra-high" development stage, with a high water cut and a high recovery degree, and the oil field output continues to decline. In the early stages of development, the main oil layer contributes to the main output of the reservoir, while the thin oil layer is labeled as "poor physical property" and "poor production" due to its own physical conditions [1], and is often not given priority in a development plan. In order to ensure the stable production of old oilfield areas, many oil fields transfer to explore potential objects and change the development mode. The potential of non-major oil reservoirs, such as thin and differential oil reservoirs, cannot be ignored [2–4]. For example, the proven geological reserves of medium-thin and low-margin oil reservoirs in Sazhong Development Zone of Daqing Changyuan are more than 300 million tons, accounting for 2/5 of the total geological reserves. The exploitation and utilization of the medium-thin and low-margin oil reservoirs have contributed to the production of Changyuan and provided a guiding direction for the production growth of old block. The exploitation and utilization of the medium-thin and low-margin oil reservoirs have contributed to the production of Changyuan and provided a guiding direction for the production growth of old block.

This paper takes the thin oil layer of the third member of the Shahejie Formation in the Guan187 area as the research object. The main oil-producing reservoirs in the Guan187 area are the first member of the Shahejie Formation and the third member of the Shahejie Formation of Paleogene. In recent years, the water content of the block has been high and liquid production has been low, and most of the oil wells have been shut down at low energy. There are a lot of potential reservoirs in the three oil formations of the third member of the Shahejie Formation, which can be interpreted as a dry reservoir, a low yield reservoir, and a poor oil reservoir, which are suitable for the research aim of this paper. In this paper, the identification method of a thin differential reservoir is established by comprehensively using various logging curves. This research can play a key role in the future development of unutilized reserves of similar reservoirs.

For most thin oil layers, the 2.5 m and 4.4 m apparent resistivity and spontaneous potential charts can distinguish between oil and water layers. Liu Jiang [5] adopted the conventional identification method to establish the four relationships between the oil layer, the water layer, and the dry layer, to establish the interpretation template, determine the lower limit value of thin differential oil layer, and identify it. The identification results have a high coincidence rate. Guo Hongyan [6] proposed an effective method for the fluid identification of thin differential reservoirs. Carbon-oxygen ratio energy spectrum logging was used to improve the interpretation accuracy of thin differential reservoirs. The interpretation chart was drawn combined with spontaneous potential, and the interpretation results were highly consistent with the oil test results. Shan Xuguang [7] used the Fourier spectrum method, the resolution matching method, and other methods for comparative processing to improve the resolution of thin layer identification, more accurately restore the real logging value of thin oil layer, and make the prediction results more in line with reality. Tang Hong [8] used variance functions to correct logging curves, improve logging identification resolution, and accurately interpret thin differential oil formations. Hou Jun [9] used wave impedance inversion to simulate the deep lateral resistivity prediction reservoir sand body and effectively identify the thin sand body with a thickness of 0.5 m.

For thin reservoirs, the logging response value of the target layer is greatly affected by the surrounding rock. It is difficult to quantitatively identify oil and water layers by conventional logging interpretation methods. Under the conditions of very limited sample data, it is necessary to seek a method that can integrate various logging and geological information to identify oil and water. Because the lithology of the reservoir is complex, its shale content is high, and the reservoir is mainly thin interbedded, resulting in the logging response being distorted and affected to different degrees. The identification of oil and water layers is difficult, the log interpretation coincidence rate is low, and the traditional empirical log interpretation is gradually unable to meet the production needs. Therefore, this paper proposes the artificial intelligence method of support vector machine (SVM) to identify thin layers.

Support vector machine (SVM) is a machine learning algorithm proposed by VAPNIK in the mid 1990s [10,11]. It is a pattern classifier based on VC dimension theory of statistical learning theory and structural risk minimization principle [12]. It has the advantages of complete theory, strong adaptability, global optimization, a short training time, and great generalization performance. It can successfully solve the "dimensionality disaster" problem in traditional learning methods, and has been widely used in pattern recognition, regression estimation, reservoir prediction, and other fields, which are research hotspots in the field of machine learning.

2. Overview of Research Area

The study area is Guan187 area in Wangguantun oilfield, south area of Huanghua Depression. Huanghua Depression is located in the central part of Bohai Bay Basin with a total area of 1.7×10^4 km^2. It is adjacent to Yanshan Fold in the north, Cangxian Uplift in the west, and Chengning Uplift and Bozhong Depression in the southeast. It is spread in a long strip in the direction of NEE-SW, and the width of the depression is up to 70 km. Wangguantun oilfield is located on the Kongdian tectonic belt in the southern area of the Huanghua Depression, and is divided into two parts by the Kongdong fault zone [13–15]. It is adjacent to Cangdong Depression in the north, the Liupu tectonic belt in the Xiaoji fault in the south, and the Changzhuang Depression in the east. The Guan187 area is located in the middle of Wangguantun oilfield and is on the east side of the Kongdong fault zone. It contains two major development fault blocks, Guan187 and Guan913, with an area of about 9.6 km^2 (Figures 1 and 2). Among them, the third member of Shasan-3 contains a large number of non-major formations, such as low-producing and thin oil formations, which are the main research objects of this paper.

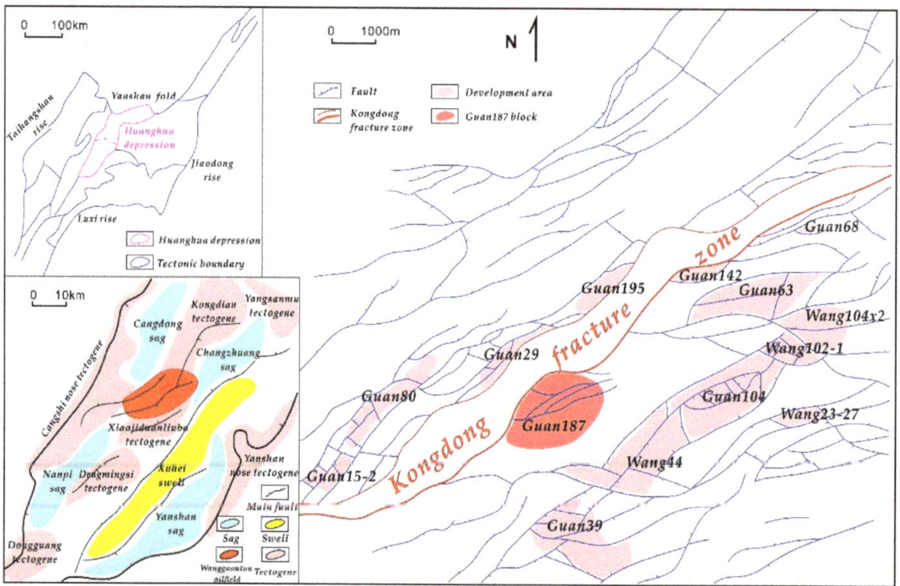

Figure 1. The tectonic location of study area.

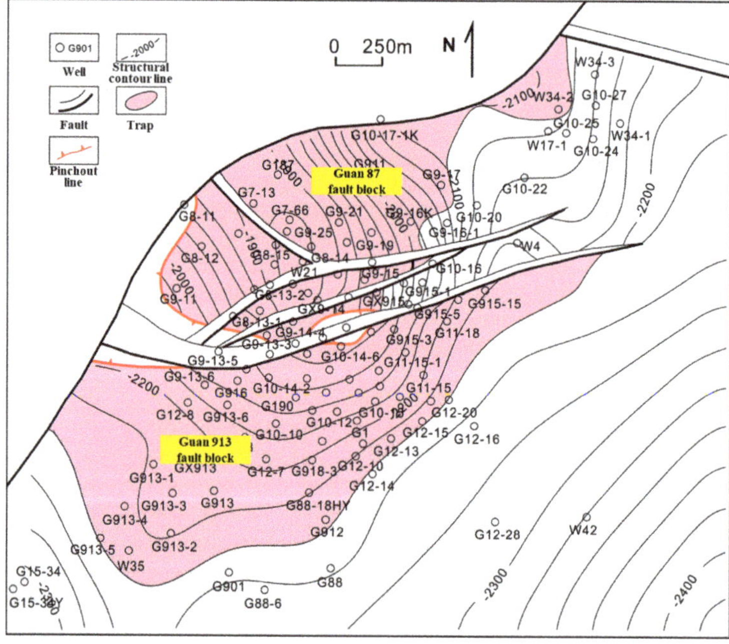

Figure 2. Top structural map of Es33 in Guan187.

Located in the east of the Kongdong fault zone and controlled by the Kongdong fault, the Guan187 area is high in the north and low in the south, and its interior is divided by several faults [16,17]. The whole study area is divided into the northern Guan187 fault block and the southern Guan913 fault block by the central fault. The highest point of the structure in the whole area is near well G187, which gradually decreases to the four sides, forming an anticlinal trap. There is a secondary high near Wang 34-2, which gradually decreases to the four sides (Figure 2).

The target horizon of this paper is the third oil formation of the third member of Shahejie Formation, which belongs to the Shahejie Formation. The lower strata are the Paleogene Kongdian Formation, and the upper strata are the Dongying Formation, Guantao Formation, and Minghuazhen Formation. In the Shahejie Formation, the lower part of the first member of the Shahejie Formation and the third member of the Shahejie Formation are the main oil-producing reservoirs in this area. The third member of the Shahejie Formation can be subdivided into three oil groups. Affected by paleotopography, strata in the study area show a trend of thickening in the south and thinning in the north, with large thicknesses in low parts and thin thicknesses in high parts as a whole. The thickness of sandstone in the reservoir also has a certain thinning trend. There are a large number of thin layers of light green fine sandstone and argillaceous siltstone in the third oil formation of the third member of the target formation, which are mainly characterized by "mud-coated sand". The upper part of the Sha32 oil Formation contains a set of stable volcanic rock sedimentary layers, and the lower part of the Zao 0 oil Formation of the first member of the Kongdian Formation is lake deposition, with a set of stable paste rock layers. Therefore, the target interval can be accurately identified and divided. Most of the wells can be drilled into the third oil group of the third member of the Shahejie Formation. In some areas, due to the influence of the central fault, the target strata have a formation loss phenomenon to varying degrees (Table 1).

Table 1. Sedimentary characteristics of Guan187 of Wangguantun oilfield.

System	Series	Group	Section	Oil Group	Lithologic Character
		Stratigraphic System		**Oil Group**	**Lithologic Character**
Neogene	Pliocene	Minghuazhen Formation			Light gray green, gray green sandstone, brown, brown red mudstone
	Miocene	Guantao Formation			Relatively thick light green, gray white sandstone, mixed with gray green, purple mudstone
	Oligocene	Dongying Formation			Gray argillaceous siltstone, mudstone, the lower mudstone is rich in ostracoda fossils
		Shahejie Formation	Sha1		It is mainly composed of biological limestone and dolomitic limestone, with oil shale and mudstone
			Sha2		Light green and gray sandstone interbedded with purple red and gray mudstone
			Sha3	Sha31	Biolithite limestone
				Sha32	Thick layer volcanic rock segment, dark basalt
				Sha33	Gray mudstone, mixed with thin layer of light green fine sandstone, medium sandstone
Paleogene	Eocene	Kongdian Formation	Kong1	Zao 0	Huge thick layer of paste rock
				Zao I	Brown red mudstone, mixed with brown siltstone, fine sandstone
				Zao II	It is mainly composed of gray-brown coarse sandstone and pebbled sandstone, mixed with gray-green and purplish red mudstone
				Zao III	It is mainly composed of brown fine sandstone, coarse sandstone and pebbled sandstone, mixed with gray-green and purple-red mudstone, and the bottom is mainly purple-red mudstone
				Zao IV	Grey sandstone, brown red mudstone
				Zao V	Grey sandstone, brown red mudstone

3. SVM Classification Principle

Support Vector Machine (SVM), first proposed by Vapnik, is a new machine learning method based on statistical theory [18]. The basic approach is to map the nonlinear problem (input space) to a new high-dimensional feature space by introducing a kernel function, and then construct the optimal decision surface in the high-dimensional feature space and conduct sample classification (Figure 3). Support vector machines (SVM) have the advantages of high accuracy, fast speeds, strong versatility, and perfect theory when solving nonlinear problems related to research targets and multiple uncertain features [19].

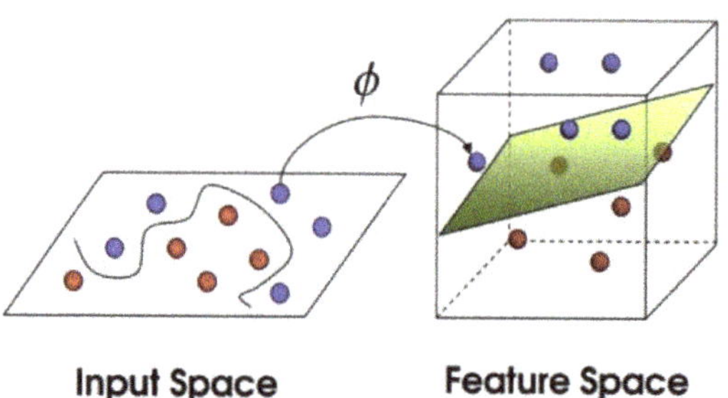

Figure 3. Support vector machine (SVM) classification principle [19].

SVM's core idea is that, for a given learning task with a limited number of training samples, there is a trade-off between the accuracy of the given training set and machine capacity with the preferable generalization capability [20]. Since the final solution of the SVM is a convex optimization problem, the obtained solution must be the global optimal solution, which is not found in other algorithms, including neural networks.

This paper uses the support vector machine (SVM) algorithm with strong nonlinear processing ability to classify and identify thin layers. The structure diagram of the support vector machine algorithm and the thin layer division process are shown in Figures 4 and 5. The implementation process of the algorithm is as follows: the known samples are selected to form learning samples to train the model, so as to establish the thin layer quantitative recognition model; the prediction model is verified by the test samples; and the thin layer of unknown samples is predicted by the verified prediction model.

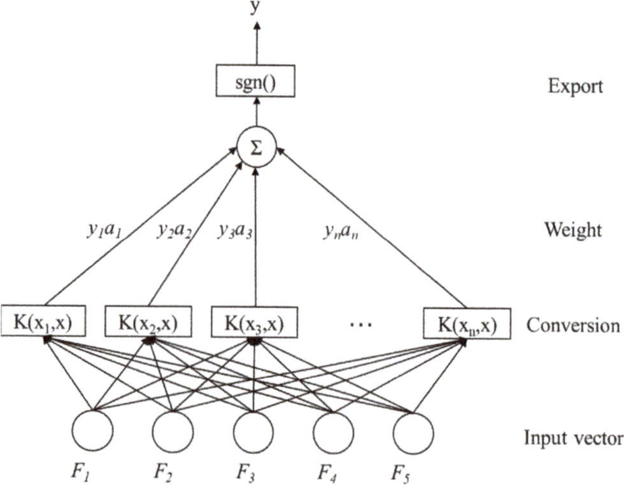

Figure 4. Structure chart of the support vector machine (SVM) method.

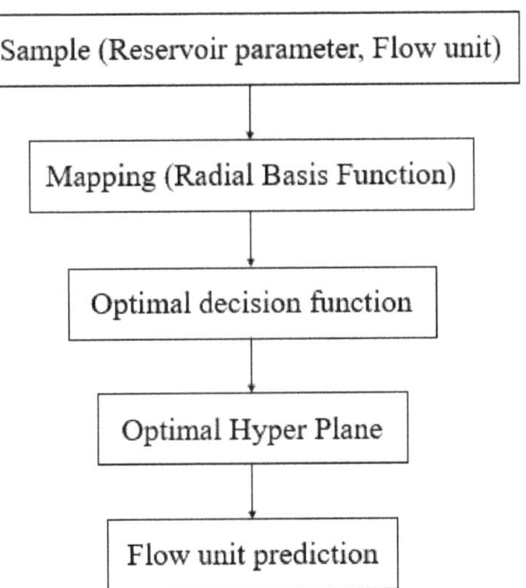

Figure 5. Flowchart of the support vector machine (SVM) method.

3.1. Two-Class SVM

Assume that sample set $(x_i, y_i), i = 1, 2, \cdots, n, x_i \in \mathbf{R}^d, y_i \in y = \{+1, -1\}$. For the case of a linearly separable sample set separated by a hyperplane. Remember the hyperplane $w \cdot x + b = 0$, where w is the normal line of the classification surface, b is the outlier that represents the position of the modified normal with respect to the origin [21]. The optimal hyperplane not only separates the two types of samples error-free, but also maximizes the classification interval. The problem of optimal hyperplane construction can be translated to calculate the minimum of the formula

$$\begin{cases} \varphi(w) = \frac{1}{2} \| w \|^2 \\ s, t, y_i(x_i \cdot w + b) \geqslant 1 \; i = 1, 2, \cdots, n \end{cases} \tag{1}$$

The problem can also be transformed into a simpler dual problem to calculate the maximum of the formula:

$$\begin{cases} Q(\alpha) = \sum\limits_{i=1}^{n} a_i - \frac{1}{2} \sum\limits_{i=1}^{n} \sum\limits_{j=1}^{n} \alpha_i \alpha_j y_i y_j (x_i \cdot x_j) \\ s, t, \sum\limits_{i=1}^{n} y_i \alpha_i = 0 \; \alpha_i \geqslant 0 \end{cases} \tag{2}$$

where, α_i is the Lagrange multiplier for each sample. According to the condition of Kuhn-Tucher, the optimal solution must satisfy the following conditions:

$$\alpha_i[y_i(w \cdot x_i + b) - 1] = 0 \quad i = 1, 2, \cdots, n \tag{3}$$

Therefore, only the support vector coefficients α_i^* are a non-zero value. If α_i is the optimal solution,

$$w^* = \sum_{i=1}^{n} \alpha_i^* y_i x_i \tag{4}$$

By choosing the corresponding i when $\alpha_i \neq 0$, we can obtain the value of b using Formula (3). Then, the optimal classification function can be obtained by substituting w^* and b into the formula $\mathrm{sgn}(w^* \cdot x + b)$.

For a linearly inseparable case, we can introduce a slack vector ξ, which satisfies:

$$\begin{cases} y_i[(w \cdot x_i) + b] - 1 \geq \xi_i \\ \xi_i \geq 0 \; i = 1, 2, \cdots, n \end{cases} \tag{5}$$

The generalized optimal classification surface can be obtained by changing the objective to find the minimum value of formula $\varphi(w, \xi) = \frac{1}{2}\|w\|^2 + C \sum_{i=1}^{n} \xi_i$, where C stands for penalty function, it shows the penalty for misclassification.

For nonlinear problems, it can be transformed into a linear problem in a higher dimensional space by nonlinear transformation and obtain the optimal classification surface. The linear classification of a nonlinear problem after transformation can be realized by using the proper kernel function. In this case, the classification function named SVM is:

$$f(x) = \text{sgn}\left\{ \sum \alpha_i^* y_i K(x_i, x) + b^* \right\} \tag{6}$$

The function's remarkable feature is that data only appear in the inner product. It is not necessary to know the specific nonlinear mapping process before calculation, only to select an appropriate model to replace the inner product, so as to economize the complex calculation. It should be noted that the model here must satisfy the conditional kernel parameters.

3.2. Multi-Class SVM

SVM technology was originally proposed for the two-class problem, but the oil-water-dry layer identification problem belongs to the multi-class problem. In order to effectively use the SVM method to divide the oil-water-dry layers, it is necessary to extend the SVM and build a reasonable multi-class coding scheme [22,23]. At present, there are two main methods for constructing SVM multi-class: one is the direct method, represented by the multi-class algorithm proposed by Weston [22], which has a high degree of complexity and is difficult to implement. The other is the indirect method, which mainly includes "one-to-one", "one-to-many", and a SVM decision tree. This paper mainly adopts the "one-to-one" method and uses the Libsvm classifier for training [24]. Libsvm is a simple, practical, fast, and effective SVM pattern recognition and regression software package. The algorithm combines the ideas of SMO and SVM-Light, and adopts a voting strategy to support multi-class. By training $k(k-1)/2$ classifiers, the samples are labeled with the highest votes. The principle of the "one-to-one" algorithm is as follows: if there are k types of data, select the i-th type of data and the j-th type of data to construct a classifier, where $i < j$, so that $k(k-1)/2$ classifiers need to be trained. For the i-th and j-th types of data, a two-class problem needs to be solved:

$$\min_{w^{ij}, b^{ij}, \xi^{ij}} \frac{1}{2}\left(w^{ij}\right)^T w^{ij} + C \sum_t \xi_t^{ij} \tag{7}$$

If $y_t = i$,

$$\left(w^{ij}\right)^T \varphi(x_i) + b^{ij} \geq 1 - \xi_t^{ij} \tag{8}$$

If $y_t = j$, $\xi_t^{ij} \geq 0$,

$$\left(w^{ij}\right)^T \varphi(x_i) + b^{ij} \leq -1 + \xi_t^{ij} \tag{9}$$

Solve this problem by using the voting method: If $\text{sign}\left[\left(w^{ij}\right)^T, \varphi(x_i) + b^{ij}\right]$, consider x as the i-th type, i-th type plus one vote; else j-th type plus one vote. Finally, x belongs to the type with the most votes.

4. Application of SVM in Thin Reservoir Identification

The identification of an oil layer, a water layer, and a dry layer belongs to the problem of multi-class discriminant pattern recognition, so it can be completed by using the SVM method to establish a fluid identification model. The basic idea is to collect modeling samples and perform data preprocessing to generate feature quantities first, and then perform parameter optimization to determine the best combination of parameters for modeling, and finally use the established model to predict targets and identify oil-water-dry layers.

4.1. Model Building

Model construction mainly includes the determination of the kernel function and penalty factor C. The most common kernel functions in SVM mainly include the Gaussian radial basis kernel function, the multi-layer perceptron kernel function, and the polynomial kernel function. In this section, the well logging curve is optimized, the sample points are collected, and then the kernel function is optimized to establish a model to identify the thin layer.

4.1.1. Sample Set Selection

The identification of an oil-water-dry layer is an important feature of logging evaluation. The logging curve indirectly reflects the properties of the fluid in the reservoir. Different logging curves will show a certain degree of difference and regularity for different fluid characteristics, such as oil-water-dry layers [25]. Through the description of logging characteristics, integrate the interpretation experience of experts and the correlation analysis and comparison of coring wells. Finally, we select the logging curves closely related to reservoir fluid, such as: acoustic (AC), true formation resistivity (RT), neutron (CNL), density (DEN), natural gamma ray (GR), spontaneous potential (SP); porosity (POR), used as input eigenvalues for the sample. The output positive integer represents the fluid identification result, where 1 represents the oil layer, 2 represents the water layer, and 3 represents the dry layer. By stratifying the logging curve and combining it with the oil test data, the typical characteristics of oil, water, and dry layers of seven wells were selected as the training objects and the remaining four wells were selected as the testing objects in the study area. Several reliable and representative logging data from each study interval were selected as training samples for this interval. Finally, a total of 203 logging data from 42 characteristic layers in seven wells were selected as the training sample dataset. According to the oil test data, the test sample dataset was obtained from the remaining four wells using a logging curve from twenty-one layers when building test samples.

4.1.2. Normalization of Sample Data

There is no standardized format data obtained from logging data. Therefore, the data should be normalized first [26] in order to avoid the difficulty of calculating the inner product of the kernel function caused by the difference of each parameter dimension and improve the prediction accuracy. This can avoid some eigenvalue ranges that are too large and other eigenvalue ranges that are too small, resulting in large numbers drowning the decimal.

The normalization formula adopted is: $X = (x - x_{min})/(x_{max} - x_{min})$, where x is the actual logging value, x_{max}, x_{min} are the maximum and minimum values among all sampling points of the logging curve, X is the log value after normalization, $X \in [0, 1]$. The training and test samples are located in the normalized interval, which ensures the reliability of the classification results [27–29].

4.1.3. Model Selection

The identification of oil, water, and dry layers belongs to the problem of multi-class discriminant pattern recognition. In theoretical analysis, the determination of the classification function is mainly the determination of the kernel function r and penalty coefficient C.

These two parameters have great influence on the prediction results, and their reasonable determination directly affects the accuracy and generalization ability of the model. The Gaussian radial basis function (GRBF) is usually used to establish the identification model of reservoir fluid. The cross-validation method is used to optimize the C and r [30–34]. The Gaussian radial basis kernel function has a wide convergence domain, has a high applicability for a variety of sample cases, and has only one kernel parameter r. It has a high flexibility, is currently the most widely used and the best effect of the classification kernel function. Therefore, the Gaussian radial basis kernel function is used to build the prediction model of support vector machine. The Gaussian radial basis kernel function formula is:

$$K(x_i, x) = \exp \frac{-\|x - x_i\|^2}{r^2} \tag{10}$$

Finally, the optimal parameter combination is calculated as $C = 4.1541$ and $r = 0.7218$.

4.2. Application Effect and Analysis

Using MATLAB R2019a software, with forty-two logging data of eight single layers in seven wells of Wangguantun oilfield as training samples, the SVM model for identifying oil-water-dry layers in this area was established. The model was used to identify twenty-one layers in four other wells in the area. Finally, the identification results of inspection with the production testing results were compared, where 18 of the 21 layers were correctly identified; recognition accuracy was 85.71% (Figure 6), which was better than conventional cross plot identification results at 80.95% (Figure 7).

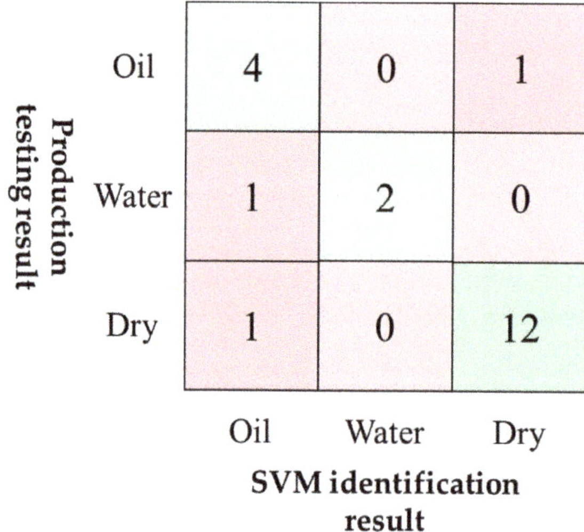

Figure 6. Confusion matrix of oil and water classification of training samples. Green means accurate prediction, pink means wrong prediction.

In order to evaluate the reliability of the SVM method for layers fluid identification, the identified results were compared with the cross-plot method. The comparison results are shown in Table 2. Through comparison and analysis, the SVM method has the highest accuracy (85.71%) in identifying oil-water-dry layers, which is higher than the cross-plot identification accuracy (80.95%). The identification results show that the SVM method based on the principle of structural risk minimization has a more stable performance when solving thin reservoirs and small sample problems.

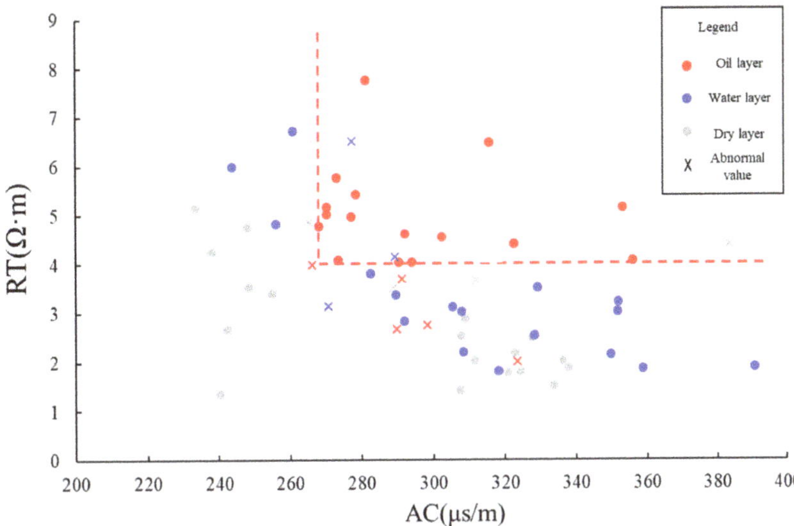

Figure 7. Crossplot of AC and RT of thin oil layer of Es33.

Table 2. Comparison table of identification results of test samples.

Well	Layer	Production Testing Depth/m	Production Testing Result	SVM Identification Result	Cross-Plot Identification Result
	311	2235.6~2236.2	Dry	Dry	Dry
	312	2243.7~2245.6	Oil	Oil	Oil
	313	2250.4~2251.9	Oil	Oil	Oil
G913-1	321	2259.0~2259.5	Dry	Dry	Dry
	322	2263.0~2264.2	Dry	Dry	Dry
	323	2270.7~2272.9	Dry	Dry	Dry
	324	2284.1~2286.3	Water	Oil	Oil
	311	2228.5~2229.5	Dry	Dry	Dry
	312	2237.6~2238.6	Oil	Oil	Oil
	313	2243.6~2245.4	Oil	Oil	Oil
G913-2	321	2252.3~2258.8	Dry	Dry	Dry
	322	2262.9~2264.8	Dry	Dry	Dry
	323	2272.8~2275.1	Dry	Dry	Dry
	324	2280.4~2282.1	Water	Water	Water
	311	2203.1~2205.0	Dry	Dry	Dry
G918-2	312	2226.1~2228.2	Dry	Oil	Oil
	313	2230.0~2231.9	Dry	Dry	Dry
	323	2239.3~2240.5	Dry	Dry	Dry
	321	2248.7~2251.0	Oil	Dry	Dry
G12-13	323	2260.9~2262.3	Dry	Dry	Dry
	324	2265.7~2268.0	Water	Water	Water
Accuracy				85.71%	80.95%

The method was applied to other wells in the study area to identify thin layer fluids, and well G9-14-4 was used as an example (Figure 8).

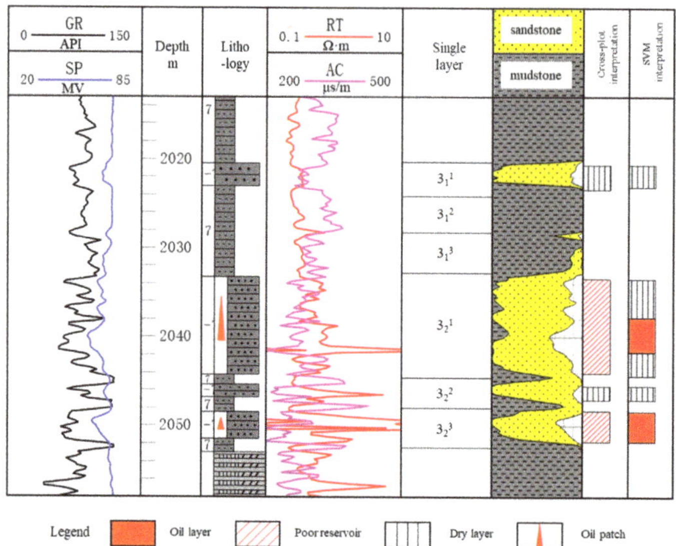

Figure 8. Identification of thin oil layer of well G9-14-4.

5. Conclusions

(1) The logging curves indirectly reflect the properties of the fluid in the reservoir. Well log data can be used to comprehensively identify the thin layers.

(2) The accuracy of the SVM method for reservoir fluid identification is obviously higher than that of the conventional cross-plot identification method.

(3) The SVM-based reservoir fluid identification model has high convergence accuracy and strong generalization ability, and can make full use of limited logging data information to obtain the optimal identification results. Especially in areas where the test data are lacking or the oil-water system is complex, this method can improve the identification accuracy of the oil-water dry layer. It has good reference values in actual logging reservoir evaluation and can be extended to lithology identification and reservoir parameter prediction.

Author Contributions: X.Z.: Writing—original draft; Y.L.: Data curation; X.S.: Writing—review and editing; L.J.: Methodology; X.W.: Visualization. All authors have read and agreed to the published version of the manuscript.

Funding: This research was funded by Science Foundation of China University of Petroleum, Beijing, grant number 2462020YXZZ022 and CNPC Innovation Found (2021DQ02-0106).

Data Availability Statement: The data that support the findings of this study are available on request from the corresponding author.

Conflicts of Interest: The authors declare no conflict of interest.

References

1. Yu, H. Study on remaining oil in the north of Daqing Oilfield. *Acta Pet. Sin.* **1993**, *14*, 72–80.
2. Qadri, S.T.; Ahmed, W.; Haque, A.E.; Radwan, A.E.; Hakimi, M.H.; Abdel Aal, A.K. Murree Clay Problems and Water-Based Drilling Mud Optimization: A Case Study from the Kohat Basin in Northwestern Pakistan. *Energies* **2022**, *15*, 3424. [CrossRef]
3. Haque, A.E.; Qadri, S.T.; Bhuiyan, M.A.H.; Navid, M.; Nabawy, B.S.; Hakimi, M.H.; Abd-El-Aal, A.K. Integrated wireline log and seismic attribute analysis for the reservoir evaluation: A case study of the Mount Messenger Formation in Kaimiro Field, Taranaki Basin, New Zealand. *J. Nat. Gas Sci. Eng.* **2022**, *99*, 104452. [CrossRef]
4. Osinowo, O.O.; Ayorinde, J.O.; Nwankwo, C.P.; Ekeng, O.M.; Taiwo, O.B. Reservoir description and characterization of Eni field offshore Niger Delta, southern Nigeria. *J. Pet. Explor. Prod. Technol.* **2018**, *8*, 381–397. [CrossRef]

5.   Liu, J.; Wang, A.; Lang, F.; Zhang, J. A new technique for identifying the fliud in thin, poor and low resistivity pay zone. *Well Logging Technol.* **2000**, *24*, 515–517.

6.   Guo, H.; Huang, D.; Pan, G. Fluid identification and interpretation method of thin differential oil-water layer in low permeability reservoir. *China Pet. Explor.* **2001**, *6*, 31–33.

7.   Shan, X. The Research to Log Recognition Technology for Thin Oil Layer in Qilicun Oilfield. Bachelor's Thesis, Xi'an Shiyou University, Xi'an, China, 2014; pp. 34–54.

8.   Tang, H.; Luo, M.; Yan, Q. Recognition and interpretation of water encroaching in thin and poor-quality pay zone. *J. Southwest Pet. Inst.* **2003**, *25*, 1–3.

9.   Hou, J.; Wang, J.; Wang, C.; Cheng, L. Application of thin & poor reservoir predicted technology to the Punan oilfield. *Southwest Pet. Inst.* **2006**, *28*, 53–56.

10.   Vapnik, V.N. *The Nature of Statistical Learning Theory*; Springer: New York, NY, USA, 1995; pp. 54–96.

11.   Luo, N. A New Method in Data Mining-Support Vector Machine. *Softw. Guide* **2008**, *7*, 30–31.

12.   Vapnik, V.; Levin, E.; Le, C.Y. Measuring the VC-Dimension of a Learning Machine. *Neural Comput.* **1994**, *6*, 851–876. [CrossRef]

13.   Yang, C. Genesis and accumulation of non-type natural gases in Huanghua depression, Dagang oilfield. *Pet. Explor. Dev.* **2006**, *33*, 335–339.

14.   Xia, R.; Tang, J. The development and evaluated patterns of Ordovician palaeo karst in the Huanghua depression. *Pet. Explor. Dev.* **2004**, *31*, 51–53.

15.   Jiao, Q.; Hou, J.; Xing, H. Anastomosing river sediment of the Zao 0 reservoir group in the Duanliubo oilfield, Huanghua depression. *Pet. Explor. Dev.* **2004**, *31*, 72–74.

16.   Ren, W. Fault structure characteristics of Guan-3 block in Wangguantun Oilfield. *Petrochem. Ind. Technol.* **2017**, *8*, 127.

17.   Zhang, X.; Hou, J.; Hu, C.; Liu, Y.; Wang, X.; Ji, L. Evaluation of reservoir permeability heterogeneity by principal component analysis—Taking Wangguantun oil field Wang 23–27 block for example. *J. East China Univ. Technol. Nat. Sci.* **2018**, *41*, 41–45.

18.   Vapnik, V.N. *The Nature of Statistical Learning Theory*, 2nd ed.; Springer: New York, NY, USA, 1999; pp. 12–15.

19.   Lu, W.; Guo, J.; Dong, H.; Zhang, Y.; Lin, L. Evaluating Mine Geology Environmental Quality Using Improved SVM Method. *J. Jilin Univ. Earth Sci. Ed.* **2016**, *46*, 1511–1519.

20.   Peng, T.; Zhang, X. Review of support vector machine and its applications in petroleum exploration and development. *Prog. Explor. Geophys.* **2007**, *30*, 91–95.

21.   Yi, Z.; Lv, M. Intrusion Detection Method Based on Multi-class Support Vector Machines. *Comput. Eng.* **2007**, *33*, 167–169.

22.   Weston, J.; Watkins, C. *Multi-Class Support Vector Machines*; CSD-TR-98-04; Royal Holloway College: London, UK, 1998.

23.   Wang, Z.; Xue, X. *Multi-Class Support Vector Machine*; Springer: Berlin/Heidelberg, Germany, 2014; pp. 23–48.

24.   Ren, L.; Li, W.; Ci, X.; Shi, X.; Sun, Z.; Zheng, R. A Method for Identification of Cuttings in Petroleum Logging by LIBSVMs. *Period. Ocean. Univ. China* **2010**, *40*, 131–136.

25.   Xu, D.; Li, T.; Huang, B.; Li, N. Research on the identification of the lithology and fluid type of foreign M oilfield by using the cross-plot method. *Prog. Geophys.* **2012**, *27*, 1123–1132.

26.   Zhang, Y.; Tong, K.; Zheng, J.; Wang, D. Application of Support vector machine method in fluid identification of low resistivity reservoir. *Geophys. Prospect. Pet.* **2008**, *47*, 306–310.

27.   Tao, S.; Xiao, C.; Yang, B.; Cai, Y. The application of the artificial neural network in the log interpretation. *Geophys. Prospect. Pet.* **1995**, *34*, 90–102.

28.   Zhu, G.; Liu, S.; Yu, J. Support vector machine and its applications to function approximation. *J. East China Univ. Sci. Technol.* **2002**, *28*, 555–559.

29.   Yu, D.; Sun, J.; Zhang, Z.; Wu, J. Reservoir Fluid Property Identification with Support Vector Machine Method. *Xinjiang Pet. Geol.* **2005**, *26*, 675–677.

30.   Ivšinović, J.; Malvić, T. Comparison of Mapping Efficiency for Small Datasets using Inverse Distance Weighting vs. Moving Average, Northern Croatia Miocene Hydrocarbon Reservoir. *Geologija* **2022**, *65*, 47–57. [CrossRef]

31.   Zhang, X.; Liu, Y. Performance Analysis of Support Vector Machines with Gauss Kernel. *Comput. Eng.* **2003**, *29*, 22–25.

32.   Yue, Y.; Yuan, Q. Application of SVM method in reservoir prediction. *Geophys. Prospect. Pet.* **2005**, *44*, 388–392.

33.   Wang, X.; Li, Z. Parameter determination of kernel function of support vector Machine based on grid search. *Period. Ocean. Univ. China* **2005**, *35*, 859–862.

34.   Chen, K.; Wu, L.; Chen, Y.; Wang, G. Classification and Recognition of Polyhalite in Chuanzhong Based on Support Vector Machine. *Adv. Earth Sci.* **2016**, *31*, 1041–1046.

Article

Effects of Inorganic Minerals and Kerogen on the Adsorption of Crude Oil in Shale

Yanyan Zhang, Shuifu Li *, Shouzhi Hu and Changran Zhou

School of Earth Resources, China University of Geosciences (Wuhan), Wuhan 430074, China
* Correspondence: lishf@cug.edu.cn

Abstract: Shale oil stored in the shale system occurs mainly in adsorbed and free states, and ascertaining the amount of adsorbed crude oil in shale is a method of ascertaining its free oil content, which determines the accuracy of shale oil resource evaluation. Both inorganic minerals and kerogen have the ability to adsorb crude oil, but there is controversy surrounding which plays the greatest part in doing so; clarifying this would be of great significance to shale oil resource evaluation. Therefore, in this study, the evolution states of inorganic minerals and kerogen in shale were changed using pyrolysis, and the adsorbents were prepared for crude oil adsorption experiments, to explore the effects of inorganic minerals and kerogen on the crude oil adsorption of shale. The results showed that the differences in kerogen's structural units and content in organic-rich shale (TOC = 1.60–4.52%) had no obvious effects on its crude oil adsorption properties. On the contrary, inorganic minerals, as the main body of shale, played a dominant role in the adsorption of crude oil. The composition and evolution of the inorganic minerals controlled the surface properties of shale adsorbents, which is the main reason for the different crude oil adsorption properties of the different types of adsorbents. The results of this study are helpful in improving our understanding of the performance and mechanisms of shale in adsorbing crude oil and promoting the development of shale oil resource evaluation.

Keywords: oil; adsorption experiment; inorganic minerals; kerogen; thermal simulation of adsorption

Citation: Zhang, Y.; Li, S.; Hu, S.; Zhou, C. Effects of Inorganic Minerals and Kerogen on the Adsorption of Crude Oil in Shale. *Energies* **2023**, *16*, 2386. https://doi.org/10.3390/en16052386

Academic Editors: Reza Rezaee, Shu Tao and Dameng Liu

Received: 15 January 2023
Revised: 12 February 2023
Accepted: 27 February 2023
Published: 2 March 2023

1. Introduction

Shale oil is a liquid petroleum resource that is retained and enriched in the shale system after hydrocarbon generation and expulsion from source rocks, and it will be an essential replacement resource in the future. Shale oil stored in the shale system occurs mainly in its free and adsorption states; because the former is the main contributor to productivity, the accurate evaluation of free shale oil content is an important prerequisite to shale oil exploration and development [1,2]. However, there is usually some deviation in the results upon directly quantifying the free oil content, so the indirect calculation of free oil content through adsorbed and total oil content is attracting increasing attention [3–6]. Previous studies indicated that the liquid hydrocarbon adsorption properties of kerogen per unit mass are much higher than those of inorganic minerals; thus, previous studies mainly considered the adsorption of total organic carbon (TOC) on liquid hydrocarbons [7–11], among which the most intuitive and widely accepted method for evaluating the potential producibility of shale oil is the oil saturation index (OSI) (S_1/TOC) [1,10–13]. However, when evaluating the potential of continental shale oil resources in some areas of China, the results of OSI do not precisely match the actual exploration results, which indicates that it is not sufficient to consider only the adsorption of liquid hydrocarbons on organic carbon [4,14,15].

Actually, the inorganic minerals (such as clay, quartz, feldspar, pyrite, carbonate, etc.) in shale still have a certain effect on the adsorption of liquid hydrocarbon [16–18]. Despite the adsorption properties of organic matter being much higher than those of inorganic minerals, with inorganic minerals forming the main part of shale, their relative content is much higher than the content of kerogen; so, the liquid hydrocarbon adsorption capacity of inorganic minerals cannot be ignored [19–21]. Meanwhile, the changes in the pore characteristics and surface properties of shale will affect its hydrocarbon adsorption properties because of the process of sedimentation and the evolution of inorganic minerals and kerogen in shale [22–24]; thus resulting in the capacity of shale to adsorb liquid hydrocarbons not being equivalent to summing the capacity of purified kerogen and different types of purified inorganic minerals to adsorb liquid hydrocarbons [21,25]. Therefore, to research the overall crude oil adsorption capacity of organic-rich shale, the adsorption properties of inorganic minerals and kerogen must be considered simultaneously. At present, there are many reports of different types and maturities of purified kerogen adsorbing pure hydrocarbon liquids [9,26,27], and the purified inorganic mineral (or mixed purified inorganic mineral) adsorption of liquid hydrocarbons has also been explored [17,28,29]. However, at present, the results of these studies cannot completely or effectively guide the exploration of shale oil [25,30]. Therefore, using actual shale as an adsorbent and actual crude oil as an adsorbate in crude oil adsorption experiments may have more practical significance and facilitate guidance.

In order to clarify the influence of kerogen's structural and inorganic mineral composition and evolution in shale on its crude oil adsorption properties, this experiment used actual shale and crude oil from the same region. Under the condition of inorganic minerals and kerogen existing simultaneously, the shale samples underwent pyrolysis, and extracts were prepared of the shale adsorbents that contained different kerogen content and structures and different transformation degrees of inorganic minerals. After performing the crude oil adsorption experiments, according to the organic carbon content and the composition characteristics of inorganic minerals in the shale adsorbents, we analyzed the effects of inorganic minerals and kerogen on crude oil adsorption in shale.

2. Samples

The shale samples were collected from the Cheng 2 well in the central deep sag area of the Biyang Depression. A total of five shale samples (①~⑤ in Figure 1) were selected from the fifth shale layer of the third member of the Hetaoyuan formation (Eh_3), at a vertical depth of $2771-2797$ m. This depression is a secondary structural unit of the Nanxiang Basin, located in Tanghe County and Biyang County, Henan Province, China. Covering an area of about 1000 km^2, it is a Meso-Cenozoic (Cretaceous–Paleogene–Neogene) rift depression [31]. According to the structure, the depression can be further divided into three subunits, including the northern slope belt, central deep sag, and southern steep slope belt. The central deep sag sedimented the thickness strata of the Paleogene (which is the main enrichment area for lithologic and shale oil reservoirs [32]), includes the Dacangfang, Yuhuangding, Hetaoyuan (Eh_1, Eh_2, Eh_3) and Liaozhuang formations from the bottom to the top. Moreover, the third member of the Hetaoyuan formation (Eh_3) can be divided into eight sand formations (Eh_3^{I}~Eh_3^{VIII}). There are six layers of organic-rich shale developed in Eh_3^{III}~Eh_3^{VI} in the central deep depression [5,33]. Based on wireline logs and core sampling analysis (Rock-Eval, GC-MS, etc.), the fifth shale layer was the thickest and most widely distributed, was abundant in retained hydrocarbons, and was the main target layer of shale oil exploration in this area [5,32].

Previous studies have shown that the third member of the Hetaoyuan formation (Eh_3) in the deep sag of the Biyang Depression is mainly composed of dark-gray and gray shale interbedded with dolomite and sandstone [32]. The types of kerogen are mainly type I and type II$_1$, with little type II$_2$ [34]. The maturity of organic matter is at the low stage ($Ro\% \approx 0.7\%$; average $T_{max} \approx 446.35$ °C) [5]. The crude oil samples used in the adsorption experiment were from the same layer in the same area, and the characteristics of the crude oil group were as follows: saturated hydrocarbon (Sat.): 60.7 wt%, aromatic hydrocarbon (Aro.): 13.5 wt%, non-hydrocarbon and asphaltene (Res. + Asp.): 25.7 wt%; this is normal crude oil.

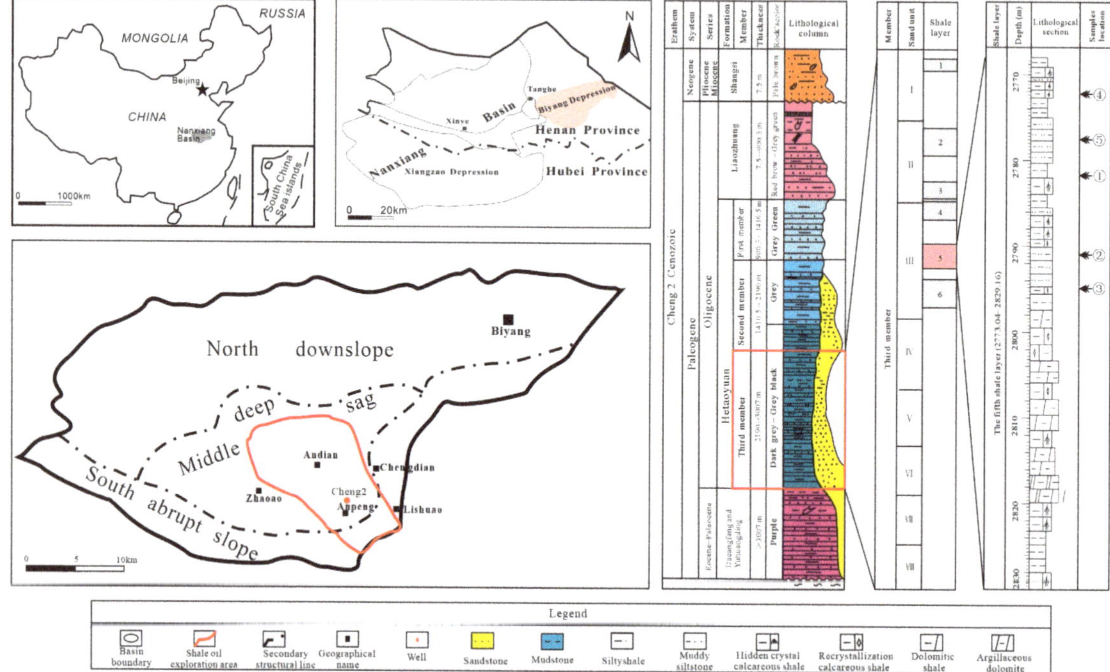

Figure 1. Location and generalized stratigraphic columns of the Biyang Depression and sample locations (Adapted with permission from Ref. [5]. 2016, Elsevier, adapted with permission from Ref. [33]. 2019, Elsevier).

3. Methodology

3.1. Preparation of Adsorbents

Organic carbon in shale includes insoluble kerogen and soluble asphaltene. Kerogen is the organic adsorptive carrier (adsorbent) for soluble asphaltene, and soluble asphaltene is the adsorbed object (adsorbate); so, it is more scientific to determine the amount of oil adsorbed by kerogen than by total organic matter [27,35,36]. According to the principle of the hydrocarbon generation of kerogen and the theory of the kerogen structure model [37], in the process of hydrocarbon generation, part of the reactive kerogen can be directly converted into oil and gas (this is called effective carbon); meanwhile, the other part, despite participating in the process of hydrocarbon generation, is not converted into hydrocarbon but into residual carbon (RC). However, the inert kerogen neither participates nor is affected by hydrocarbon generation, so it can be called dead carbon (in this paper, the residual carbon and dead carbon are collectively called ineffective carbon, dividing the organic carbon in kerogen into two parts: effective carbon and ineffective carbon) [10,38]. Because reactive kerogen and inert kerogen represent a theoretical model, we could not obtain separate kerogen with these structural units, but we could measure the effective carbon (TOC > RC) and ineffective carbon (TOC ≈ RC) in kerogen via pyrolysis to measure the reactivity and inertia of kerogen. In this way, we could explore the oil adsorption properties of different kerogen structural units for crude oil.

Before preparing the adsorbents, the carbonate minerals in shale needed to be removed using hydrochloric acid (HCl). The main reason for this was that the upper layer of Eh_3 in the Biyang Depression was formed in an alkaline environment, which limited the dissolution of carbonate minerals in the early period of diagenesis and resulted in the precipitation and cementation of carbonate minerals in the late period of diagenesis [39]. However, during the process of preparing the adsorbents via thermal treatment, carbonate minerals tended to decompose after 650 °C. This will lead to a significant increase in the specific surface area (SSA) of the samples, and the CO_2 generated will affect the measurement of TOC [40–42]. The former will seriously affect the adsorption performance of shale adsorbents [19,21,25,29], while the latter will affect the evaluation of kerogen's adsorption properties in shale adsorbents. Therefore, in order to keep the inorganic mineral composition of shale adsorbents obtained via pyrolysis treatment as consistent as possible, it was necessary to remove carbonate minerals in shale via hydrochloric acid treatment before pyrolysis treatment of the shale samples.

The temperature conditions for preparing the adsorbents were those of the Rock-Eval analysis method. A shale sample treated with hydrochloric acid (abbreviated as T_{HCl}) was heated in a N_2 flow at 300 °C for 1 h to remove the pyrolytic hydrocarbon S_1. Then, on the pyrolysis products, we used DCM (dichloromethane) and MeOH (methanol) (93:7, v/v) for extraction for 72 h; after that, extraction was conducted for 72 h using methanol, acetone, and benzene (MAB) (2:5:5, $v/v/v$) to remove as much of the soluble organic matter in the pyrolysis products of shale as possible. After finishing the above process, the Type-A (abbreviated as T_A) adsorbent that contained the effectiveness of carbon in kerogen (containing available carbon+ and ineffective carbon+ inorganic minerals) was prepared. To prepare the Type-B adsorbent (abbreviated as T_B) that contained the carbon in kerogen ineffective (containing ineffective carbon+ inorganic minerals), a fresh shale sample of T_{HCl} was heated in a N_2 flow at 650 °C for 1 h to remove the pyrolytic hydrocarbons S_1 and S_2; then, we treated the pyrolysis products with the same extraction process as the extract for T_A, removing as much of the soluble organic matter as possible. To prepare the Type-C (represented by T_C) adsorbent that removed kerogen (containing only inorganic minerals), a fresh shale sample of T_{HCl} was heated in an O_2 flow at 900 °C for 1 h to remove all organic carbon; then, we treated the pyrolysis products with the same extraction process as the extract for T_A, removing the soluble organic matter. The adsorbents and T_{HCl} were subsequently analyzed using XRD and Rock-Eval to verify the quality.

X-ray diffraction (XRD) analysis of the adsorbents was performed using a German Bruker AXS D8 Advance, with a Cu target; a ceramic X-ray tube; operation at 40 kV and 40 mA; a focal spot size of 0.4 × 12 mm; a LynxEye XE detector; test mode: wide angle of $5°-90°$; and a rate of 5°/min. XRD analysis was completed in the Analysis and Test Center of the School of Materials and Chemistry, China University of Geosciences (Wuhan).

Rock-Eval analysis of the adsorbents was performed according to Lafargue et al. (1998) [6]. The powder samples (20 g and about 80 mesh, ≤ 178 μm in grain size) were analyzed using Rock-Eval 6 (manufactured by Vinci Technologies in France). Rock-Eval analysis and data processing were performed by the Guangzhou Institute of Geochemistry, Chinese Academy of Sciences (GIG, Guangzhou, China). Subsequently, we obtained the pyrolysis data of S_1, S_2, total organic carbon (TOC), residual organic carbon (RC), mineral carbon (MinC), Tmax (°C), and the hydrogen index (HI) of the adsorbents.

3.2. Adsorption Experimental Method

Generally, an adsorbate solution with a low or high concentration can reach adsorption equilibrium with the adsorbent within a certain period of time (many liquid adsorption experiments measure the concentration of the adsorbate solution in the adsorption equilibrium stage by spectrophotometry to estimate the adsorption capacity of the adsorbent [17,43]. Therefore, there is still fluidity of the unadsorbed adsorbate molecules in the adsorbate solution at the adsorption equilibrium state); when the concentration of adsorbate reaches a certain level at which the adsorbent reaches saturated adsorption equilibrium, the adsorbent will no longer receive adsorbate. Even if the concentration of the adsorbate increases further, it will not affect the adsorption capacity of the adsorbent [21,28]. In other words, the crude oil (adsorbate) in the formation fluid (such as the oil–water solution containing salt and other soluble substances), under geological conditions, does not need to satisfy saturated adsorption by inorganic minerals and kerogen (adsorbents). After the crude oil in the formation fluid and the inorganic minerals and kerogen reach adsorption equilibrium, the unadsorbed crude oil still has mobility. This principle is often used to reduce the adsorption equilibrium concentration of crude oil and improve its mobility by injecting hot water, gas, and other polymer solutions to enhance oil recovery (EOR) in the secondary and tertiary oil recovery stages of the oilfield [44,45]. Therefore, the setting of the adsorbate content (crude oil) in the experiment should be close to the content of chloroform bitumen A in geological samples. At the same time, in many studies on the organic matter adsorption properties of inorganic minerals, most choose single or mixed standard samples as adsorbates or adsorbents [17,20,46,47]. However, when crude oil is adsorbing on shale, different types of compounds and multifunctional carbon compounds in crude oil compete for the chemical bonds of the adsorbent for adsorption, so using crude oil as an adsorbate could more effectively reflect practical geological problems.

Ertas et al. [8] tested kerogen swelling under high-pressure and closed conditions, and found that the average swelling ratio at 150 °C was slightly higher than that at 30 °C and 90 °C, indicating that temperature affects kerogen swelling [26]. Therefore, the simulated temperature of this adsorption experiment was set at 90 °C, which was close to the actual formation temperature of the sample. Pernyeszi et al. conducted the adsorption of asphaltenes on clay and reservoir rocks by shaking them at 298 K for 24 h, and adsorption equilibrium was reached after the reaction [17]. Daughney [28] reported that adsorption equilibration was achieved within 24 h for the sorption of oil onto powdered quartz in both the presence and absence of an aqueous phase. Ertas et al. (2006) also verified that an adsorbate solution could reach adsorption equilibrium within 24 h at 30 °C, 90 °C, and 150 °C [8]. In fact, in many reports, it was found that the adsorption equilibrium time of organic matter on inorganic minerals only took tens to hundreds of minutes [43,48,49]. Furthermore, previous studies have shown that an increase in temperature reduced the time taken to reach adsorption equilibrium [50,51]. Therefore, because the simulated temperature of this experiment was close to the temperature of the formation where the shale sample was located (90 °C), combined with the above experience, continuing the experiment at 90 °C for 24 h was enough to reach adsorption equilibrium.

The conditions for each adsorption experiment are shown in Table 1. To ensure the adsorbate was evenly distributed in the adsorbents, it was necessary to dilute the crude oil with DCM and fully mix it with the adsorbent in the lining of the autoclave (25 mL), and then, wait for the DCM to evaporate completely. The adsorption experiment continued at 90 °C for 24 h, and then, the unadsorbed hydrocarbons were recovered via extraction.

Table 1. Conditions used in the adsorption experiments.

Sample No.	Chloroform Bitumen C (wt.%)	Temperature of Adsorbent Preparation (°C)	Adsorbent No.	Weight of Adsorbent/Adsorbate (g)		
				Adsorbent Content	Calculate Adsorbate Content	Adsorbate Content
1	0.597	300	1-T_A	10.0534	0.0601	0.0603
		650	1-T_B	10.0457	0.0600	0.0601
		900	1-T_C	10.0285	0.0599	0.0600
2	0.589	300	2-T_A	10.1000	0.0595	0.0591
		650	2-T_B	10.3014	0.0607	0.0606
		900	2-T_C	10.3354	0.0609	0.0606
3	0.479	300	3-T_A	10.0139	0.0480	0.0478
		650	3-T_B	10.2368	0.0490	0.0490
		900	3-T_C	10.3310	0.0495	0.0498

Table 1. Cont.

Sample No.	Chloroform Bitumen C (wt.%)	Temperature of Adsorbent Preparation (°C)	Adsorbent No.	Weight of Adsorbent/Adsorbate (g)		
				Adsorbent Content	Calculate Adsorbate Content	Adsorbate Content
4	0.986	300	4-T_A	9.6927	0.0956	0.0989
		650	4-T_B	10.0054	0.0987	0.0975
		900	4-T_C	10.0219	0.0988	0.0989
5	0.575	300	5-T_A	9.9344	0.0571	0.0574
		650	5-T_B	10.1426	0.0583	0.0582
		900	5-T_C	10.3579	0.0596	0.0601

Here, in Table 1, 1-T_A represents the Type-A adsorbent of sample 1 that was treated with hydrochloric acid to remove the pyrolytic hydrocarbon S_1 (heated in a N_2 flow at 300 °C for 1 h) and the soluble organic matter; 1-T_B represents the Type-B adsorbent of sample 1 that was treated with hydrochloric acid to remove the pyrolytic hydrocarbons S_1 and S_2, (heated in a N_2 flow at 650 °C for 1 h) and the soluble organic matter; 1-T_C represents the Type-C adsorbent of sample 1 that was treated with hydrochloric acid to remove all of the organic matter (heated in an O_2 flow at 900 °C for 1 h). Furthermore, 2-T_A, 2-T_B and 2-T_C represent shale adsorbents prepared by the same thermal decomposition and extraction process as 1-T_A, 1-T_B and 1-T_C, respectively, for shale powder samples after hydrochloric acid treatment. The other adsorbent codes also represent the corresponding treatment procedures for the corresponding samples.

4. Results

4.1. Adsorbent Properties

The quality of the adsorbent is the key to a successful adsorption experiment; so, it was necessary to test the quality of the adsorbent before the adsorption experiment. For the detection of adsorbent properties, we mainly used XRD analysis to detect the composition and content of inorganic minerals, and Rock-Eval analysis to detect various indicators of organic matter.

T_{HCl} and the adsorbents of T_A (heated in a N_2 flow at 300 °C), T_B (heated in a N_2 flow at 650 °C), and T_C (heated in an O_2 flow at 900 °C) were analyzed via XRD (Figure 2), and the relative contents of the main inorganic minerals were statistically analyzed (Figure 3). The result showed that T_{HCl} and the adsorbents mainly contained quartz (Q), orthoclase (Or), plagioclase (Pl), illite (I), illite/smectite (I/S), and a little chlorite (Ch), among which quartz had the highest relative content. Pyrolysis treatment under different conditions had a great influence on the composition and content of inorganic minerals in shale.

With the increased temperature in the adsorbent preparation, the illite diffraction peak intensity increased significantly (Figure 2), which was due to the transformation of illite/smectite to illite [52–54]. At the same time, the diffraction peak intensity of plagioclase also increased. This was due to the consumption of K^+ in the transformation of illite/smectite to illite; as long as the conversion exists, the dissolution of orthoclase will continue until exhausted [55]. Moreover, plagioclase will be difficult to dissolve, even when precipitated, because of the buffering of Na^+ produced by the transformation of illite/smectite to illite [55,56]. Subsequently, when the temperature of the adsorbent preparation reaches about 650 °C, the illite/smectite is completely transformed, and the plagioclase gradually decomposes and its diffraction peak intensity is reduced [57]. Meanwhile, the dehydroxylation reaction of illite will also reduce its diffraction peak intensity. When the temperature of the adsorbent preparation reaches about 900 °C, the diffraction peak intensity of illite decreases further due to the high degree of the dehydroxylation reaction and a certain degree of amorphous transformation (sintering phenomenon) [58–63].

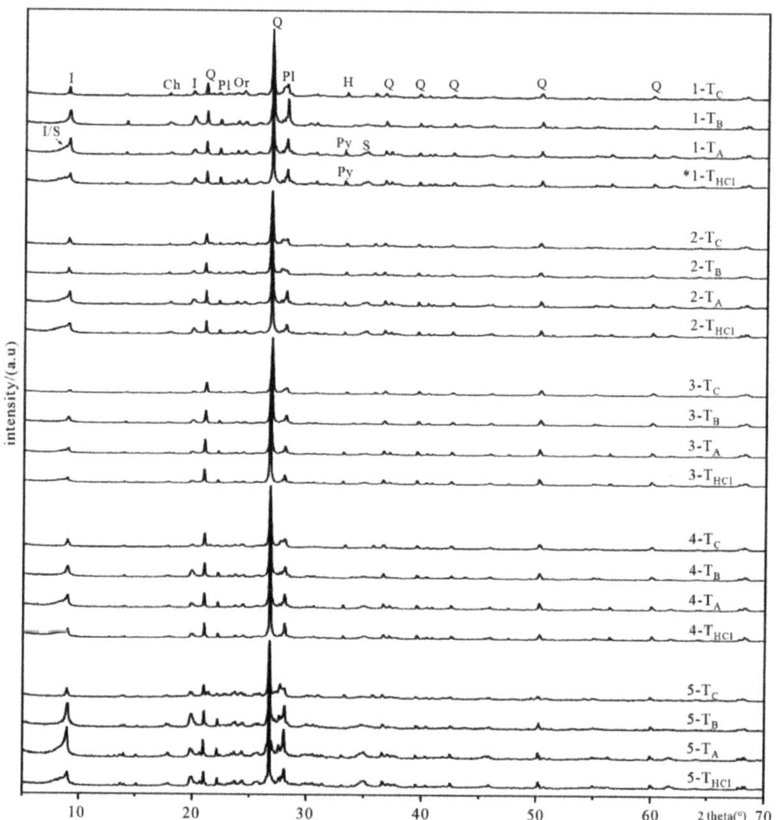

Figure 2. XRD pattern of T_{HCl} and adsorbents. The identified minerals are Q: quartz; Or: orthoclase; Pl: plagioclase; Py: pyrite; H: hematite; I: illite; S: smectite; I/S: illite/smectite; and Ch: chlorite. * T_{HCl} represents sample treated with hydrochloric acid.

For the samples treated with hydrochloric acid (T_{HCl}) and the adsorbents of T_A (heated in a N_2 flow at 300 °C), T_B (heated in a N_2 flow at 650 °C), and T_C (heated in an O_2 flow at 900 °C), after being analyzed via Rock-Eval (Table 2), it was found that the mineral carbon (MinC%) content was very low (0–0.23 wt.%, average of 0.1 wt.%), which indicated that the measurements of TOC, S_2, and RC were little affected by the thermal decomposition of carbonate minerals. The TOC content in T_A (containing effective carbon) was high (2.01–4.84 wt%), and S_1 almost disappeared, indicating that pyrolysis and solvent extraction substantially removed free hydrocarbons from T_A. T_B (including ineffective carbon) contained almost no S_1 and S_2, and the TOC was close to the RC, which indicated that the free hydrocarbons and effective carbon in T_B were substantially removed by pyrolysis and solvent extraction. While T_B had a higher content of RC (1.60–3.29 wt%) than of T_A, indicating that part of the reactive kerogen was converted into inert kerogen through hydrocarbon generation, T_C contained no organic carbon (TOC ≈ RC ≈ MinC ≈ 0), indicating that all organic carbon was removed by oxidation at 900 °C, and only inorganic minerals remained. The results of the pyrolysis data show that the three types of adsorbents satisfied the requirements of the adsorption experiment and eliminated the interference factors of the experiment.

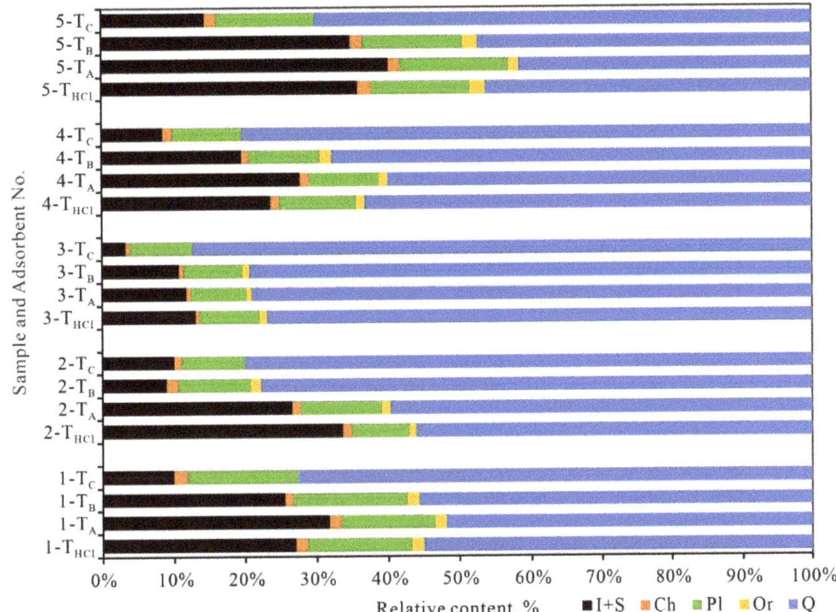

Figure 3. Relative contents of main inorganic minerals in T_{HCl} and adsorbents. I + S: illite + smectite + illite/smectite.

Table 2. Rock-Eval analysis data of T_{HCl} and adsorbents.

No.	S_1 (mg/g)	S_2 (mg/g)	TOC (wt.%)	RC (wt.%)	MinC (wt.%)	Tmax (°C)	HI (mg/g)
1-T_{HCl}	2.58	21.18	4.81	2.83	0.15	446	440
1-T_A	0.04	18.44	4.49	2.94	0.15	445	411
1-T_B	0.01	0	3.29	3.28	0.07	486	0
1-T_C	0	0	0	0	0	486	0
2-T_{HCl}	0.06	6.55	2.41	1.84	0.16	440	272
2-T_A	0.08	4.45	2.03	1.64	0.09	440	219
2-T_B	0.06	0.01	1.76	1.75	0.12	297	1
2-T_C	0	0	0.01	0.01	0.01	399	0
3-T_{HCl}	0.26	22.42	4.82	2.92	0.07	444	465
3-T_A	0.04	20.44	4.52	2.8	0.14	445	452
3-T_B	0.07	0.06	3.03	3.01	0.13	305	2
3-T_C	0.03	0	0.01	0	0.01	327	0
4-T_{HCl}	0.47	14.19	3.98	2.75	0.23	443	357
4-T_A	0.06	10.52	3.51	2.62	0.14	443	300
4-T_B	0.06	0.01	2.71	2.7	0.17	293	0
4-T_C	0.01	0	0	0	0.01	392	0
5-T_{HCl}	0.2	4.89	2.01	1.57	0.16	439	243
5-T_A	0.04	1.91	1.72	1.54	0.07	440	111
5-T_B	0.06	0.01	1.6	1.59	0.05	290	0
5-T_C	0.01	0	0	0	0	376	0

4.2. Results of Adsorption Experiment

After the adsorption experiment, we referred to the calculation method of adsorbed hydrocarbon by Zhang et al. [25] and took the volatility of crude oil into account. The product of the adsorption experiment was extracted with a mixed solvent of DCM and

MeOH (93:7, v/v) for 72 h, and the unadsorbed hydrocarbons (adsorbate) were recovered. After weighing it, we performed a calculation using the following formula:

$$q = [m_{oil} \times (1 - K_{oil}) - m_{ext}]/m_a \tag{1}$$

$$Q = [1 - m_{ext}/(m_{oil} - m_{oil} \times K_{oil})] \times 100\% \tag{2}$$

Here, q (mg/g) is the crude oil adsorption capacity; m_{oil} (mg) is the weight of the adsorbate (crude oil); K_{oil} (%) is the volatilization ratio of the adsorbate (crude oil) during the process of the adsorption experiment, which was 14.27% (the date is the average value that measured multiple times in under the experimental conditions of continued the crude oil comparison group at 90 °C for 24 h); m_{ext} (mg) is the unadsorbed hydrocarbons that were extracted in the adsorption experiment; m_a (g) is the wight of adsorbent; and Q (wt.%) is the crude oil adsorption ratio. Then, we obtained the adsorption capacity (Equation (1)) and the crude oil adsorption ratio (Equation (2)) of different types of adsorbents (Table 3).

Table 3. The date of adsorption capacity and adsorption ratio.

No.	Adsorption Capacity (mg/g)	Adsorption Ratio (wt.%)
1-T_A	1.14	19.02
1-T_B	1.96	32.76
1-T_C	0.97	16.23
2-T_A	1.32	22.34
2-T_B	1.86	31.55
2-T_C	1.00	17.02
3-T_A	0.84	17.55
3-T_B	1.27	26.48
3-T_C	0.54	11.36
4-T_A	2.23	22.64
4-T_B	3.31	33.51
4-T_C	1.15	11.67
5-T_A	1.47	25.52
5-T_B	2.24	38.98
5-T_C	0.93	16.23

The date of the adsorption capacity and adsorption ratio (Table 3) showed that the different samples of three types of adsorbents had the same trend of adsorption ratio for crude oil, among which T_B had the highest crude oil adsorption ratio and adsorption capacity per unit mass, followed by T_A, and then, T_C. This indicated that the evolution state of the adsorbent had a great influence on its own adsorption performance of crude oil. Meanwhile, the crude oil adsorption ratio of the same types of adsorbent were obviously different; it was evident that the difference of the samples had an influence on the adsorption properties of the same type of adsorbent. So, it was necessary to analyze the influence of kerogen and inorganic minerals on the crude oil adsorption properties.

Based on Table 3, the adsorbate content with the adsorption capacity (Figure 4a) and the adsorbents with the adsorption ratio (Figure 4b) were drawn. Upon combining Figure 4 with Figure 2, it was found that although T_C contained only inorganic minerals and lost more clay minerals than T_A and T_B, the crude oil adsorption ratio of T_C was not much lower than that of T_A and T_B, which implies that the more clay minerals contained in T_A and T_B is more important for its adsorption properties. Meanwhile, after the pyrolysis treatment, the changes in the kerogen content and structure, inorganic mineral content, and surface properties of T_B were much greater than those of T_A, which may be the main reason that T_B had the highest crude oil adsorption capacity. Therefore, further analysis of the crude oil adsorption capacity of T_B is needed.

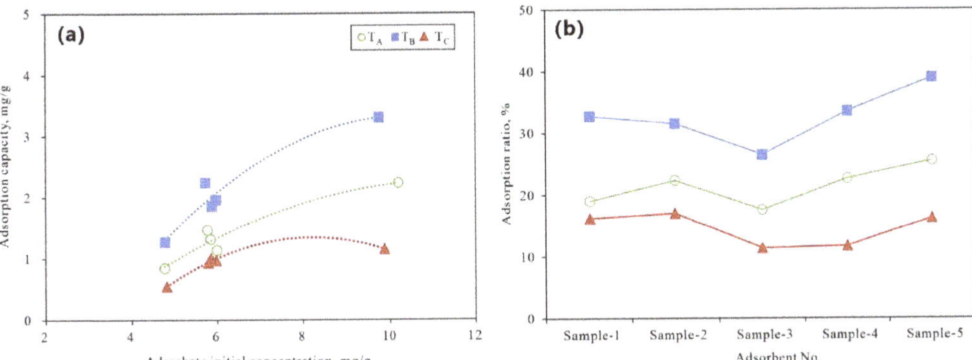

Figure 4. Adsorbate content with adsorption capacity (**a**), and adsorbents with adsorption ratio (**b**).

Figure 4a shows that, with an adsorbate content increase, the adsorption capacity gradually approached saturation. In this process, and with the adsorbents in an unsaturated adsorption equilibrium state, the adsorbate content (corresponding to the chloroform bitumen C content of T_{HCl}) determined the adsorption capacity in the adsorption equilibrium stage. This result conformed to the original intention of this experiment's design: to observe the adsorption properties of the different shale adsorbents of crude oil under the condition of the content of adsorbate being close to the oil content of geological samples. Figure 4b shows that when the T_A of samples 1, 2, and 4, the T_B of samples 2 and 4, and the T_C of samples 3 and 4 had different adsorbate contents, the adsorption ratio of crude oil was basically the same, indicating that adsorbate content cannot determine the crude oil adsorption ratio of shale.

The adsorbents prepared from different shale samples had obvious differences in their crude oil adsorption ratios. For example, the different types of adsorbents in sample 3 had the lowest crude oil adsorption ratios. As shown in Figure 2, among the same types of adsorbent, sample 3 had the lowest levels of clay and feldspar mineral content. Other adsorbents with higher contents of clay and feldspar minerals generally had higher crude oil adsorption ratios, indicating that the content of inorganic minerals in shale adsorbents had a significant effect on crude oil adsorption performance in shale. However, compared with T_A, T_B lost more clay minerals, but its crude oil adsorption ratio was higher than that of T_A. The reasons for this phenomenon also need to be analyzed.

5. Discussion

5.1. Crude Oil Adsorption Ratio of Kerogen

As T_C does not contain organic carbon, only the organic carbon content of T_A and T_B needs to be compared with the crude oil adsorption ratio. By comparing the organic carbon content of the same types of adsorbents with the crude oil adsorption ratio, it was found that there was a negative correlation between the crude oil adsorption ratio and the TOC of the adsorbents. The crude oil adsorption ratio generally increased with decreasing TOC (Figure 5a,b). This phenomenon is contrary to previous adsorption experiments using purified kerogen that showed that the kerogen content was the main controlling factor in the adsorption of crude oil, but similar to some studies that demonstrated that kerogen in shale was not the main adsorbent of hydrocarbon [26,64,65]. As can be seen from Figures 2 and 5b, the adsorbent prepared from sample 5 had the highest crude oil adsorption ratio and the highest I + S content among the same types of adsorbent. The adsorbent prepared from sample 3 had the lowest crude oil adsorption ratio and the lowest I + S content among the same types of adsorbent. Both of these results indicate that in actual shale, an increase in organic carbon (TOC = 1.60–4.52%) content cannot improve the crude oil adsorption ratio in shale and cannot dominate crude oil adsorption in shale. However, the crude oil adsorption ratio may be more affected by clay minerals.

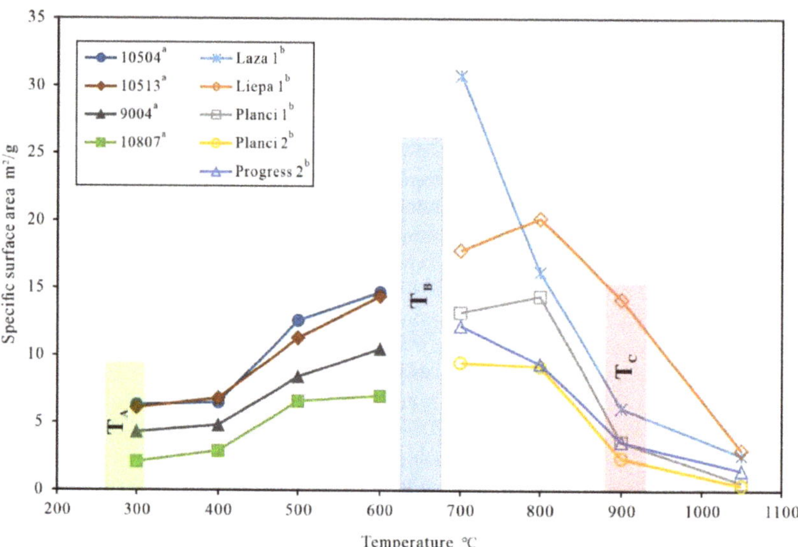

Figure 8. Effect of temperature on the specific surface area of shale samples and clay minerals. Left: Four groups of data on shale samples, referenced and adapted with permission from Ref. [42], marked with "a". These shale samples were not treated with hydrochloric acid, and the carbonate minerals were largely decomposed when the temperature was higher than 600 °C, which increased the SSA of the shale samples. Therefore, only data for temperatures lower than 600 °C can represent the pyrolysis characteristics of T_A and T_B in this experiment. Right: Five groups of data on clay minerals, referenced from Dabare and Svinka [80], marked with "b". The changes in SSA in several clay minerals (calcination at 700−1050 °C) were not affected by the thermal decomposition of carbonate. Moreover, the composition of the main inorganic minerals was basically consistent with that of T_{HCl}, so this study can approximately reflect the changes in the SSA of T_C in this experiment.

6. Conclusions

(1) Under the condition of the adsorbate content being close to the oil content of geological shale samples, the crude oil adsorption capacity of T_A (containing kerogen effective carbon) ranged from 1.39 to 3.66 mg/g, and the adsorption ratio ranged from 17.56 to 25.52%. The crude oil adsorption capacity of T_B (containing kerogen ineffective carbon) ranged from 1.77 to 4.12 mg/g, and the adsorption ratio ranged from 26.48 to 38.98%. The crude oil adsorption capacity of T_C (only inorganic minerals) ranged from 1.18 to 2.40 mg/g, and the adsorption ratio ranged from 11.36 to 17.02%. The change in the surface properties of shale adsorbents during thermal evolution was the main reason for the different crude oil adsorption properties of different types of adsorbent.

(2) In the adsorbents prepared from general organic-rich shale samples (TOC = 1.60−4.52%), because of the wide content difference between kerogen and inorganic minerals, resulting in a change in kerogen's structural units and content, it cannot dominate the crude oil adsorption of shale. On the contrary, the composition and evolution of inorganic minerals are closely related to the crude oil adsorption properties of shale, and they play a dominant role in shale's adsorption of crude oil. Among them, the content and evolution characteristics of illite + smectite in shale had the most significant effect on the adsorption of crude oil, and orthoclase can indirectly affect the crude oil adsorption properties of shale by affecting the conversion process of illite/smectite to illite.

Author Contributions: Conceptualization, S.L.; methodology, S.L. and Y.Z.; validation, Y.Z.; investigation, S.L., Y.Z. and C.Z.; resources, S.L.; data curation, Y.Z.; writing—original draft preparation, Y.Z.; writing—review and editing, S.L. and S.H.; supervision, S.L. and C.Z.; project administration, S.L. All authors have read and agreed to the published version of the manuscript.

Funding: This research was funded by the National Natural Science Foundation of China, grant number 42073067.

Data Availability Statement: Not applicable.

Acknowledgments: This research was financially supported by the National Natural Science Foundation of China (grant no. 42073067). We sincerely appreciate all anonymous reviewers and the handling editor for their comments and suggestions.

Conflicts of Interest: The authors declare no conflict of interest.

References

1. Jarvie, D.M. Shale Resource Systems for Oil and Gas: Part2-Shale-Oil Resource Systems. *Shale Reserv.-Giant Resour. 21st Century* **2012**, *97*, 89–119.
2. Zou, C.N.; Yang, Z.; Cui, J.W.; Zhu, R.K.; Hou, L.H.; Tao, S.Z.; Yuan, X.J.; Wu, S.T.; Lin, S.H.; Wang, L.; et al. Formation mechanism, geological characteristics and development strategy of nonmarine shale oil in China. *Pet. Explor. Dev.* **2013**, *40*, 15–27. [CrossRef]
3. Modica, C.J.; Lapierre, S.G. Estimation of kerogen porosity in source rocks as a function of thermal transformation: Example from the Mowry Shale in the Powder River Basin of Wyoming. *AAPG Bull.* **2012**, *96*, 87–108. [CrossRef]
4. Jiang, Q.G.; Li, M.W.; Qian, M.H.; Li, Z.M.; Li, Z.; Huang, Z.K.; Zhang, C.M.; Ma, Y.Y. Quantitative characterization of shale oil in different occurrence states and its application. *Pet. Geol. Exp.* **2016**, *38*, 842–849.
5. Li, S.F.; Hu, S.Z.; Xie, X.N.; Lv, Q.; Huang, X.; Ye, J.R. Assessment of shale oil potential using a new free hydrocarbon index. *Int. J. Coal Geol.* **2016**, *156*, 74–85. [CrossRef]
6. Lafargue, E.; Marquis, F.; Pillot, D. Rock-Eval 6 applications in hydrocarbon exploration, production, and soil contamination studies. *Rev. L'institut Français Pétrole* **1998**, *53*, 421–437. [CrossRef]
7. Ritter, U. Fractionation of petroleum during expulsion from kerogen. *J. Geochem. Explor.* **2003**, *78–79*, 417–420. [CrossRef]
8. Ertas, D.; Kelemen, S.R.; Halsey, T.C. Petroleum Expulsion Part 1. Theory of Kerogen Swelling in Multicomponent Solvents. *Energy Fuels* **2006**, *20*, 295–300. [CrossRef]
9. Kelemen, S.R.; Walters, C.C.; Ertas, D.; Kwiatek, L.M.; Curry, D.J. Petroleum Expulsion Part 2. Organic Matter Type and Maturity Effects on Kerogen Swelling by Solvents and Thermodynamic Parameters for Kerogen from Regular Solution Theory. *Energy Fuels* **2006**, *20*, 301–308. [CrossRef]
10. Pepper, A.S.; Corvi, P.J. Simple Kinetic-Models of Petroleum Formation.1. Oil and Gas Generation from Kerogen. *Mar. Pet. Geol.* **1995**, *12*, 291–319. [CrossRef]
11. Pepper, A.S.; Corvi, P.J. Simple kinetic models of petroleum formation. Part III: Modelling an open system. *Mar. Pet. Geol.* **1995**, *12*, 417–452. [CrossRef]
12. Jarvie, D.M. Components and processes affecting producibility and commerciality of shale resource systems. *Geol. Acta* **2014**, *12*, 307–325.
13. Jarvie, D.M.; Hill, R.J.; Ruble, T.E.; Pollastro, R.M. Unconventional shale-gas systems: The Mississippian Barnett Shale of north-central Texas as one model for thermogenic shale-gas assessment. *AAPG Bull.* **2007**, *91*, 475–499. [CrossRef]
14. Li, Z.M.; Tao, G.L.; Li, M.W.; Jiang, Q.G.; Cao, T.T.; Liu, P.; Qian, M.H.; Xie, M.M.; Li, Z. Favorable interval for shale oil prospecting in coring Well L69 in the Paleogene Es3L in Zhanhua Sag, Jiyang Depression, Bohai Bay Basin. *Oil Gas Geol.* **2019**, *40*, 236–247.
15. Cai, Y.L.; Zhang, X.; Zou, Y.R. Solvent swelling: A new technique for oil primary migration. *Geochimica* **2007**, *36*, 351–356.
16. Collins, S.H.; Melrose, J.C. Adsorption of Asphaltenes and Water on Reservoir Rock Minerals. *SPE Oilfield Geotherm. Chem. Symp.* **1983**, 249–254. [CrossRef]
17. Pernyeszi, T.; Patzko, A.; Berkesi, O.; Dekany, I. Asphaltene adsorption on clays and crude oil reservoir rocks. *Colloids Surf. A Physicochem. Eng. Asp.* **1998**, *137*, 373–384. [CrossRef]
18. Schettler, P.D., Jr.; Parmely, C.R. Contributions to Total Storage Capacity in Devonian Shales. In Proceedings of the SPE Eastern Regional Meeting, Lexington, KY, USA, 23 October 1991.
19. Cao, Z.; Jiang, H.; Zeng, J.H.; Hakim, S.; Lu, T.Z.; Xie, X.M.; Zhang, Y.C.; Zhou, G.G.; Wu, K.Y.; Guo, J.R. Nanoscale liquid hydrocarbon adsorption on clay minerals: A molecular dynamics simulation of shale oils. *Chem. Eng. J.* **2021**, *420*, 127578. [CrossRef]
20. Acevedo, S.; Ranaudo, M.A.; Escobar, G.; Gutiérrez, L.; Ortega, P. Adsorption of asphaltenes and resins on organic and inorganic substrates and their correlation with precipitation problems in production well tubing. *Fuel* **1995**, *74*, 595–598. [CrossRef]
21. Li, Z.; Zou, Y.R.; Xu, X.Y.; Sun, J.N.; Li, M.W.; Peng, P.A. Adsorption of mudstone source rock for shale oil—Experiments, model and a case study. *Org. Geochem.* **2016**, *92*, 55–62. [CrossRef]

22. Guo, S.B.; Mao, W.J. Division of diagenesis and pore evolution of a Permian Shanxi shale in the Ordos Basin, China. *J. Pet. Sci. Eng.* **2019**, *182*, 106351. [CrossRef]

23. Mastalerz, M.; Schimmelmann, A.; Drobniak, A.; Chen, Y. Porosity of Devonian and Mississippian New Albany Shale across a maturation gradient: Insights from organic petrology, gas adsorption, and mercury intrusion. *AAPG Bull.* **2013**, *97*, 1621–1643. [CrossRef]

24. Li, F.L.; Wang, M.Z.; Liu, S.B.; Hao, Y.W. Pore characteristics and influencing factors of different types of shales. *Mar. Pet. Geol.* **2019**, *102*, 391–401. [CrossRef]

25. Zhang, L.Y.; Bao, S.Y.; Li, J.Y.; Li, Z.; Zhu, R.F.; Zhang, L.; Wang, Y.R. Hydrocarbon and crude oil adsorption abilities of minerals and kerogens in lacustrine shales. *Pet. Geol. Exp.* **2015**, *37*, 776–780.

26. Fan, E.P.; Tang, S.H.; Zhang, C.L.; Guo, Q.L.; Sun, C.H. Methane sorption capacity of organics and clays in high-over matured shale-gas systems. *Energy Explor. Exploit.* **2014**, *32*, 927–942. [CrossRef]

27. Tian, S.S.; Xue, H.T.; Lu, S.F.; Zeng, F.; Xue, Q.Z.; Chen, G.H.; Wu, C.Z.; Zhang, S.S. Molecular Simulation of Oil Mixture Adsorption Character in Shale System. *J. Nanosci. Nanotechnol.* **2017**, *17*, 6198–6209. [CrossRef]

28. Daughney, C.J. Sorption of crude oil from a non-aqueous phase onto silica: The influence of aqueous pH and wetting sequence. *Org. Geochem.* **2000**, *31*, 147–158. [CrossRef]

29. Dudášová, D.; Simon, S.; Hemmingsen, P.V.; Sjöblom, J. Study of asphaltenes adsorption onto different minerals and clays. *Colloids Surf. A Physicochem. Eng. Asp.* **2008**, *317*, 62–72. [CrossRef]

30. Wei, Z.F.; Zou, Y.R.; Cai, Y.L.; Wang, L.; Luo, X.R.; Peng, P.A. Kinetics of oil group-type generation and expulsion: An integrated application to Dongying Depression, Bohai Bay Basin, China. *Org. Geochem.* **2012**, *52*, 1–12. [CrossRef]

31. Chen, J.H.; Philp, R.P. Porphyrin distributions in crude oils from the Jianghan and Biyang basins, China. *Chem. Geol.* **1991**, *91*, 139–151. [CrossRef]

32. Dong, Y.L.; Zhu, X.M.; Xian, B.Z.; Hu, T.H.; Geng, X.J.; Liao, J.J.; Luo, Q. Seismic geomorphology study of the Paleogene Hetaoyuan Formation, central-south Biyang Sag, Nanxiang Basin, China. *Mar. Pet. Geol.* **2015**, *64*, 104–124. [CrossRef]

33. Song, Y.; Li, S.F.; Hu, S.Z. Warm-humid paleoclimate control of salinized lacustrine organic-rich shale deposition in the Oligocene Hetaoyuan Formation of the Biyang Depression, East China. *Int. J. Coal Geol.* **2018**, *202*, 69–84. [CrossRef]

34. Qing, J.Z. *Source Rocks in China*; Science Press: Beijing, China, 2005.

35. Sun, J.N. Evalution of Movable Oil and Retention Machanics of Dongying Depression Shales. Ph.D. Thesis, University of Chinese Academy of Sciences, Guangzhou, China, 2021.

36. Sert, M.; Ballice, L.; Yüksel, M.; Saglam, M.; Reimert, R.; Erdem, S. Effect of solvent swelling on pyrolysis of kerogen (type-I) isolated from Göynük oil shale (Turkey). *J. Anal. Appl. Pyrolysis* **2009**, *84*, 31–38. [CrossRef]

37. Cooles, G.P. Calculation of petroleum masses generated and expelled from source rocks. *Org. Geochem.* **1986**, *10*, 235–245. [CrossRef]

38. Dow, W.G. Kerogen studies and geological interpretations. *J. Geochem. Explor.* **1977**, *7*, 79–99. [CrossRef]

39. Qiu, L.W.; Jiang, Z.X.; Cao, Y.C.; Qiu, R.H.; Chen, W.X.; Tu, Y.F. Alkaline diagenesis and its influence on a reservoir in the Biyang depression. *Sci. China (Earth Sci.)* **2002**, *45*, 643–653. [CrossRef]

40. Wang, Q.; Xu, F.; Bai, J.R.; Sun, B.Z.; Li, S.H.; Sun, J. Study on the Basic Physicoche mical Characteristics of the Huadi an Oil Shales. *J. Jilin Univ. (Earth Sci. Ed.)* **2006**, *36*, 1006–1011.

41. Dembicki, J.H. The effects of the mineral matrix on the determination of kinetic parameters using modified Rock Eval pyrolysis. *Org. Geochem.* **1992**, *18*, 531–539. [CrossRef]

42. Bai, J.R.; Song, K.T.; Chen, J.B. The migration of heavy metal elements during pyrolysis of oil shale in Mongolia. *Fuel* **2018**, *225*, 381–387. [CrossRef]

43. Yu, Z.; Ding, K.; Han, C.; Wu, Y.; Guan, F. Adsorption of diesel oil onto natural mud shale: From experimental investigation to thermodynamic and kinetic modelling. *Int. J. Environ. Anal. Chem.* **2021**, *8*, 1–15. [CrossRef]

44. Hao, Y.M.; Lu, M.J.; Dong, C.S.; Jia, J.P.; Su, Y.L.; Sheng, G.L. Experimental Investigation on Oil Enhancement Mechanism of Hot Water Injection in tight reservoirs. *Open Phys.* **2016**, *14*, 703–713.

45. Shakeel, M.; Pourafshary, P.; Hashmet, M.R. Hybrid Engineered Water-Polymer Flooding in Carbonates: A Review of Mechanisms and Case Studies. *Appl. Sci.* **2020**, *10*, 6087. [CrossRef]

46. Zhang, J.; Lu, S.F.; Li, J.Q.; Zhang, P.F.; Xue, H.T.; Zhao, X.; Xie, L.J. Adsorption Properties of Hydrocarbons (n-Decane, Methyl Cyclohexane and Toluene) on Clay Minerals: An Experimental Study. *Energies* **2017**, *10*, 1586. [CrossRef]

47. Zeng, T.Z.; Kim, K.T.; Werth, C.J.; Katz, L.E.; Mohanty, K.K. Surfactant Adsorption on Shale Samples: Experiments and an Additive Model. *Energy Fuels* **2020**, *34*, 5436–5443. [CrossRef]

48. Rajak, V.K.; Kumar, H.; Mandal, A. Kinetics, equilibrium and thermodynamic studies of adsorption of oil from oil-in-water emulsion by activated charcoal. *Int. J. Surf. Sci. Eng.* **2016**, *10*, 600–621. [CrossRef]

49. Acar, E.T.; Ortaboy, S.; Atun, G. Adsorptive removal of thiazine dyes from aqueous solutions by oil shale and its oil processing residues: Characterization, equilibrium, kinetics and modeling studies. *Chem. Eng. J.* **2015**, *276*, 340–348. [CrossRef]

50. Zhang, H.X.; Wang, X.Y.; Liang, H.H.; Tan, T.S.; Wu, W.S. Adsorption behavior of Th(IV) onto illite: Effect of contact time, pH value, ionic strength, humic acid and temperature. *Appl. Clay Sci.* **2016**, *127*, 35–43.

51. Li, S.G.; Bai, Y.; Lin, H.F.; Yan, M.; Liu, J.B. Experimental study on the effect of temperature on the kinetics characteristics of gas adsorption on coal. *J. Xi'an Univ. Sci. Technol.* **2018**, *38*, 181–186+272.

52. Berger, G.; Lacharpagne, J.C.; Velde, B.; Beaufort, D.; Lanson, B. Kinetic constrains on illitization reactions and effects of organic diagenesis in sandstone/shale sequences. *Appl. Geochem.* **1997**, *12*, 23–35. [CrossRef]

53. Velde, B.; Vasseur, G. Estimation of the diagenetic smectite to illite transformation in time-temperature space. *Am. Mineral.* **1992**, *77*, 967–976.

54. Boles, J.R.; Franks, S.J. Clay diagenesis in Wilcox sandstones of Southwest Texas; implications of smectite diagenesis on sandstone cementation. *J. Sediment. Res.* **1979**, *49*, 55–70.

55. Huang, S.J.; Huang, K.K.; Feng, W.L.; Tong, H.P.; Liu, L.H.; Zhang, X.H. Mass exchanges among feldspar, kaolinite and illite and their influences on secondary porosity formation in clastic diagenesis—A case study on the Upper Paleozoic, Ordos Basin and Xujiahe Formation, Western Sichuan Depression. *Geochimica* **2009**, *38*, 498–506.

56. Aagaard, P.; Egeberg, P.K.; Saigal, G.C.; Morad, S.Y.; Bjorlykke, K. Diagenetic albitization of detrital K-feldspars in Jurassic, Lower Cretaceous and Tertiary clastic reservoir rocks from offshore Norway; II, Formation water chemistry and kinetic considerations. *J. Sediment. Res.* **1990**, *60*, 575–581. [CrossRef]

57. Wei, W.W.; Huang, S.J.; Huan, J.L. Thermodynamic Calculation of Illite Formation and Its Significance on Research of Sandstone Diagenesis. *Geol. Sci. Technol. Inf.* **2011**, *30*, 20–25.

58. Hollanders, S.; Adriaens, R.; Skibsted, J.; Cizer, Ö.; Elsen, J. Pozzolanic reactivity of pure calcined clays. *Appl. Clay Sci.* **2016**, *132–133*, 552–560. [CrossRef]

59. Fernandez, R.; Martirena, F.; Scrivener, K.L. The origin of the pozzolanic activity of calcined clay minerals: A comparison between kaolinite, illite and montmorillonite. *Cem. Concr. Res.* **2011**, *41*, 113–122. [CrossRef]

60. He, C.L.; Osbaeck, B.; Makovicky, E. Pozzolanic reactions of six principal clay minerals: Activation, reactivity assessments and technological effects. *Cem. Concr. Res.* **1995**, *25*, 1691–1702. [CrossRef]

61. Li, G.H.; Jiang, T.; Fan, X.H.; Qiu, G.Z. Thermochemical Activation of Silicon in Illite and Its Removal. *Met. Mine* **2004**, *5*, 18–24.

62. Zhou, Z.J.; Chen, D.Z. The properties and evolution of illite at high temperature. *J. Mineral. Petrol.* **1999**, *19*, 20–22.

63. He, C.L.; Osbaeck, B.; Makovicky, E. Thermal stability and pozzolanic activity of calcined illite. *Appl. Clay Sci.* **1995**, *9*, 337–354. [CrossRef]

64. Zhu, X.J.; Cai, J.G.; Wang, G.L.; Song, M.S. Role of organo-clay composites in hydrocarbon generation of shale. *Int. J. Coal Geol.* **2018**, *192*, 83–90. [CrossRef]

65. Ross, D.; Bustin, R.M. The importance of shale composition and pore structure upon gas storage potential of shale gas reservoirs. *Mar. Pet. Geol.* **2009**, *26*, 916–927. [CrossRef]

66. Tian, S.S.; Xue, H.T.; Lu, S.F.; Chen, G.H. Discussion on the Mechanism of Oil and Gas Retention of Different Types of Kerogen, Symposium 57: Basin Dynamics and Unconventional Energy. In Proceedings of the 2014 China Geoscience Union Academic Conference, Beijing, China, 20–23 October 2014.

67. Rahman, H.M.; Kennedy, M.; LöHr, S.; Dewhurst, D.N. Clay-organic association as a control on hydrocarbon generation in shale. *Org. Geochem.* **2017**, *105*, 42–55. [CrossRef]

68. Tian, S.S.; Erastova, V.; Lu, S.F.; Greenwell, H.C.; Underwood, T.R.; Xue, H.T.; Zeng, F.; Chen, G.H.; Wu, C.Z.; Zhao, R.X. Understanding Model Crude Oil Component Interactions on Kaolinite Silicate and Aluminol Surfaces: Toward Improved Understanding of Shale Oil Recovery. *Energy Fuels* **2018**, *32*, 1155–1165. [CrossRef]

69. Jada, A.; Debih, H. Hydrophobation of Clay Particles by Asphaltenes Adsorption. *Compos. Interfaces* **2012**, *16*, 219–235. [CrossRef]

70. Chu, X.Z.; Xu, J.M. Relationship between the Adsorption Capacity of Hydrogen Isotopes and Specific Surface Area of Adsorbents. *Chem. J. Chin. Univ.-Chin. Ed.* **2008**, *29*, 775–778.

71. Nijkamp, M.G. Hydrogen Storage using Physisorption: Modified Carbon Nanofibers and Related Materials. Ph.D. Thesis, Utrecht University Repository, Utrecht, The Netherlands, 2001.

72. Almon, W.; Davies, D.K.; Longstaffe, F. Formation damage and the crystal chemistry of clays. *Short Course Clays Resour. Geol. Montr. Mineral. Assoc. Can.* **1981**, *7*, 81–102.

73. Olphen, V.H.; Fripiar, J.J. *Data Handbook for Clay Materials and Other Non-Metallic Minerals*; Pergamon Press: Oxford, UK; Elmsford, NY, USA, 1979; Volume 131, pp. 62–63.

74. Zhao, X.Y.; Zhang, Y.Y.; Song, J. Some mineralogical characteristics of clay minerals in China oil-bearing basins. *Geoscience* **1994**, *8*, 254–272.

75. Zhang, W.; Ren, Y.N.; Zhao, X.L.; Yan, T.G.; Sun, Q. Investigations of the adsorption of fluoride on natural sodium feldspar and potassium feldspar in aqueous solution. *J. Yunnan Univ. Nat. Sci. Ed.* **2020**, *42*, 977–984.

76. Yazdani, R.; Tuutijrvi, T.; Bhatnagar, A.; Vahala, R. Adsorptive removal of arsenic(V) from aqueous phase by feldspars: Kinetics, mechanism, and thermodynamic aspects of adsorption. *J. Mol. Liq.* **2016**, *214*, 149–156. [CrossRef]

77. Wu, S.T.; Zhu, R.K.; Cui, J.G.; Cui, J.W.; Bai, B.; Zhang, X.X.; Jin, X.; Zhu, D.S.; You, J.C.; Li, X.H. Characteristics of lacustrine shale porosity evolution, Triassic Chang 7 Member, Ordos Basin, NW China. *Pet. Explor. Dev.* **2015**, *42*, 167–176. [CrossRef]

78. Liu, B.; Wang, Y.; Tian, S.S.; Guo, Y.L.; Wang, L.; Yasin, Q.; Yang, J.G. Impact of thermal maturity on the diagenesis and porosity of lacustrine oil-prone shales: Insights from natural shale samples with thermal maturation in the oil generation window. *Int. J. Coal Geol.* **2022**, *261*, 104079. [CrossRef]

79. Goodman, C.; Vahedifard, F. Micro-Scale Characterization of Clay at Elevated Temperatures. *Géotechnique Lett.* **2019**, *9*, 225–230. [CrossRef]
80. Dabare, L.; Svinka, R. Characterization of porous ceramic pellets from Latvian clays. *Chemija* **2014**, *25*, 82–88.

Article

Heterogeneity of a Sandy Conglomerate Reservoir in Qie12 Block, Qaidam Basin, Northwest China and Its Influence on Remaining Oil Distribution

Qingshun Gong [1], Zhanguo Liu [1], Chao Zhu [1,*], Bo Wang [2], Yijie Jin [3], Zhenghao Shi [2], Lin Xie [2] and Jin Wu [1]

[1] PetroChina Hangzhou Research Institute of Geology, Hangzhou 310023, China; gongqs_hz@petrochina.com.cn (Q.G.); wuj_hz@petrochina.com.cn (J.W.)
[2] Exploration and Development Institute of Qinghai Oilfield Company, CNPC, Dunhuang 736202, China
[3] China National Logging Company, Beijing 100101, China
* Correspondence: zhuc_hz@petrochina.com.cn; Tel.: +86-531-8522-4934

Academic Editor: Dameng Liu

Received: 14 February 2023
Revised: 11 March 2023
Accepted: 22 March 2023
Published: 24 March 2023

Abstract: In view of the key geological factors restricting reservoir development, the reservoir heterogeneity of an alluvial fan sandy conglomerate reservoir in the Qie12 block of Qaidam Basin, Northwest China, and its influence on remaining oil distribution, were studied according to geology, wireline logging data, and dynamic production data. This study illustrates that the difference in pore structures, which are controlled by different sedimentary fabrics, is the main cause of reservoir microscopic heterogeneity. Besides, the temporal and spatial distribution of architectural units in the alluvial fan controls reservoir macroheterogeneity. Our results show that the thick sandy conglomerate develops two types of pores, two types of permeability rhythms, two types of interlayers, two types of interlayer distribution, two types of effective sand body architecture, and four types of sand body connecting schemes. The strongest plane heterogeneity is found in the composite channel unit formed by overlapping and separated stable channels of the middle fan, and the unit's permeability variation coefficient is >0.7. However, the variation coefficient in the range of 0.3–0.5 is found in the extensively connected body unit sandwiched with intermittent channels of the inner fan. The distributions of the remaining oil vary significantly in different architectural units because of the influence of reservoir heterogeneity, including distribution patterns of flow barriers, permeability rhythm, and reservoir pore structures. The composite channel unit formed by overlapping and separated stable channels, or the lateral alternated unit with braided channel and sheet flow sediment of the middle fan, is influenced by the inhomogeneous breakthrough of injection water flowing along the dominant channel in a high-permeability layer. The microscopic surrounding flow and island-shaped remaining oils form and concentrate mainly in the upper part of a compound rhythmic layer. Meanwhile, in the extensively connected body unit sandwiched with intermittent channels of the inner fan, poor injector–producer connectivity and low reservoir permeability lead to a flake-like enrichment of the remaining oil.

Keywords: reservoir heterogeneity; remaining oil; sandy conglomerate reservoir; alluvial fan; Qaidam Basin

1. Introduction

Reservoir heterogeneity is the term used to describe the uneven alternations in spatial distribution and internal attributes, which are caused by the influences of sedimentary environments, diagenetic processes, and tectonism during the formation of oil and gas reservoirs [1–3]. It is an important content of reservoir characterization manifested in both macroscale (plane, intrabedding, and interbedding) and microscale (pore space and structure). Additionally, it is the main geological factor determining the reservoir quality, which constricts the adjustment of a reservoir development scheme and affects the distribution of the remaining oil [4–6].

Conglomerate reservoirs are widely distributed, and they include the Permian–Triassic succession on the northwest edge of Junggar Basin, China; Paleogene Shahejie Formation of Dongying Depression, China; Yingcheng Formation of Songliao Basin, China; Barrancas Formation of Mendoza area, Argentina; and Upper Morrow in Anadarko Basin and Bend Conglomerate in Boonsville Gas Field of Fort Worth Basin, USA [7–12]. In general, they are strongly heterogeneous because of depositional and diagenetic alternations; however, their main controlling factors and specific characteristics on the macrolevel and microlevel are different. In this study, the research area is the Qie12 block in Qaidam Basin, which is a typical sandy conglomerate reservoir. Qie12 block was put into production according to the "thick massive reservoir" geological model during the initial stage of its development; however, the model caused prominent problems such as rapidly declining productivity, quickly declining water volume fraction, uneven injection production, and unclear law of remaining oil distribution. Although research on the characteristics of its sandy conglomerate reservoir has been reported [13], studies on its reservoir heterogeneity have not been conducted. Therefore, this study deeply analyzed the heterogeneous characteristics of Qie12 block's sandy conglomerate reservoir according to geology, logging data, and dynamic production data and determined their influence on the distribution of remaining oil in the research area. On the basis of the results, we provide a theoretical basis for establishing the potential of remaining oil and comprehensive reservoir management.

2. Geological Background

The research area is the Qie12 block, which is located in the Kunbei oilfield of the Qaidam Basin, northwest China. Kunbei oilfield is composed of several tertiary oil-bearing structures, and it is located specifically in the Kunbei fault step zone on the western depression of the Qaidam Basin. Meanwhile, the Kunbei fault step zone is distributed in front of the East Kunlun structural belt, and it is developed by compression from the East Kunlun Mountain to the basin. It has a large piedmont compression torsional thrust structural belt and shows a structural pattern of north–south zoning and east–west blocking [14,15]. In particular, the Qie12 block reservoir is distributed on the west side of the step zone, whereas the block's structural form is an anticline, and its axis is NNE (Figure 1). The lower member of the Ganchaigou Formation (E_3^1), comprising four oil groups, I, II, III, and IV (from top to bottom of Figure 1), was deposited during the Paleogene. In this study, the target layer is oil group IV, which has an average thickness of 55 m and is subdivided into seven small layers.

In the research area, the dominant colors of sediments are brown and brown–red, which are oxidation colors reflecting the characteristics of near-source and rapid accumulation. Meanwhile, the area's sandy conglomerate reservoir shows low composition and structure maturity. The two main types of hydrodynamic mechanisms affecting the area's bedding structure, which can be further used for representing the sedimentary environment, are traction current and gravity flow. The traction current resulted in tabular crossbedding, trough crossbedding, scouring–filling structure, parallel bedding, and gravel orientation structures (Figure 2a–d), whereas the massive bedding and matrix-supported suspended gravel structure were formed by the gravity flow (Figure 2e–g). The lithology types in the block include conglomerate, glutenite, pebbly coarse–fine sandstone, siltstone, pebbly sandstone, and mudstone. We defined eight lithofacies within this study area according to bedding structures, and their characteristics and genesis are summarized in Table 1.

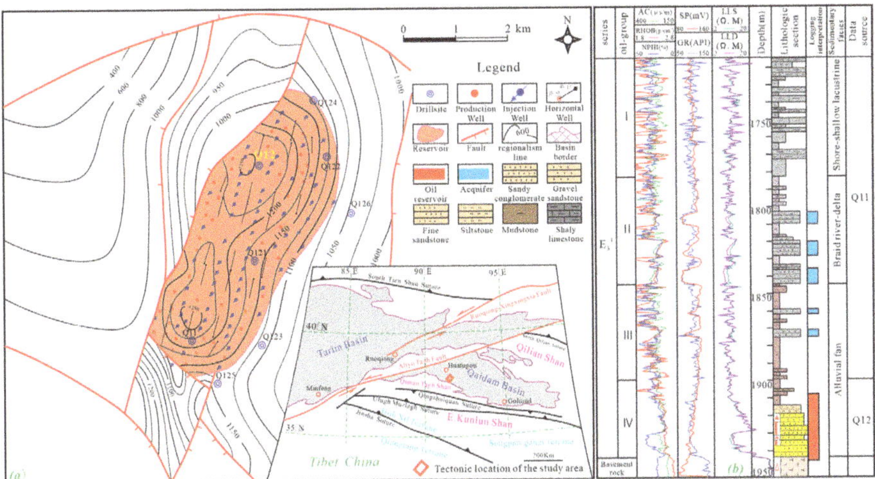

Figure 1. (**a**) Tectonic location of the research area, which is marked with a red box, is the west block of the Kunbei thrust structural belt. (**b**) Diagram showing the composite columnar of the reservoir, where oil layers are concentrated at the bottom of the target stratum. There are six lithofacies and two types of logging interpretation conclusions along the target stratum.

Figure 2. Typical sedimentary structures charts found in the Qie12 block: (**a**) Q12−10−8, 2059.1 m, tabular crossbedding; (**b**) Q12−7−28, 1829.4 m, tabular crossbedding; (**c**) Q12p1, 1809 m, tabular crossbedding; (**d**) Q125, 2026.5 m, trough crossbedding, gravel orientation arrangement; (**e**) Q12−23−6, 1917 m, massive bedding, gravel upright; (**f**) Q12−7−28, 1842.4 m, massive bedding, gravel upright; and (**g**) Q121, 1842.7 m, massive bedding. *T* is the top surface, and *B* is the bottom surface of the core in these figures.

Table 1. Characteristics and genesis of the lithofacies in the Qie12 block.

Code	Lithology Type	Bedding Structure	Genesis	Amplitude Difference of SP Curve	RHOB (g/cm³)	GR (API)
Gmm	Argillaceous conglomerate–glutenite	Massive bedding, matrix support	Debris flow/sheet flow	moderate	2.36–2.57	85–117
Gei	Sandy conglomerate–glutenite	Massive bedding, grain support, gravel orientation arrangement	Braided channel floor lag	large	2.36–2.57	85–100
Gt	Sandy conglomerate–glutenite	Trough crossbedding	Braided channel-filled deposit	large	2.36–2.57	73–105
Gp	Sandy conglomerate–glutenite	Tabular crossbedding	Braided channel-filled deposit	large	2.36–2.57	73–105
St	Pebbly sandstone	Trough crossbedding	Braided channel-filled deposit	large	2.23–2.44	73–105
SSh	Siltstone	Parallel bedding	Abandoned channel/silted channel deposit/runoff channel	large	2.23–2.44	73–105
Sm	Anisometric sandstone	Massive bedding	Sheet flow/runoff channel	moderate	2.23–2.44	80–125
Mm	Sandy mudstone	Massive bedding, flat bedding	Flood plain	small	1.75–2.41	100–125

By integrating the aforementioned geological features, we believe that the Qie12 block is an alluvial fan, which is a genetic type of thick sandy conglomerate. In the alluvial fan, there are five sedimentary microfacies: a fourth-order architectural element [16] that includes a braided channel, debris flow, sheet flow, runoff channel, and flood plain. According to the vertical assemblage of different architectural elements and the distribution pattern of interlayers [17,18], we also identify four fifth-order architectural units: the extensively connected body unit sandwiched with intermittent channels, the composite channel unit formed by overlapping and separated stable channels, the lateral alternated unit with braided channel and sheet flow sediment, and the runoff channel inlaid in flood plain mudstone.

Based on the theory of high-resolution sequence stratigraphy [19], we established the sequence stratigraphy framework in this study area. We divided 1 long-term, 7 medium-term, and 24 short-term base-level cycles in the target stratum. The time span of the medium and short-term cycles is approximately 0.23 and 0.07 Ma, respectively. The vertical assemblage of different sedimentary microfacies constitutes the sedimentary sequence of the alluvial fan. Its sediments show an upward-fining grain-size distribution, which suggests that the sedimentary sequence is a retrogradation sequence. The physical properties of the reservoir structures include low porosity and ultra-low permeability, and they have average porosity and permeability values of 10.5% and 7.8 mD, respectively.

3. Data and Methods

The reservoir data in the Qie12 area contain various data types: geology, wireline logging, seismic, and production dynamic. The area houses 13 coring wells, including 3 systematic coring wells, and the total core length of the target layer is 309 m. After more than 1000 analyses, the resulting test data, including microreservoir analysis, reservoir sensitivity, and microseepage test, are corrected using core and ground gamma data. These data are detailed and reliable, and they provide data support for the microheterogeneity evaluation of reservoirs. All 127 wells are well-logged, and logging interpretation was performed on these wells. By evaluating the four properties of the reservoir, we established the identification charts of effective reservoirs and interlayers, interpreted and divided the effective reservoirs and interlayers, and clarified the structures of effective reservoirs and distribution styles of interlayers. The three-dimensional seismic work area is approximately 100 km². The quality of seismic data obtained at main frequency, frequency band range, and sampling rate values of 30 Hz, 10–50 Hz, and 2 ms, respectively, is good. The research area has been in development and has rich production dynamic data. Its geology, wireline logging, seismic data, and dynamic production data provided a basis for our reservoir macroheterogeneity evaluation.

There are various methods to evaluate reservoir heterogeneity [20–24]. In this study, we used geological origin, geostatistics, logging interpretation, and analysis of production performance to evaluate the macroheterogeneity of the studied reservoir. Additionally, an experimental method was used to analyze the microheterogeneity of the reservoir. The geological origin affecting reservoir heterogeneity was studied initially using a geological origin method, and the results revealed the main factors that influence heterogeneity and depicted the strength of reservoir heterogeneity. We found that sedimentary factors are the main geological causes of reservoir heterogeneity. Different configuration units of alluvial fans directly determined the strength of macroheterogeneity. Geostatistics is a widely used method for studying reservoir heterogeneity, and this method uses statistical analysis of reservoir permeability to judge the strength of reservoir heterogeneity. It also includes the statistical analysis of the thicknesses and frequencies of reservoir interlayers and the scales and distributions of effective reservoir sand bodies. Meanwhile, the logging interpretation method provides continuous reservoir permeability data through elaborate logging interpretation. The method provides effective reservoir and interlayer identification charts with four-property evaluation, interpretation of effective reservoirs and interlayers on a single well, development positions of interlayers and intralayers, and qualitative description of macroheterogeneity. Continuous reservoir permeability data are also basic data for geostatistics. An analysis of production performance provides a verifiable and effective way to determine reservoir heterogeneity using various dynamic data that directly reflect the strength of reservoir heterogeneity. Experimental analysis is an important tool for studying microheterogeneity. The dynamic conditions of alluvial fans are complex. The lithofacies are diverse, and the microheterogeneity of reservoirs is strong. Thus, with the help of different experimental analysis methods, the following can be achieved: characteristics of the pore structure of the reservoir, insights into the microseepage mechanism of the reservoir, and sensitivity characteristics of the reservoir. These achievements will provide a geological basis for reservoir reconstruction, remaining oil potential tapping, and reservoir management.

4. Results

4.1. Reservoir Microheterogeneity

In this study area, reservoir petrology is characterized by low compositional and structural maturity. The compositional maturity index $(Q/(F + R))$ is 0.23. The reservoir is poorly sorted, and the main contact method among its particles is the line–point contact relationship. The average mass fraction of its mud base is approximately 8.4%, and its average cement content of 2.1% is relatively low. We utilized various experimental techniques, including thin-section observation, scanning electron microscopy (SEM), mercury injection, and computed tomography (CT) scanning, and their results confirmed the dual porosity media characteristics of the space-type sandy conglomerates in this study area [25,26]. Four types of pores were found in the reservoir: primary intergranular (Figure 3a–d), intragranular dissolution (Figure 3h–j), diagenetic fracture (Figure 3e–g) [27], and argillaceous micropores (Figure 3k,l), and the proportion of primary intergranular pores is >60%. In this study, the diagenesis affecting the sand conglomerate reservoir includes compaction, cementation, and dissolution [28]. Compaction is the main diagenesis to reduce the primary intergranular pores, the average compaction pore reduction and compaction pore reduction rates are 19% and 53%, respectively. The compaction intensity is medium. Additionally, the diagenetic fractures developed due to compaction increased the reservoir pore space and partially improved the reservoir's permeability; this had a certain constructive effect on the reservoir transformation. Cementation is another important factor contributing to the reduction of the primary intergranular pores, with a pore reduction of 3% and a pore reduction rate of 8%. Calcite, anhydrite, and dolomite are mostly distributed in porphyry pores; moreover, they are only partially enriched in the unconformity surface between the sand conglomerate and the basement rock. Dissolution plays a constructive role in modifying reservoir performance. In this study, dissolution either selectively dissolved feldspar and igneous rock debris to form intragranular pores or spread along diagenetic fractures to create dissolution pores. The dissolution pore enhancement and dissolution pore enhancement rates are 2% and 5%, respectively.

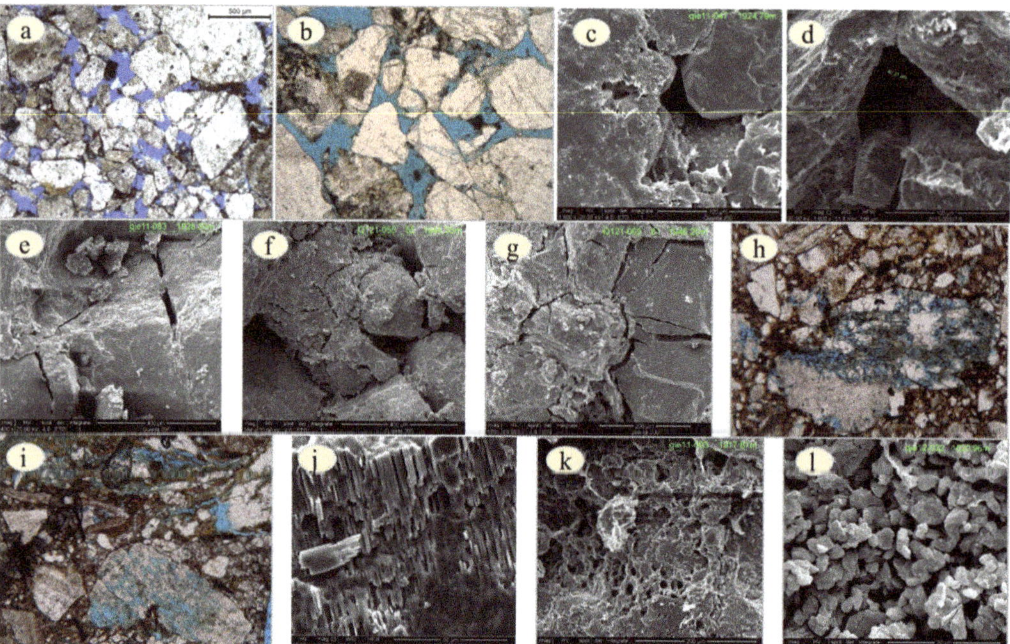

Figure 3. Characteristics of reservoir storage space in the Qie12 block: (**a**) Well Qie12−10−8, 1857.01 m, medium-coarse-grained feldspar lithic sandstone, primary intergranular pores; (**b**) Well Qie11, 1920.13 m, gravel-bearing coarse sandstone, primary intergranular pores, (-) × 25; (**c**) Well Qie11, 1924.79 m, sandy conglomerate, primary intergranular pores; (**d**) Well Qie12−10−8, 1850.82 m, gravel-bearing coarse sandstone, primary intergranular pores; (**e**) Well Qie11, 1928.3 m, sandy conglomerate, cracked fracture; (**f**) Well Qie121, 1944.36 m, cracked fracture; (**g**) Well Qie121, 1946.2 m, cracked fracture; (**h**) Well Qie12, 1818.6 m, gravel-bearing sandstone, detritus dissolution pores, (-) × 25; (**i**) Well Qie12, 1820.5 m, pebbly sandstone, feldspar dissolution pores, (-) × 25; (**j**) Well Q12−7−28, 1831.72 m, sandy conglomerate, feldspar dissolves along the joint; (**k**) Well Qie11, 1917.87 m, sandy conglomerate, argillaceous micropores; and (**l**) Well Qie12, 1820.96 m, argillaceous micropores.

The microheterogeneity of the reservoir is mainly due to various pore structures, which are formed by different sedimentary fabrics [29,30]. Under the same geological background and identical dynamic diagenetic conditions within the reservoir, the grain sizes and cement contents of different sedimentary microfacies are the same. The mud content is the most intuitive reflection that describes the differences in control sedimentary fabrics found in various facies belts of alluvial fans. The high or low contents of the mud controlled the pore structure of the reservoir, and it also determines the strength of the reservoir's microheterogeneity. SEM images of thin sections and particle size data showed two types of pore structures in the reservoir of this study area; furthermore, these structures are bimodal and multimodal.

The bimodal pore structure of the reservoir is supported by gravel, forming a skeleton that is filled with sandy debris. Alternatively, the gravel is suspended in the sandy debris, the grain-size histogram is bimodal, and the main grain-size interval is fine gravel and giant sand–fine sand with low clay content (Figure 4a,b). The reservoir space is characterized by combinations of primary intergranular pores, intergranular dissolved pores, and diagenetic fractures. The proportion of primary intergranular pores is >70%, and the pores observed in thin sections are well developed. We performed two experimental analyses, namely conventional and constant-rate mercury intrusions, to characterize the pore structure of the reservoir. The conventional method mainly reflects the change in the pore volume during mercury intrusion, while the constant-rate method highlights the number distribution of pores and throat tracts. In pore structure analysis and statistics, the quantity distribution is more accurate than the volume distribution, particularly the quantity distribution of the throat tract, which can better characterize the reservoir's seepage characteristics. The results of mercury intrusion experiments revealed pore structures in the reservoir of coarse skewness, well sorted, low drainage pressure, and median pressure. Macropores and mesopores are the dominant pore sizes, and their main peak value is approximately 125 μm, which suggests a unimodal distribution. The distribution of throats, which are mainly fine throats, is not significant, and the radii of the mainstream throats are in the range of 3.05–10.22 μm. Local large throats are also found, and the average pore–throat ratios obtained for the reservoir are in the range of 58.51–130.94. CT scanning revealed that the percentage of pore–throat volume is 52%. This modal type is developed in most braided channel microfacies and the minority of runoff channel microfacies that are effective reservoirs.

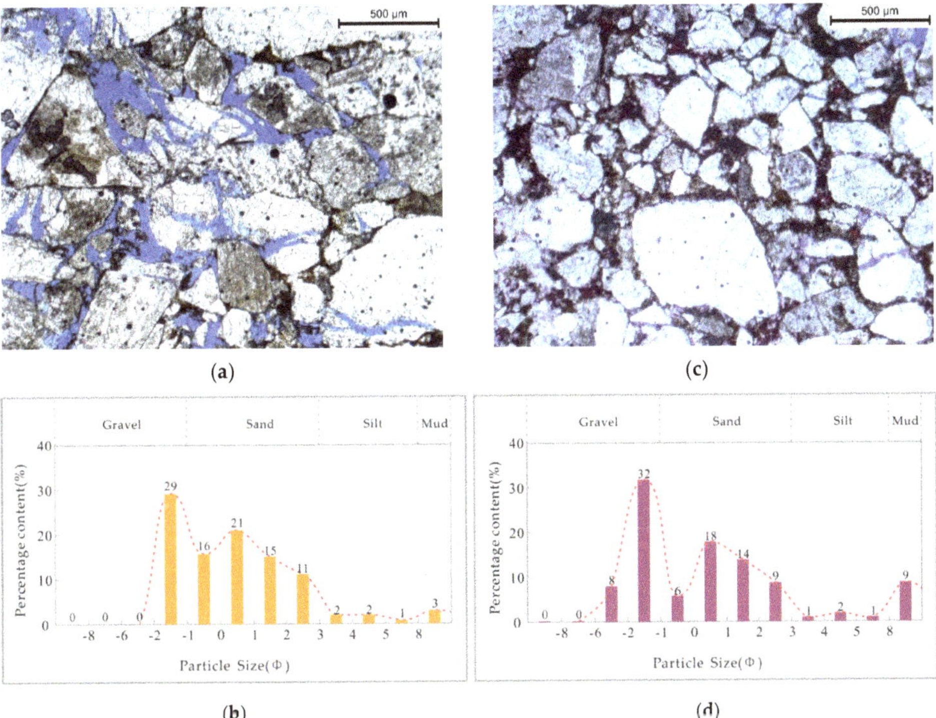

Figure 4. A chart of the reservoir structure model for the Qie12 block: (**a**) Well Qie12−10−8, 1858.06 m, gravel-bearing medium-coarse feldspar sandstone, primary intergranular pores; (**b**) Well Qie12−10−8, 1858.06 m, histogram of particle size distribution; (**c**) Well Qie12−10−8, 1859.4 m, argillaceous sandy conglomerate, dissolution pores, and argillaceous pores are the main pore types, rarely with visible pores; and (**d**) Well Qie12−10−8, 1859.4 m, histogram of particle size distribution.

The multimodal pore structure of the reservoir is complex, and it is composed of gravel, sand, and clay. The clay content is high, and the histogram of the grain size distribution shows multiple peaks (Figure 4c,d). The reservoir space is characterized by a combination of intergranular dissolved pores and argillaceous micropores. Although the primary intergranular pores under thin sections are poorly developed, they seem to be filled mostly with argillaceous interstitial fillers. The pore structures of the reservoir are characterized by fine skewness, poorly sorted, high displacement pressure, pore–throat distribution, multi-peak forms, and small pore–throat radius. This mode type is mainly developed in debris flow, diffuse deposition, and the majority of runoff channel microfacies, and it is manifested in poor reservoir and nonreservoir.

4.2. Reservoir Macroheterogeneity

Reservoir macroheterogeneity is controlled primarily by the spatial and temporal distribution of different sedimentary configuration units of the alluvial fan. This heterogeneity determines the depositional architecture, rhythm, geometry, and connection mode of sand bodies of different origins; thus, it directly affects the characteristics and strength of macroheterogeneity [31,32].

4.2.1. In-Layer Heterogeneity

In-layer heterogeneity refers to the variation rule of vertical reservoir parameters in small layers (e.g., rhythmic characteristics of vertical permeability, interlayer types, and distribution styles) [33,34], and it is a key geological factor controlling the swept volume of vertical injectors and the remaining oil distribution in small layers.

Characteristics of Rhythms

Based on the data obtained from the core physical property analysis, the permeability rhythmic characteristics of the single sand body in this study area have two rhythm types: compound and homogeneous. The former is a typical characteristic of the sand body in the braided channel. The median grain size of sands in the sand body shows positive and compound positive rhythms, and the wellbore logging curve characteristics show bell and box shapes. WellQ11, with a depth range of 1929–1932 m, has a typical permeability compound rhythm that is composed of a single-granularity positive rhythm (Figure 5a). The bottom of a rhythm is controlled by hydrodynamic conditions, and it has coarse grains that are mainly composed of gravel with poorly sorted sedimentary fabric and poor permeability. In the middle section of the rhythm, the hydrodynamic energy is moderate, and the sand is mainly coarse-medium in grain size and well-sorted with low mud content and good permeability. In the upper section of the rhythm, the channel energy is attenuated, which turns the grain sizes to fine (mainly fine-silty sand), and the fabric sorting is medium. The mud content in this section is high, and its permeability is poor. Its permeability rush coefficient and ratio are 3.3 and 126, respectively. Finally, this section has strong heterogeneity.

WellQ11, with a depth range of 1923.0–1926.5 m, has a composite permeability rhythm composed of two or more single-granularity positive rhythms that are superposed on top of each other (Figure 5b). It is formed by aqueduct accretion, and it has strong heterogeneity. The change in its permeability in the longitudinal direction is very complicated. Its permeability rush coefficient and ratio are 3.6 and 175, respectively. The primary rhythm type in its sheet flow and debris flow sand body is homogeneous. It has poor fabric sorting, mixed base support, massive structure, and no obvious grain rhythm. Its reservoir permeability is poor, but the reservoir is relatively homogeneous.

WellQ12-10-8, with a depth range of 1872–1875 m, is a homogeneous rhythmic debris flow deposit (Figure 5c). Its permeability rush coefficient and ratio are 1.7 and 18, respectively.

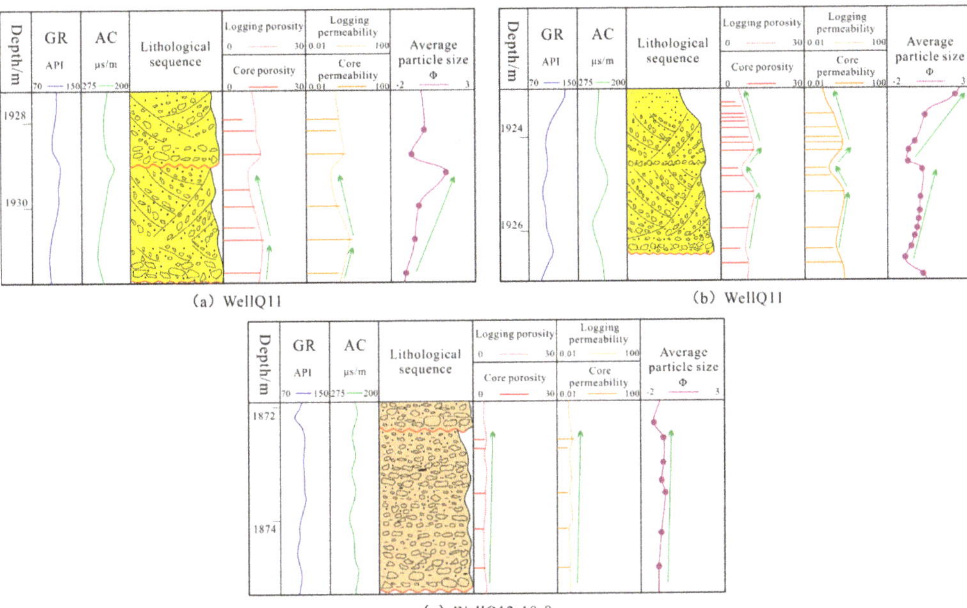

(a) WellQ11

(b) WellQ11

(c) WellQ12-10-8

Figure 5. Rhythm characteristics of the Qie12 block in the Kunbei oilfield: (**a**) Well Qie11, 1929–1932 m, single-granularity positive rhythm; (**b**) Well Qie11, 1923.0–1926.5 m, composite permeability rhythm; (**c**) Well Qie12−10−8, 1872–1875 m, homogeneous rhythm.

Quantitative Evaluation of the Range of Permeability Difference

Quantitative evaluation of the intralayer permeability difference in this study area includes the following parameters: average permeability, permeability ratio, and permeability rush coefficient. The distribution histograms of permeability difference and permeability rush coefficient are shown in Figure 6. For IV-6 and IV-7 layers, the permeability difference, rush coefficient, and average permeability ranges are 33–43, 2.5–2.7, and 0.10–1.56 mD, respectively. These values reflect the lack of a high-permeability braided channel reservoir in the extensively connected body unit sandwiched with intermittent channels of the inner fan. They also suggest poor reservoir storage performance and medium heterogeneity. For IV-4 and IV-5 single layers, the permeability difference, rush coefficient, and mean permeability ranges are 72–74, 3.8–4.1, and 4.71–25.41 mD, respectively. These values suggest good reservoir performance in the composite channel unit formed by overlapping and separated stable channels in the alluvial fan. This unit has a high-permeability section that intensifies reservoir heterogeneity, and it is the most heterogeneous configuration unit in the alluvial fan. The IV-3 single layer has values for permeability difference, rush coefficient, and average permeability of 45, 3.3, and 3.2 mD, respectively. These values reflect poor reservoir performance and low heterogeneity in the lateral alternated unit with braided channel and sheet flow sediment of the middle fan; however, the overall heterogeneity is still high. In the IV-1 and IV-2 single layers, mudstone in their flood plain is regional cap rock, and the reservoir is poorly developed.

In-Layer Type and Distribution Style

Systematic coring wells reveal the absence of mudstone interlayers in the thick sandy conglomerate section. However, the characteristics of the oil-bearing sections, oil infiltration, oil spot, and oil trace, are discontinuous, showing the interbedded distribution of these sections and the oil-free section. On the basis of the oil test results, the oil-free section is not a water layer. The main reason for the vertical difference in oil content is the presence of muddy interlayers and part of partial calcareous interlayers, which are controlled by the different sedimentary fabrics or cements. The pore structure of the muddy interlayer is multimodal. This interlayer has a high content of clay mineral matrix and tight physical characteristics. The wellbore logging curve suggests the following features: high natural gamma-ray, high acoustic time, low resistivity, high nuclear magnetic total porosity, and mainly argillaceous micropores. The pore structure of the calcareous interlayers is also multimodal. These interlayers have high cement carbonate content and medium clay mineral matrix. In contrast to the muddy interlayer, the wellbore logging curve of a calcareous interlayer suggests the following features: low natural gamma-ray, low acoustic time, high resistivity, and low nuclear magnetic total porosity. We determined the lower limit of physical property, the upper limit of muddy content, and the logging curve parameters of the effective reservoir and interlayer using a four-property relation analysis. These enabled us to establish the quantitative identification plate (Table 2) of the effective reservoir and interlayer.

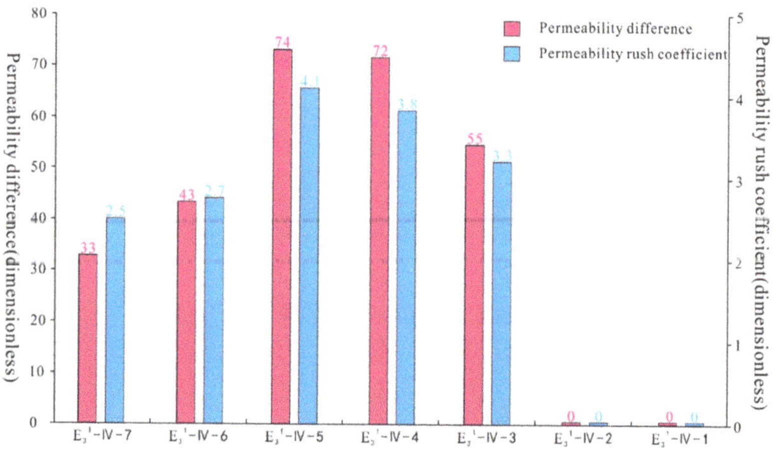

Figure 6. Distribution histograms of permeability difference and permeability rush coefficient.

Table 2. Identification plate of effective reservoir and interlayer in the Qie12 block.

Type	Permeability (mD)	Porosity (%)	Shale Content (%)	AC (μs·m^{-1})	ΔGR	LLD (Ω·m)	Sedimentary Facies
Effective reservoir	>0.6	>8.5	<5	>230	<0.55	>8	Braided channel
Muddy interlayer	<0.6	<8.5	>8	>230	>0.55	5–8	Debris flow/sheet flow/abandoned channel/silted channel deposit/runoff channel
Calcareous interlayer	<0.6	<8.5	5–8	<230	<0.6	>8	Riverbed detention deposit of braided channel/runoff channel/glutenite above the unconformity surface

The identification and division of effective reservoir and interlayer are performed for all wells using the quantitative identification plates. After comparing the effective reservoir and interlayer, we conclude the presence of two interlayer distribution patterns: (1) layer-cake architecture and (2) generally scattered and locally interlaced (Figure 7). The former pattern is formed primarily at the extensively connected body unit sandwiched with intermittent channels of the inner fan. This pattern is laminar with stacked interlayers that divide the lenticular effective reservoir sand body, and its seepage barrier has the strongest shading capacity among all the alluvial fan architectural units. The latter pattern develops in the composite channel unit formed by overlapping and separated stable channels and the lateral alternated unit with braided channel and sheet flow sediment of the middle fan. It is manifested as partial discontinuities in the grid of "plywood" high-permeability reservoirs. As the channels gradually shrink and die out, the pattern's sheet flood deposits increase, and the scale of its interlayer development increases gradually.

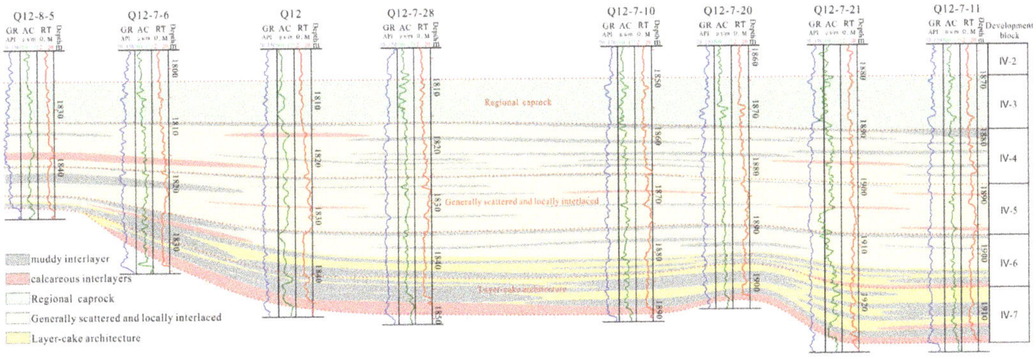

Figure 7. The interlayer distribution of the Qie12 block.

The isolation ability of an interlayer mainly depends on its permeability, which increases variably with increasing interlayer thickness. Parameters such as thickness, density, and frequency are often used to quantitate the distribution pattern of an interlayer [35]. On the basis of our measurements, the thickness and density of the interlayer in this study area from bottom to the top showed a characteristic change from large to small and then to large, whereas its frequency increased gradually. The highest and lowest density values are found in layers IV-7 (0.78%) and IV-5 (0.38%), respectively. Meanwhile, the highest and lowest frequency values are found in layers IV-3 (0.23 bars/m) and IV-7 (0.09 bars/m), respectively.

4.2.2. Interlayer Heterogeneity

Interlayer heterogeneity, which refers to the differences between sand bodies, including cyclicity of strata system, distribution of sand barrier, and characteristics of interlayer fractures, is an important cause of interlayer interference and water displacement differences during water injection development. We focused on the characteristics of interlayer barriers and found that barriers between layer IV-7 and its overlying basement rock are the most developed. These 1.1–2.8-m-thick barriers showed a stable lateral distribution and good continuity. The barriers between layers IV-6 and IV-7 have a more continuous lateral distribution, and their thickness is in the range of 0.5–2.3 m. The barriers between layers IV-5 and IV-6 and layers IV-3 and IV-4 have a discontinuous lateral distribution, and their thickness in the range of 0.4–1.6 m is small. Meanwhile, the barriers between layers IV-4 and IV-5 are developed on a small scale, and their thickness is in the range of 0.3–0.5 m. They are also poorly continuous, and their isolation ability is limited. The development scale and distribution pattern of the interval layers are controlled by the architectural unit type; thus, serious heterogeneity in the development of water injection is present in different architectural units, which limits the effectiveness of reservoir waterflood.

4.2.3. Horizontal Heterogeneity

Reservoir horizontal heterogeneity is caused by the geometry, scale, and continuity of the sand body and its permeability planar variation. It directly affects the waterflood-swept area and planar water displacement efficiency.

Sand Body Geometry

Sedimentary facies determine the geometry of sand body distribution. The 5.8–62.0-m-thick alluvial fan sandy conglomerate in this study area is distributed with a thin western to thick eastern distribution. Controlled by the development scale of the braided channels in different sedimentary architectural units, the effective sand body developed two types of reservoir architecture. One is jigsaw-puzzle reservoir architecture, which developed in the composite channel unit formed by overlapping and separated stable channels of the middle fan. The plane geometry is fan-shaped, but with the channel shrinkage, the plane geometry becomes dendritic. The other is the labyrinth architecture, which is developed in the extensively connected body unit sandwiched with intermittent channels. The lenticular effective sand body is distributed sporadically, and its geometric shape is similar to a potato.

Sand Body Connectivity

Through a good correlation comparison of effective sand bodies with interlayers in this study area, we concluded the presence of four sand body connectivity mode types: disconnected, weakly connected, connected multilayer, and connected multilateral type, which are shown in Figure 8a–d, respectively. The disconnected, weakly connected, and two-connected types account for 20%, 35%, and 45% of the total, respectively. The disconnected type mainly develops in the extensively connected body unit sandwiched with intermittent channels. The weakly connected type develops mainly in the lateral alternated unit with braided channel and sheet flow sediment. Meanwhile, the connected types mainly develop in the composite channel unit formed by overlapping and separated stable channels, and they play an important role in controlling water injection and well pattern arrangement.

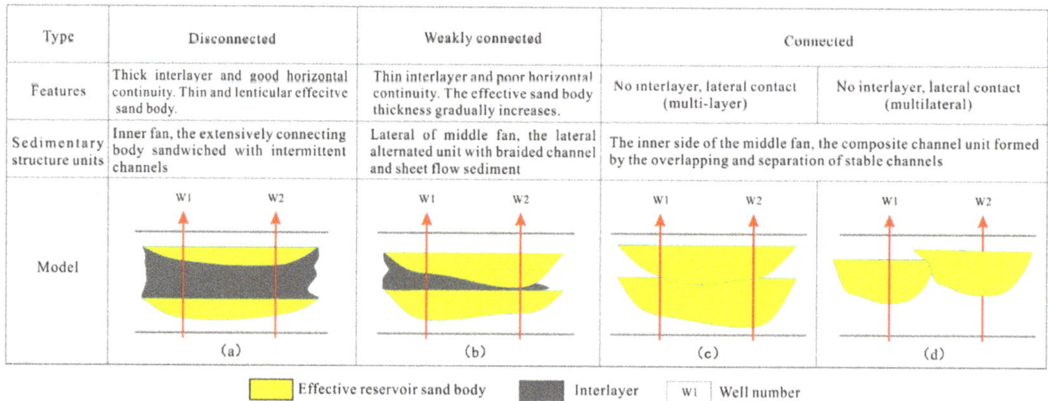

Type	Disconnected	Weakly connected	Connected	
Features	Thick interlayer and good horizontal continuity. Thin and lenticular effecitve sand body.	Thin interlayer and poor horizontal continuity. The effective sand body thickness gradually increases.	No interlayer, lateral contact (multi-layer)	No interlayer, lateral contact (multilateral)
Sedimentary structure units	Inner fan, the extensively connecting body sandwiched with intermittent channels	Lateral of middle fan, the lateral alternated unit with braided channel and sheet flow sediment	The inner side of the middle fan, the composite channel unit formed by the overlapping and separation of stable channels	
Model	(a)	(b)	(c)	(d)

Figure 8. Sand body connecting type of the Qie12 block: (**a**) disconnected; (**b**) weakly connected; (**c**) connected multilayer; (**d**) connected multilateral.

Distribution of Reservoir Permeability and Permeability Variation Coefficient

We analyzed the horizontal distribution characteristics of reservoir parameters using a geostatistics method with E_3^1-IV-4 as an example. The average permeability, mean variation coefficient, mean range difference, and rush coefficient of the selected single reservoir are 7.8 mD, 0.86, 66.3, and 3.4, respectively. A comprehensive evaluation of reservoir heterogeneity is strong. The permeability distribution map of the reservoir is shown in Figure 9a. The long axis of the isoline is in the direction of northwest to southeast. The permeability changes from low to high and then to low from west to east, and the composite channel in the middle fan has the best permeability. The permeability variation coefficient is distributed in a circular band, as shown in Figure 9b. The composite channel unit heterogeneity in the inner side of the middle fan is the strongest, and its variation coefficient is >0.7. This unit is followed by the lateral alternated unit with braided channel and sheet flow sediment on the lateral side of the middle fan, and its variation coefficients are in the range of 0.3–0.9. The extensively connected body unit sandwiched with intermittent channels in the inner fan showed moderate reservoir heterogeneity, with its variation coefficients in the range of 0.3–0.5.

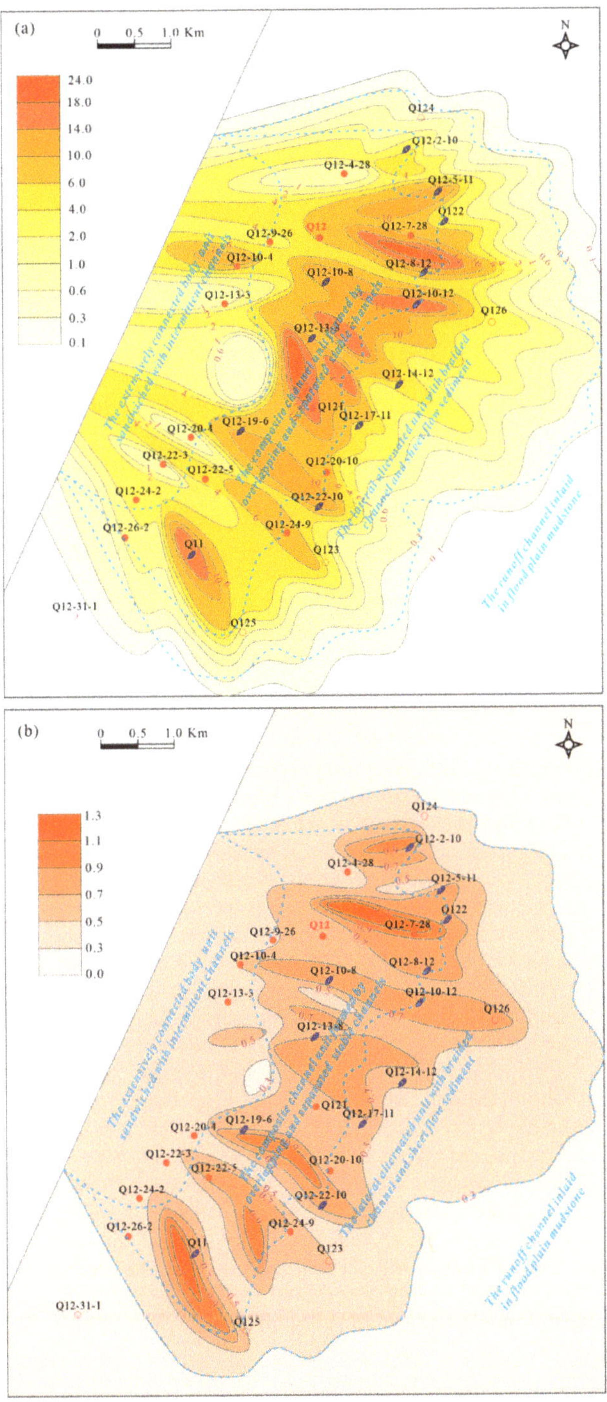

Figure 9. (**a**) Permeability and (**b**) variation coefficient distribution of the E31-IV-4 formation in the Qie12 block.

5. Discussion

Reservoir heterogeneity from macroscopic to microscopic and from intralayer to interlayer and plane is the main factor that controls the distribution of remaining oil [36–38]. Thus, the geological factors that control the distribution of the remaining oil in alluvial fan glutenite reservoirs include the distribution pattern of flow barriers, permeability rhythm characteristics, and reservoir microscopic pore structure of the sedimentary architectural unit.

5.1. Influence of Flow Barrier Pattern on Remaining Oil Distribution in Architectural Units

The hydrocarbon distribution in this study area is uneven and segmented according to the description of the systematic coring of Well Q12-10-8. There are four oil segments from bottom to top. The segment at the bottom contains the least oil and a significant amount of massive barren argillaceous conglomerates. The middle segment contains a relatively continuous oil distribution segment, which is regarded as the best oil-bearing interval in the alluvial fan. This segment is mainly composed of oil-immersion, oil-bearing, and oil-patch sandy conglomerates or pebbly coarse-medium sandstones. It also contains thin barren muddy glutenite. The upper segment has a discontinuous property of oiliness and is composed of oil-patch and oil-immersion sandy conglomerates, pebbly sandstone, barren muddy glutenite, and anisometric sandstone. The topmost segment is barren, and it contains massive mudstones that do not have oil and gas.

After analyzing the relationship between hydrocarbon enrichment characteristics and the liquid production profile of architectural units, we found that the effective reservoir architecture and flow barrier distribution pattern of different architectural units control the vertical "four-segmentation" enrichment law of hydrocarbon and affect the macroscopic distribution of remaining oil (Figure 10).

In the extensively connected body unit sandwiched with intermittent channels, the effective reservoir with a lenticular shape distribution is isolated by "layer-cake architecture" interlayers, resulting in poor oiliness properties. The reservoir has poor permeability and injection–production connectivity, and it has a high starting water pressure that slows waterline advancement. It manifests as under-injection or no injection and has a relative water absorption of <20%. It has poor sweep efficiency of injected water and degree of water driving and produces flake-like remaining oil.

In the composite channel unit formed by overlapping and separated stable channels, some interlayers have a "generally scattered and locally interlaced" distribution in the unit's effective reservoir, having a plate-like synthesis distribution. Therefore, the best oiliness property and most continuous hydrocarbon distribution are found in the proximal part of the middle fan. The injection–production connectivity rate, sweeping efficiency of water injection, and degree of water driving in this unit are high. A total of 90% of daily oil production and 80% of relative water absorption occur in this unit. In the lateral alternated unit with braided channel and sheet flow sediment, there is gradual shrinkage of the braided channel and a gradual increase of sheet flow sediments. In addition, the thickness of the effective reservoir reduces, whereas the thickness and the frequency of interlayers increase. Meanwhile, hydrocarbon distribution becomes discontinuous in the distal part of the middle fan, and both liquid production and water absorption significantly decrease. In the aforementioned two units, according to the perforation plan of "large diameter casing pipe and mixed injection," the injected water passes quickly along large channels in the high-permeability reservoir, resulting in water flooding, and the unswept area of injected water forms plaque-like remaining oil.

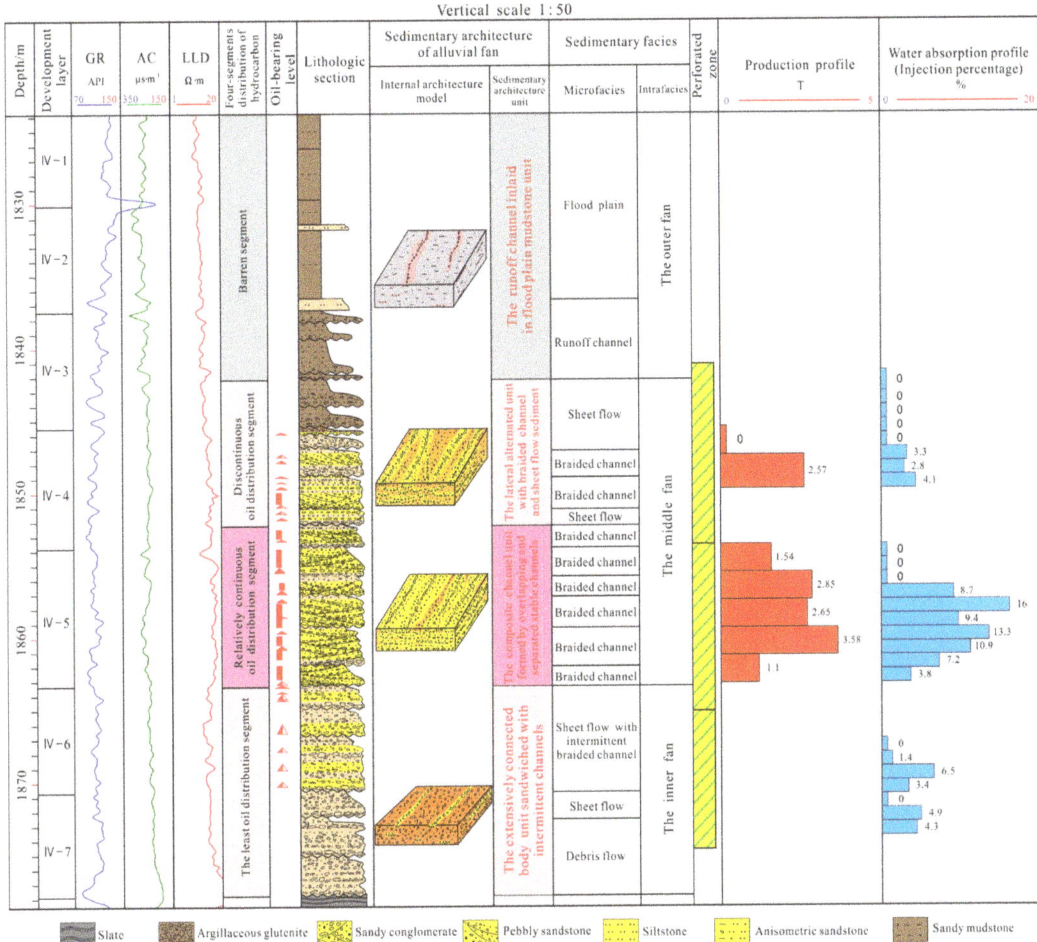

Figure 10. Relationship between production profile and sedimentary architectural units.

5.2. Influence of Permeability Rhythm on Macroscopic Remaining Oil Distribution

The water absorption profile of the 1922- to 1935-m braided channel sand body in Well Qie11 is shown in Figure 11. We used this profile as an example to analyze the influence of reservoir permeability rhythm characteristics on the macroscopic remaining oil distribution. There are four permeability composite rhythms developed vertically in this section reservoir. In the middle and lower parts of the rhythm, the injected water flows easily along the large pores of the high-permeability layer because of the superposition of gravity and intralayer heterogeneity effects. Relative water absorption accounts for approximately 85% of the reservoir's total in Well Qie11 water absorption profile. After the water injection front breaks through, the water cut of the production well rises rapidly, and the injected water circulates ineffectively. In the upper part of the rhythm layer, the water breakthrough flow is slow, relative water absorption accounts for approximately 15%, oil displacement efficiency is low, and the remaining oil is enriched.

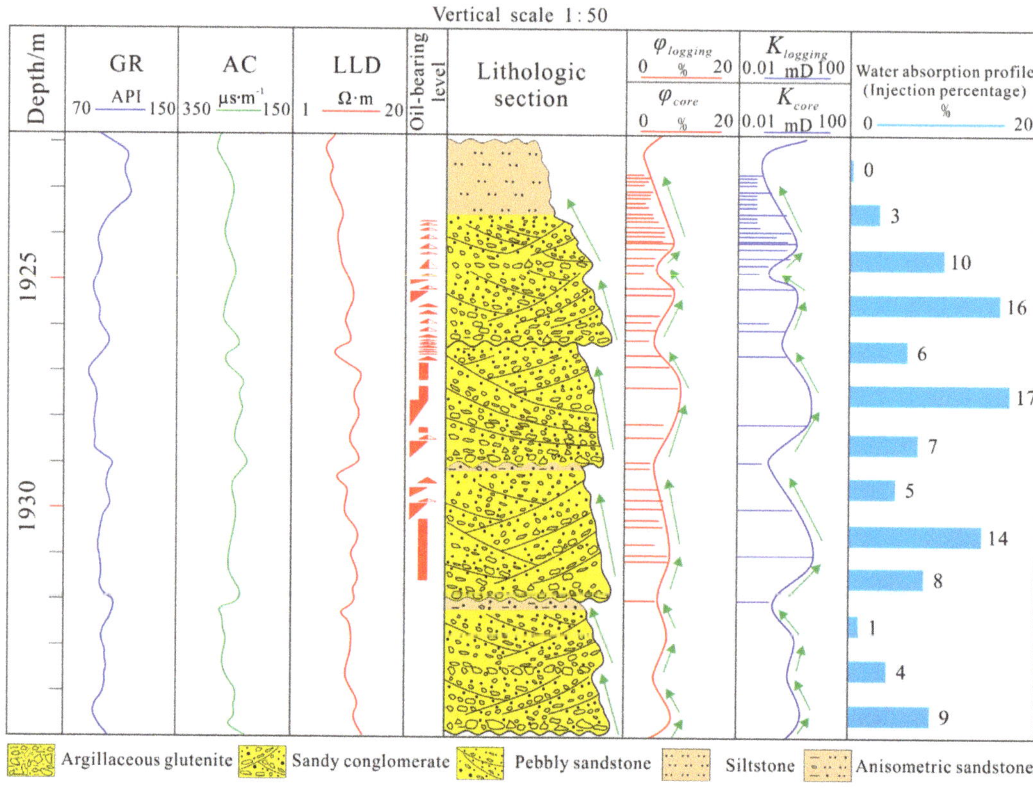

Figure 11. Relationship between permeability compound rhythmicity and water absorption.

5.3. Influence of Reservoir Microheterogeneity on Microscopic Remaining Oil Distribution

The flow in the large pores of the bimodal reservoir pore structure is the main factor that leads to the strong heterogeneity of microscopic seepage [39,40]. The microscopic flow experiment shows the following results: the water displacement process is characterized by heterogeneous flooding, the water-free oil recovery period is short, and oil displacement efficiency is relatively low. In the process of water displacement, the heterogeneous intrusion and flow around the water flooding front form a large area of remaining oil that is not displaced by the surrounding flow and island-shaped remaining oil, as shown in Figure 12a,b, respectively. Water phase permeability is low in the presence of remaining oil; thus, the injected water is more likely to penetrate along the pore channels that have been broken through. Therefore, it is difficult to displace oil with the injected water, resulting in low oil displacement efficiency and a rapid increase in water content.

In summary, reservoir heterogeneity plays an important role in controlling the distribution of remaining oil. We note that deep profile control and oil displacement technique are key to the comprehensive management of the composite channel unit formed by overlapping and separated stable channels or the lateral alternated unit with braided channel and sheet flow sediment. Meanwhile, to tap the potential of the remaining oil in the extensively connected body unit sandwiched with intermittent channels, we need to depressurize, increase the injection, and improve the quality of the reservoir. The study effectively guided the deep profile control of the test well group, and it achieved practical results. Under the premise of an unchanged production system and after profile control, the daily fluid production of a single test well (i.e., Q12H13-9) was stable (Figure 13). The test well's daily oil production increased by approximately 40%, whereas its water content decreased by 35%.

(a)   (b)

Figure 12. Distribution model of the remaining oil in the Qie12 block. (**a**) Surrounding flow remaining oil, Well Qie12−7−8, 1814.92 m; (**b**) island-shaped remaining oil, Well Qie12−7−28, 1844.12 m.

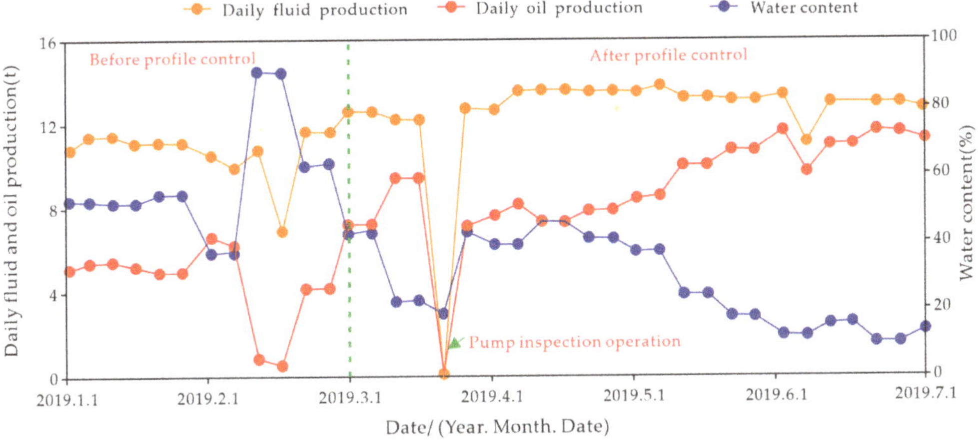

Figure 13. Oil production rate curve of the Q12H13−9 well in the Qie12 block.

6. Conclusions

The results of this study showed that the alluvial fan reservoir exhibited strong heterogeneity. By considering sedimentary architectural units as the research object, we qualitatively and quantitatively investigated the microheterogeneity and macroheterogeneity of a sandy conglomerate reservoir. Based on these results, we discussed the effect of heterogeneity on the remaining oil distribution and highlighted the main measures required to tap the remaining oil in different architectural units.

Note that reservoir heterogeneity is the geological basis of reservoir management. This is the first study to propose the concept based on sedimentary architectural units. These are important for the comprehensive reservoir management in the Qie12 block, but would also be relevant in other alluvial fan reservoirs.

(1) The difference in reservoir pore structures that are controlled by the different sedimentary fabrics is the main cause of reservoir microheterogeneity, whereas the spatial and temporal distribution of alluvial fan sedimentary architecture units is the main factor that controls reservoir macroheterogeneity.

(2) Reservoir heterogeneity affects the distribution of remaining oil through the flow barrier distribution pattern of sedimentary architecture units, permeability rhythm, reservoir pore structure, and other aspects. The remaining oil distribution of different

structural units is different. The composite channel unit formed by overlapping and separated stable channels or the lateral alternated unit with braided channel and sheet flow sediment is affected by the inhomogeneous inching of injected water along the large pore–throat channel. The remaining oil is formed in patchy distributions, such as flow around an island, and it is enriched in the upper part of the composite rhythmic layer. Therefore, deep profile control and oil displacement technologies are keys to the comprehensive management of these units. In the extensively connected body unit sandwiched with intermittent channels, poor injection–production connectivity and low reservoir permeability caused flake-like remaining oil distribution. To fully realize the potential of the remaining oil, the fundamental requirements are depressurization, increase in injection, and improvement of reservoir quality.

(3)   The research results have guiding significance for comprehensive reservoir management. Daily oil production increased by approximately 40%, whereas water content decreased by 35%.

Author Contributions: Conceptualization and writing–original draft preparation, Q.G. and C.Z.; writing–review and editing, Y.J.; reservoir microscopic heterogeneity, Z.L. and Q.G.; reservoir macro-heterogeneity, Q.G., B.W. and L.X.; resources, Z.S. and L.X.; data curation, J.W. and Y.J. All authors have read and agreed to the published version of the manuscript.

Funding: This study was financially supported by PetroChina's 14th Five-year Forward-looking Fundamental Scientific and Technology Projects: Research on Reservoir Formation Law and Key Technology of Lithologic Stratigraphic Oil and Gas Reservoirs (2021DJ0402) and Research on Oil and Gas Enrichment Law and Exploration Evaluation Technology in Continental Deep and Ultra-deep Layers (2021DJ0202).

Data Availability Statement: Not applicable.

Acknowledgments: We would like to thank the PetroChina Hangzhou Research Institute of Geology and the PetroChina Qinghai Oilfield Company for permission to release their data. The first author would like to express warm gratitude to all those who supported this research. This paper describes objective technical results and analysis. Comments from anonymous reviewers greatly improved the manuscript.

Conflicts of Interest: The authors declare no conflict of interest.

References

1.   Qiu, Y. The methodology of petroleum development geology. *Pet. Explor. Dev.* **1996**, *23*, 43–47.
2.   Weber, J.K. Influence of Common Sedimentary Structures on Fluid Flow in Reservoir Models. *Int. J. Rock Mech. Min. Sci. Geomech. Abstr.* **1982**, *34*, 665–672. [CrossRef]
3.   Douglas, S.; Hamilton. Approaches to identifying reservoir heterogeneity in barrier/strandplain reservoirs and the opportunities for increased oil recovery: An example from the prolific oil-producing Jackson-Yegua trend, south Texas. *Mar. Pet. Geol.* **1995**, *12*, 273–290.
4.   Yang, S. A new method for quantitatively studying reservoir heterogeneity. *J. Univ. Pet.* **2000**, *24*, 53–56.
5.   Zhang, W.; Zhang, J.; Xie, J. Research on reservoir bed heterogeneity, interlayers and seal layers and controlling factors of 2+3 sands of upper second member, Shahejie formation, in the west of the Pucheng Oilfield. *Pet. Sci.* **2008**, *21*, 135–144. [CrossRef]
6.   Kaviani, D.; Valkò, P.; Jensen, J. Analysis of injection and production data for open and large reservoirs. *Energies* **2011**, *4*, 1950–1972. [CrossRef]
7.   Zheng, Z.; Wu, S.; Xu, C.; Yue, D.; Wang, W.; Zhang, F. Lithofacies and reservoirs of allluvial fan in the Lower Keramay formation in the block-6 of Karamay Oilfield, the Junggar Basin. *Oil Gas Geol.* **2010**, *31*, 463–471.
8.   Kong, F. Exploration technique and practice of sandy-conglomeratic fans in the northern part of Dongying depression. *Acta Pet. Sin.* **2000**, *21*, 27–31.
9.   Lei, Z.; Bian, D.; Du, S.; Wei, Y.; Feng, H. Characteristics of fan forming and oil-gas distribution in west-north margin of Junggar Basin. *Acta Pet. Sin.* **2005**, *26*, 8–12.
10.   Simlote, V.N.; Ebanks, W.J.; Eslinger, E.V. Synergistic evaluation of a complex conglomerate reservoir for enhanced oil recovery, barrancas formation, Argentina. *J. Pet. Technol.* **1982**, *37*, 295–305. [CrossRef]
11.   Shelby, J.M. Geologic and Economic Significance of the Upper Morrow Chert Conglomerate Reservoir of the Anadarko Basin. *J. Pet. Technol.* **1980**, *32*, 489–495. [CrossRef]

12. Carr, D.L.; Oliver, K.L. Surface-bounded reservoir compartmentalization in the Caddo Conglomerate, Boonsville (Bend Conglomerate) gas field, Fort Worth Basin, Texas. *AAPG Bull.* **1996**, 5. [CrossRef]
13. Lin, L.; Mu, Z.; Ma, D.; Kou, F.; Jing, H. Characteristics and control factors of the E31 clastic rock reservoir in the Q12 block of the Kunbei oilfield. *Spec. Oil Gas Reserv.* **2011**, *18*, 26–29.
14. Fu, S.; Ma, D.; Wang, L.; Feng, Q.; Xue, J.; Chen, Y.; Zhang, X.; Kou, F. Characteristics and accumulation conditions of paleo-uplift reservoirs in Kunbei thrust belt, Qaidam Basin. *Acta Pet. Sin.* **2013**, *34*, 675–682.
15. Gong, Q.; Shou, J.; Huang, G.; Li, S.; Wang, Y. Sedimentary characteristic of braided delta in Lulehe Formation of Kunbei oilfield in Qaidam Basin. *Chin. J. Geol.* **2012**, *47*, 116–128.
16. Miall, A.D. *The Geology of Fluvial Deposits*; Springer: Berlin/Heidelberg, Germany, 1996; pp. 75–178.
17. Decelles, P.G.; Gray, M.B.; Ridgway, K.D.; Cole, R.B.; Pivnik, D.A.; Pequera, N.; Srivastava, P. Controls on synorogenic alluvial-fan architecture, Beartooth Conglomerate (Paleocene), Wyoming and Montana. *Sedimentology* **1991**, *38*, 567–590. [CrossRef]
18. Wu, S.; Fan, Z.; Xu, C.; Yue, D.; Zheng, Z.; Peng, S.; Wang, W. Internal architecture of alluvial fan in the Triassic Lower Karamay Form ation in Karamay Oilfield, Xinjiang. *J. Paleogeography* **2012**, *14*, 331–340.
19. Cross, T.A. Stratigraphic controls on reservoir attributes in continental strata. *Earth Sci. Front.* **2000**, *7*, 322–350.
20. Yazynina, I.V.; Shelyago, E.V.; Abrosimov, A.A.; Yakushev, V.S. New method of oil Reservoir rock heterogeneity quantitative estimation from X-ray MCT data. *Energies* **2021**, *14*, 5103. [CrossRef]
21. Peter, J.R.F.; Mike, A.L.; Sarah, J.D. An integrated and quantitative approach to petrophysical heterogeneity. *Mar. Pet. Geol.* **2015**, *63*, 82–96.
22. Chen, H.; Wang, J.; Du, Y. Advances of research methods on reservoir heterogeneity. *Geol. J. China Univ.* **2017**, *23*, 104–116.
23. Zhang, Y.; Tang, Y. A review of reservoir heterogeneity research. *Mar. Geol. Front.* **2011**, *27*, 17–22.
24. Jiao, Y.; Li, S.; Li, Z.; Wen, X. Heterogeneity of porosity and permeability in clastic rock reservoirs. *Oil Gas Geol.* **1998**, *19*, 89–92.
25. Su, P.; Xia, Z.; Wang, P.; Ding, W.; Hu, Y.; Zhang, W.; Peng, Y. Fractal and multifractal analysis of pore size distribution in low permeability reservoirs based on mercury intrusion porosimetry. *Energies* **2019**, *12*, 1337. [CrossRef]
26. Gao, H.; Wang, M.; Shang, S. Quantitative evaluation of micro-pore throat heterogeneity in extra-low permeability sandstone using constant rate mercury penetration: Taking the Chang 8 reservoir of Xifeng oilfield in Ordos Basin. *Prog. Geophys.* **2013**, *28*, 1900–1907.
27. Guo, M.; Zhu, G.; Shou, J.; Xu, X. Feature origin and petroleum explorative significance of crushed fracture in clastic rock. *ACTA Sedimentol. Sin.* **2006**, *24*, 331–340.
28. Giles, M.R. *Diagenesis: A Quantitative Perspective: Implications for Basin Modelling and Rock Property Prediction*; Kluwer Academic Publishers: Dordrecht, The Netherlands, 1997.
29. Luo, M.G. Quantitative models for pore structures of clastic sedimentary rocks. *Acta Pet. Sin.* **1991**, *12*, 27–38.
30. Liu, Y.; Ma, K.; Hou, J.; Yan, L.; Chen, F. Diagenetic controls on the quality of the Middle Permian lucaogou formation tight reservoir, southeastern Junggar Basin, northwestern China. *J. Asian Earth Sci.* **2019**, *178*, 139–155. [CrossRef]
31. Tian, J.; Liu, W.; Wang, F.; Chen, R.; Lin, X. Heterogeneity of the Paleozoic tight sandstone reservoirs in Gaoqiao area of Ordos Basin. *Oil Gas Geol.* **2014**, *35*, 183–189.
32. Cui, J.; Zhu, R.; Wu, S.; Wang, T. Heterogeneity and lower oily limits for tight sandstones: A case study on Chang 7 oil layers of the Yanchang Formation, Ordos Basin. *Acta Pet. Sin.* **2013**, *34*, 877–882.
33. Yue, D.; Lin, C.; Wu, S.; Hou, L. Application of quantitative method for characterizing reservoir heterogeneity to the development of reef limestone reservoir. *Acta Pet. Sin.* **2004**, *25*, 75–79.
34. Feng, C.; Shan, Q.; Shi, W.; Zhu, S. Reservoirs heterogeneity and its control on remaining oil distribution of K1q4, Fuyu oilfield. *J. China Univ. Pet.* **2013**, *37*, 1–7.
35. Hu, R.; Ma, D.; Ma, S.; Yan, B. Identification of flow units and distribution of remaining oil controlled by the architectural structure of point bar. *J. China Univ. Min. Technol.* **2016**, *45*, 135–140.
36. Rasmussen, L.; Fan, T.; Rinehart, A.; Luhmann, A.; Ampomah, W.; Dewers, T.; Heath, J.; Cather, M.; Grigg, R. Carbon storage and enhanced oil recovery in Pennsylvanian morrow formation clastic reservoirs: Controls on oil–brine and oil–Co2 relative permeability from diagenetic heterogeneity and evolving wettability. *Energies* **2019**, *12*, 3663. [CrossRef]
37. Yin, Z.; Lu, G.; Zou, X.; Yang, Z. Heterogeneity of non-marine reservoirs and its influences on recovery factor: Take Gaoshangpu and Yonganzhen oil reservoirs in Jidong and Shengli oilfields as examples. *Oil Gas Geol.* **2006**, *27*, 106–117.
38. Ren, D.; Sun, W.; Zhao, J.; Qu, X.; Zhang, Q. Microscopic waterflooding characteristics of lithologic reservoirs in Ordos Basin and its influence factors: Taking the Chang 81 reservoir in Huaqing oilfield as an example. *J. China Univ. Min. Technol.* **2015**, *44*, 1043–1052.
39. Charbi, R.; Alajmi, A.; Algharaib, M. The potential of a surfactant/polymer flood in a middle eastern reservoir. *Energies* **2012**, *5*, 58–70. [CrossRef]
40. Lee; Sang, K. Performance of a polymer flood with shear-thinning fluid in heterogeneous layered systems with crossflow. *Energies* **2011**, *4*, 1112–1118. [CrossRef]

Article

Current Status of Helium Resource Research and Prediction of Favorable Areas for Helium Reservoir in China

Ye Xiong [1], Shan Jiang [1,*], Jingjing Yi [2] and Yi Ding [2]

1   College of Earth Sciences, Yangtze University, Wuhan 430000, China; 2021710373@yangtzeu.edu.cn
2   Jianghan Oilfield Exploration and Development Research Institute, Sinopec Group, Wuhan 430082, China; ciciyi1986@126.com (J.Y.); xxldy123@sohu.com (Y.D.)
*   Correspondence: jiangshan0712@126.com; Tel.: +86-139-7127-8916

Abstract: As an unconventional oil and gas reservoir, helium gas reservoirs have gradually become a focus of attention. In recent years, with the continuous increase in demand for helium gas, the uneven distribution of global helium resources has attracted China's attention to helium resources. In this study, a method for predicting favorable areas of helium gas was proposed based on the natural gas exploration theory and the idea of "finding gas in enrichment areas". We conducted an in-depth study and analysis of the types of helium gas formations in China by comprehensively using geochemical and isotope-testing data, identifying the distribution of helium source rocks in China. Based on this, we conducted directed analyses of the transport channels and caprock conditions for helium gas, and summarized the enrichment modes of helium gas. Using this method, we predicted five favorable areas for the enrichment of helium gas in China, providing an important basis for the future exploration and development of helium resources in China.

Keywords: helium resources; unconventional oil and gas reservoirs; resource distribution; helium formation; favorable area forecasts

Citation: Xiong, Y.; Jiang, S.; Yi, J.; Ding, Y. Current Status of Helium Resource Research and Prediction of Favorable Areas for Helium Reservoir in China. *Energies* **2024**, *17*, 1530. https://doi.org/10.3390/en17071530

Academic Editor: Nikolaos Koukouzas

Received: 23 January 2024
Revised: 10 March 2024
Accepted: 15 March 2024
Published: 22 March 2024

1. Introduction

Unconventional oil and gas reservoirs are currently a key area of oil and gas development worldwide, and in recent years, helium gas reservoirs have become a new hot spot in this field. Research on helium gas dates back to the 1930s, and there have been various opinions regarding its source. Ruedemann [1] believed that the helium gas in the Panhandle, which is a large and rich helium gas field in the United States, primarily originated from the decay of U and Th in granite and pegmatite in the reservoir. Pierce [2] believed that helium gas originates from source rocks, while Katz [3] believed that helium gas comes from sedimentary rocks containing ammonium minerals and organic matter. However, Nikonov [4] believed that helium gas originates from the crust, and Maione [5] believed that helium gas comes from the basement. Gold [6] and others inferred that helium gas is generated jointly from basement rocks and basement source rocks. Currently, scholars believe that helium gas primarily originates from deep mantle-source fluids and the decay of U and Th in granitic rocks.

In addition to identifying the source of helium, it is also necessary to determine the transport pathways and migration modes of helium gas during the accumulation process. Different scholars have proposed various viewpoints regarding the transport modes and pathways of helium gas. Some scholars believe that helium gas is transported through diffusion, and can also be upwardly transported with fluids [7]. Other scholars propose that the main pathways for helium gas are basement faults and fractures [4,6,8]. Studies by Ballentine [9] and Brown [10] suggest that the helium gas in the Panhandle gas field is transported via groundwater. Qin [11] and colleagues suggest that in the Weiyuan gas field, helium gas is dissolved in water before degassing and accumulation. In the research conducted by Zhang [12] and his team, they suggest that helium gas generated from radioactive minerals initially dissolves in water, and then dissolves in a carrier gas before being transported into the reservoir.

For a long time, China has relied heavily on imports to obtain sufficient amounts of helium gas, as its own production and resource reserves are very inadequate. This situation is mainly due to the past underestimation and neglect of helium resources in China, which prevented adequate exploration and development work from being carried out. In addition, the incomplete theories on the formation of helium reservoirs have further hindered the exploration and development of China's helium resources. Therefore, it is necessary to predict the favorable areas for helium gas in China to provide a sufficient theoretical basis for finding and developing helium resources.

2. Global Helium Situation

2.1. Global Distribution of Helium Resources

The global distribution of helium gas is uneven, and according to a report from the US Geological Survey in 2023 [13], the average recoverable helium reserves in known natural gas reservoirs in the United States is 8.49×10^9 m^3. Except for the United States, the estimated global helium resources are about 3.13×10^9 m^3. They are in Qatar with 10.1×10^9 m^3, Algeria with 8.2×10^9 m^3, Russia with 6.8×10^9 m^3, Canada with 2.0×10^9 m^3, and China with 1.1×10^9 m^3, indicating the extremely uneven distribution of global helium resources. Overall, the world's proven and remaining helium reserves are gradually decreasing, but at the same time, some large helium fields have been discovered. For example, the "world-class" helium field discovered in the East African Rift Valley in Tanzania is estimated to have reserves of up to 2.8×10^9 m^3 [14].

According to the US Geological Survey and Danabalan's article on the distribution of helium gas in the United States, the helium resources of the United States are mainly distributed in the Panhandle-Hugoton gas field in Kansas, Oklahoma, and Texas, the Riley Ridge gas field in Wyoming, the Greenwoods gas field in Kansas, the Keyes gas field in Oklahoma, and the Cliffside gas field in Texas. In addition, some helium-rich natural gas resources have also been discovered in sedimentary basins to the east of the Rocky Mountains and in New Mexico (Figure 1) [8,15,16].

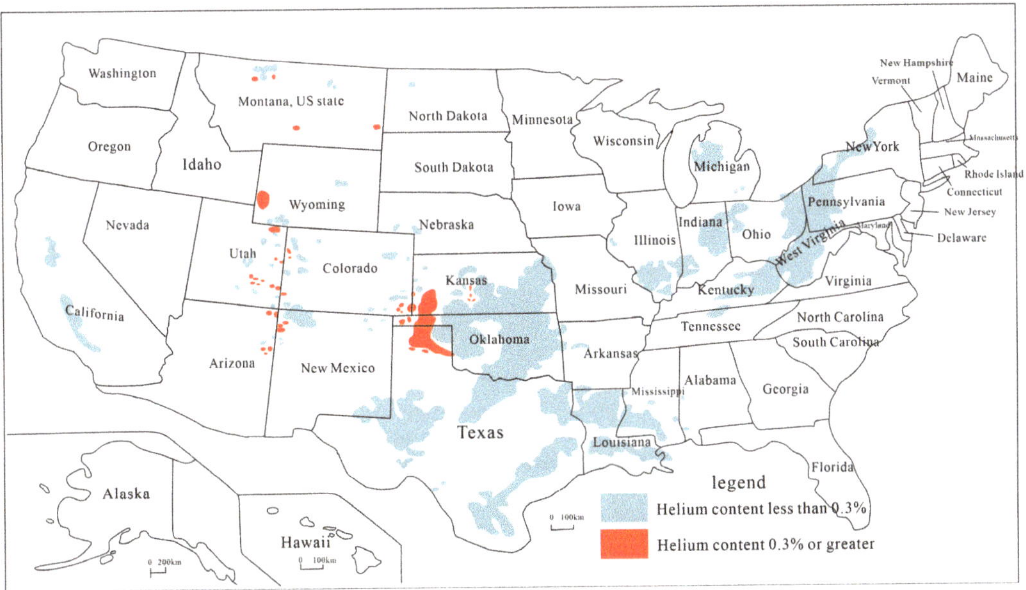

Figure 1. Distribution of helium resources in the United States (from Danabalan D. 2017 [16]).

According to Akutsent's research in 2014, it was found that the helium resources discovered in Russia are mainly distributed in natural gas fields on the West Siberian Plateau. In addition, there are also some helium-rich natural gas fields in the Orenburg field in the Northern Caspian region, the Chayanda gas field in the Yakutia region, and some areas in the Irkutsk region and Komi Republic (Figure 2) [17]. From the distribution maps of the helium contents in the United States and Russia, uneven distributions also occur within these countries.

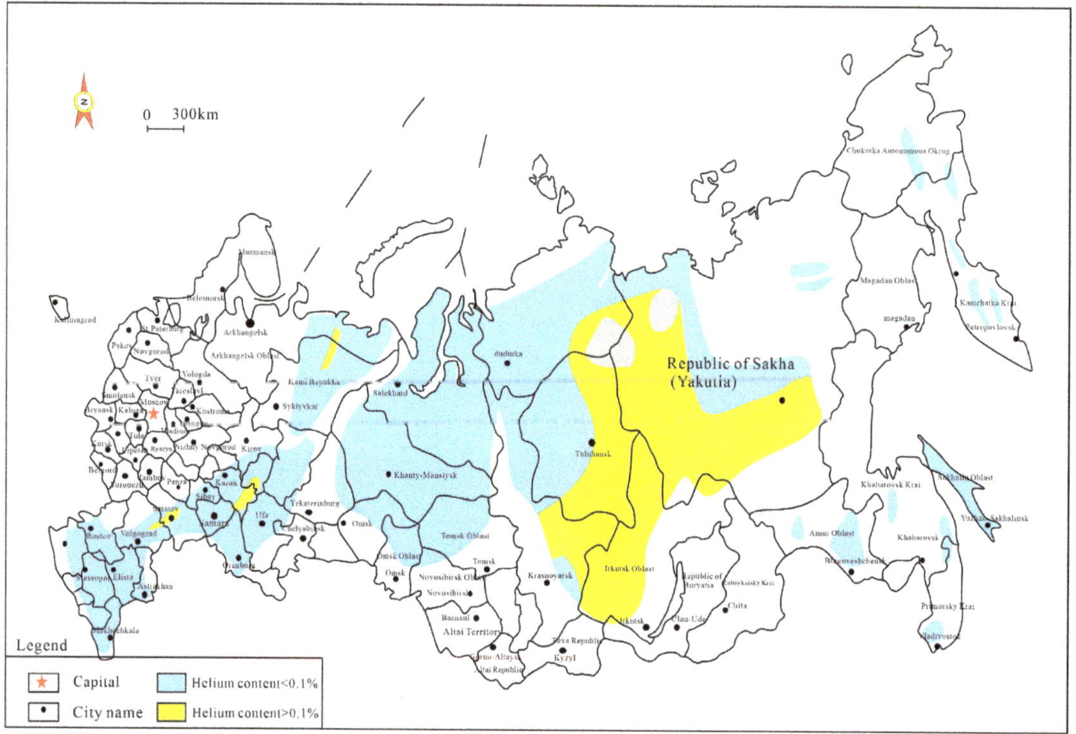

Figure 2. Distribution of helium resources in Russia (from Akutsent V.P. 2014 [17]).

2.2. Global Helium Supply and Demand

In 2021, the helium production capacity in the United States was about 77 million cubic meters, accounting for about 48.13% of the global production, while Qatar's production capacity was about 51 million cubic meters, accounting for about 31.88%. Together, their production capacity accounts for almost 80% of the global production, while China only accounts for about 0.63%. Helium resources are almost completely owned by a few countries such as the United States and Qatar [18].

The United States is the world's largest producer and supplier of helium. Prior to 1996, 90% of the world's helium production came from the United States. After 2000, most of the helium-rich natural gas fields in the United States entered a depletion phase of development. Since 2012, US helium production has been declining at an annual rate of about 10%. Since 2005, the helium production in Qatar has rapidly increased to 5×10^7 m^3 after being put into operation. As Qatar, Algeria, and other countries increased their helium production, the proportion of US helium production in the global total decreased from 90% in the 1990s to 55% in 2016. Among the global helium production proportions, Qatar accounts for 32%, Algeria for 6.5%, Australia for 2.6%, Russia for 1.9%, and Poland for 1.3% (Figure 3) [14].

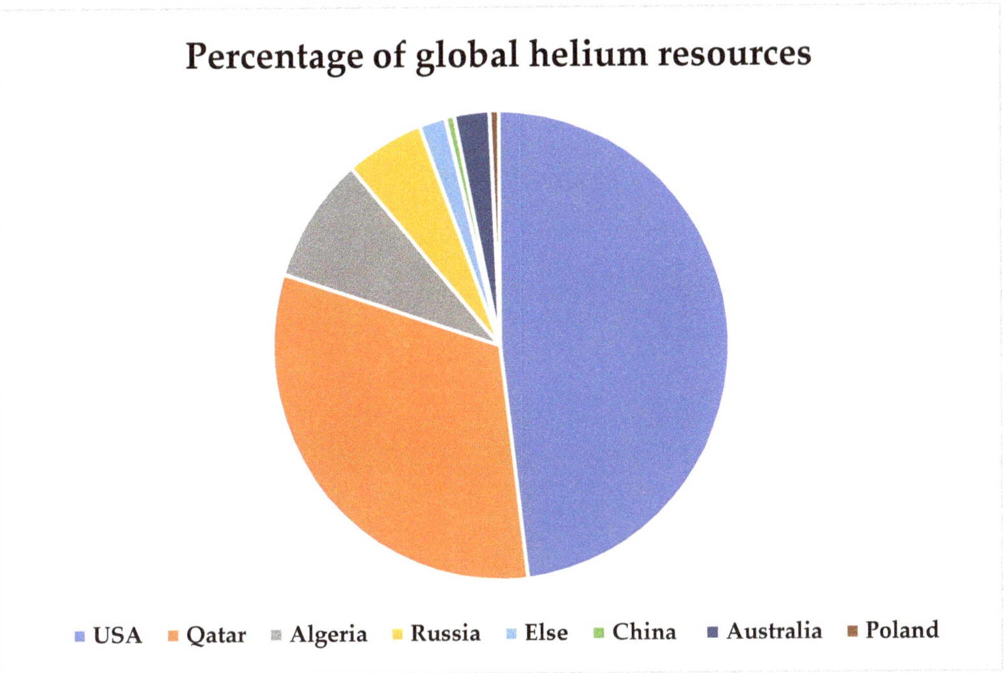

Figure 3. Pie chart showing the proportion of global helium production capacity.

Due to the impact of the reduction in US helium production, the global helium supply has been in an overall downward trend since 2013. As the economy and high-tech industries continue to develop rapidly, the application scope of helium has gradually expanded, and global helium demand is growing at an annual rate of 4% to 6% [19]. This has led to a current supply shortage of helium, which is expected to last for a long time.

In 2016, the global helium demand was 2.3×10^8 m^3, but the annual production was only 1.54×10^8 m^3. According to the United States Geological Survey, the global helium resource amount was approximately 520×10^8 m^3 in 2020, of which the United States accounts for about 206×10^8 m^3, followed by Qatar with a helium resource reserve of about 10×10^8 m, while China's helium resource amount is about 11×10^8 m^3, accounting for only 2% [13].

China's demand for helium was very small prior to 1990, and the highest production of helium before 1989 was only 3×10^4 m^3 [20]. In 2005, China imported 2×10^6 m^3 of helium, and by 2012, the total amount of imported helium had reached 5×10^6 m. In 2017, China imported approximately 20×10^6 m^3 of helium [21], which means that the amount of imported helium has increased by 10 times that amount in just 10 years. Currently, China's helium supply is heavily dependent on imports, mainly from Qatar, the United States, and Australia, so the situation for helium resources in China is very severe.

3. Materials and Methods

Based on natural gas exploration theory and the "finding gas in enrichment zones" concept, this article proposes a method for predicting favorable areas for helium gas exploration. The Combination of Functional Elements to Control Reservoir Formation Method (CPH) involves using three functional elements, including the caprock (C), effective migration pathway (P), and helium source rock (H). With the distribution and development of CPH, this method predicts favorable areas for the enrichment of helium in China. Using the helium content of discovered oil and gas reservoirs, the rationality of the subdivision scope of the predicted favorable areas has been confirmed.

We reprocessed the geological, geochemical, and isotopic data of nine samples collected from seven regions, conducted an in-depth analysis, and studied the types of helium gas geneses in China. Through this analysis, we discovered the laws governing the sources of helium gas and the distribution patterns of helium source rocks in China. We also conducted directional analyses of the migration pathways and caprock conditions for helium gas, and summarized the enrichment patterns of helium gas. Based on these findings, we predicted favorable areas for helium gas exploration and verified their rationality using the helium content of discovered oil and gas reservoirs. This study provides an important foundation for the future exploration and development of China's helium resources. The research outcomes of this study hold significant practical and theoretical implications, which will promote the development and utilization of helium resources.

4. Results

4.1. Genetic Types and Distribution Characteristics of Helium in China

4.1.1. Types of Helium Geneses

Helium gas in natural gas reservoirs has various sources and origins. The origin and source of helium can be effectively identified using the composition of helium and rare gas isotopes. Helium has two stable isotopes, ^{3}He and ^{4}He. ^{3}He is mainly derived from mantle degassing, while ^{4}He is mainly derived from the decay of radioactive elements like ^{238}U, ^{235}U, and ^{232}Th. The main radioactive decay reactions are as follows [19]:

$$^{238}U \rightarrow 8\,^4He + 6\beta + ^{206}Pb \tag{1}$$

$$^{235}U \rightarrow 7\,^4He + 4\beta + ^{207}Pb \tag{2}$$

$$^{232}Th \rightarrow 6\,^4He + 4\beta + ^{208}Pb \tag{3}$$

The classification of helium gas origin types is determined by isotopic ratios. The isotopic signature of the sample helium is expressed as the ^{3}He/^{4}He ratio (R) relative to the atmospheric helium ^{3}He/^{4}He ratio (Ra), to determine the helium isotope characteristics of the gas sample.

$$R/Ra = (^3He/^4He)\ sample/(^3He/^4He)\ atmosphere \tag{4}$$

In natural gas reservoirs, the composition of atmospheric helium is so small that it can be neglected. Therefore, the source of atmospheric helium can be ignored, and a binary composite model can be used to calculate the proportions of crustal and mantle helium in natural gas samples. When the isotopic ratio R/Ra is used to express the isotopic distribution characteristics of the gas sample, if (R/Ra) > 3.94, then the proportion of mantle helium is greater than 50%; if (R/Ra) > 0.1, then the proportion of mantle helium is greater than 1%; if (R/Ra) < 0.1, then it is from crustal sources [11]. Mantle-source helium gas will ultimately enter crustal reservoirs and become accumulated. Therefore, mantle-source helium-rich natural gas reservoirs are always mixed with crustal-sourced helium gas.

Cao, Z.X. (2001), Guo, N.F. (1999), Dai, J.X. (2003), Yu, Q.X. (2013), and Zhang, X.B. (2020) [12,22–28] collected natural gas samples from the Hankou Gas Field and the Huangqiao Gas Field on the Tan-Lu Fault, the Wanjinta Gas Field in the Songliao Basin, the Weiyuan Gas Field in the Upper Yangtze Plate, the Yakra Gas Field and the Hetianhe Gas Field in the Tarim Plate, and the Mabei, Dongping, and Niudong areas in the Qaidam Block, respectively, and analyzed their natural gas components using mass spectrometry. We reprocessed the test result data (Table 1). We found that helium gas is produced along with natural gas in gas reservoirs. Helium-rich gas reservoirs with a crustal origin are mainly composed of organic (CH_4) gas reservoirs. Helium-rich gas reservoirs with a mantle origin are characterized by a high content of nitrogen gas. The helium-rich natural gas can be classified into organic helium-rich gas reservoirs and inorganic helium-rich gas reservoirs based on its geochemical composition. From a geological structural perspective, the mantle-origin helium-rich gas reservoirs are mainly distributed in fault zones, where there is no supply of hydrocarbon source rocks for the reservoirs, but the tectonic activity is strong and the faults are well developed. The helium gas reservoirs with a crustal origin are generally associated with natural gas and are distributed along the edges of tectonic plates.

Table 1. Composition and Isotopic Characteristic Data Table of Natural Gas Reservoirs.

Classification of Causes	Geotectonic Position	Component Characteristics (%)					^{3}He/^{4}He (10^{-8})	R/Ra	Mantle Source Share (%)	Reference
		CH_4	C_2^+	N_2	CO_2	He				
Crustal-source helium	Southwest edge of the Upper Yangtze Plate	83.97	2.3	9.6	4.3	0.21~0.34	2.9~3.0	0.21~0.00	0.08~0.09	Dai et al., 2003 [29]
	Northern margin of the Tarim Plate	70.37	3.1	6.88	19.34	0.22	21.6	0.15	1.79	Yu et al., 2013 [22]
	Western margin of the Tarim Plate	80.35	3.64	10.39	1.36	0.249	11.6~12.7	0.08~0.09	0.87~0.97	Liu et al., 2009 [23]
	Northern margin of the Tsaidam plate	77~79		7.41~8.87	0.10~0.53	0.12~0.20	4.90~6.85	0.04~0.05	0.3~0.4	Zhang et al., 2019 [12]
	Northern margin of the Tsaidam plate	62~95	1~13	3~30	0.01~2.02	0.08~1.07	1.01~2.21			Zhang et al., 2020 [28]
	Fenwei graben of the North China Plate	71.70	72.38	27.20	0.567	0.395	17.64	0.13	1.42	Zhang et al., 2019 [12]
Mantle-source helium	Songliao Basin	9.69	0.39		89.92	0.1	687	4.91	62.38	Cao et al., 2001 [26]
	Tanlu Fault Zone	1.77		68.86	34.27	2.08	434	3.10	39.3	Guo et al., 1999 [27]
	Tanlu Fault Zone	27.057	2.825	64.529	4.241	1.33	371	2.65	33.61	Dai et al., 2017 [25]

4.1.2. Genetic Types and Distribution Characteristics of Helium in China

We conducted an investigation and analysis of eight basins in China. Xu, Y.C. (1997) [24], Yu, Q.X. (2013) [22], DAIJ. X. (2005), He, F.Q (2022) [30], Cao, Z.X. (2001) [26], Guo, N.F. (1999) [27], Zhang, Y.P (2016) [31], HAN W. (2020) [32] and others collected samples from oil and gas reservoirs in the Tarim Basin, Qaidam Basin, Junggar Basin, Sichuan Basin, Songliao Basin, Bohai Bay Basin, Subei Basin, and San-shui Basin, respectively, and conducted a helium isotope analysis of their natural gas using a mass spectrometer. We have summarized their analysis results in Table 2. There are significant differences in the helium isotopic ratios and mantle-source helium proportions among the helium-bearing basins in China. According to the research data analyzed in this paper, the ^{3}He/^{4}He values were found to range between 1.01×10^{-8} and 7.2×10^{-6}, and the R/Ra values were found to range between 0.01 and 5.14. In China, the helium-bearing basins with high mantle-source helium proportions are mainly distributed in the eastern part of the country. Among them, the Sanshui Basin shows a strong correlation between its helium content and mantle-source helium proportion. The input of mantle-source helium is the main reason for the high helium content in the Sanshui Basin and other eastern basins in China. Except for some samples from the Junggar Basin and Tarim Basin, the R/Ra values of samples from the central and western basins in China were mostly lower than 0.1, indicating that the mantle-source helium proportions were less than 1.1%, and almost all the helium gas was derived from crustal sources.

The Tarim Basin is the largest onshore oil and gas basin in China, and is a large cratonic basin developed on the Precambrian basement. The ^{3}He/^{4}He values in the basin were found to range between 1.92×10^{-8} and 2.6×10^{-7}, and the R/Ra values were found to range between 0.014 and 0.186, indicating that the helium gas is predominantly derived from crustal sources. The Qaidam Basin is a large-scale cratonic-type oil and gas basin in western China, with a complex fault system and Precambrian metamorphic rocks and Hercynian granites as the basement of its northern margin. Helium isotope ratios in the basin range from 1.01×10^{-8} to 1.3×10^{-6} for ^{3}He/^{4}He values and from 0.007 to 0.93 for R/Ra values, indicating that most of the helium in the basin originates from a crustal source. The Junggar Basin is a large oil and gas basin in northwest China that has undergone multiple tectonic events. The ^{3}He/^{4}He values in the basin were found to range from 1.96×10^{-8} to 5.4×10^{-7}, while the R/Ra values were found to range from 0.014 to 0.386. The helium in the basin is mostly derived from crustal sources, with only a small amount of mantle-source helium mixing occurring in the southern sag area. The Ordos Basin is a large-scale oil and gas basin developed on the craton basement. The ^{3}He/^{4}He values in the basin range from 2.07×10^{-8} to 13.64×10^{-8}, and the R/Ra values vary between 0.01 and 0.1, indicating that the helium gas in the basin is mostly derived from crustal sources. The basement of the Sichuan Basin is composed of magmatic and metamorphic rocks, and the Weiyuan gas field, which belongs to the Cambrian gas reservoirs, was the first commercial helium production area in China. The ^{3}He/^{4}He values in the gas field range from 1.4×10^{-8} to ~5.74×10^{-8}, while the R/Ra values are between 0.01 and 0.04, indicating that crustal helium is the main source of helium in the Weiyuan gas field. The Songliao Basin is a Cenozoic continental extensional rift basin with a large-scale intrusion of granites in the basin. The ^{3}He/^{4}He values in the basin are distributed within the range of 1.01×10^{-6} to 7.2×10^{-6}, while the R/Ra values are between 0.72 and 5.14, indicating that the helium gas in the basin is mainly derived from a crustal source. The Bohai Bay Basin is a typical inland rift basin that contains six sag basins and has experienced multiple magmatic activities. The ^{3}He/^{4}He values in the basin are distributed in the range of 8.97×10^{-7} to 5.22×10^{-6}, while the R/Ra values are between 0.64 and 0.73. Due to the influence of magmatic and fault activities, a large amount of mantle-derived helium has entered the basin, making it a mixed type of crust-mantle helium. The main tectonic pattern of the Subei Basin consists of "one uplift and two sag basins". The ^{3}He/^{4}He values in the basin are distributed in the range of 3.71×10^{-6} to 5.54×10^{-6}, while the R/Ra values are distributed between 2.63 and 3.96, indicating that the helium gas in the basin is a typical mixture of crust-mantle sources.

Table 2. Statistical table of helium characteristics in basins in China.

Area	Basin	^{3}He/^{4}He	R/Ra	Helium Genetic Type	Reference
Midwest	Tarim	1.92×10^{-8}–2.60×10^{-7}	0.014–0.186	crustal-based	Yu et al. (2013) [22]
	Qaidam	1.01×10^{-8}–1.30×10^{-6}	0.007–0.930	crustal-based	Zhang et al. (2016) [31] Han et al. (2020) [32]
	Junggar	1.96×10^{-8}–5.40×10^{-7}	0.014–0.386	crustal-based	Xu et al., 1997 [24]. Xu et al. (1995) [33]
	Ordos	2.07×10^{-8}–13.64×10^{-8}	0.010–0.100	crustal-based	Dai et al. (2017) [25], Xu et al. (1997) [24], He et al. (2022) [30]
	Sichuan	1.40×10^{-8}–5.74×10^{-8}	0.010–0.040	crustal-based	Xu et al., 1997 [24], Dai et al. (2003) [29]
East	Songliao	1.01×10^{-6}–7.20×10^{-6}	0.720–5.140	crust-mantle mixing	Feng et al. (2001) [34]
	Bohai Bay	8.97×10^{-7}–5.22×10^{-6}	0.640–0.730	crust-mantle mixing	Sun et al. (1996) [35], Cao et al. (2001) [26]
	Subei	3.71×10^{-6}–5.54×10^{-6}	2.630–3.960	crust-mantle mixing	Xu et al. [24]. Guo et al. (1999) [27]
	Sanshui	1.60×10^{-6}–6.39×10^{-6}	1.140–4.560	crust-mantle mixing	Zhang et al. (2014) [21]

Based on the above data, we have found that the helium reservoirs in western China are mainly located in the Sichuan, Tarim, Qaidam, and Junggar basins, as well as south of the Weihe River Fault, with the crustal source being the primary source of helium. In the eastern region, helium is mainly located in the oil and gas-bearing basins on both sides of the Tan-Lu fault zone, such as the Subei, Songliao, Hailar, Bohai Bay, and Sanshui basins, and is mostly of a mixed crust-mantle origin, with the mantle source being the dominant source (Figure 4).

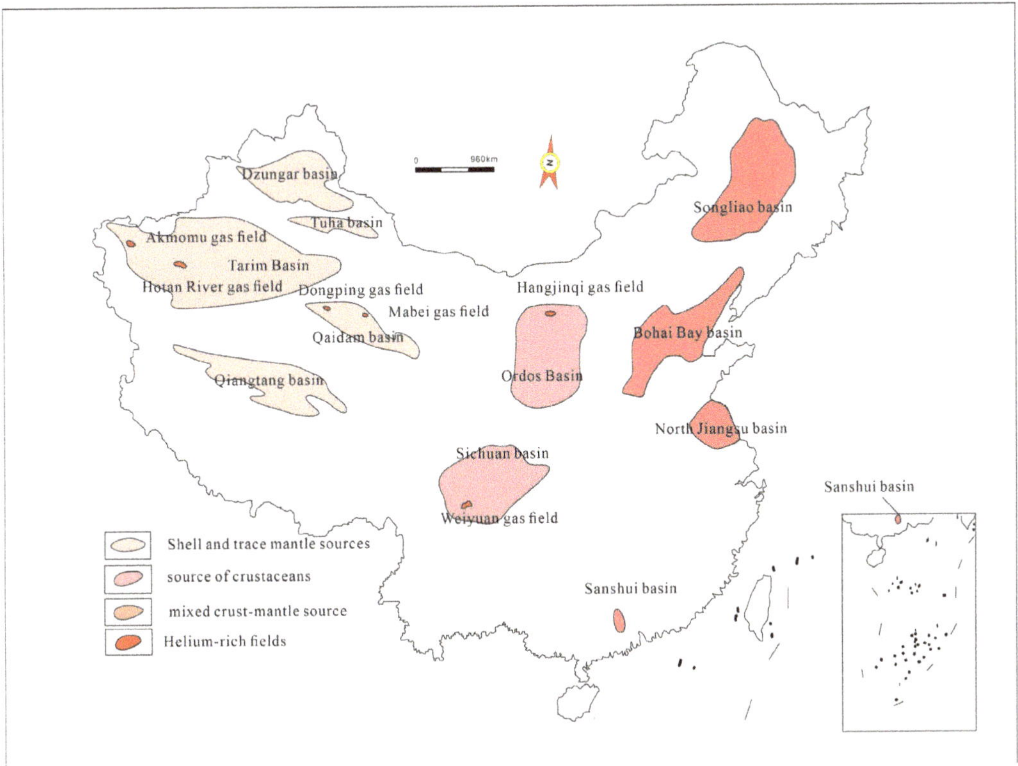

Figure 4. Genetic types of helium gas in major petroliferous basins in China.

4.2. China's Helium Accumulation Conditions and Enrichment Patterns

Under the guidance of the theoretical and enrichment zone concepts in natural gas exploration, we believe that research on the accumulation conditions of helium in China is essential. By studying the helium source rocks, migration pathways, and sealing layers in China, we can understand how helium gas is enriched in different regions of the country. This research also provides the basis for predicting the distribution of favorable areas for helium in China.

4.2.1. Distribution Characteristics of Helium Source Rocks

Research on the genesis of helium types has shown that most basins in China have crustal sources of helium, and the helium source rocks in China are mainly rich in uranium-containing granites in the crust. These types of rocks are mainly distributed in tectonic-magmatic activation zones on the edges of cratons, fold belts in orogenic zones, and active continental margins. The fold belts on the edges of cratons generally developed along the margins of the Archean continental nucleus and later underwent strong folding, metamorphisms, and granite transformations to form these rocks. After multiple tectonic movements in large orogenic belts, basement rocks and granites with a high uranium content are the main helium source rocks. Based on the above conditions, we conducted a survey of the strata in China and drew a distribution map of the helium source rocks in China.

According to the distribution of helium source rocks in China (Figure 5), the main distribution areas of helium source rocks in China are mainly in the southeast of South China, the northern part of the Qinling Mountains, the eastern part of Tibet, the Tianshan Mountains, the Yin Mountains, and the northern part of the Greater Khingan Mountains.

They are also distributed in Western Sichuan, the southern part of the Qaidam Basin, the western Kunlun Mountains, Qilian Mountains, Greater Khingan Mountains, Eastern Liaoning, and Erguna. This also indicates that the research on the distribution of helium source rocks is the basis for predicting the distribution of favorable areas for helium in China.

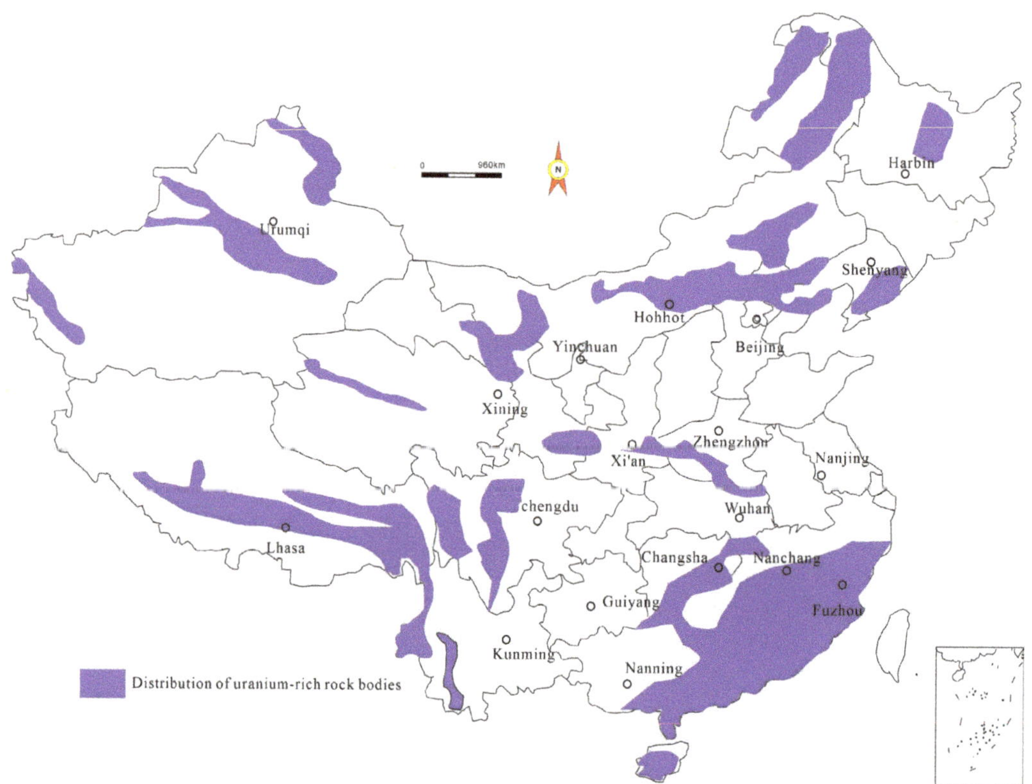

Figure 5. The distribution map of helium source rocks in China.

4.2.2. Effective Migration Pathway in China

Transport pathways are an important part of the formation of a helium reservoir. There are two main transport pathways for helium in China. One is through major deep-seated faults, and the other is through an ancient formation of water that serves as the carrier.

The development of fault structures plays a dual controlling role in the formation of helium deposits. On the one hand, it can serve as a pathway for the upward migration of crustal-sourced helium, and on the other hand, it can serve as a pathway for the input of mantle-sourced helium [36]. The process of helium transferring from mineral particles that undergo decay to pore water is the first stage of helium transport. Then, the helium in the pore water is fractionated and extracted by the migrating gas, which contains helium gas or helium groundwater. This migrates through fractures, faults, and undergoes secondary transport until it accumulates in a structural trap, forming a helium reservoir in the geological stratum. In the crust-mantle mixed-type helium-bearing basins in eastern China, there is a strong correlation between helium gas and the Tan-Lu fault zone [37]. Research has found that areas where helium accumulates are near the Tan-Lu fault zone. According to survey data (Table 2), high helium samples from the Songliao Basin, Bohai Bay Basin, and northern Jiangsu Basin are distributed on both sides of the Tan-Lu fault zone (Figure 6), exhibiting a strong correlation. The Tan-Lu fault zone is a crustal fault that cuts through at a thickness of 30–40 km.

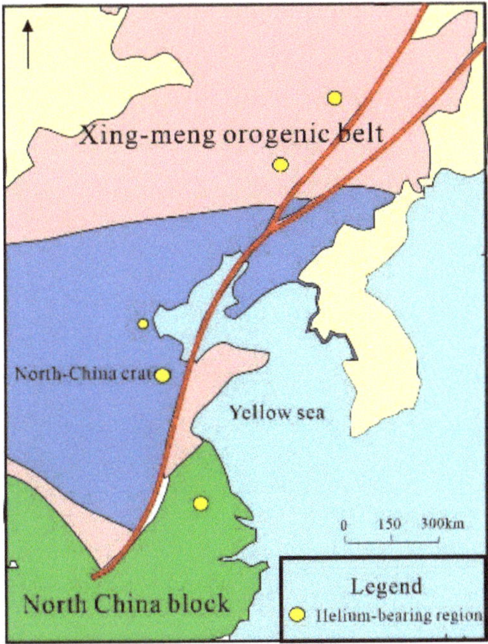

Figure 6. Schematic diagram of the Tan-Lu fault zone and the distribution of high helium gas blocks on both sides.

Water-carried gas is the main transport pathway in the central and western regions, mainly distributed in the Sichuan Basin, the Qaidam Basin, and the Tarim Basin. Helium source rocks provide the source for helium gas generation, and the development of structural faults provides the pathway for helium gas to migrate upwards. However, the movement of helium in pore water is very slow [38,39]. Moreover, helium cannot form a separate gas stream and cannot enter a trap driven by buoyancy alone. Therefore, the large-scale migration of helium relies on groundwater and gas-bearing layers as carriers. As mentioned earlier, helium gas often accompanies the enrichment of natural gas. An analysis of the gas composition in Table 2 shows that the primary transport pathway for helium in Central and Western China is through the migration of helium in natural gas, mainly distributed in the Sichuan Basin, Qaidam Basin, and Tarim Basin.

4.2.3. Cap Rock Distribution

After helium enters a gas reservoir, if the cap rock conditions of the overlying strata are well-developed, it has the potential to become a helium-bearing gas reservoir and continuously receives replenishment of helium from the groundwater in the subsequent process. However, if the overlying strata of the gas reservoir lack cap rock conditions or the gas-bearing layer is affected by tectonic activity, the helium-bearing gas layer may migrate laterally along the geological stratum and eventually form a helium-enriched gas reservoir under suitable trap conditions.

Helium gas molecules are the smallest known chemical substance in nature, so their storage requires even stricter cap rock conditions. Helium-rich gas reservoirs are usually capped by tight anhydrite, salt rock, and shale formations [40]. Based on the distribution of helium source rocks and the lithology of cap rocks, we conducted directional and qualitative searches for cap rocks in Chinese gas reservoirs, and drew up a map showing the types and distribution of cap rocks (Figure 7).

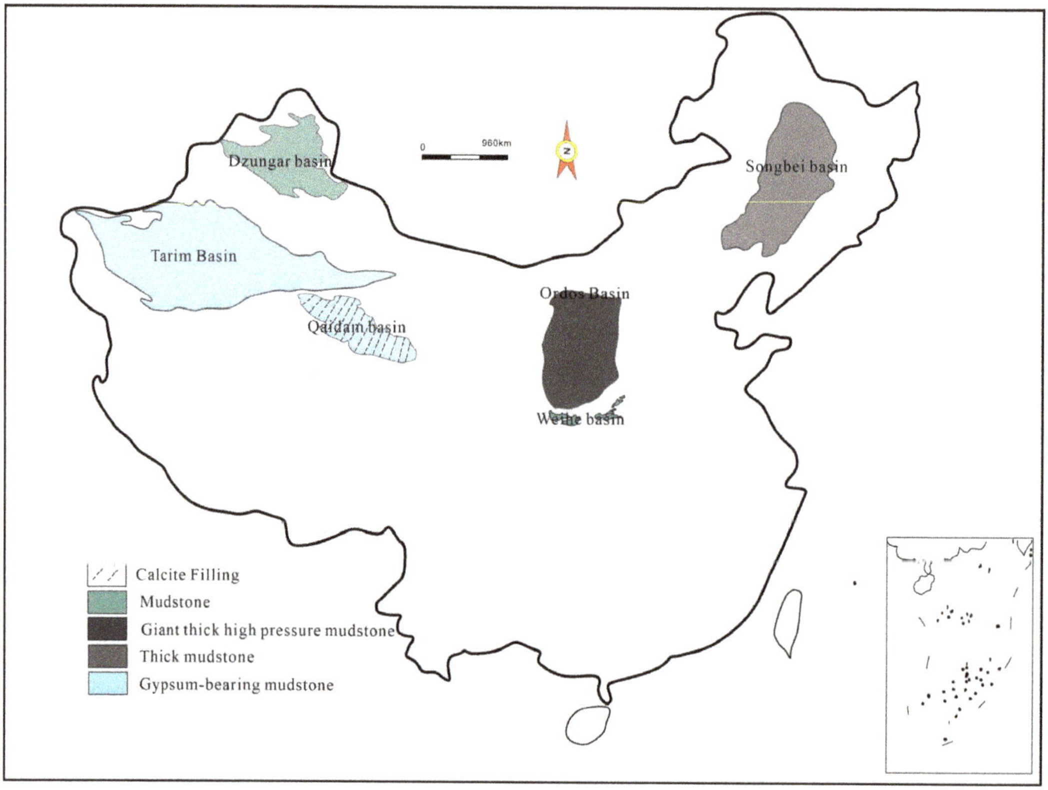

Figure 7. Types and distribution of cap rocks in China.

4.2.4. Chinese Model of Helium Enrichment

In the helium-rich reservoirs in central and western Chinese basins, shell-source helium is the main type of helium, and the enrichment model is mainly associated with the migration of groundwater and natural gas in the strata. Helium gas is generated by the basal helium source rocks and dissolved in ancient formation water. Due to tectonic movements, ancient formation water migrates upward along faults, releasing free helium gas along the way. The helium gas then migrates to the gas reservoirs as it moves upward. Natural gas can also serve as a carrier for helium migration. During the upward migration process of the helium-containing groundwater, if it encounters natural gas, the helium in the groundwater is displaced into the natural gas due to Henry's Law and is carried along with the natural gas until it forms a gas reservoir in a suitable trap position (Figure 8).

In the helium-rich reservoirs in the eastern Chinese basins, the helium is of a mantle-crust mixing origin. The thinning of the lithosphere, upwelling of the mantle, and degassing caused by magmatic activities produces mantle-source helium gas and non-organic gas (CO_2, N_2). More shell-source helium is released through the thermal effects of magma. The mantle-origin fluids containing helium, N_2, and CO_2 migrate upward along deep faults. At the same time, helium generated from other strata is also trapped in the gas reservoirs. Eventually, helium gas from different sources and natural gas combine to form a helium-rich natural gas reservoir in an appropriate trap position (Figure 9).

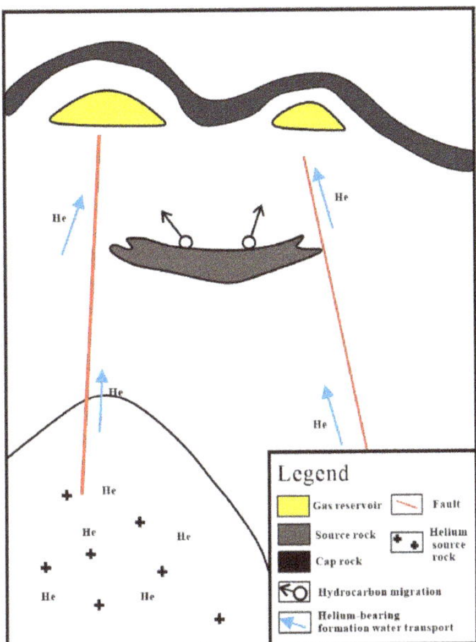

Figure 8. Helium enrichment model in western China.

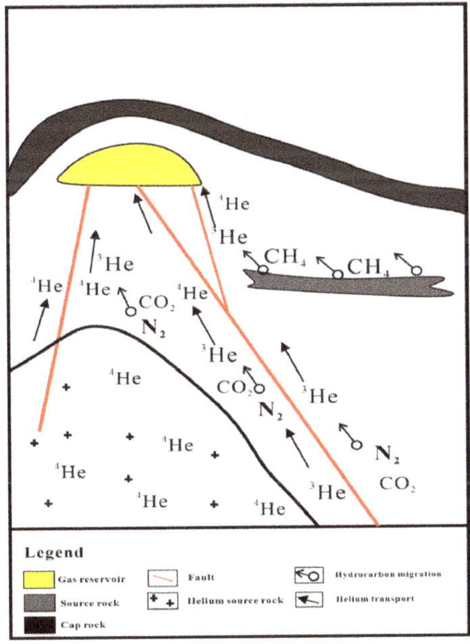

Figure 9. Helium enrichment model in eastern China.

4.3. Prediction of Favorable Areas for Helium Exploration in China

Based on the distribution of helium source rocks, migration channels, and cap rocks in China, we have predicted the favorable areas for helium exploration in China (Figure 10).

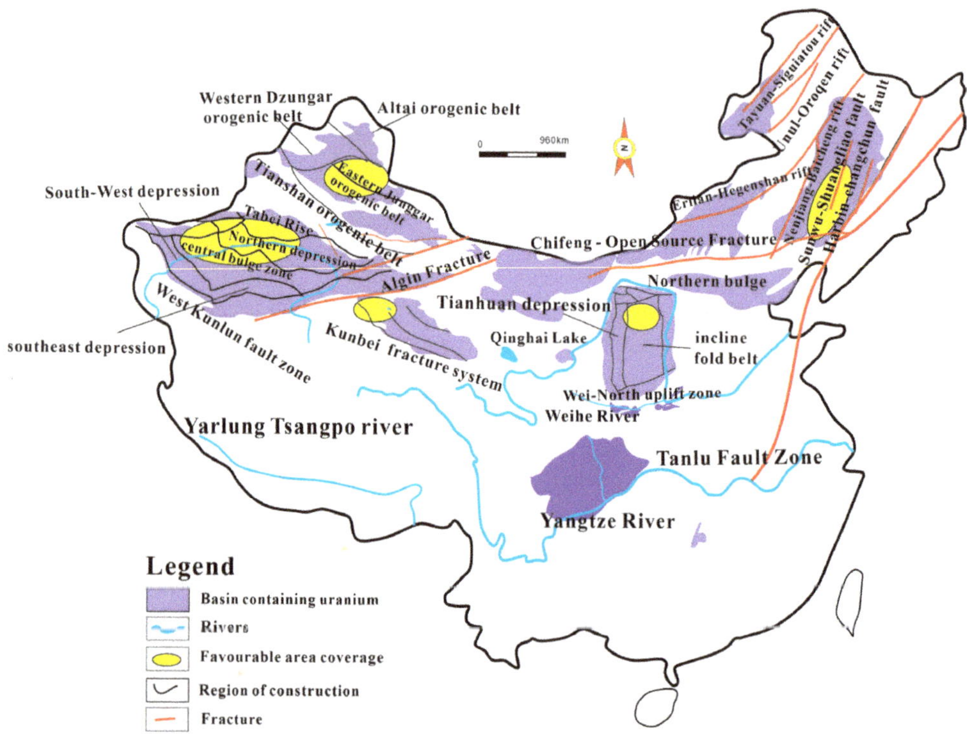

Figure 10. Distribution of Favorable Areas for Helium Resources in China.

Permian helium source rocks are widely distributed in the Tarim Basin, appearing in areas such as Shaya, Tangguba, Shuntuoguole, and the Kongque River. In the northern part of the Tarim Basin, the Yakela structure is covered by the gypsum-mudstone segments of the Mesozoic and Cenozoic periods, which constitute the best cap rocks. The Baomashaya Uplift area in the northern Tarim Basin has good helium sources, various structural units, and cap rocks, providing favorable conditions for helium enrichment and accumulation. Therefore, it is a favorable area for helium exploration in the Tarim Basin.

Helium source rocks are well developed in both the eastern and southern parts of the Junggar Basin, mainly from the Permian and Carboniferous systems, especially in the areas of Kamsut and Kuposu. The Junggar Basin has complex tectonic activity and a NNW trending basement-deep fault zone, providing good conduit systems. Overlying the basement is Cretaceous mudstone, which forms a lithologic trap that causes the natural gas carrying the helium gas to accumulate. The Junggar Basin has excellent conditions for the generation of helium gas and has high exploration potential.

In the frontal zone of the southern Altyn Mountains in the southwestern margin of the Qaidam Basin, there are widely developed helium source rocks. The Dongping-Niudong Slope Belt has multiple north–south deep faults that serve as effective channels for the vertical transport of the helium-bearing fluid from the deep to the shallow layers. Moreover, regional gypsum salt rocks have been discovered in the northwest part of the basin, which consist mainly of gypsum-mudstone and gypsum-bearing mudstone, with a small amount of pure gypsum salt rock layers, forming excellent cap rocks. We believe that the western part of the basin is an area of excellent potential for helium gas.

The Ordos Basin experienced a large-scale granite intrusion in the Paleoproterozoic era, and the Paleoproterozoic granite in the northern region is a good source rock. There are several faults that penetrate the basement in the basin, and fractures mainly developed from the Jurassic to the early Cretaceous period, providing good conduit systems for the transport of helium from the basement. A thick layer of mudstone has developed in the northern part of the Ordos Basin, forming a high-pressure mudstone cap rock and providing conditions for the accumulation of helium. This indicates that the northern part of the basin has great potential for helium exploration and should be given special attention.

The genesis of helium gas reservoirs in eastern China is mostly from a crust-mantle hybrid source, and this method leads to errors in predicting favorable helium areas. The Songliao Basin is a continental extensional basin with active faults that have developed beneath the depression, and numerous helium sources have been discovered in this basin. A Neogene mudstone layer has developed as a cap rock in the southeastern part of the basin. Many faults have developed in the central depression and southeastern uplift zone of the basin, extending to the basement and deep mantle. These faults have a good transport capacity for helium gas and are also an important reason for the crust-mantle hybrid genesis. The central and southern parts, the southern margin, and the eastern margin of the Songliao Basin are favorable areas for helium gas accumulation.

5. Discussion

The validation method used in this paper is to confirm whether the location of the discovered helium natural gas reservoirs is within the favorable areas predicted by our method. If the location of the discovered helium natural gas reservoirs is within the predicted areas, it indicates that our method is feasible (Table 3).

Table 3. Chinese Helium-Bearing Oil and Gas Reserve Statistical Table.

Basin	Field/Regions	Stratum	Helium Content (%)	Resource Volume (m^3)	Reference
Tarim	Donghetang, Hudson, Tazhong	C	0.16–2.19%	1.959×10^8 m^3	Zhang et al. (2005) [41]
Qaidam	Qunjishan, Tuanyushan	J	0.47–1.14%	Projected 4.000×10^8 m^3	Zhang et al. (2020) [28]
Junggar	Central Depression, Southern Depression	J	About 0.2%	1.172×10^8 m^3	Xu et al. (2016) [42]
Ordos	Dongsheng, Sugeli	T, P, J	0.016–0.478%	2.444×10^8 m^3	Liu et al. (2007) [43]
Songliao	Wanjinta, Shuangcheng-Taiping chuan	Mz	0.1.2–0.404%	5.588×10^8 m^3	Zhao et al. (2023) [44]

The Donghetang Oilfield in the Tarim Basin and the Hudson Oilfield have relatively a high helium content in their Carboniferous crude oil associated gas, reaching 0.40% and 0.219%, respectively [41]. The Triassic reservoirs in the Lunan and Jiefangqu oil and gas fields have a helium content of up to 0.93%. The helium content in the Carboniferous deposits in the Tazhong region ranges from 0.16% to 0.23%, while the overall helium content in the Ordovician and Silurian strata is low. The estimated helium reserve in the Tarim Basin is 1.959×10^8 m^3.

The Qunjishan Gas Field in the Qaidam Basin has a tested helium content of up to 1.10%. An analysis of the dark mud shale interval of the Jurassic Daheishan Formation in the Tuanyushan Gas Field shows a helium content ranging from 0.47% to 1.14%. The estimated helium reserve in the Qaidam Basin is 4×10^8 m^3.

The helium content in the central and southern depressions of the Junggar Basin is around 0.2%, specifically 0.203% and 0.236%, mainly found in the Jurassic strata. The estimated helium reserve in the Junggar Basin is 1.172×10^8 m^3 [42].

The Dongsheng Gas Field and Sulige Gas Field, both located in the central part of the Ordos Basin, have indications of helium. The helium content ranges from 0.045% to 0.478% in the Dongsheng Gas Field and from 0.016% to 0.035% in the Sulige Gas Field. The estimated helium reserve in the Ordos Basin is 2.444×10^8 m^3 [43].

The results of our studies have confirmed the rationality of the predicted favorable areas for helium gas exploration. The research outcomes have important implications for the exploration and development of helium gas resources in China.

6. Conclusions

We have found that the distribution of helium resources is not only uneven globally, but also imbalanced within a country. The formation process of helium is due to the decay of the U-element. In China, there are two main types of helium formation, one is the crust-mantle mixture source and the other is the crustal source. The distribution of these sources has distinct characteristics, with the mixture source located in the east and the crustal source located in the west. Based on the analysis of geology, the distribution of helium source rocks, effective transport, and cap rock distribution, we have summarized two enrichment modes. One mode involves multi-source enrichment in the east due to the crust-mantle mixture, and the other involves crustal enrichment in the west due to the accumulation of formations of water and natural gas. We predicted five favorable areas for helium accumulation, namely in the northern Tarim Basin, eastern and southern Junggar Basin, western Qaidam Basin, western Ordos Basin, and southern and eastern Songliao Basin, which were verified by the helium contents in discovered gas fields.

Author Contributions: Investigation, analysis, conceptualization, writing-original draft, Y.X.; supervision, review, S.J.; data processing, methodology, J.Y.; graphical processing, validation Y.D. All authors have read and agreed to the published version of the manuscript.

Funding: This research is funded by the Open Project of State Key Laboratory of Shale Oil and Gas Enrichment Mechanisms and Effective Development (GSYKY-B09-33).

Data Availability Statement: All data and materials are available on request from the corresponding author. The data are not publicly available due to ongoing research using a part of the data.

Conflicts of Interest: The authors declare no conflicts of interest.

References

1. Ruedeman, N.P.; Oles, L.M. Helium-its probable origin and concentration in the Amarillo fold, Texas. *AAPG Bull.* **1929**, *13*, 799–810. [CrossRef]
2. Pierce, A.P.; Gott, G.B.; Mytton, J.W. Uranium and Helium in the Panhandle Gas Field, Texas, and Adjacent Area. US Geological Survey Professional Papers, 1964, 454-G:1-57. Available online: https://pubs.usgs.gov/publication/pp454G (accessed on 1 February 2024).
3. Katz, D.L. *Source of Helium in Natural Gases: USBM Information Circular 8417*; Helium Research Center Library: Amarillo, TX, USA, 1969; pp. 242–255. [CrossRef]
4. Nikonov, V.F. Formation of helium-bearing gases and trends in prospecting for them. *Int. Geol. Rev.* **1973**, *15*, 534–541. [CrossRef]
5. Maione, S.J. Helium exploration: A 21st century challenge. *Houst. Geol. Soc. Bull.* **2004**, *46*, 27–28,30.
6. Gold, T.; Held, M. Helium-nitrogen-methane systematics in natural gases of Texas and Kansas. *J. Pet. Geol.* **2007**, *10*, 415–420. [CrossRef]
7. Ballentine, C.J.; Burnard, P. Production, release, and transport of noble gases in the continental crust. *Rev. Mineral. Geochem.* **2002**, *47*, 481–538. [CrossRef]
8. Broadhead, R.F. Helium in New Mexico: Geological distribution, resource demand, and exploration possibilities. *New Mex. Geol.* **2005**, *27*, 93–100. [CrossRef]
9. Ballentine, C.J.; Lollar, B.S. Regional groundwater focusing of nitrogen and noble gases into the Hugoton-Panhandle giant gas field, USA. *Geochem. Cacochymical Acta* **2002**, *66*, 2483–2497. [CrossRef]
10. Brown, A. Origin of helium and nitrogen in the Panhandle–Hugoton field of Texas, Oklahoma, and Kansas, UnitedStates. *AAPG Bull.* **2019**, *103*, 369–403. [CrossRef]
11. Qin, S.; Yuan, M.; Zhou, Z.; Yang, J. Distribution law of helium in Leshan-Longevism paleo-uplift in Sichuan Basin, China. *IOP Conf. Ser. Earth Environ. Sci.* **2019**, *360*, 012031. [CrossRef]

12. Zhang, W.; Li, Y.; Zhao, F.; Han, W.; Li, Y.; Wang, Y.; Holland, G.; Zhou, Z. Using noble gases to trace groundwater evolution and assess helium accumulation in Weihe Basin, central China. *Geochem. Cosmochim. Acta* **2019**, *251*, 229–246. [CrossRef]

13. USGS. Mineral Commodity Summaries 2022. U.S. Geological Survey. 2022. Available online: http://minerals.usgs.gov/minerals/pubs/commodity/heliumn/;2022-01 (accessed on 3 February 2024).

14. Tao, X.; Li, J.; Zhao, L.; Li, L.; Zhu, W.; Xing, L.; Su, F.; Shan, X.; Zheng, H.; Zhang, L. Current status of helium resources in China and the discovery of the first extra-large helium-rich reserve: Hotanhe gas field. *Earth Sci.* **2019**, *44*, 1024–1041. [CrossRef]

15. Lueth, V.; Lucas, S.G.; Chamberlin, R.M. (Eds.) Geology of Geology of the Chupadera Mesa. In *The 60th Annual Fall Field Conference Guidebook*; New Mexico Geological Society, New Mexico Institute of Mining & Technology: Socorro, NM, USA, 2009; pp. 359–374. [CrossRef]

16. Danabalan, D. Helium: Exploration Methodology for a Strategic Resource. Ph.D. Thesis, Durham University, Durham, UK, 2017.

17. Akutsent, V.P. World helium resources and the perspectives of helium industry development. *Pet. Geol.* **2014**, *9*, 11. [CrossRef]

18. Zhao, H.X.; Zhang, Y.; Li, C.L. Analysis of global helium supply and price system. *Chem. Propellants Polym. Mater.* **2012**, *10*, 91–96. [CrossRef]

19. Wang, X.; Liu, W.; Li, X.; Tao, C.; Borjigin, T.; Liu, P.; Luo, H.; Li, X.; Zhang, J. Application of noble gas geochemistry to the quantitative study of the accumulation and expulsion of lower Paleozoic shale gas in southern China. *Appl. Geochem.* **2022**, *146*, 105446. [CrossRef]

20. Feng, W.F. Economic benefit analysis of helium extraction from natural gas in Weiyuan. *Nat. Gas Ind.* **1989**, *9*, 69–71.

21. Zhang, L.L.; Sun, Q.G.; Liu, Y.Y.; Qiu, X.L.; Jiang, Y.M. Global market of helium and suggestions for helium supply security in China. *Cryog. Spec. Gas* **2014**, *32*, 106935.

22. Yu, Q.X.; Shi, Z.; Wang, D.G. Analysis on helium enrichment characteristics and reservoir forming condition in Northwest Tarim Basin. *Northwest. Geol.* **2013**, *46*, 215–222.

23. Liu, Q.Y.; Dai, J.X.; Jin, Z.J.; Li, J. Geochemical and genesis of natural gas in the foreland and platform of the Tarim Basin. *J. Geol.* **2009**, *83*, 107–114.

24. Xu, Y.C. Helium isotope distribution and tectonic environment in natural gas. *Geol. Front.* **1997**, *4*, 185–190.

25. Dai, J.; Ni, Y.; Qin, S.; Huang, S.; Gong, D.; Liu, D.; Feng, Z.; Peng, W.; Han, W.; Fang, C. Geochemical characteristics of He and CO_2 from the Ordos (cratonic) and Bohaibay (rift) basins in China. *Chem. Geol.* **2017**, *469*, 192–213. [CrossRef]

26. Cao, Z.; Che, Y.; Li, J.; Li, H. Accumulation analysis on a helium-enriched gas reservoir in Huagou area, the Jiyang depression. *Pet. Exp. Geol.* **2001**, *23*, 395–399.

27. Guo, N.; You, X.; Xu, J. Geological characteristics and prospect of helium-bearing natural gas exploration in the Xiqiao helium-bearing natural gas field, North Jiangsu Basin. *Pet. Explor. Dev.* **1999**, *26*, 24–26.

28. Zhang, X.; Zhou, F.; Cao, Z.; Liang, M. Finding of the Dongping economic Helium gas field in the Qaidam Basin, and Helium source and exploration prospect. *Nat. Gas Geosci.* **2020**, *31*, 1585–1592.

29. Dai, J.X. The formation period and gas source of Weiyuan gas field. *Pet. Exp. Geol.* **2003**, *25*, 473–480. [CrossRef]

30. He, F.Q.; Wang, G.; Wang, J.; Zou, Y.; An, C.; Zhou, X.; Ma, L.; Zhao, Y.; Zhang, J.; Liu, D.; et al. Helium distribution pattern of Dongsheng gas field in Ordos Basin and discovery of extra-large helium-rich field. *Pet. Exp. Geol.* **2022**, *44*, 1–10.

31. Zhang, Y.P.; Li, Y.H.; Lu, J.; Li, Y.; Song, B.; Guo, W. Discovery of helium-rich natural gas in the northern margin of the Qaidam Basin—Another consideration of geological conditions of formation. *Geol. Bull.* **2016**, *35*, 364–371.

32. Han, W.; Liu, W.; Li, Y.; Zhou, J.; Zhang, W.; Zhang, Y.; Chen, X. Characteristics of rare gas isotopes and main controlling factors of helium enrichment in the northern margin of the Qaidam Basin, China. *J. Nat. Gas Geosci.* **2020**, *5*, 299–306. [CrossRef]

33. Xu, S.; Nakai, S.I.; Wakita, H.; Xu, Y.; Wang, X. Helium isotope compositions in sedimentary basins in China. *Appl. Geochem.* **1995**, *10*, 643–656. [CrossRef]

34. Feng, Z.H.; Huo, Q.L.; Wang, X. A syudy of helium reservolr formation characteristic in the north part of Songliao Basin. *Nat. Gas Ind.* **2001**, *21*, 27–30. [CrossRef]

35. Sun, M.L.; Chen, T.F.; Liao, Y.S. Helium isotope characteristics, genesis of CO2 in natural gases and distribution of tetradymite in the Jiyang depression. *Geochemistry* **1996**, *25*, 475–480. [CrossRef]

36. Wan, T.F.; Zhu, H.; Zhao, L. Formation and evolution of Tancheng-Lujiang fault zone: A review. *Geoscience* **1996**, *10*, 159–168. Available online: https://cpfd.cnki.com.cn/Article/CPFDTOTAL-ZGDW199810001329.htm (accessed on 26 February 2024).

37. Zhu, G.; Liu, C.; Gu, C.; Zhang, S.; Li, Y.; Su, N.; Xiao, S. Oceanic plate subduction history in the western Pacific Ocean: Constraint from Late Mesozoicevolution of the Tan-Lu fault zone. *Sci. China Earth Sci.* **2018**, *48*, 415–435. [CrossRef]

38. Sathaye, K.J.; Larson, T.E.; Hesse, M.A. Noble gas fractionation during subsurface gas migration. *Earth Planet. Sci.* **2016**, *450*, 1–9. [CrossRef]

39. Qin, S.; Li, J.; Liang, C.; Zhou, G.; Yuan, M. Mechanism of helium enrichment in helium-rich reservoirs in west-central China—Helium enrichment by de-heating of ancient formation water. *Nat. Gas Geosci.* **2022**, *33*, 1203–1217.

40. Zhang, C.; Guan, P.; Zhang, J.; Song, D.; Ren, J. Characteristics of helium resource zoning and reservoir formation pattern in China. *Nat. Gas Geosci.* **2023**, *34*, 656–671. [CrossRef]

41. Zhang, D.W.; Liu, W.; Zheng, J. Helium and argon isotope compositions of natural gas in Tazhong area, Tarim Basin. *Pet. Explor. Dev.* **2005**, *32*, 38–41.

42. Xu, S.; Zheng, G.; Zheng, J.; Zhou, S.; Shi, P. Mantle-derived helium in foreland basins in Xinjiang, NorthwestChina. *Tectonophysics* **2017**, *694*, 319–331. [CrossRef]

43. Liu, Q.Y.; Liu, W.H.; Xu, Y.C.; Li, J.; Chen, M.J. Geochemistry of natural gas and crude computation of gas-generated contribution for various source rocks in Sulige Gas Field, Ordos Basin. *Nat. Gas Geosci.* **2007**, *18*, 697–702.
44. Zhao, H.; Liang, K.; Wei, Z. Differential enrichment of helium-rich gas reservoirs in Songliao Basin and favorable area forecast. *Nat. Gas Geosci.* **2023**, *34*, 628–646.

 *energies*

Article

Pre-Stack Fracture Prediction in an Unconventional Carbonate Reservoir: A Case Study of the M Oilfield in Tarim Basin, NW China

Bo Liu [1,2], Fengying Yang [1,2], Guangzhi Zhang [1,*] and Longfei Zhao [3]

1   School of Geosciences, China University of Petroleum (East China), Qingdao 266580, China; liubo.swty@sinopec.com (B.L.); yangfy.swty@sinopec.com (F.Y.)
2   SINOPEC Geophysical Research Institute Co., Ltd., Nanjing 211103, China
3   PetroChina Tarim Oilfield Company, Korla 841000, China
*   Correspondence: zhanggz@upc.edu.cn

Abstract: The reservoir of the M oilfield in Tarim Basin is an unconventional fracture-cave carbonate rock, encompassing various reservoir types like fractured, fracture-cave, and cave, exhibiting significant spatial heterogeneity. Despite the limited pore space in fractures, they can serve as seepage pathways, complicating the connectivity between reservoirs. High-precision fracture prediction is critical for the effective development of these reservoirs. The conventional post-stack seismic attribute-based approach, however, is limited in its ability to detect small-scale fractures. To address this limitation, a novel pre-stack fracture prediction method based on azimuthal Young's modulus ellipse fitting is introduced. Offset Vector Tile (OVT) gather is utilized, providing comprehensive information on azimuth and offset. Through analyzing azimuthal anisotropies, such as travel time, amplitude, and elastic parameters, smaller-scale fractures can be detected. First, the original OVT gather data are preprocessed to enhance the signal-to-noise ratio. Subsequently, these data are partially stacked based on different azimuths and offsets. On this basis, pre-stack inversion is carried out for each azimuth to obtain the Young's modulus in each direction, and, finally, the ellipse fitting algorithm is used to obtain the orientation of the long axis of the ellipse and the ellipticity, indicating the fracture orientation and density, respectively. The fracture prediction results are consistent with the geological structural features and fault development patterns of the block, demonstrating good agreement with the imaging logging interpretations. Furthermore, the results align with the production dynamics observed in the production wells within the block. This alignment confirms the high accuracy of the method and underscores its significance in providing a robust foundation for reservoir connectivity studies and well deployment decisions in this region.

Keywords: fracture prediction; OVT; azimuthal anisotropy; unconventional carbonate reservoir; Tarim Basin

Citation: Liu, B.; Yang, F.; Zhang, G.; Zhao, L. Pre-Stack Fracture Prediction in an Unconventional Carbonate Reservoir: A Case Study of the M Oilfield in Tarim Basin, NW China. *Energies* **2024**, *17*, 2061. https://doi.org/10.3390/en17092061

Academic Editor: Pavel A. Strizhak

Received: 15 March 2024
Revised: 19 April 2024
Accepted: 19 April 2024
Published: 26 April 2024

1. Introduction

Fractures are widely distributed underground, and they can serve as both oil and gas storage spaces and percolation channels for oil and gas migration, playing an important role in the formation and distribution of oil and gas reservoirs. Especially in recent years, with the continuous discoveries of unconventional reservoirs, such as shale, tight sandstone, and complex carbonate rocks, fracture prediction has become a hot topic for geophysical researchers.

The M oilfield is located in the Tabei uplift of the Tarim Basin and is a fractured-vuggy carbonate oil reservoir with an anticline structural background. Overall, the oil–water interface is relatively uniform, but it also exhibits strong "one cave, one reservoir" heterogeneity. Studies have suggested that, due to the influence of tectonic and fault evolution and karstification, fractured, fractured-vuggy, and cave reservoirs have generally developed in the

M oilfield. Although the fractures are usually small in scale and have limited pore space, they have a significant impact on the connectivity between oil reservoir units. Therefore, high-precision fracture prediction is crucial for the later development and adjustment of the oil reservoir. At present, there are numerous seismic fracture prediction methods in the geophysical field. Based on the type of reflected wave, they can be divided into shear wave fracture prediction methods and compressional wave fracture prediction methods. Of these, the shear wave or converted shear wave methods, which exploit the shear wave splitting phenomenon induced by fractures, are theoretically the most direct and efficient approach for fracture prediction [1–5]. However, despite its theoretical superiority, the application of this type of method has not become widespread due to the stringent requirements it imposes on field data acquisition and seismic processing, particularly in multi-wave or multi-component exploration settings. This has limited its utilization as a mainstream fracture prediction method. Utilizing compressional waves for fracture prediction is currently a more commonly used approach among major oil and gas fields. Based on the source data and methodological principles employed, it can be categorized as post-stack and pre-stack fracture prediction. For post-stack fracture prediction, commonly used methods involve calculating geometric attributes such as coherence, curvature, and variance from post-stack seismic data. The fundamental principle is that, when faults and fractures develop underground, seismic events may appear discontinuous or curved. Detecting these geological features helps identify fractures. To highlight the seismic response characteristics of fractures and enhance the accuracy of post-stack attribute fracture prediction, numerous scholars have conducted extensive research on seismic data preprocessing. This includes techniques such as structure-oriented filtering [6–12], seismic spectral decomposition processing [13–16], and curvelet transform multi-scale decomposition [17]. Concurrently, research on attributes such as maximum likelihood, ant tracking, and AFE has further diversified the methods for post-stack attribute fracture prediction [18–27]. However, different attributes have their own advantages and disadvantages in characterizing various types and scales of fractures. A single attribute is often insufficient for comprehensively representing the development patterns of fractures. Therefore, methods utilizing the fusion of multiple attributes have been proposed to improve the accuracy of post-stack attribute fracture prediction [28,29]. Nevertheless, due to the limitations of seismic resolution, post-stack attribute fracture prediction is primarily suitable for identifying large-scale fractures associated with structures or faults, and the prediction results cannot be quantified. In recent years, with the promotion and application of wide-azimuth seismic acquisition techniques and the advancements in OVT processing techniques, pre-stack fracture prediction based on OVT-domain seismic data has become a research hotspot [30,31]. The primary advantage of this methodology is its ability to detect fractures of smaller scale and quantitatively delineate their orientation and density, thereby providing significant technical assistance in oil and gas exploration and development. In terms of the theoretical study for fracture prediction, the azimuthal anisotropy characteristics of physical quantities such as travel time, velocity, amplitude, frequency, and phase exhibited by compressional waves (P-waves) propagating through fractured media play a pivotal role [32–34]. Grechka et al. [35] derived the travel time expression for compressional waves propagating in HTI media, revealing that travel time exhibits periodic variations with azimuth in HTI media. Rüger et al. [36–38] conducted a study based on the theory of weak anisotropy and derived formulas for the azimuthal variation of compressional wave reflection coefficients in anisotropic media, laying a theoretical foundation for predicting fractures using dynamic parameters such as amplitude. Mallick et al. [39] observed a cosine-like variation in seismic amplitudes with respect to azimuth in fractured strata, enabling the accurate indication of fracture orientation. These theoretical studies have provided crucial guidance for subsequent fracture prediction practices. In practical field applications, numerous scholars have successfully utilized the azimuthal anisotropy characteristics of compressional waves to predict fractures. Qu et al. [40] introduced a quantitative approach for predicting fracture orientation and density through analyzing the variation in P-wave impedance with azimuth, and they successfully applied it in practical field blocks. Li et al. [41] proposed an anisotropic gradient inversion method based on zero constraints, utilizing the maximum likelihood solution of anisotropic gradients to predict fracture development. The prediction results match well with the drilling data. Zhang et al. [42] simplified and linearized the Rüger approximation, obtaining anisotropic strength parameters through maximum likelihood inversion. Model tests demonstrated that the algorithm exhibited good noise resistance. Wang et al. [43] conducted a comprehensive prediction of fractures in practical field blocks by utilizing travel time anisotropy and AVO gradient attributes. Zhou et al. [44] conducted an analysis of amplitude azimuthal anisotropy and frequency azimuthal anisotropy based on OVT-domain gather data and predicted the intensity and orientation of small-scale fractures. Chen et al. [45] introduced a statistics-based anisotropic strength prediction technique, achieving good results in predicting fractured reservoirs in the subsurface of the Bongor Basin in Chad. These application cases fully demonstrate the practical value of the azimuthal anisotropy of P-waves in fracture prediction. Moreover, in addition to directly utilizing the azimuthal anisotropy characteristics of P-waves, the Young's modulus—an essential physical parameter for evaluating rock brittleness and fracturing capability—also plays a significant role in fracture prediction [46–48]. Sayers [49] conducted studies and revealed the anisotropy of the Young's modulus in fractured media. Specifically, the Young's modulus along the axis of symmetry was consistently smaller than that along the strike direction of the fractures. Zong et al. [50] and Wang et al. [51] used pre-stack inversion to obtain the Young's modulus and ellipse fitting to predict fractures, and they achieved good results. These studies provide novel ideas and methodologies for utilizing Young's modulus in fracture prediction.

High-angle single-group fractures are extensively developed within the target layer of the M oilfield in the study area, serving as significant permeability channels and storage spaces for hydrocarbons within the block. For this study, we first optimized the pre-tack OVT gather data to suppress random noise, eliminate travel time disparities, and formulate an appropriate stacking scheme based on azimuth angle and offset. Subsequently, we applied pre-stack inversion techniques to the processed pre-stack OVT data to obtain the Young's modulus for each azimuth. Then, ellipse fitting was carried out, and the ellipticity and direction of the major axis obtained from the fitting indicated the density and orientation of the fractures, thus realizing the pre-stack fracture prediction of the M oilfield in Tarim Basin, facilitating the subsequent enhanced oil recovery study of unconventional carbonate reservoirs [52,53].

2. Geological Setting

The primary exploration and development stratum in the study area is the Yijianfang Formation of the Ordovician, exhibiting a nearly northeast–southwest trending anticline structure. This is a typical fractured-vuggy carbonate reservoir (Figure 1). The study area has experienced multi-stage tectonic movements, leading to the development of diverse fault types and an intricate faulting pattern. Within the Ordovician strata, three primary groups of faults are particularly prominent. The first group comprises strike-slip faults and their associated branch faults, which were formed during the middle Caledonian. These faults trend northwest and are arranged in a nearly parallel fashion within the study area. The second group of faults was mainly developed in the Hercynian period, which was affected by the lateral intrusion of the Mana igneous rock on the northwest side of the study area into the Middle Cambrian plastic strata and was formed by the arching of the supra-salt strata. This group of normal faults presents a northwest–southeast radial distribution pattern on the plane. Under the influence of tensile stress, this group of normal faults developed numerous fractures, thereby improving the quality of the reservoir and further enhancing the connectivity of the fractured-vuggy reservoir. The third group consists of thrust faults that formed during the late Hercynian to early Indosinian, which control the formation of the NE-trending asymmetric long-axis anticline of the M oilfield.

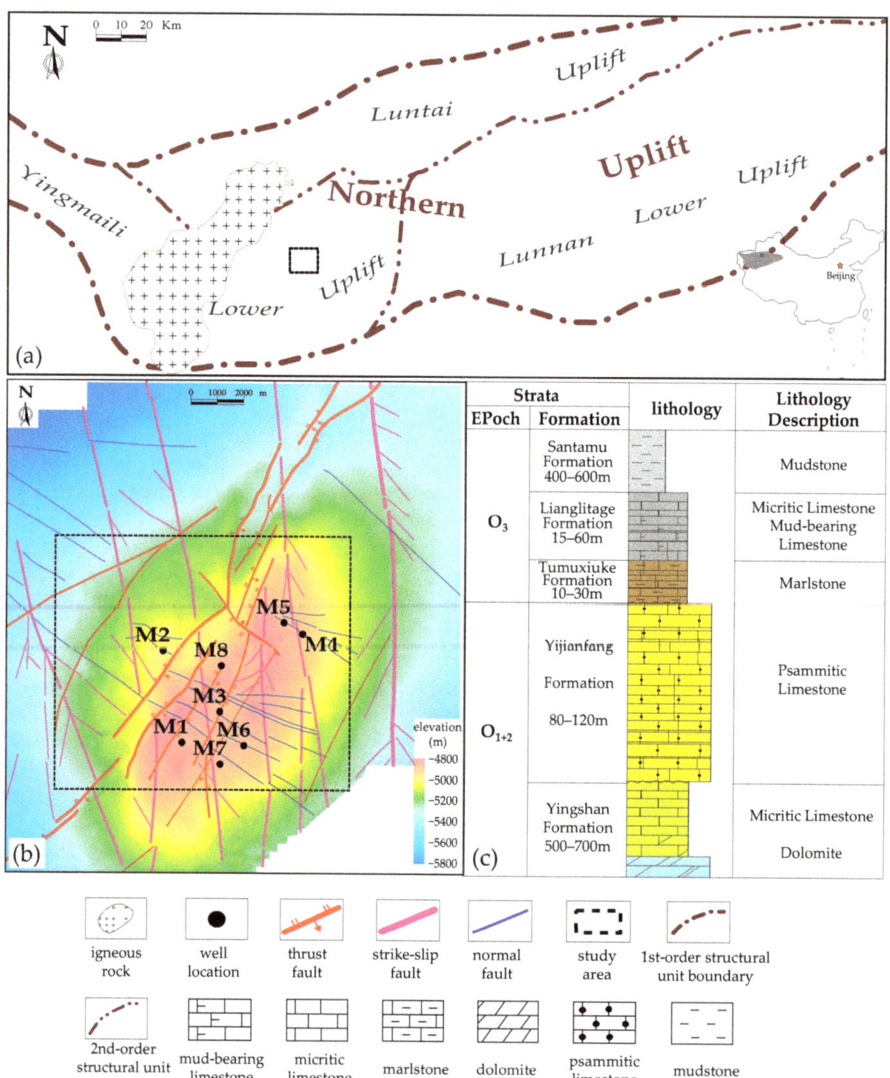

Strata		lithology	Lithology Description
EPoch	Formation		
O₃	Santamu Formation 400–600m		Mudstone
	Lianglitage Formation 15–60m		Micritic Limestone Mud-bearing Limestone
	Tumuxiuke Formation 10–30m		Marlstone
O₁₊₂	Yijianfang Formation 80–120m		Psammitic Limestone
	Yingshan Formation 500–700m		Micritic Limestone Dolomite

Figure 1. (**a**) Structural location map of the study area; (**b**) structural map of the top surface of the Yijianfang Formation in the target layer; (**c**) stratigraphic column of the Ordovician series.

The sedimentary sequence of the Ordovician strata in the M oilfield is relatively complete, with the development of the Santamu Formation, Lianglitage Formation, Tumuxiuke Formation, Yijianfang Formation, and Yingshan Formation in sequential order. Among them, the Yijianfang Formation and the upper part of the Yingshan Formation are the main reservoir development intervals. The lithology of the Yijianfang Formation is mainly composed of sparry oolitic limestone and sparry bioclastic limestone. The lithology of the upper part of the Yingshan Formation is dominated by micritic limestone and psammitic limestone (Figure 1c). The formation of high-quality reservoirs in the Yijianfang–Yingshan Formation is primarily controlled by the combined effects of high-energy facies belts, faulting, and karstification [54–56]. The reservoir spaces are dominated by fractures and small-scale dissolved pores and fractures, with a small number of cave-type reservoirs. The rock matrix has poor physical properties, with porosity mostly less than 1%, mainly concentrated around 0.4% to 0.7%, and an average porosity of 0.5%. The permeability is generally less than 0.1 mD. Drilling core and thin section observations reveal the presence of high-angle fractures and dissolved pores developed along these fractures (Figure 2). Based on the interpretation results of micro-resistivity imaging logging in the study area, fractures in this region are primarily composed of single-set high-angle oblique and vertical fractures. The fracture widths generally range from 0.01 to 0.1 mm. Nearly 52% of the fractures are filled, primarily with calcite. However, the overall fracture aperture is relatively good, with an average aperture of 77 μm [57]. From the regression relationship between fracture aperture, permeability, and single-well production, it is evident that single-well production positively correlates with both factors (Figure 3). Therefore, high-precision fracture prediction is very important for efficient well deployment and later reservoir development technology policy formulation.

Figure 2. Core photos and a thin section of the target layer of typical wells in the study area: (**a–c**) core photos of M1, M2, and M3 wells, respectively; (**d**) thin section of M8 well. M1: residual dissolution pores developed along high-angle fractures; M2: crude oil extravasation along high-angle unfilled fractures; M3: crude oil extravasation along vertical unfilled fractures; M8: bright crystal psammitic limestone with visible fractures.

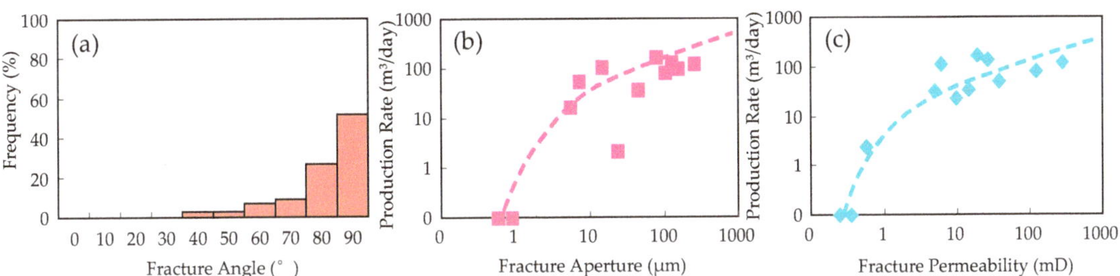

Figure 3. (**a**) Statistical diagram of fracture angle; (**b**) relationship between fracture development and production; and (**c**) relationship between fracture permeability and production in the study area. The scatter points presented in (**b**) depict the measured fracture aperture and production rate data for a well within the study area. Analogously, the scatter points in (**c**) illustrate the measured fracture permeability and production rate data for a well (potentially the same or a different one) within the study area. The dashed curves approximate the underlying trends within these measurements.

3. Data and Methods

3.1. Data Set

In the study area, high-density 3D seismic data were acquired with a wide azimuth, a bin size of 15 m $\times$ 15 m, full coverage of 320 times, and a shot-trace density of 1.4222 million traces per square kilometer. The data exhibited a favorable aspect ratio of 0.8, indicating good spatial coverage. Furthermore, the data quality was excellent, having undergone targeted amplitude-preserving and fidelity-enhancing processing in the OVT domain. This ensured the data's suitability for subsequent pre-stack fracture prediction. After preprocessing, the OVT gather data were fully stacked and partially stacked based on azimuth and incident angles, generating multiple sets of 3D seismic data volumes.

The fully stacked seismic data volume Is employed for conducting detailed horizon interpretation, from which post-stack attributes are extracted to facilitate the analysis of fault systems and reservoir development characteristics. On the other hand, the partially stacked data volumes are utilized for azimuthal pre-stack reservoir inversion, serving as a basis for estimating the Young's modulus and fracture prediction. To comprehensively assess the accuracy of the fracture prediction results, imaging logging interpretation data from four wells, along with drilling, logging, and production data from one well, were collected. Subsequently, these data sets were rigorously compared and analyzed with the prediction results, ensuring robust validation of the methodologies and enhancing the reliability of the predictions.

3.2. Methods

3.2.1. Azimuthal Young's Modulus Calculation

Geophysicists generally agree that the anisotropy of media in the Earth's crust is primarily caused by oriented fractures and thin interbeds. Specifically, the formation of HTI media (horizontally transverse isotropic media) is often associated with vertically aligned and parallel fractures. Based on the results of core observation and imaging logging statistics from the M oilfield, it is evident that the block primarily exhibits the development of high-angle fractures, so it can be approximately regarded as HTI media. The formula for calculating the equivalent Young's modulus of HTI media is as follows [50]:

$$E = \rho v_s^2 \left(3v_p^2 - 4v_s^2\right) / \left(v_p^2 - v_s^2\right) \tag{1}$$

where ρ represents density, v_s represents shear wave velocity, and v_p represents compressional wave velocity.

To calculate the Young's modulus at any given azimuth, pre-stack inversion techniques are applied to obtain density and compressional and shear wave velocities. These values are then substituted into Formula (1) so as to obtain the Young's modulus of the corresponding azimuth.

3.2.2. Principle of Fracture Prediction Based on Elliptical Fitting of Young's Modulus

Sayers' experimental research revealed the directional characteristics of the Young's modulus in fractured media: along the direction of the fracture, the Young's modulus reaches its maximum, whereas it reaches its minimum along the symmetrical axis of the fracture [49]. In terms of theoretical research, Zong et al. [50] constructed a series of layered models with varying fracture densities based on Thomsen's fracture theory. Through model analysis, it was found that the Young's modulus exhibits periodic fluctuations similar to a cosine curve as the azimuth angle changes. This variation pattern is consistent with Sayers' experimental results. Furthermore, when the Young's modulus is projected onto a polar coordinate system based on the azimuth angle, it exhibits an elliptical shape. The long axis of the ellipse aligns with the orientation of the fracture, and as the fracture density increases, the ellipse becomes flatter. Based on this pattern, the equivalent Young's modulus at different azimuths can be calculated by using Formula (1), and subsequently, the ellipticity and the direction of the ellipse's long axis can be obtained through ellipse fitting algorithms such as least squares. This approach enables not only quantitative prediction of the orientation of fractures but also qualitative assessment of their density.

3.2.3. Fracture Prediction Workflow

The research is completed in five steps (Figure 4).

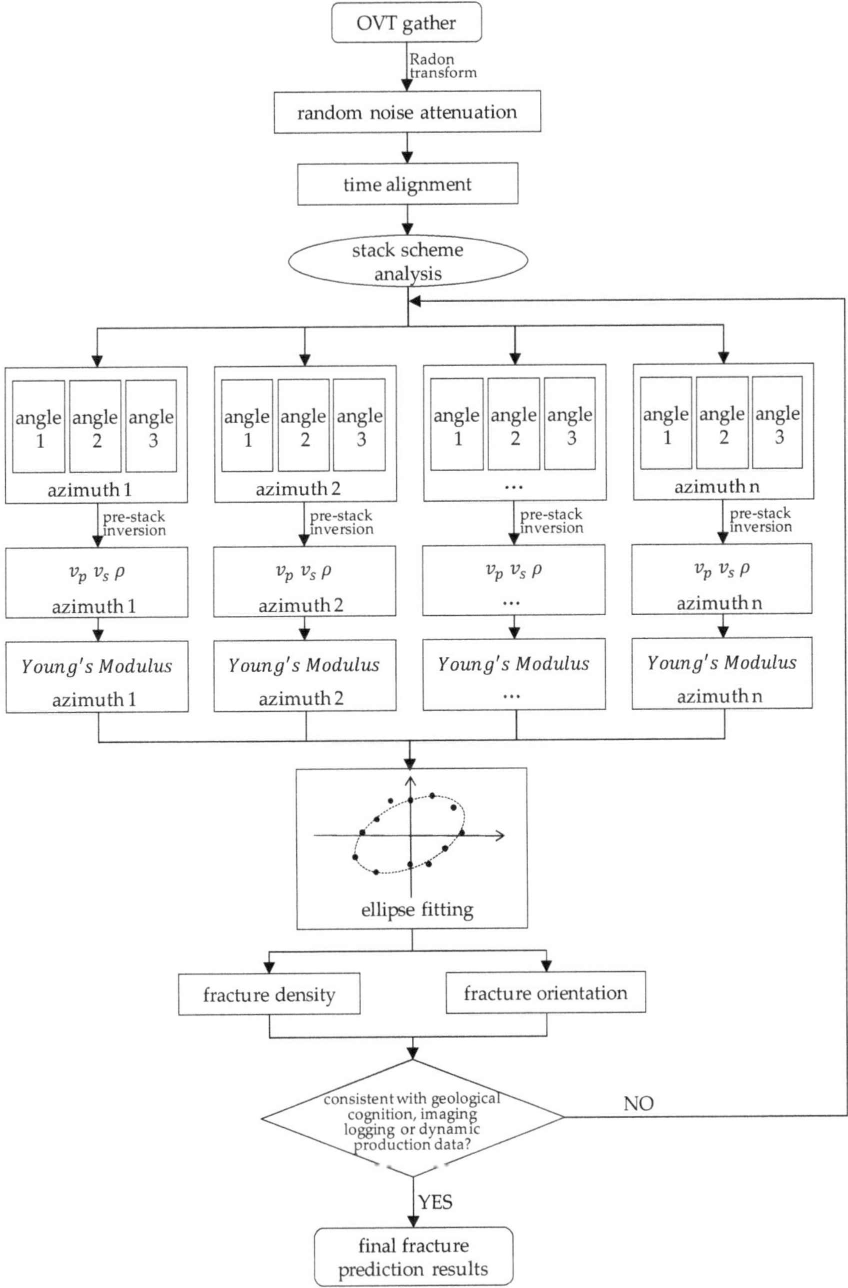

Figure 4. Workflow of pre-stack fracture prediction based on OVT gather.

Step 1: Initially, the original OVT gather data are optimized, which mainly includes two pivotal steps: random noise attenuation and anisotropic time alignment. This optimization aims to enhance the signal-to-noise ratio of the data, mitigate the travel time disparities arising from anisotropy, maintain azimuthal amplitude variations, and ensure the robustness of pre-stack elastic parameter inversion and Young's modulus ellipse fitting. Consequently, this process facilitates the generation of reliable fracture prediction outcomes.

Step 2: The OVT gather data are analyzed to formulate a reasonable stacking scheme, followed by partial stacking based on azimuth and incident angle (offset).

Step 3: Utilizing pre-stack inversion techniques, the data obtained from each azimuth in Step 2 are processed to derive Young's modulus data volumes for each azimuth.

Step 4: A least-squares ellipse fitting algorithm is utilized to individually fit each trace and sample within the Young's modulus data volumes derived from Step 3. The orientation of fractures is represented by the long axis of the fitted ellipse, while the ellipticity provides an indication of the fracture density.

Step 5: The fracture prediction results obtained from Step 4 are compared and validated against geological cognition, imaging logging data, and dynamic production data of the block. If the fracture prediction results do not align, return to Step 3 for modifications. Otherwise, output the final fracture prediction results.

4. Results and Discussion

4.1. OVT Gather Optimization and Processing, and Stacking Scheme Analysis

One of the key steps in pre-stack fracture prediction is to perform azimuthal stacking of the OVT gather data, followed by stacking based on incident angles for each azimuth. Although the OVT gathers have been denoised during the seismic processing, compared with the full stack data, partial stack data are more sensitive to noise. Therefore, it is necessary to optimize the original OVT gather data to improve the signal-to-noise ratio. Figure 5 shows the typical OVT gather data in the study area. Figure 5b shows the original gather record. It can be seen from the extracted observation system that the OVT gather contains both azimuth and offset information. These data are systematically organized in a spiral pattern, as shown in Figure 5a. From Figure 5b, it can be seen that there are obvious random noise and residual time differences in the original data, which need to be further optimized. Firstly, the Radon transform is applied to attenuate the random noise, as depicted in Figure 5c. Analysis of the residual profile in Figure 5d indicates that the denoising process preserves most of the effective signals while effectively eliminating random noise components. Subsequently, time alignment is conducted on the denoised gather data to mitigate travel time variations arising from velocity anisotropy while maintaining amplitude differences associated with azimuth, as illustrated in Figure 5e.

As the fundamental elliptic equation encompasses five variables, it is imperative to input at least five azimuthal Young's modulus data points for elliptic fitting. Given the central symmetry of the OVT gather, it is partitioned into six azimuthal sectors, as depicted in Figure 6a: 0° to 30°, 30° to 60°, 60° to 90°, 90° to 120°, 120° to 150°, and 150° to 180°. The selection of the maximum offset must strike a balance between ensuring relatively uniform coverage across all azimuths and maximizing the incident angle near the target layer, crucial for preserving the accuracy of pre-stack inversion. In this study, we conducted azimuth-specific stacking tests with maximum offsets of 4500 m, 5000 m, and 6500 m. The results of these tests are presented in Figure 6b–d.

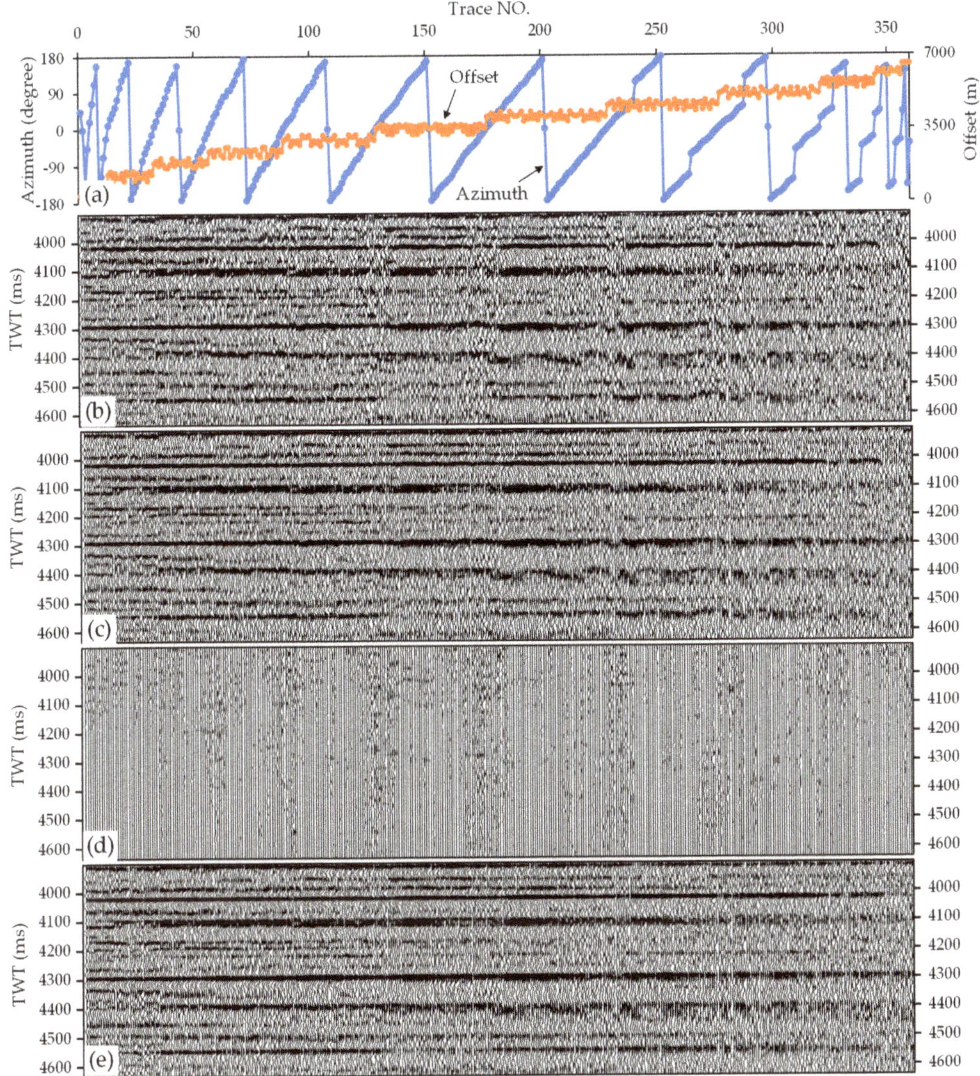

Figure 5. Typical OVT gather display: (**a**) spiral arrangement display of azimuth and offset for OVT gather; (**b**) original OVT gather; (**c**) gather after Radon transform denoising; (**d**) display of random noise; (**e**) gather after time alignment.

Upon analysis, it is evident that when the maximum offset is set to 6500 m (Figure 6b), there is a significant inhomogeneity in coverage across various azimuths. Although certain azimuths, such as 45°, achieve a maximum incident angle of 40°, azimuths like 75° and 105° only reach a maximum incident angle of approximately 25°, indicating a lack of far-offset data. This inhomogeneity in coverage can introduce spurious anisotropies. On the other hand, when the maximum offset is reduced to 4500 m (Figure 6d), while the coverage across azimuths becomes more uniform, the maximum incident angle is limited to approximately 25°, significantly compromising the accuracy of pre-stack inversion. This, in turn, has a substantial impact on the estimation of the Young's modulus and subsequent elliptic fitting. Finally, when the maximum offset is set to 5000 m (Figure 6c), the maximum incident angle near the target layer is approximately 30°, and the coverage across various azimuths is relatively uniform. This configuration satisfies the requirements for subsequent pre-stack inversion and elliptic fitting of the Young's modulus. Therefore, 5000 m was ultimately chosen as the optimal maximum offset.

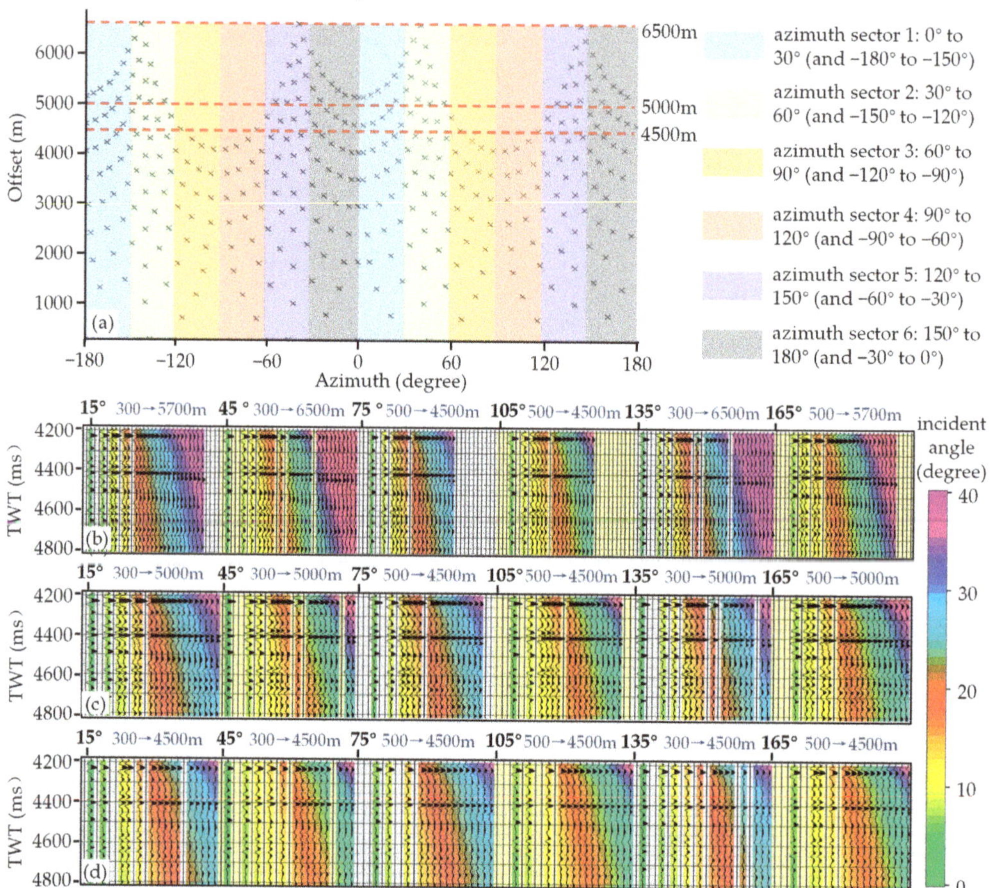

Figure 6. Analysis of azimuth angle and offset division scheme: (**a**) azimuth and offset distribution of OVT gather; (**b**–**d**) gather on each azimuth when the maximum offset is 6500 m, 5000 m, and 4500 m, respectively.

4.2. Fracture Prediction and Analysis

Based on the determined stacking scheme, the processed OVT gathers were stacked, generating partially stacked seismic data volumes for small, medium, and large incident angles across six azimuths. Using pre-stack inversion techniques, the equivalent Young's moduli for each of the six azimuths were obtained. Subsequently, the least-squares ellipse fitting algorithm was applied to fit the azimuthal Young's moduli on a trace-by-trace and sample-by-sample basis, yielding the direction of the ellipse's major axis and ellipticity for each sample point. The direction of the ellipse's major axis indicates the orientation of fractures, while the ellipticity reflects the difference in Young's moduli along and perpendicular to the fracture orientation. Although the absolute value of ellipticity does not directly correspond to the linear or volumetric density of fractures, it does represent the strength of anisotropy induced by fractures [50]. Therefore, ellipticity can indirectly reflect the density of fractures, with higher ellipticity values indicating a higher density of fractures.

The formation and distribution of fractures are usually closely related to structures and faults. Curvature attributes, as an effective post-stack analytical tool, are often used to characterize the morphological features of small-scale faults and large-scale fractures associated with structures and faults. To verify the rationality and accuracy of the fracture prediction method proposed in this study, the curvature attribute of the target layer's top surface was extracted and compared with the ellipticity attribute. Figure 7a shows a structural map of the top surface of the Yijianfang Formation in the study area (partially enlarged from the black dashed box in Figure 1b). The study area is located at the high part of an anticline structure, where three main fault systems are developed. Figure 7b–d present the curvature attribute map, the ellipticity attribute map, and a superimposed plan view of both attributes for the top surface of the Yijianfang Formation, respectively. It can be observed that the curvature attribute effectively characterizes the faults and the associated fracture development zones in the study area. From the superimposed image of ellipticity and curvature attributes, it can be seen that the predicted results of fracture density have a good correlation with the faults, generally showing a trend that the more developed the faults are, the higher the fracture density is. Different types of faults exert varying degrees of control over fracture development. Specifically, the fracture development zones influenced by thrust faults are relatively narrow and tend to concentrate near the fault planes. This phenomenon may be related to the local compressive stress environment. In contrast, strike-slip faults produce broader fracture zones, which extend a significant distance from the main fault plane, aligning with field observations of strike-slip fault geological structures. On the plane, the fractures surrounding the radially distributed tensile normal faults exhibit the most significant development, forming complex networks of fractures. By incorporating the tectonic evolution background of the block, it is speculated that this system of normal faults and fractures is closely related to the deformation and arching mechanisms of the brittle limestone strata overlying the salt layer. Overall, the prediction results of fracture density are consistent with the tectonic geological background of this area, which proves that the method adopted in this study is applicable and effective under the geological conditions of this area.

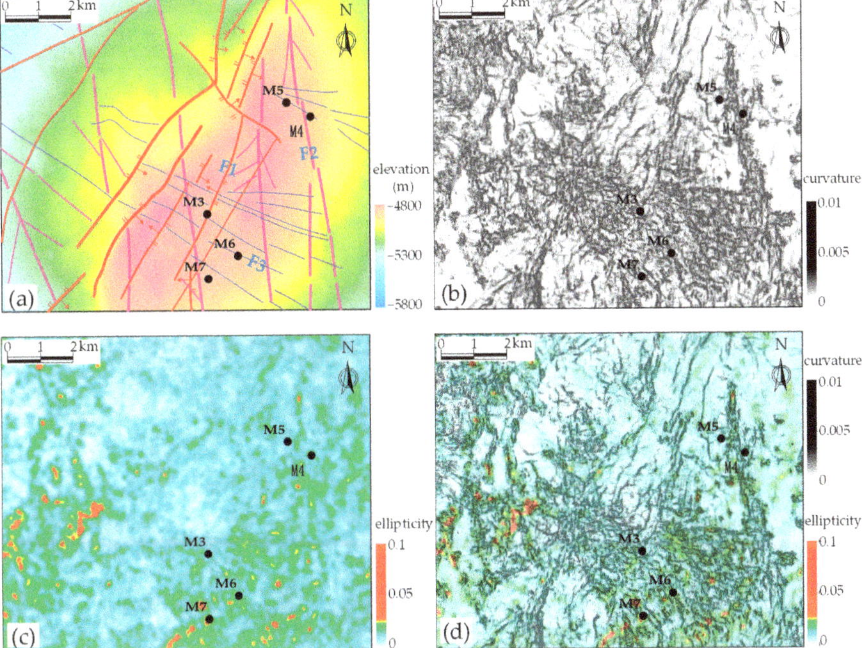

Figure 7. Sub-maps of the top surface of the Yijianfang formation: (**a**) structural map; (**b**) curvature attribute map; (**c**) ellipticity attribute map; (**d**) superimposed plan view displaying both ellipticity and

curvature attributes. Black dots represent the well locations, red lines indicate thrust faults, purple lines indicate strike-slip faults, and blue lines indicate normal faults, which are consistent with Figure 1.

Imaging logging technology is capable of clearly projecting the structural features of rock on the wellbore surface, exhibiting directionality and high resolution. It has been widely applied in the interpretation and evaluation of reservoirs such as fractures and dissolved pores in carbonate formations [58]. To further assess the reliability of the fracture prediction results in this study, we collected and compared the predicted results with fracture interpretation conclusions from imaging logging of four wells in the study area, as shown in Figure 8 and Table 1.

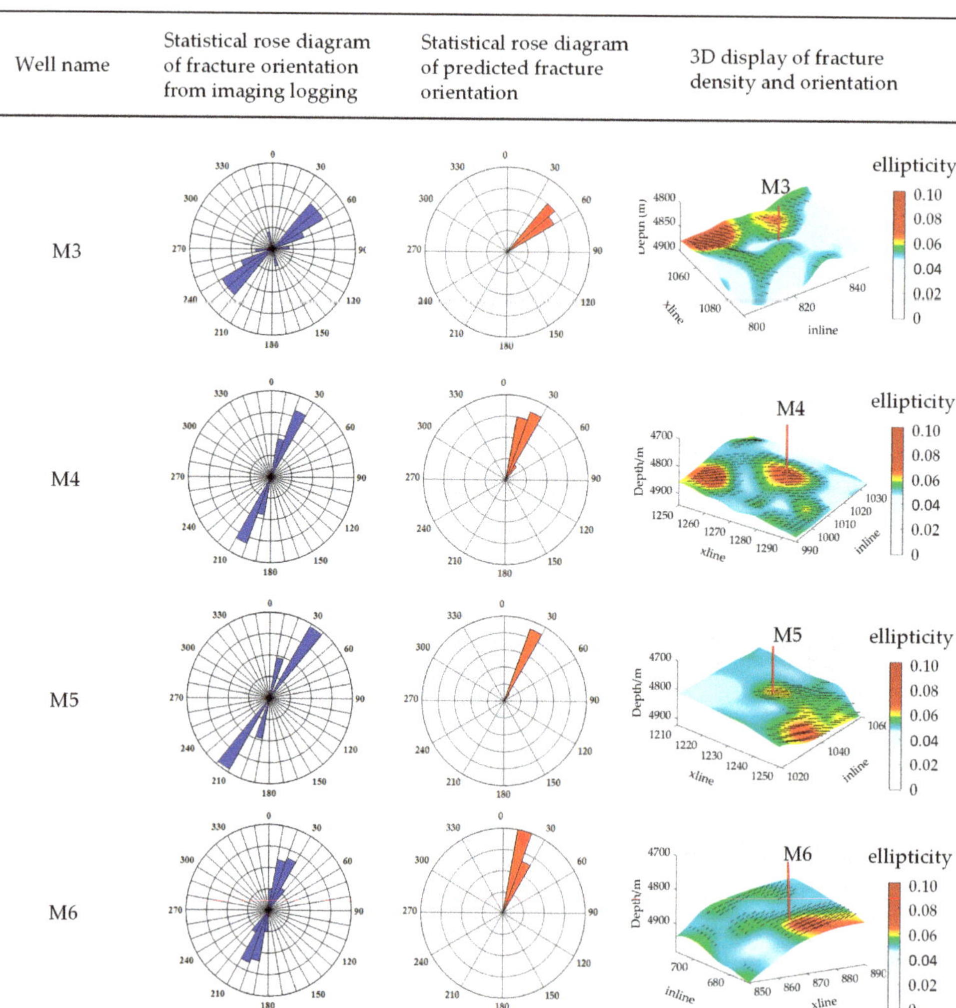

Figure 8. Comparison between imaging logging fracture interpretation conclusions and prediction results for wells such as M3 within the study area. For a given well, the sub-diagrams presented from left to right are the statistical rose diagram of fracture orientation from imaging logging, the statistical rose diagram of predicted fracture orientation, and the superimposed three-dimensional display of the geological structure, fracture orientation, and density around the well area. The locations of the wells are shown in Figure 7a.

Table 1. Statistical table of imaging logging fracture orientations and predicted fracture orientations for wells M3, M4, M5, and M6.

Well Name	Orientations from Imaging Logging	Orientations from Prediction	Match
M3	N40°E~N60°E	N40°E~N60°E	Matching
M4	N20°E~N30°E	N20°E~N30°E	Matching
M5	N10°E~N20°E N30°E~N40°E	N20°E~N30°E	Not matching well
M6	N10°E~N30°E	N10°E~N30°E	Matching

Well M3 is located near the thrust fault F1. The main orientation of fractures within the target layer is predicted to range from N40 °E to N60 °E using this method, and this prediction aligns well with the observed fracture orientation. Wells M4 and M5 are located near the strike-slip fault F2. The main fracture orientation in the target layer of well M4, interpreted from well logging, is N20 °E to N30 °E, which aligns with our prediction. For well M5, the well logging interpretation of its target layer reveals two primary fracture sets with orientations ranging from N10 °E to N20 °E and from N30 °E to N40 °E. However, our prediction for well M5 indicates a fracture orientation of N20 °E to N30 °E. Given the limited resolution of seismic data compared to imaging logging, it is possible that the prediction represents a combined response from both fracture sets. Therefore, the prediction result is considered to be relatively consistent. Well M6 is situated near the normal fault F3. The primary fracture orientation in its target layer, interpreted from well logging, ranges from N10 °E to N30 °E. Our method predicts a similar orientation of N10 °E to 30 °E, indicating a consistent result. Consequently, there is a high degree of agreement between the predicted fracture orientations at the wellbore location using our method and the interpretations from imaging logging. A comprehensive analysis of multiple wells reveals that the primary fractures in the target layer generally trend towards the northeast, which aligns with the northeast-oriented principal stress during the middle-to-late Ordovician in the Tabei uplift [59]. Additionally, this corroborates the effectiveness of our method.

The actual drilling data from the study area indicate that the strong reflections presenting as "beaded" patterns on seismic profiles represent high-quality fractured-vuggy reservoirs, while non-beaded reflections are typically associated with dense surrounding rocks, which function as lateral barriers or capping layers, enclosing the fractured-vuggy reservoirs. The magnitude of Young's modulus serves as an indicator of the stiffness of the medium; the higher its value, the less susceptible it is to deformability, resulting in a denser and more stable rock formation. Statistics from drilled wells within the study area indicate that the Young's modulus of high-quality fractured-vuggy reservoirs typically ranges from 6.7 to 7.2 $\times$ 10^{10} N/m^2, whereas the Young's modulus of compact surrounding rocks falls within the range of 7.2 to 9.0 $\times$ 10^{10} N/m^2. Figure 9a presents the seismic profile of well M7 located within the study area, while Figure 9b–g depict the corresponding Young's modulus profiles along six different azimuths. The specific locations of these profiles are shown in Figure 10a. As can be observed from Figure 9a, three "beaded" reflections are observed around the wellbore, which manifest as three fractured-vuggy systems on the six corresponding Young's modulus profiles (indicated by the black dashed boxes). Although there are differences in local details and numerical values across the six azimuths of the Young's modulus profiles, the overall morphologies are generally similar. The reservoirs are located in regions with low Young's modulus values. The drilling target of well M7 was bead II. While drilling at depths ranging from 5914.93 to 5915.58 m and from 5917.14 to 5927.1 m, the well experienced void drilling of 0.65 m and 9.96 m, respectively, resulting in a cumulative loss of 142 cubic meters of drilling fluid. This indicated that high-quality fractured-vuggy reservoirs were uncovered, which aligned with the inversion results of the Young's modulus.

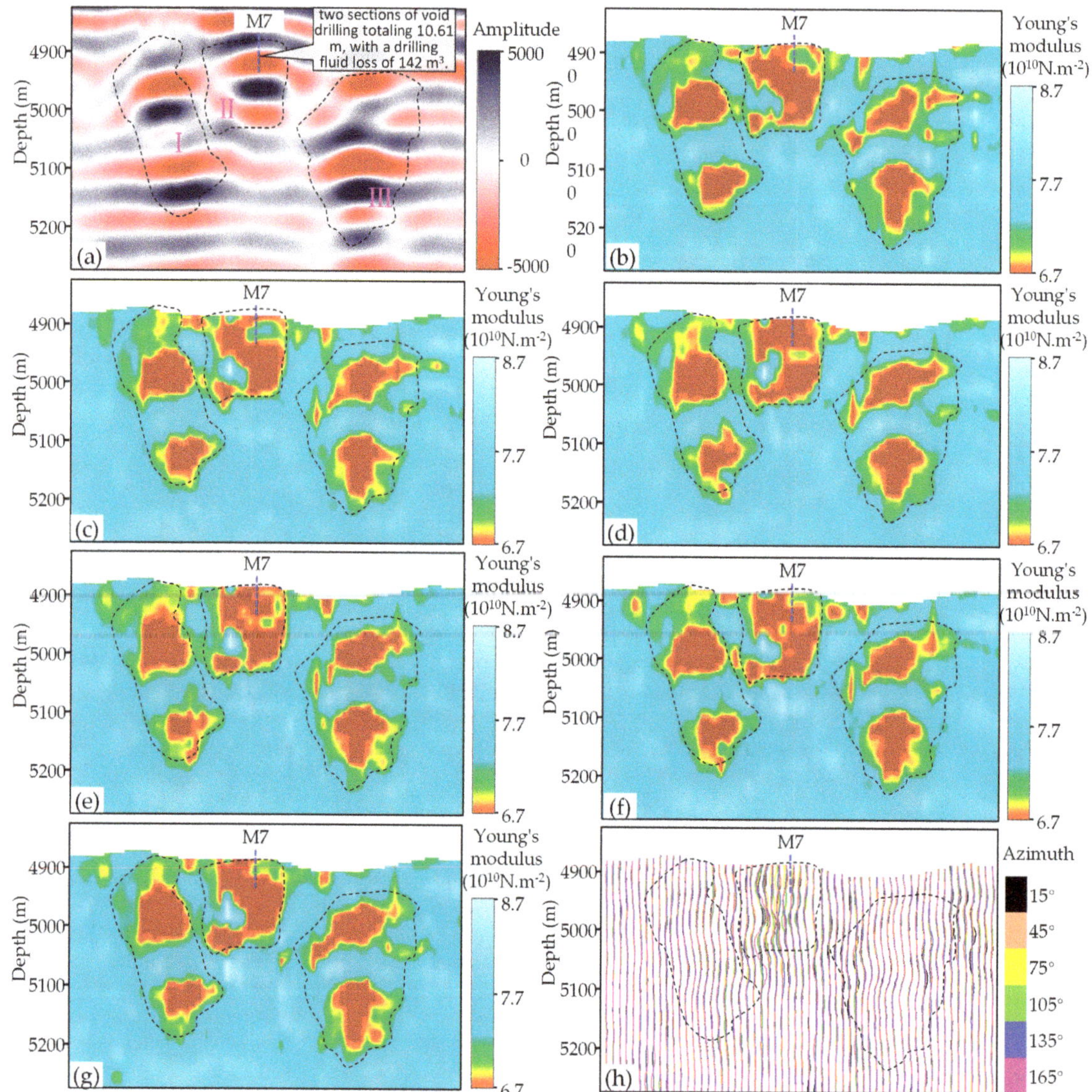

Figure 9. (**a**) Seismic profile; (**b–g**) azimuthal Young's modulus profiles; and (**h**) superimposed waveform display of azimuthal Young's modulus for well M7 in the study area. The locations of the diagrams (**a–h**) are indicated as shown by the dashed arrows in Figure 10a.

Figure 9h displays a superimposed waveform profile of the Young's modulus for the six azimuths. It is evident that the variations in Young's modulus among different azimuths are relatively minor near bead I and bead III, whereas a significant difference is observed near bead II. Figure 10c–e provide enlarged views of the Young's modulus data for the six azimuths at the centers of the three beads. It is apparent that the Young's modulus values for different azimuths are generally comparable and do not exhibit significant variations for bead I and bead III. However, notable changes are observed in the Young's modulus curves for different azimuths near bead II. Figure 10f displays a box plot comparing the Young's modulus values across six azimuths at the centers of three beads. For bead I, the Young's modulus ranges from a minimum of 5.82×10^{10} N/m² to a maximum of

6.19×10^{10} N/m^2, with a median value of 6.0×10^{10} N/m^2. At bead II, the Young's modulus exhibits a wider variation, with a maximum of 6.73×10^{10} N/m^2, a minimum of 5.32×10^{10} N/m^2, and a median of 6.32×10^{10} N/m^2. Finally, for bead III, the Young's modulus varies from a minimum of 6.33×10^{10} N/m^2 to a maximum of 6.56×10^{10} N/m^2, with a median of 6.47×10^{10} N/m^2. Notably, the dispersion of Young's modulus data across different azimuths is more pronounced for bead II, indicating significant differences in the Young's modulus among various azimuths. Based on the experimental conclusions of Sayers [49] and Zong et al. [50], a greater difference between the long and short axes of the azimuthal Young's modulus ellipse indicates stronger anisotropy and a higher density of fractures. Therefore, it is speculated that the anisotropy intensity is relatively low for bead I and bead III, whereas it is significantly high for bead II, suggesting a greater density of fracture within bead II.

Figure 10a displays a plan view of the maximum trough attribute surrounding well M7. The warm colors represent the bead-like geobodies. Figure 10b presents a plan view of the extracted fracture density and fracture orientation attributes around well M7. It can be observed that the fractures surrounding bead II, where well M7 is located, are more developed than those around beads I and III, albeit with limited extension lengths, approximately 100 to 160 m from the center of the bead. Relatively less developed fractures are observed around beads I and III. Based on the fracture prediction results, it is evident that these three beads are not interconnected. Figure 11 depicts the production profile of well M7. Immediately after the commencement of production, the well exhibited a sharp decline in oil pressure and output, with a brief self-flowing period lasting only 11 days. Subsequently, a total of 29 rounds of water injection were executed to stimulate oil displacement, resulting in a cumulative water injection volume of 0.51 million tons and a cumulative oil production of 0.49 million tons, indicating excellent oil enhancement through water injection. The production characteristics of well M7 closely resemble those exhibited by a single-cavity, constant-volume well [60], thereby reinforcing the conclusion of limited connectivity between well M7 and its surrounding reservoirs. This observation aligns with the previously mentioned fracture prediction results. Based on a comprehensive analysis of static fracture prediction results and dynamic production characteristics, it is concluded that bead II, where well M7 is located, is not connected to beads I and III. Consequently, a sidetrack drilling operation was designed to target bead I. After sidetracking, the wellbore was tested with a 5 mm choke at the target layer, achieving an oil pressure of 5.7 MPa and a daily oil production equivalent of 133 cubic meters, confirming the disconnectedness of bead I and bead II. This demonstrates that the reservoir associated with bead I remains untapped and validates the high precision and reliability of the fracture prediction method employed in this study.

According to the fracture prediction results, a secondary sidetrack drilling operation can be considered in the future to tap into the reserves associated with bead III. Employing the volumetric method based on carving techniques, the geological reserves of bead III are estimated to be approximately 86,000 tons, with an expected cumulative oil production of 12,900 tons, indicating significant economic potential. The fracture prediction methodology introduced in this study can be further promoted and utilized to guide the deployment of infill wells and the exploitation of untapped potential in fractured-vuggy carbonate reservoirs, ultimately maximizing the exploitation of geological reserves.

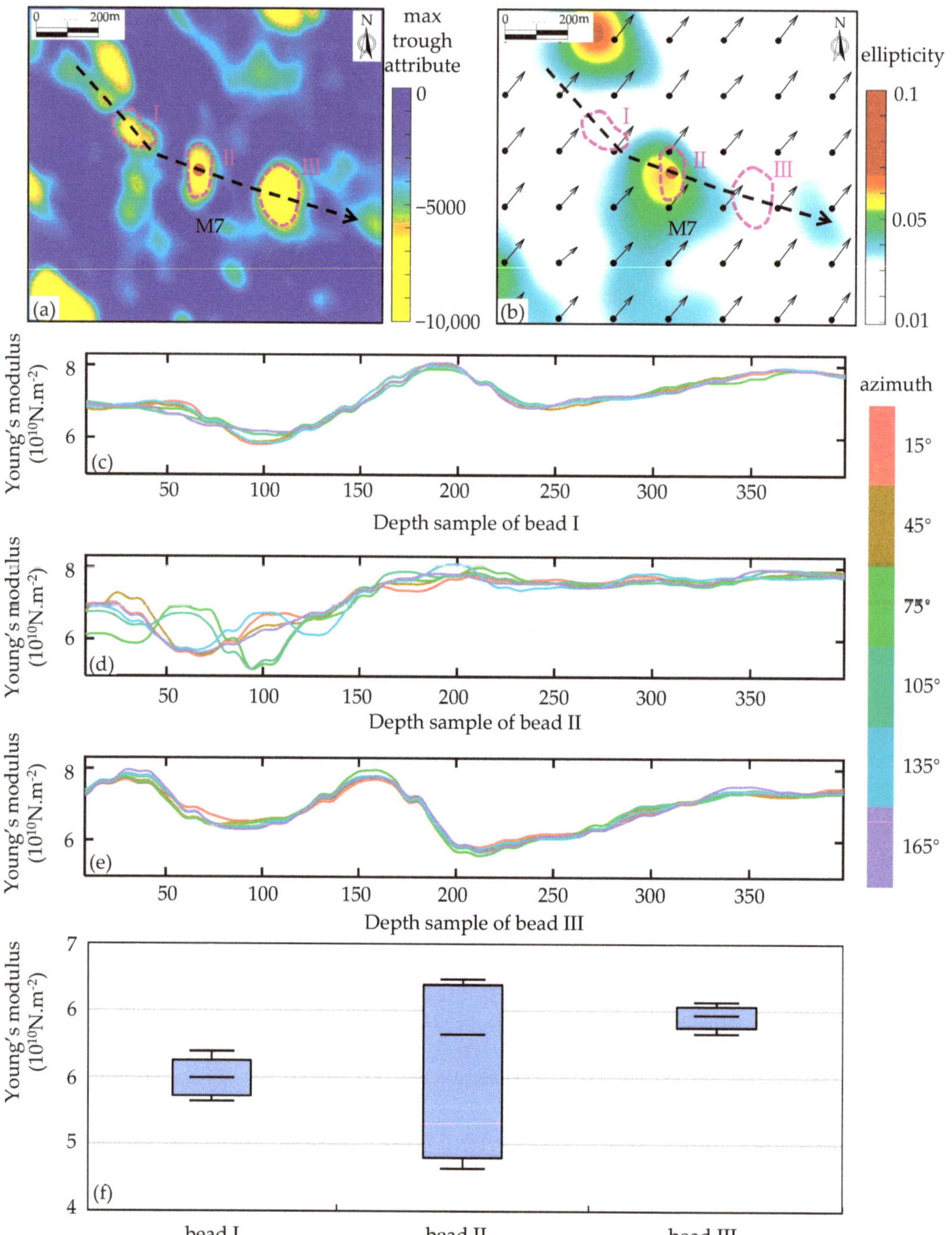

Figure 10. (**a,b**) Maximum trough attribute and ellipticity attribute plans for the M7 well area, respectively. The purple dashed line boxes indicate the development zones of beaded reservoirs; (**c–e**) enlarged local views of the 6-azimuthal Young's modulus data extracted from the centers of three beaded reservoirs; (**f**) a box plot depicting the statistical distribution of the Young's modulus values measured in six azimuths at the centers of the three beaded reservoirs.

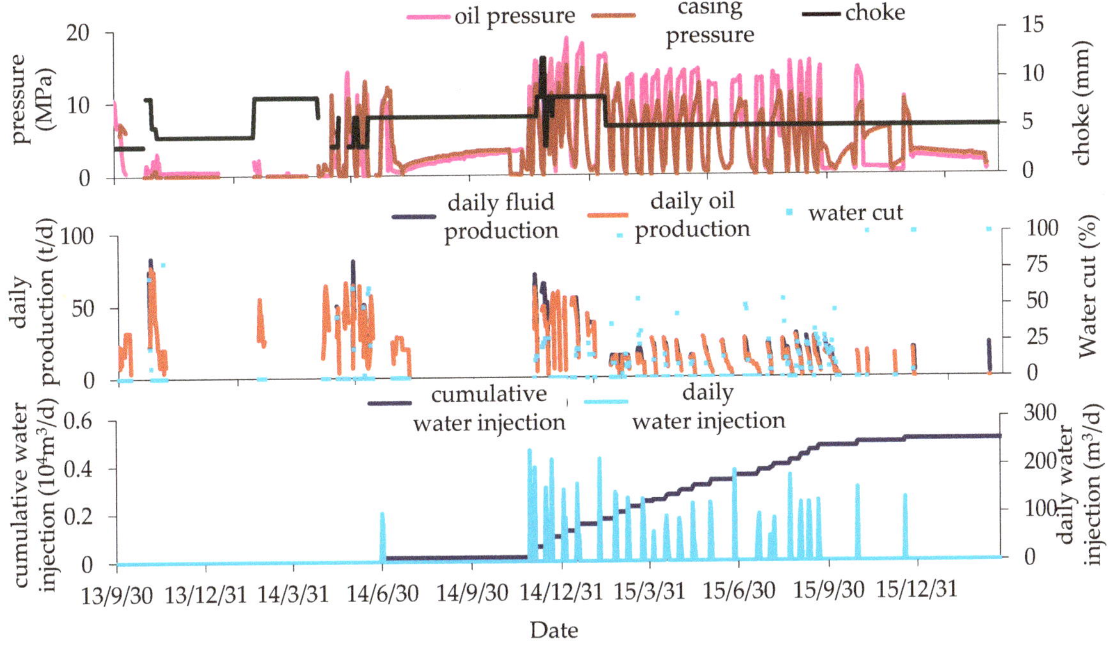

Figure 11. Production curves of well M7.

5. Conclusions

In this study, we conducted research on a pre-stack fracture prediction method based on azimuthal Young's modulus ellipse fitting for the M oilfield. Based on the obtained results, the following conclusions were drawn:

(1) Compared with traditional post-stack attribute-based fracture prediction methods, such as curvature analysis, the fracture prediction method utilizing OVT (Offset Vector Tile) gather data can predict fractures of smaller scales and quantitatively characterize fracture development.

(2) The pre-stack technology employed in this study primarily relies on azimuthal variations in Young's modulus. Consequently, it has higher requirements for amplitude-preserving and fidelity-enhancing processing in the seismic processing stage, as well as optimized preprocessing of OVT gather data in the interpretation stage.

(3) The method adopted in this study is primarily suitable for scenarios involving the development of a single set of high-angle fractures. In cases where two or more sets of high-angle fractures exist, such as in well M5, the prediction results may manifest as a combined response from multiple sets of fractures due to resolution limitations. For complex areas with the simultaneous development of low-angle and multiple sets of fractures, further research is needed on the azimuthal response characteristics of parameters such as the Young's modulus.

(4) A pre-stack fracture prediction technical workflow in the OVT domain for ultra-deep unconventional fractured-vuggy carbonate reservoirs is established in this paper. The fracture prediction results were subsequently tested against the geological cognition, imaging logging data, and dynamic production data of the block. This validation process confirms the applicability and reliability of the technique in unconventional fractured-vuggy carbonate reservoirs, providing valuable insights for future fracture prediction in similar geological settings.

Author Contributions: Conceptualization, G.Z.; methodology, B.L. and F.Y.; software, B.L.; investigation, B.L. and F.Y.; data curation and visualization, L.Z. and F.Y.; writing, B.L.; supervision, G.Z. All authors have read and agreed to the published version of the manuscript.

Funding: This research received no external funding.

Data Availability Statement: The original contributions presented in the study are included in the article, further inquiries can be directed to the corresponding author.

Acknowledgments: The authors thank the editor and reviewers for their comments regarding manuscript improvement. We also thank Kuanzhi Zhao, Suo Cheng and Yongjian Zeng for their help.

Conflicts of Interest: Author Fengying Yang was employed by the company SINOPEC Geophysical Research Institute Co., Ltd. Author Longfei Zhao was employed by the PetroChina Tarim Oilfield Company. The remaining authors declare that the research was conducted in the absence of any commercial or financial relationships that could be construed as a potential conflict of interest.

References

1. Garotta, R.; Granger, P.Y. Acquisition and Processing of 3C × 3-D Data Using Converted Waves. In *SEG Technical Program Expanded Abstracts 1988*; SEG Technical Program Expanded Abstracts; Society of Exploration Geophysicists: Anaheim, CA, USA, 1988; pp. 995–997.
2. Garotta, R. Detection of Azimuthal Anisotropy. In *SEG Technical Program Expanded Abstracts 1989*; SEG Technical Program Expanded Abstracts; Society of Exploration Geophysicists: Dallas, TX, USA, 1989; pp. 861–863.
3. Gaiser, J.E. Enhanced PS-Wave Images and Attributes Using Prestack Azimuth Processing. In *SEG Technical Program Expanded Abstracts 1999*; SEG Technical Program Expanded Abstracts; Society of Exploration Geophysicists: Houston, TX, USA, 1999; pp. 699–702.
4. Bale, R.A.; Li, J.; Mattocks, B. Robust Estimation of Fracture Directions from 3-D Converted Waves. In *SEG Technical Program Expanded Abstracts 2005*; SEG Technical Program Expanded Abstracts; Society of Exploration Geophysicists: Houston, TX, USA, 2005; pp. 889–892.
5. Mattocks, B.; Li, J.; Roche, S.L. Converted-Wave Azimuthal Anisotropy in a Carbonate Foreland Basin. In *SEG Technical Program Expanded Abstracts 2005*; SEG Technical Program Expanded Abstracts; Society of Exploration Geophysicists: Houston, TX, USA, 2005; pp. 897–900.
6. Fehmers, G.C.; Höcker, C.F.W. Fast Structural Interpretation with Structure-oriented Filtering. *Geophysics* **2003**, *68*, 1286–1293. [CrossRef]
7. Lu, Y.; Lu, W. Edge-Preserving Polynomial Fitting Method to Suppress Random Seismic Noise. *Geophysics* **2009**, *74*, V69–V73. [CrossRef]
8. AlBinHassan, N.M.; Luo, Y.; Al-Faraj, M.N. 3D Edge-Preserving Smoothing and Applications. *Geophysics* **2006**, *71*, P5–P11. [CrossRef]
9. Ashraf, H.; Mousa, W.A.; Al Dossary, S. Sobel Filter for Edge Detection of Hexagonally Sampled 3D Seismic Data. *Geophysics* **2016**, *81*, N41–N51. [CrossRef]
10. Guo, L.; Liu, Y.; Liu, C.; Zheng, Z. Structure-Oriented Filtering for Seismic Images Using Nonlocal Median Filter. In *SEG Technical Program Expanded Abstracts 2018*; SEG Technical Program Expanded Abstracts; Society of Exploration Geophysicists: Anaheim, CA, USA, 2018; pp. 4623–4627.
11. Chopra, S.; Marfurt, K.J. Structure-Oriented Filtering and Image Enhancement. In *Seismic Attributes for Prospect Identification and Reservoir Characterization*; Geophysical Developments Series; Society of Exploration Geophysicists and European Association of Geoscientists and Engineers: Denver, CO, USA, 2007; pp. 187–218.
12. Steklain, A.F.; Ganacim, F.; Adames, M.R.; Gonçalves, J.L.; Oliveira, D.S. Structure-Oriented Filtering in Unsupervised Multiattribute Seismic Facies Analysis. *Lead. Edge* **2022**, *41*, 366–436. [CrossRef]
13. Barbato, U.; Castagna, J.; Portniaguine, O.; Fagin, S. Composite Attribute from Spectral Decomposition for Fault Detection. In *SEG Technical Program Expanded Abstracts 2014*; SEG Technical Program Expanded Abstracts; Society of Exploration Geophysicists: Denver, CO, USA, 2014; pp. 2542–2546.
14. Yuan, Z.; Huang, H.; Jiang, Y.; Tang, J.; Li, J. An Enhanced Fault-Detection Method Based on Adaptive Spectral Decomposition and Super-Resolution Deep Learning. *Interpretation* **2019**, *7*, T713–T725. [CrossRef]
15. Miao, X.; Todorovic-Marinic, D.; Klatt, T. Enhancing Seismic Insight by Spectral Decomposition. In *SEG Technical Program Expanded Abstracts 2007*; SEG Technical Program Expanded Abstracts; Society of Exploration Geophysicists: San Antonio, TX, USA, 2007; pp. 1437–1441.
16. Jahan, I.; Castagna, J. Spectral Decomposition Using Time-Frequency Continuous Wavelet Transforms for Fault Detection in the Bakken Formation. In *SEG Technical Program Expanded Abstracts 2017*; SEG Technical Program Expanded Abstracts; Society of Exploration Geophysicists: Houston, TX, USA, 2017; pp. 2190–2194.

17. Liu, S.; Wen, X.; Li, L.; Yang, J.; Chen, X. Fault Analysis of Azimuth Curvature Attribute Based on Curvelet Transform. In *SEG 2018 Workshop: Reservoir Geophysics, Daqing, China, 5–7 August 2018*; SEG Global Meeting Abstracts; Society of Exploration Geophysicists and the Chinese Geophysical Society: Daqing, China, 2018; pp. 1–4.

18. Brito, L.S.B.; Alaei, B.; Torabi, A.; Leopoldino-Oliveira, K.M.; Vasconcelos, D.L.; Bezerra, F.H.R.; Nogueira, F.C.C. Automatic 3D Fault Detection and Characterization—A Comparison between Seismic Attribute Methods and Deep Learning. *Interpret.-J. Subsurf. Charact.* **2023**, *11*, T793–T808. [CrossRef]

19. Islam, M.M.; Babikir, I.; Elsaadany, M.; Elkurdy, S.; Siddiqui, N.A.; Akinyemi, O.D. Application of a Pre-Trained CNN Model for Fault Interpretation in the Structurally Complex Browse Basin, Australia. *Appl. Sci.* **2023**, *13*, 11300. [CrossRef]

20. Rodriguez-Pradilla, G.; Verdon, J.P. Quantifying the Variability in Fault Density across the UK Bowland Shale with Implications for Induced Seismicity Hazard. *Geomech. Energy Environ.* **2024**, *38*, 100534. [CrossRef]

21. Hale, D. Methods to Compute Fault Images, Extract Fault Surfaces, and Estimate Fault Throws from 3D Seismic Images. *Geophysics* **2013**, *78*, O33–O43. [CrossRef]

22. Basir, H.M.; Javaherian, A.; Yaraki, M.T. Multi-Attribute Ant-Tracking and Neural Network for Fault Detection: A Case Study of an Iranian Oilfield. *J. Geophys. Eng.* **2013**, *10*, 015009. [CrossRef]

23. Xie, Q.; Zhao, C.; Rui, Z.; Guan, S.; Zheng, W.; Fan, H. An Improved Ant-Tracking Workflow Based on Divided-Frequency Data for Fracture Detection. *J. Geophys. Eng.* **2022**, *19*, 1149–1162. [CrossRef]

24. Li, S.; Zhao, Y.; Xian, C.; Liang, X.; Zhang, J.; Qiao, Q.; Yan, L.; Shen, Y.; Cao, H. Small-Scale Fracture Detection via Anisotropic Bayesian Ant-Tracking Colony Optimization Driven by Azimuthal Seismic Data. *IEEE Trans. Geosci. Remote Sens.* **2023**, *61*, 1–12. [CrossRef]

25. Ge, X.; Guo, T.; Ma, Y.; Wang, G.; Li, M.; Zhao, P.; Yu, X.; Li, S.; Fan, H.; Zhao, T. Fracture Development and Inter-Well Interference for Shale Gas Production from the Wufeng-Longmaxi Formation in a Gentle Syncline Area of Weirong Shale Gas Field, Southern Sichuan, China. *J. Pet. Sci. Eng.* **2022**, *212*, 110207. [CrossRef]

26. Luo, J.; Yang, X.; Wang, L.; Li, X.; Xu, M.; Guo, H.; Wang, Z.; Deng, W. Finely Description for Fractured Reservoir and Comprehensive Evaluation of Seismic. In Proceedings of the 2021 International Petroleum and Petrochemical Technology Conference, Beijing, China, 8–10 June 2021; Lin, J., Ed.; Springer: Singapore, 2022; pp. 166–173.

27. Wang, R.; Liu, T.; Zhang, C. Fault Interpretation for Carbonate Reservoir and Its Application for Reservoir Connectivity. In *SEG Technical Program Expanded Abstracts 2019*; Society of Exploration Geophysicists: San Antonio, TX, USA, 2019; pp. 3479–3482.

28. Li, K.; Zong, J.; Fei, Y.; Liang, J.; Hu, G. Simultaneous Seismic Deep Attribute Extraction and Attribute Fusion. *IEEE Trans. Geosci. Remote Sens.* **2022**, *60*, 1–10. [CrossRef]

29. Jiang, W.; Zhang, D.; Hui, G. A Dual-Branch Fracture Attribute Fusion Network Based on Prior Knowledge. *Eng. Appl. Artif. Intell.* **2024**, *127*, 107383. [CrossRef]

30. Yin, X.; Zhang, H.; Zong, Z. Research status and progress of 5D seismic data interpretation in OVT domain. *Geophys. Prospect. Pet.* **2018**, *57*, 155–178. (In Chinese with English Abstract)

31. Pei, J.; Guo, X.; Hu, Y.; Guo, F.; Liu, B.; Hao, L.; Zhang, R.; Zheng, F. Research and Application of 5D Seismic Prediction Technology. *Interpretation* **2023**, *11*, T189–T197. [CrossRef]

32. Williams, M.; Jenner, E. Interpreting Seismic Data in the Presence of Azimuthal Anisotropy; or Azimuthal Anisotropy in the Presence of the Seismic Interpretation. *Lead. Edge* **2002**, *21*, 771–774. [CrossRef]

33. Pan, X.; Li, L.; Zhang, G. Multiscale Frequency-Domain Seismic Inversion for Fracture Weakness. *J. Pet. Sci. Eng.* **2020**, *195*, 107845. [CrossRef]

34. Pan, X.; Zhang, D.; Zhang, P. Fracture Detection from Azimuth-Dependent Seismic Inversion in Joint Time–Frequency Domain. *Sci. Rep.* **2021**, *11*, 1269. [CrossRef] [PubMed]

35. Grechka, V.; Tsvankin, I. 3-D Description of Normal Moveout in Anisotropic Inhomogeneous Media. *Geophysics* **1998**, *63*, 1079–1092. [CrossRef]

36. Rüger, A. *P*-wave Reflection Coefficients for Transversely Isotropic Models with Vertical and Horizontal Axis of Symmetry. *Geophysics* **1997**, *62*, 713–722. [CrossRef]

37. Rüger, A.; Tsvankin, I. Using AVO for Fracture Detection: Analytic Basis and Practical Solutions. *Lead. Edge* **1997**, *16*, 1429–1434. [CrossRef]

38. Rüger, A. *Reflection Coefficients and Azimuthal AVO Analysis in Anisotropic Media*; Society of Exploration Geophysicists: Tulsa, OK, USA, 2002; ISBN 1-56080-10-7.

39. Mallick, S.; Craft, K.L.; Meister, L.J.; Chambers, R.E. Chambers Determination of the Principal Directions of Azimuthal Anisotropy from *P*-Wave Seismic Data. *Geophysics* **1998**, *63*, 692–706. [CrossRef]

40. Qu, S.L.; Ji, Y.X.; Wang, X.; Wang, X.L.; Chen, X.R.; Shen, G.Q. Seismic method for using full-azimuth P wave attribution to detct fracture. *Oil Geophys. Prospect.* **2021**, *36*, 390–397. (In Chinese with English Abstract)

41. Li, C.P.; Yin, X.Y.; Liu, Z.G.; Li, A.S.; Yuan, F. An anisotropic gradient inversion for fractured reservoir prediction. *Geophys. Prospect. Pet.* **2017**, *56*, 835–840. (In Chinese with English Abstract)

42. Zhang, G.Z.; Chen, H.Z.; Yin, X.Y.; Li, N.; Yang, B.Y. Method of Fracture Elastic Parameter Inversion Based on Anisotropic AVO. *J. Jilin Univ. (Earth Sci. Ed.)* **2012**, *42*, 845–851. (In Chinese with English Abstract)

43. Wang, H.Q.; Yang, W.Y.; Xie, C.H.; Zheng, D.M.; Wang, H.L.; Zhang, X.M.; Jiang, C.L. Azimuthal anisotropy analysis of different seismic attributes and fracture prediction. *Oil Geophys. Prospect.* **2014**, *49*, 925–931. (In Chinese with English Abstract)

44. Zhou, L.; Zou, J.H.; Dai, R.X.; Zhang, Y.; Lan, X.M.; Wu, Y.; Wang, H.Q.; Liu, S.M. Application of OVT-domain 5-dimensional seismic attributes in fracture prediction in the Qixia Formation of the Shuangyushi area. *Earth Sci. Front.* **2023**, *30*, 213–228. (In Chinese with English Abstract)

45. Chen, Z.G.; Li, F.; Wang, X.; Wu, R.K.; Sun, X.; Zhao, Q.; Song, D.C.; Ma, H. Application of prestack anisotropic intensity attribute in prediction of P Buried hill fractured reservoir in Bongor Basin, Chad. *Chin. J. Geophys.* **2018**, *61*, 4625–4634. (In Chinese with English Abstract)

46. Zong, Z.Y.; Yin, X.Y.; Wu, G.C. Model Parameterization and EVA-DSVD Inversion with Young's Modulus and Poisson's Ratio. In *SEG Technical Program Expanded Abstracts 2013*; SEG Technical Program Expanded Abstracts; Society of Exploration Geophysicists: Houston, TX, USA, 2013; pp. 408–412.

47. Zong, Z.Y.; Yin, X.Y.; Wu, G.C. Elastic Impedance Parameterization and Inversion with Young's Modulus and Poisson's Ratio. *Geophysics* **2013**, *78*, N35–N42. [CrossRef]

48. Yaojie, C.; Shulin, P. Calculation Example of Brittleness Index of Tight Glutenite Based on Prestack Seismic Inversion. In *SEG Integration of Geophysics, Geology, and Engineering Workshop, Chengdu, China, 26–28 June 2023*; SEG Global Meeting Abstracts; Society of Exploration Geophysicists: Houston, TX, USA, 2023; pp. 18–21.

49. Sayers, C.M. The Effect of Anisotropy on the Young's Moduli and Poisson's Ratios of Shales. In *SEG Technical Program Expanded Abstracts 2010*; Society of Exploration Geophysicists: Denver, CO, USA, 2010; pp. 2606–2611.

50. Zong, Z.; Sun, Q.; Li, C.; Yin, X. Young's Modulus Variation with Azimuth for Fracture-Orientation Estimation. *Interpretation* **2018**, *6*, T809–T818. [CrossRef]

51. Wang, J.H.; Zhang, J.M.; Wu, G.C. Wide-azimuth Young's modulus inversion and fracture prediction: An example of H structure in Bozhong sag. *Oil Geophys. Prospect.* **2021**, *56*, 593–602. (In Chinese with English Abstract)

52. Vakhin, A.V.; Khelkhal, M.A.; Tajik, A.; Ignashev, N.E.; Krapivnitskaya, T.O.; Peskov, N.Y.; Glyavin, M.Y.; Bulanova, S.A.; Slavkina, O.V.; Schekoldin, K.A. Microwave Radiation Impact on Heavy Oil Upgrading from Carbonate Deposits in the Presence of Nano-Sized Magnetite. *Processes* **2021**, *9*, 2021. [CrossRef]

53. Al-Mishaal, O.F.; Suwaid, M.A.; Al Muntaser, A.A.; Khelkhal, M.A.; Varfolomeev, M.A.; Djimasbe, R.; Zairov, R.R.; Saeed, S.A.; Vorotnikova, N.A.; Shestopalov, M.A.; et al. Octahedral Cluster Complex of Molybdenum as Oil-Soluble Catalyst for Improving In Situ Upgrading of Heavy Crude Oil: Synthesis and Application. *Catalysts* **2022**, *12*, 1125. [CrossRef]

54. Zhao, Y.; Wu, G.; Zhang, Y.; Scarselli, N.; Yan, W.; Sun, C.; Han, J. The Strike-Slip Fault Effects on Tight Ordovician Reef-Shoal Reservoirs in the Central Tarim Basin (NW China). *Energies* **2023**, *16*, 2575. [CrossRef]

55. Zhu, Y.; Zhang, Y.; Zhao, X.; Xie, Z.; Wu, G.; Li, T.; Yang, S.; Kang, P. The Fault Effects on the Oil Migration in the Ultra-Deep Fuman Oilfield of the Tarim Basin, NW China. *Energies* **2022**, *15*, 5789. [CrossRef]

56. Chen, L.; Jiang, Z.; Sun, C.; Ma, B.; Su, Z.; Wan, X.; Han, J.; Wu, G. An Overview of the Differential Carbonate Reservoir Characteristic and Exploitation Challenge in the Tarim Basin (NW China). *Energies* **2023**, *16*, 5586. [CrossRef]

57. He, G.Z. Study on Controlling Effectes of Faults on Ordovician Carbonate Hydrocarbon Accumulation in the YM2 Area. Master's Thesis, China University of Petroleum, Beijing, China, 2017. (In Chinese with English Abstract)

58. Zhang, J.; Nie, X.; Xiao, S.; Zhang, C.; Zhang, C.; Zhang, Z. Generating Porosity Spectrum of Carbonate Reservoirs Using Ultrasonic Imaging Log. *Acta Geophys.* **2018**, *66*, 191–201. [CrossRef]

59. Zhao, R.; Deng, S.; Yun, L.; Lin, H.; Zhao, T.; Yu, C.; Kong, Q.; Wang, Q.; Li, H. Description of the reservoir along strike-slip fault zones in China T-Sh oilfield, Tarim Basin. *Carbonates Evaporites* **2020**, *36*, 2. [CrossRef]

60. Deng, X.L.; Luo, X.S.; Liu, Y.F.; Huang, L.M.; Xiong, C. Forming Mechanism and Development Measures of Constant-Volume Fractured-Vuggy Carbonate Oil Reservoirs. *Xinjiang Pet. Geol.* **2019**, *40*, 79–83. (In Chinese with English Abstract)

MDPI AG
Grosspeteranlage 5
4052 Basel
Switzerland
Tel.: +41 61 683 77 34

Energies Editorial Office
E-mail: energies@mdpi.com
www.mdpi.com/journal/energies